**STATISTICAL
ANALYSIS IN
PSYCHOLOGY
& EDUCATION**

Adams, *Human Memory*
Beach, Hebb, Morgan, and Nissen, *The Neuropsychology of Lashley*
Von Békésy, *Experiments in Hearing*
Berkowitz, *Aggression: A Social Psychological Analysis*
Berlyne, *Conflict, Arousal, and Curiosity*
Blum, *Psychoanalytic Theories of Personality*
Brown, *The Motivation of Behavior*
Brown and Ghiselli, *Scientific Method in Psychology*
Butcher, *MMPI: Research Developments and Clinical Applications*
Campbell, Dunnette, Lawler, and Weick, *Managerial Behavior, Performance, and Effectiveness*
Cofer, *Verbal Learning and Verbal Behavior*
Cofer and Musgrave, *Verbal Behavior and Learning: Problems and Processes*
Crafts, Schneirla, Robinson, and Gilbert, *Recent Experiments in Psychology*
Crites, *Vocational Psychology*
D'Amato, *Experimental Psychology: Methodology, Psychophysics, and Learning*
Davitz, *The Communication of Emotional Meaning*
Deese and Hulse, *The Psychology of Learning*
Dollard and Miller, *Personality and Psychotherapy*
Edgington, *Statistical Inference: The Distribution-free Approach*
Ellis, *Handbook of Mental Deficiency*
Epstein, *Varieties of Perceptual Learning*
Ferguson, *Statistical Analysis in Psychology and Education*
Forgus, *Perception: The Basic Process in Cognitive Development*
Franks, *Behavior Therapy Appraisal and Status*
Ghiselli, *Theory of Psychological Measurement*
Ghiselli and Brown, *Personnel and Industrial Psychology*
Gilmer, *Industrial Psychology*
Gilmer, *Industrial and Organizational Psychology*
Gray, *Psychology Applied to Human Affairs*
Guilford, *Fundamental Statistics in Psychology and Education*
Guilford, *The Nature of Human Intelligence*
Guilford, *Personality*
Guilford, *Psychometric Methods*
Guilford and Hoepfner, *The Analysis of Intelligence*
Guion, *Personnel Testing*
Haire, *Psychology in Management*
Hirsch, *Behavior-genetic Analysis*
Hirsh, *The Measurement of Hearing*
Hurlock, *Adolescent Development*
Hurlock, *Child Development*
Hurlock, *Developmental Psychology*
Jackson and Messick, *Problems in Human Assessment*
Karn and Gilmer, *Readings in Industrial and Business Psychology*

George A. Ferguson *Chairman, Professor of Psychology, McGill University*

STATISTICAL ANALYSIS IN PSYCHOLOGY & EDUCATION

THIRD EDITION

McGraw-Hill Book Company New York St. Louis San Francisco
Düsseldorf Johannesburg Kuala Lumpur London Mexico Montreal
New Delhi Panama Rio de Janeiro Singapore Sydney Toronto

This book was set in Bodoni Book by Progressive Typographers, printed on permanent paper by Halliday Lithograph Corporation, and bound by The Book Press, Inc. The designer was Elliot Epstein; the drawings were done by John Cordes, J. & R. Technical Services, Inc. The editors were Walter Maytham, John Hendry, and Barry Benjamin. Peter D. Guilmette supervised production.

CONTENTS

PREFACE

The object of this book, like that of the second edition, is to introduce students and research workers in psychology and education to the concepts and applications of statistics. Emphasis is placed on the analysis and interpretation of data resulting from the conduct of experiments. Students and investigators in experimental medicine, psychiatry, sociology, and other disciplines should also find the book useful.

This book has been designed as a text for both one-semester and full-year courses in statistics. In either case the instructor has some freedom of choice in the selection of material. For a one-semester course the selection will usually include most of Chapters 1 to 13 and possibly sections of some of the remaining chapters.

I have attempted not only to introduce the student to the practical technology of statistics but also to explain in a nonmathematical and frequently intuitive way the nature of statistical ideas. This is not always easy. Obviously, the extent to which an understanding of statistics can be communicated without some mathematical knowledge is limited. Skill in high school or first-year college algebra will prove most helpful.

The writing of a book of this type demands numerous compromises between a tidy, logical arrangement of material, sound pedagogy, and common usage, which are not always compatible. The desire for completeness has led to the inclusion of occasional sections which are not essential in an introductory text. The instructor can readily identify these sections and omit them if he chooses.

Many changes have been made in this edition, some important and some trivial. The book has been divided into four parts: Basic Statistics, The Design of Experiments, Nonparametric Statistics, and Psychological Test and Multivariate Statistics. Four new chapters have been added: Analysis of Variance: Three-way Classification, Multiple Comparisons, Test Construction Statistics, and Factor Analysis. The section on the analysis of variance for repeated measurement designs has been expanded. The material on nonparametric statistics has been modified. Some material on now-obsolete computational methods has been deleted, and some rarely used statistical procedures have been dropped.

The usefulness of this book is enhanced by the kindness of authors and publishers who have permitted the adaptation and reproduction of tables and other materials published originally by them. I should like to express my gratitude to Francis G. Cornell, Allen L. Edwards, J. P. Guilford, Harry H. Harman, H. Leon Harter, R. W. B. Jackson, S. K. Katti, M. G. Kendall, John F. Kenny, Don Lewis, Quinn McNemar, Edwin G. Olds, George W. Snedecor, Herbert Sorenson, L. R. Verdooren, James E. Wert, and Frank Wilcoxon; and to the Scottish Council for Research in Education, the University of London Press Ltd., Charles Griffin & Company, Ltd., Prentice-Hall, Inc., John Wiley & Sons, Inc., Van Nostrand Reinhold Company, Rinehart & Company, Inc., Iowa State College Press, The University of Chicago Press, *Biometrika*, and the *Annals of Mathematical Statistics*. I owe special thanks to the Literary Executor of the Late Sir Ronald A. Fisher, F.R.S., to Dr. Frank Yates, F.R.S., and to Oliver and Boyd, Edinburgh, for permission to reprint Tables III, IV, and VII of their book *Statistical Tables for Biological, Agricultural, and Medical Research.*

I should like to express here my indebtedness to the late Sir Godfrey H. Thomson and to W. G. Emmett and D. N. Lawley, all of the University of Edinburgh. These three are responsible for my persisting interest in the applications of statistical method to psychological problems. In particular, I should like to express my gratitude to Lady Thomson for permission to reproduce certain tables from Sir Godfrey's work.

This book has benefited greatly from the many constructive criticisms of the manuscript given by Julian C. Stanley of the University of Wisconsin, Lyle Jones and Stanley A. Mulaik of the University of North Carolina, and Philip H. DuBois of Washington University. I regret that limitations of time have prevented me from incorporating some of their more insightful recommendations.

George A. Ferguson

STATISTICAL
ANALYSIS IN
PSYCHOLOGY
& EDUCATION

BASIC STATISTICS

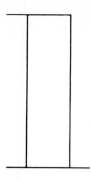

BASIC IDEAS IN STATISTICS

1.1 INTRODUCTION

This book is concerned with the elementary statistical treatment of experimental data in psychology, education, and related disciplines. The data resulting from any experiment are usually a collection of observations or measurements. The conclusions to be drawn from the experiment cannot be reliably ascertained by simple direct inspection of the data. Classification, summary description, and rules of evidence for the drawing of valid inference are required. Statistics provides the methodology whereby this can be done.

Implicit in any experiment is the presumption that it is possible to argue validly from the particular to the general and that new knowledge can be obtained by the process of inductive inference. The statistician does not assume that such arguments can be made with certainty. On the contrary, he assumes that some degree of uncertainty must attach to all such arguments; that some of the inferences drawn from the data of experiments are wrong. He further assumes that the uncertainty itself is amenable to precise and rigorous treatment, that it is possible to make rigorous statements about the uncertainty which attaches to any particular inference. Thus in the uncertain milieu of experimentation he applies a rigorous method.

A knowledge of statistics is an essential part of the training of all students in psychology. There are many reasons for this. *First*, an understanding of the modern literature of psychology requires a knowledge of statistical method and modes of thought. A high proportion of current books and journal articles either report experimental findings in statistical form or present theories or arguments involving statistical concepts.

3

These concepts play an increasing role in our thinking about psychological problems, quite apart from the treatment of data. The student need only consider, for example, the role of statistical concepts in current lines of theorizing in the field of learning to grasp the force of this argument. *Second*, training in psychology at an advanced level requires that the student himself design and conduct experiments. The design of an experiment is inseparable from the statistical treatment of the results. Experiments must be designed to enable the treatment of results in such a way as to permit clear interpretation, and to fulfill the purposes which motivated the experiment in the first place. If the design of an experiment is faulty, no amount of statistical manipulation can lead to the drawing of valid inferences. Experimental design and statistical procedures are two sides of the same coin. Thus not only must the advanced student conduct experiments and interpret results, he must plan his experiments in such a way that the interpretation of results can conform to known rules of scientific evidence. *Third*, training in statistics is training in scientific method. Statistical inference is scientific inference, which in turn is inductive inference, the making of general statements from the study of particular cases. These terms are for all practical purposes, and at a certain level of generality, synonymous. Statistics attempts to make induction rigorous. Induction is regarded by some scholars as the only way in which new knowledge comes into the world. While this statement is debatable, the role in modern society of scientific discovery through induction is obviously of the greatest importance. For this reason no serious student of psychology, or any other discipline, can afford not to know something of the rudiments of the scientific approach to problems. Statistical procedures and ideas play an important role in this approach.

1.2   THE BROAD ROLE OF QUANTIFICATION IN PSYCHOLOGY

While this book is largely concerned with elementary statistical procedures and ideas, some mention may be made of the broad role of quantitative method in psychology.

The attempt to quantify has a long and distinguished history in experimental psychology, which indeed may be regarded as synonymous with the history of that science itself. Since the experimental work in psychophysics of E. H. Weber and Gustav Fechner in the nineteenth century, determined attempts have been made to develop psychology as an experimental science. The early psychophysicists were concerned with the relationship between the "mind" and the "body" and developed certain mathematical functions which they held to be descriptive of that relationship. While much of their thinking on the mind-body problem has been discarded, their methods and techniques with development and elaboration are still used. Shorn of its philosophical and theoretical encumbrances, the work of the early psychophysicists was reduced in effect to the study of the relationship between measurements, obtained in two different ways, of what were presumed to be the same property. Thus, for example, they studied the relationship between weight, length, and temperature, defined by the responses of human subjects as instruments, and weight, length, and temperature, defined by other measuring instruments, scales, foot rules, and thermometers. A psychophysical law, so called, is a statement of the relationship between measurements obtained by these two methods. Modern psychophysics is concerned to some considerable extent with the scaling of the responses of the human subject as instrument and with the use of the human subject as instrument in dealing with a wide variety of practical problems. It may perhaps be referred to as human instrumentation.

The early psychophysicists invented certain experimental methods and developed statistical procedures for handling the data obtained by these methods. It is of interest to note that one method, the *constant process*, developed by G. E. Müller and F. M. Urban, has, with modification, found application in biological-assay work in assessing the potency of hormones, toxicants, and drugs of all types. It is currently known in biology as the *method of probits* (Finney, 1944, 1947).

Statistical methods have found extensive application in the psychological and educational testing field and in the study of human ability. Since the time of Binet, who developed the first extensively used test of intelligence and whose thinking was influenced by the early psychophysicists, a comprehensive body of theory and technique has been developed which is primarily statistical in type. This body of theory and technique is concerned with the construction of instruments for measuring human ability, personality characteristics, attitudes, interests, and many other aspects of behavior; with the nature and magnitude of the errors involved in such measurement; with the logical conditions which such measuring instruments must satisfy; with the quantitative prediction of human behavior; and with other related topics. Some of these topics are discussed in Part IV of this book.

The use of psychological tests stimulated the development of the techniques of factor analysis, which are used to some extent in contemporary psychology. Problems arise which involve a study of the relationships between sets of variables, sometimes as many as 50 or 60 and perhaps more. Factor analysis, by identifying underlying variables, attempts to provide a simplified description of these relationships, which facilitates an interpretation and comprehension of the information in the data. Factor analysis has found a number of uses in branches of science other than psychology, including meteorology and agriculture. An elementary introduction to factor analysis is given in Chap. 27 of this book.

Within recent years frequent use has been made of statistical concepts in the construction of models designed to provide some explanation and understanding of observable phenomena. Such models are used in the field of learning. Further, many biological scientists are currently concerned with the construction of models which may possibly bear some correspondence to the functioning of certain aspects of the central nervous system. While these attempts may be premature and their success cannot at this time be evaluated, it is possible that in future the models which will prove helpful in understanding the functioning of the human brain will either implicitly or explicitly involve statistical concepts. In a system comprised of a complex network of nerve fibers, the transmission of impulses can be conceived in probabilistic terms.

The study of the topics discussed above demands a knowledge of statistical method and a comprehension of the basic ideas of statistics as a starting point. It would seem that as psychology develops, increasing emphasis will be placed on quantitative procedure and an increasing degree of statistical sophistication will be required of the student.

1.3   STATISTICS AS THE STUDY OF POPULATIONS

Statistics is a branch of scientific methodology. It deals with the collection, classification, description, and interpretation of data obtained by the conduct of surveys and experiments. Its essential purpose is to describe and draw inferences about the numerical properties of populations. The terms, population and numerical property, require clarification.

In everyday language the term *population* is used to refer to groups or aggregates of people. We speak, for example, of the population of the United States, or of the state of Texas, or of the city of New York, meaning by this all the people who occupy defined geographical regions at specified times. This, however, is a particular usage of the term population. The statistician employs the term in a more general sense to refer not only to defined groups or aggregates of people, but also to defined groups or aggregates of animals, objects, materials, measurements, or "things" or "happenings" of any kind. Thus the statistician may define, for his particular purposes, populations of laboratory animals, trees, nerve fibers, liquids, soil, manufactured articles, automobile accidents, microorganisms, birds' eggs, insects, or fishes in the sea. On occasion he may deal with a population of measurements. By this is meant an indefinitely large aggregate of measurements which, hypothetically, might be obtained under specified experimental conditions. To illustrate, a series of measurements might be made of the length of a desk. Some or all of these measurements may differ one from another because of the presence of errors of measurement. This series of measurements may be regarded as part of an indefinitely large aggregate or population of measurements which might, hypothetically, be obtained by measuring the length of the desk over and over again an indefinitely large number of times.

The general concept implicit in all these particular uses of the word population is that of group or aggregation. The statistician's concern is with properties which are descriptive of the group or aggregation itself rather than with properties of particular members. Thus measurements may be made of the height and weight of a group of individuals. These measurements may be added together and divided by the number of cases to obtain the mean height and weight. These means describe a property of the group as a whole and are not descriptive of particular individuals. To illustrate further, a child may have an IQ of 90 and belong to a high socioeconomic group. Another child may have an IQ of 120 and belong to a low socioeconomic group. These facts as such about individual children do not directly concern the statistician. If, however, questions are raised about the proportion of children in a particular population or subpopulation with IQ's above or below a specified value, or if more general questions are raised about the relationship between intelligence and socioeconomic level, then these are questions of a statistical nature, and the statistician has techniques which assist their exploration.

The distinction is sometimes made between *finite* and *infinite* populations. The children attending school in the city of Chicago, the inmates of penitentiaries in Ontario, the cards in a deck are examples of finite populations. The members of such a population can presumably be counted, and a finite number obtained. The possible rolls of a die and the possible observations in many scientific experiments are examples of infinite or indefinitely large populations. The number of rolls of a die or the number of scientific observations may, at least theoretically, be increased without any finite limit. In many situations the populations which the statistician proposes to describe are finite, but so large that for all practical purposes they may be regarded as infinite. The 200 million or so people living in the United States constitute a large but finite population. This population is so large that for many types of statistical inference it may be assumed to be infinite. This would not apply to the cards in a deck, which may be thought of as a small finite population of 52 members.

Most populations are comprised of naturally distinguishable members, as is, of course, the case with people, animals, measurements, or the rolls of a die. Some populations are not so comprised, as is the case with liquids, soils, woven fabrics, or, for that matter, human behavior. How is it possible to apply the concept of group or aggregation to populations of this latter type? This may be done by defining the population member arbitrarily as a liter, a cubic centimeter, a square yard, or some such unit. The whole population may be thought to be composed of an aggregate of such members. Likewise, in the study of human behavior, the psychologist frequently concerns himself with arbitrarily defined bits of behavior, although behavior as such may perhaps be regarded as a continuous flow or sequence.

Statistics is concerned with the numerical properties of populations, that is, with properties to which numerals can in some manner be assigned. The logical implications of the term *numerical property* are complex and need not be elaborated here. To illustrate briefly, however, in any population of mental-hospital patients some may be classed as psychoneurotic, others as schizophrenic psychotic, others as psychotic with organic brain disease, and so on. Further, some patients may come from broken homes, while others may have a normal healthy home background. Some may have a history of mental disease in the family, and others may not. We may be said to apply a statistical method when we concern ourselves with how many patients in the population fall within these various classes, that is, how many are psychoneurotic, schizophrenic psychotic, and the like, and how many come from broken homes, how many do not, and so on. Further, the flicker-fusion rates of some part or all of the population may be measured and attention directed to the numbers of patients who fall within specified ranges of flicker-fusion rate, to mean rates for various classes of patients, and to related problems. The investigation of such problems as these may be said to involve a statistical method. In general, the statistician's concern is with those properties of populations which can be expressed in numerical form.

1.4   STATISTICS AS THE STUDY OF VARIATION

Statistics is sometimes conceptualized as the study of variation, because it provides a technology for the exploration of variation in the events of nature and for the making of inferences about the causal circumstances which underlie that variation. Emphasis on the study of variation originated with Darwin in *The Origin of Species* (1859). Variation was a central concept in the theory of natural selection because evolution could not occur without it. In Darwin's words,

> The many slight differences which appear in the offspring from the same parent . . . may be called individual differences. . . . These individual differences are of the highest importance for us, for they are often inherited, as must be familiar to everyone; and they thus afford materials for natural selection to act on and accumulate.

The matter is more clearly stated in an editorial in the first issue of the journal *Biometrika* (1901), probably written by Karl Pearson:

> The starting point of Darwin's theory of evolution is precisely the existence of those differences between individual members of a race or species which morphologists for the most part rightly neglect. The first condition necessary, in order that any process of Natural Selection may begin among a race, or species, is the existence of differences among its members; and the first step in an enquiry into the possible effect of a selective process upon any character of a race must be an estimate of the frequency with which individuals, exhibiting any given degree of abnormality with respect to that character, occur. The unit, with which such an enquiry must deal, is not an individual but a race, or a statistically representative sample of a race; and the result must take the form of a numerical statement, showing the relative frequency with which the various kinds of individuals composing the race occur.

Darwin made no direct contribution to statistical method. He did, however, create a theoretical context, based on observation and report, which made the study of variation meaningful and required, as it were, the development of statistical methods for its rigorous study. Darwin's disciple Galton understood fully the concept of variation. He was responsible for the initial applications of the so-called "normal" curve, or distribution, in psychological enquiry and made important contributions to the development of methods of correlation. He greatly influenced Karl Pearson, his disciple and biographer. Pearson perceived his role as that of helping to build a mathematical basis for evolutionary theory; he did, in fact, build the foundations of modern statistics. Between 1894 and 1916 he published 19 papers and monographs on statistical subjects, some of great importance and comprehensiveness. All these papers were titled *Contributions to the Mathematical Theory of Evolution.*

Despite influences from many sources, modern statistics is largely a direct emergent of the biological revolution of the nineteenth century which Darwin helped to create. The central concept in evolutionary theory which gave rise to this line of development was the concept of variation. Since Pearson the development of statistical method has been closely associated with the attempt to find solutions to biological problems. R. A. Fisher, who developed the analysis of variance and made contributions between 1920 and 1960 which exceeded those of any other living person, devoted his life primarily to statistical problems of experimentation in the biological sciences and to the mathematical foundations of genetics.

1.5 SAMPLES AND SAMPLING

Because of the large size of many populations, it may be either impracticable or impossible for the investigator to produce statistics based on all members. If, for example, interest is in investigating the attitudes of adult Canadians toward immigrants, it would obviously be a prohibitively expensive and time-consuming task to measure the attitudes of all adult Canadians and produce statistics based on a study of the complete population. If a population is indefinitely large, it is of course impossible, *ipso facto*, to produce complete population statistics. Under circumstances such as these the investigator draws what is spoken of as a sample. A sample is any subgroup or subaggregate drawn by some appropriate method from a population, the method used in drawing the sample being important. Methods used in drawing samples will be discussed in later chapters of this book. Having drawn his sample, the investigator utilizes appropriate statistical methods to describe its properties. He then proceeds to make statements about the properties of the population from his knowledge of the properties of the sample; that is, he proceeds to generalize from the sample to the population. To return to the example above, an investigator might draw a sample of 1,000 adult Canadians, the term *adult* being assigned a precise meaning, measure their attitudes toward immigrants using an acceptable technique of measurement, and calculate the required statistics. Questions may then be raised about the attitudes of all adult Canadians from the information obtained from a study of a sample of 1,000.

The fact that inferences can be made about the properties of populations from a knowledge of the properties of samples is basic in research thinking. Such statements are of course subject to error. The magnitude of the error involved in drawing such inferences can, however, in most cases be estimated by appropriate procedures. Where no estimate of error of any kind can be made, generalizations about populations from sample data are worthless.

Information about properties of particular samples, quite apart from any generalizations about the population, is of little intrinsic interest in itself. Consider a case where the investigator's interest is in the relative effects of two types of psychotherapy when applied to patients suffering from a particular mental disorder. He may select two samples of patients, apply one type of treatment to one sample and the other type of treatment to the other sample, and collect data on the relative rates of recovery of patients in the two samples. Clearly, in this case his interest is in finding out whether the one treatment is better or worse than the other when applied to the whole class of patients suffering from the mental disorder in question. He is interested in the sample data only in so far as these data enable him to draw inferences with some acceptable degree of assurance about this general question. His experimental procedures must be designed to enable the drawing of such inferences, otherwise the experiment serves no purpose. On occasion research reports are found where the investigator states that the experimental results obtained should not be generalized beyond the particular sample of individuals who participated in the study. The adoption of this view means that the investigator has missed the essential nature of experimentation. Unless the intention is to generalize from a sample to a population, unless the procedures used are such as to enable such generalizations justifiably to be made, and unless some estimate of error can be obtained, the conduct of experiments is without point.

Statistical procedures used in describing the properties of samples, or of populations where complete population data are available, are referred to by some writers as *descriptive statistics*. If we measure the IQ of the complete population of students in a particular university and compute the mean IQ, that mean is a descriptive statistic because it describes a characteristic of the complete population. If, on the other hand, we measure the IQ of a sample of 100 students and compute the mean IQ for the sample, that mean is also a descriptive statistic, because it describes a characteristic of that sample.

Statistical procedures used in the drawing of inferences about the properties of populations from sample data are frequently referred to as *sampling statistics*. If, for example, we wish to make a statement about the mean IQ in the complete population of students in a particular university from a knowledge of the mean computed on the sample of 100 and estimate the error involved in this statement, we use procedures from sampling statistics. The application of these procedures provides information about the accuracy of the sample mean as an estimate of the population mean; that is, it indicates the degree of assurance we may place in the inferences we draw from the sample to the population.

In this section no discussion is advanced on methods of drawing samples or the conditions which these methods must satisfy to allow the drawing of valid inferences from the sample to the population. Further, no precise meaning has been assigned to the term error. These topics will be elaborated at a later stage.

1.6 PARAMETERS AND ESTIMATES

A clear distinction is usually drawn between parameters and estimates. A *parameter* is a property descriptive of the population. The term *estimate* refers to a property of a sample drawn at random from a population. The sample value is presumed to be an estimate of a corresponding population parameter. Suppose, for example, that a sample of 1,000 adult male Canadians of a given age range is drawn from the total population, the height of the members of the sample measured, and a mean value, 68.972 in., obtained. This value is an estimate of the population parameter which would have been obtained had it been possible to measure all the members in the population. Usually parameters or population values are unknown. We estimate them from our sample values. The distinction between parameter and estimate reflects itself in statistical notation. A widely used convention in notation is to employ Greek letters to represent parameters and Roman letters to represent estimates. Thus the symbol σ, the Greek letter sigma, may be used to represent the standard deviation in the population, the standard deviation being a commonly used measure of variability. The symbol s may be used as an estimate of the parameter σ. This convention in notation is applicable only within broad limits. By and large we shall adhere to this convention in this book, although in certain instances it will be necessary to depart from it. By common practice and tradition a Greek letter may be used on occasion to denote a sample statistic.

1.7 VARIABLES AND THEIR CLASSIFICATION

The term *variable* refers to a property whereby the members of a group or set differ one from another. The members of a group may be individuals and may be found to differ in sex, age, eye color, intelligence, auditory acuity, reaction time to a stimulus, attitudes toward a political issue, and many other ways. Such properties are variables. The term *constant* refers to a property whereby the members of a group do not differ one from another. In a sense a constant is a particular type of variable; it is a variable which does not vary from one member of a group to another or within a particular set of defined conditions.

Labels or numerals may be used to describe the way in which one member of a group is the same as or different from another. With variables like sex, racial origin, religious affiliation, and occupation, labels are employed to identify the members which fall within particular classes. An individual may be classified as male or female; of English, French, or Dutch racial origin; Protestant or Catholic; a shoemaker or a farmer; and so on. The label identifies the class to which the individual belongs. Sex for most practical purposes is a two-valued variable, individuals being either male or female. Occupation, on the other hand, is a multivalued variable. Any particular individual may be assigned to any one of a large number of classes. With variables like height, weight, intelligence, and so on, measuring operations may be employed which enable the assignment of descriptive numerical values. An individual may be 72 in. tall, weigh 190 lb, and have an IQ of 90.

The particular values of a variable are referred to as *variates*, or *variate values*. To illustrate, in considering the height of adult males, height is the variable, whereas the height of any particular individual is a variate, or variate value.

In dealing with variables which bear a functional relationship one to another the distinction may be drawn between *dependent* and *independent* variables. Consider the expression $Y = f(X)$. This expression says that a given variable Y is some unspecified function of another variable X. The symbol f is used generally to express the fact that a functional relationship exists, although the precise nature of the relationship is not stated. In any particular case the nature of the relationship may be known; that is, we may know precisely what f means. Under these circumstances, for any given value of X a corresponding value of Y can be calculated; that is, given X and a knowledge of the functional relationship, Y can be predicted. It is customary to speak of Y, the predicted variable, as the dependent variable because the prediction of it depends on the value of X and the known functional relationship, whereas X is spoken of as the independent variable. Given an expression of a kind $Y = X^3$ for any given value of X, an exact value of Y can readily be determined. Thus if X is known, Y is also known exactly. Many of the functional relationships found in statistics permit probabilistic and not exact prediction to occur. Such relationships may provide the most probable value of Y for any given value of X, but do not permit the making of perfect predictions.

A distinction may be drawn between *continuous* and *discrete* (or *discontinuous*) variables. A continuous variable may take any value within a defined range of values. The possible values of the variable belong to a continuous series. Between any two values of the variable an indefinitely large number of in-between values may occur. Height, weight, and chronological time are examples of continuous variables. A discontinuous or discrete variable can take specific values only. Size of family is a discontinuous variable. A family may be comprised of 1, 2, 3 or more children, but values between these numbers are not possible. The values obtained in rolling a die are 1, 2, 3, 4, 5, and 6. Values between these numbers are not possible. Although the underlying variable may be continuous, all sets of real data in practice are discontinuous or discrete. Convenience and errors of measurement impose restrictions on the refinement of the measurement employed.

Another classification of variables is possible which is of some importance and is of particular interest to statisticians. This classification is based on differences in the type of information which different operations of classification or measurement yield. To illustrate, consider the following situations. An observer using direct inspection may rank order a group of individuals from the tallest to the shortest according to height. On the other hand, he may use a foot rule and record the height of each individual in the group in feet and inches. These two operations are clearly different, and the nature of the information obtained by applying the two operations is different. The former operation permits statements of the kind: individual A is taller or shorter than individual B. The latter operation permits statements of how much taller or shorter one individual is than another. Differences along these lines serve as a basis for a classification of variables, the class to which a variable belongs being determined by the nature of the information made available by the measuring operation used to define the variable. Four broad classes of variables may be identified. These are referred to as (1) nominal, (2) ordinal, (3) interval, and (4) ratio variables. A very interesting discussion relevant to this topic is given in Torgerson (1958).

A *nominal variable* is a property of the members of a group defined by an operation which permits the making of statements only of equality or difference. Thus we may state that one member is the *same as* or *different from* another member with respect to the property in question. Statements about the ordering of members, or the equality of differences between members, or the number of times a particular member is greater than or less than another are not possible. To illustrate, individuals may be classified by the color of their eyes. Color is a nominal variable. The statement that an individual with blue eyes is in some sense "greater than" or "less than" an individual with brown eyes is meaningless. Likewise the statement that the difference between blue eyes and brown eyes is equal to the difference between brown eyes and green eyes is meaningless. The only kind of meaningful statement possible with the information available is that the eye color of one individual is the same as or different from the eye color of another. A nominal variable may perhaps be viewed as a primitive type of variable, and the operations whereby the members of a group are classified according to such a variable constitute a primitive form of measurement. In dealing with nominal variables numerals may be assigned to represent classes, but such numerals are labels, and the only purpose they serve is to identify the members within a given class.

An *ordinal variable* is a property defined by an operation which permits the rank ordering of the members of a group; that is, not only are statements of equality and difference possible, but also statements of the kind *greater than* or *less than*. Statements about the equality of differences between members or the number of times one member is greater than or less than another are not possible. If a judge is required to order a group of individuals according to aggressiveness, or cooperativeness, or some other quality, the resulting variable is ordinal in type. Many of the variables used in psychology are ordinal.

An *interval variable* is a property defined by an operation which permits the making of statements of equality of intervals, in addition to statements of sameness or difference or greater than or less than. An interval variable does not have a "true" zero point, although a zero point may for convenience be arbitrarily defined. Fahrenheit and centigrade temperature measurements constitute interval variables. Consider three objects, A, B, and C, with temperatures 12°, 24°, and 36°, respectively. It is appropriate to say that the difference between the temperature of A and B is equal to the difference in the temperature of B and C. It is appropriate also to say that the difference between the temperature of A and C is twice the difference between the temperature of A and B or B and C. It is not appropriate to say that B has twice the temperature of A, or that C has three times the temperature of A. In common usage, if the temperature yesterday was 64° and today it was 32°, we would not say that it was twice as hot yesterday, or that the temperature was twice as great, as it was today. Calendar time is also an interval variable with an arbitrarily defined zero.

A *ratio variable* is a property defined by an operation which permits the making of statements of equality of ratios in addition to all other kinds of statements discussed above. This means that one variate value, or measurement, may be spoken of as double or triple another, and so on. An absolute zero is always implied. The numbers used represent distances from a natural origin. Length, weight, and the numerosity of aggregates are examples of ratio variables. One object may be twice as long as another, or three times as heavy, or four times as numerous. Many of the variables used in the physical sciences are of the ratio type. In psychological work, variables which conform to the requirements of ratio variables are uncommon. Scales for measuring loudness, pitch, and other variables have been developed by Stevens (1957) at Harvard. These appear to satisfy all the conditions of ratio variables.

The essential difference between a ratio and an interval variable is that for the former the measurements are made from a true zero point, whereas for the latter the measurements are made from an arbitrarily defined zero point or origin. Because of this, for a ratio variable ratios may be formed directly from the variate values themselves, and meaningfully interpreted. For an interval variable, ratios may be formed from differences between the variate values. The differences constitute a ratio variable, because the process of subtraction eliminates, or cancels out, the arbitrary origin. Differences are the same regardless of the location of the zero or origin.

A variety of refinements and elaborations can be made on the distinction between nominal, ordinal, interval, and ratio variables. For a more detailed discussion the reader is referred to Torgerson (1958).

Some writers distinguish between quantitative and qualitative variables without being explicit about the nature of this distinction. In the present classificatory system nominal and ordinal variables may be spoken of as qualitative, and interval and ratio variables as quantitative.

Statistical methods exist for the analysis of data composed of nominal variables, ordinal variables, and interval and ratio variables. From the viewpoint of practical statistical work in psychology and education the distinction between interval and ratio variables is perhaps unimportant, and it is convenient to think of three, and not four, classes of variables, with three corresponding classes of statistical method. Procedures for the analysis of interval and ratio variables constitute by far the largest, and most important, class of statistical method.

In practice we frequently apply methods appropriate to one class of variable in the statistical analysis of other classes of variables. This means that we either discard information which we do in fact possess or assume that we have information which we do not possess. An example of the former situation arises where measurements of the interval or ratio type are replaced by ranks for purposes of analysis. Measurements of height or weight for a group of N subjects may be replaced by the ranks 1, 2, 3, . . . , N, and subsequent analysis based on these ranks. Further measurements may be divided into broad classes, say top third, middle third, and bottom third, and treated as a nominal variable. A currently popular class of statistical method is called *nonparametric statistics* (see Part III). Many nonparametric statistical procedures convert problems involving interval and ratio variables to problems that involve a consideration of either nominal categories or ranks.

In the analysis of statistical data in psychology and education the investigator not infrequently assumes that he has information which actually he does not have. Variables which are in fact ordinal may be treated by a method appropriate for interval and ratio variables. An example of this situation arises when the members of a group are ordered with regard to some property. The information consists of relations of "greater than" or "less than," and these are described by a set of ordinal numbers; thus one member is first, another second, and so on. It is common practice to replace such a set of ordinal numbers by the corresponding set of cardinal numbers, 1, 2, 3, . . . , N, and to proceed to apply arithmetical operations to these numbers. This means that certain assumptions are made. Information is superimposed on the data which the measuring operation did not yield; that is, for computational purposes we assume we are in possession of information which actually we do not have. In the above instance we are making an assumption about the equality of intervals when in fact the measuring operation employed does not yield information of this kind. The assumption is that the difference between the first and second individual is equal to the difference between the second and third, and so on.

In psychological work many variables are in fact ordinal, although for statistical purposes they are, quite justifiably, commonly treated as if they were interval or ratio variables. For example, scores on intelligence tests, scholastic aptitude tests, attitude tests, personality tests, and the like, are in effect ordinal variables, although they are commonly treated as if they were of the interval or ratio type. No aspect of the operation of measuring intelligence, let us say, is such as to permit the making of meaningful statements about the equality of intervals or ratios. We cannot say that the difference in intelligence between a person with an IQ of 80 and one with an IQ of 90 is in any sense equal to the difference in intelligence between a person with an IQ of 110 and one with an IQ of 120. Nor can we meaningfully assert that a person with an IQ of 120 is twice as intelligent as a person with an IQ of 60. Such statements are without meaning. Despite this, IQ's are commonly treated by statistical methods which, from a rigorously logical viewpoint, are appropriate only to interval and ratio variables. The suggestion is not made here that the practice of assuming that we have information we do not have, or the converse practice of discarding information we do in fact have, be discontinued, although a logical purist might be led to this position. Frequently practical necessity dictates a particular procedure. Nevertheless it is a matter of some importance to know the nature of the information contained in the data. We should be able to distinguish clearly between this and the information either imposed or discarded for the purpose of making some process of calculation possible. In other words, our understanding of precisely what we are doing is enriched by knowing the nature of the assumptions made at each stage in the application of any procedure.

1.8   EXPERIMENTAL AND CORRELATIONAL INVESTIGATIONS

Much scientific enquiry is concerned with an exploration of the relations between variables. Some investigations involve a study of the relations between many variables; others, of the relation between two variables only, and are of the general form $Y = f(X)$. To illustrate the two-variable situation, Y may be a measure of motor performance; and X, four different dosages of alcohol, each dosage administered to 10 different individuals. Y may be a measure of intensity of belief in the prevalence of witches; and X, three specified times before, during, and after Halloween. Y may be average first-year marks in a university; and X, scores on a scholastic aptitude test. Again Y may be the presence or absence of cancer of the lung at age 50; and X, some index or measure of the amount of cigarette smoking. In these examples the investigator is concerned with an examination of the nature of the relation between two variables. The reader should note here that both Y and X may be nominal, ordinal, interval, or ratio variables. Thus both Y and X may be nominal, Y may be nominal and X ordinal or interval, and so on.

A useful and, indeed, important distinction may be made between *experimental* and *correlational* investigations. In an experiment the values of the X variable, and the frequency of occurrence of these values, are fixed by the investigator. In the illustrative example on the relation between motor performance and alcohol, the investigator determines the number of dosages, the amount of each dosage, and the number of experimental subjects receiving each dosage. If the experiment were repeated to check the results, the same dosage of alcohol would be used. In a correlational study the particular values of the variable, and the frequency of their occurrence, are not fixed, or controlled, by the investigator. In the example on the relation between first-year averages in university and scholastic aptitude test scores, the investigator may draw a sample of students for whom both first-year averages and scholastic aptitude test scores are available. He may then proceed to study the relation between these two variables. He exerts no control over the magnitude of particular scholastic aptitude test scores or the frequency of their occurrence. If the investigation were repeated, another sample of subjects would be used, and the particular scholastic aptitude test scores, and the frequency of their occurrence, might be expected to differ in some degree from that found in the first enquiry.

In psychology and education use is made of both experimental and correlational studies. The experimentalist is himself the creator of variation. The correlationist studies the variation which already exists in nature. Cronbach (1957) has examined in detail the relation between the experimental and correlational disciplines in psychology. He writes,

The well-known virtue of the experimental method is that it brings situational variables under tight control. It thus permits rigorous tests of hypotheses and confident statements about causation. The correlational method, for its part, can study what man has not learned to control. Nature has been experimenting since the beginning of time, with a boldness and complexity far beyond the resources of science. The correlator's mission is to observe and organize the data of nature's experiments.

1.9 ON CALCULATING

Skill in the operation of a calculating machine, and, if possible, some knowledge of electronic computers, should be acquired at an early stage in studying statistics. Many of the statistical problems which present themselves in experimental work in psychology and education involve much computation, and without a calculator the arithmetical labor is prohibitive. Indeed, many forms of modern statistical analysis are so complex that their application is simply not possible without access to a high-speed electronic computer. Skill in the operation of a calculator can be readily acquired, a reasonable level of performance on simple operations being attained by most students in a few hours of practice.

Statistical tables of various kinds are frequently required in computational work. A useful aid in computing is Barlow's Tables. The 1965 paperback edition of these tables provides the squares, cubes, square roots, cube roots, and reciprocals of all integers up to 12,500. The reader may also find *Basic Mathematical and Statistical Tables for Psychology and Education* (Meredith, 1967) a convenient source of information on many numerical functions and commonly used probability distributions. *Statistical Tables for Biological, Agricultural and Medical Research*, by Fisher and Yates (1963), has been widely used by statisticians for many years and currently is available in its sixth edition.

In computing, the importance of adequate checks on the accuracy of the calculation cannot be too emphatically stressed. Every calculation should be checked by the employment of some checking device which guarantees accuracy. There is no substitute for accuracy. The conduct of an experiment serves no purpose unless correct inferences are drawn from the data. The correctness of the inferences drawn cannot be assured unless the statistical procedures employed are appropriate to the data and unless these procedures are accurately applied. Students not infrequently feel that the statistical analysis of a set of data is laborious and time-consuming, and in their haste to arrive at some kind of result they may disregard checks which are necessary to ensure the accuracy of their calculations. When tempted in this direction, the student should observe that the time spent in the proper statistical analysis of a set of data represents in most instances a small proportion of the time required to plan the experiment and gather the data. A slipshod analysis may throw in jeopardy the total investment of time and effort.

The availability of electronic computers has changed many aspects of modern statistical work. Many computational procedures which have been devised to simplify arithmetical calculation are now obsolete. Forms of calculation which were hitherto impossible can now be readily carried out in short periods of time. Many investigators now conduct elaborate analyses of such complexity that their proper interpretation demands much in the way of statistical sophistication. Not uncommonly investigators are bewildered by the deluge of calculations which computers quickly regurgitate. The computer has clearly not only enhanced the importance of an understanding of basic statistical ideas, as distinct from computational methods, but has also reemphasized the importance of simple ways of viewing complex problems. Also, the computer has led to the increased use of multivariate statistical methods and has stimulated the development of such methods. These are methods for the treatment of data comprised of many variables.

1.10   UNITS OF MEASUREMENT

When dealing with continuous variables a unit of measurement may be regarded as any defined subdivision of a scale, however fine. In measuring length the units may be inches, yards, and miles or centimeters, meters, and kilometers. In measuring weight the units may be ounces and pounds or grams and kilograms. In measuring chronological time the units may be seconds, minutes, hours, days, months, or years.

With continuous variables, although all values are theoretically possible within any range of values, we select a unit of measurement and record our observations as discrete values. All experimental observations, however obtained, are recorded as discrete values. Thus the length of a desk or the height of a man may be measured to the nearest inch, or tenth of an inch, or hundredth of an inch, the unit of measurement in each case being 1 in., $\frac{1}{10}$ in., or $\frac{1}{100}$ in., respectively, and the number of such units involved in any particular measurement must, of necessity, be recorded as a discrete number.

The fineness of the unit of measurement employed is determined by the accuracy which the nature of the situation demands or by the accuracy which the instrument of measurement allows, or both. In the measurement of time intervals, for example, great accuracy can be obtained by the use of appropriate measuring devices. In measuring the time required for a child to solve a problem it is certainly adequate for all practical purposes to record the observation in seconds. In reaction-time experiments, however, we may require a unit of measurement of a hundredth or perhaps a thousandth part of a second. Further, the unit should reflect the accuracy of the measuring operation. To illustrate, an intelligence quotient is calculated by dividing mental age by chronological age, both expressed in months, and multiplying by 100. Quite clearly, we could speak of a child's intelligence quotient as being 103.3, or 103.23, or something of the sort. Such an attempt at accuracy would be spurious because of the large error of measurement which is known to attach to the intelligence quotient. In practice, intelligence quotients are always recorded to the nearest whole number.

When we record measurements of a continuous variable as discrete numbers in so many units, we imply in most cases that had a more accurate form of measurement been used, were this possible and desirable, the value thereby obtained would fall within certain limits, these limits being defined as one-half a unit above and below the value reported. Thus when we report a measurement to the nearest inch, say, 26 in., this is assumed to mean that the observation falls within the limits 25.5 and 26.5, or more precisely that it is *greater than or equal to* 25.5 and *less than* 26.5. Likewise, a measurement made to the nearest tenth part of an inch, say, 31.7, is assumed to fall within the limits 31.65 and 31.75. In a reaction-time experiment a particular observation measured to the nearest thousandth of a second might be, say, .196 sec. This means that the measurement is taken as falling within the limits .1955 and .1965 sec.

An exception to the above is age. When we state that a person is 18 years old, we do not mean in conventional usage that his age falls within the limits 17 years 6 months and 18 years 6 months. A person is ordinarily spoken of as 18 years old until his 19th birthday. His age is greater than or equal to 18 years and less than 19 years. Similarly to state that a person is 126 months old means that he is greater than or equal to 126 months and less than 127 months. Definitions of age other than the above are used for particular purposes.

Questions of the above kind do not, of course, arise with discrete variables. The number of animals in a cage, or children in a classroom, or teeth in a child's head are discrete observations, and to imply a range of values within which any particular observation is assumed to fall is not meaningful.

1.11 SUMMATION NOTATION

Statistical notation is a language with its own grammatical rules. One of the more frequently used forms of notation is spoken of as summation notation. Some familiarity with this class of notation should be acquired as early as possible in the study of statistics.

Let X be a variable and $X_1, X_2, \ldots, X_N$ a set of variate values. To illustrate, X might refer to a measure of the activity of rats in a maze. The symbols $X_1, X_2, \ldots, X_N$ would then refer to measures of activity for individual rats, there being N rats in the group. The sum $X_1 + X_2 + \cdots + X_N$, that is, all the individual measures added together, may be written as

$$\sum_{i=1}^{N} X_i$$

Thus

[1.1] $$\sum_{i=1}^{N} X_i = X_1 + X_2 + \cdots + X_N$$

The symbol Σ is the Greek capital letter sigma and refers to the simple operation of adding things up. The symbols above and below the summation sign define the limits of the summation. Thus $\sum_{i=1}^{N}$ means the addition of all values formed by assigning to i the values of every positive integer from $i = 1$ to $i = N$, inclusive. For example, let the numbers 10, 12, 19, 21, 32 be measures of activity for a group of five rats. The sum of these five scores may be represented symbolically by $\sum_{i=1}^{5} X_i$ and in this case is equal to 94. Where the limits of the summation are clearly understood from the context, which is very frequently the case, it is customary to omit the notation above and below the summation sign and write ΣX_i, or simply ΣX.

There are a number of very simple theorems which are useful in handling problems involving summation notation.

Theorem 1   If every variate value in a group is multiplied by a constant number or factor, that factor may be removed from under the summation sign and written outside as a factor. Thus

[1.2] $$\sum_{i=1}^{N} cX_i = cX_1 + cX_2 + \cdots + cX_N$$
$$= c(X_1 + X_2 + \cdots + X_N)$$
$$= c \sum_{i=1}^{N} X_i$$

This means that if we multiply each one of the measures 10, 12, 19, 21, 32 by any constant, say, 5, the sum of the resulting measures will be given directly by 5×94.

Theorem 2   The summation of a constant over N terms is equal to Nc. Thus

[1.3] $$\sum_{i=1}^{N} c = c + c + \cdots + c$$
$$= Nc$$

If $c = 5$ and N equals 4, it is obvious that $5 + 5 + 5 + 5 = 4 \times 5 = 20$.

Theorem 3   The summation of the sum of any number of terms is the sum of the summations of these terms taken separately. Thus

[1.4] $\displaystyle\sum_{i=1}^{N} (X_i + Y_i + Z_i) = X_1 + Y_1 + Z_1 + X_2 + Y_2 + Z_2 + \cdots + X_N + Y_N + Z_N$

$$= \sum_{i=1}^{N} X_i + \sum_{i=1}^{N} Y_i + \sum_{i=1}^{N} Z_i$$

Theorem 4   The sum of the first N integers is

[1.5] $\dfrac{N(N+1)}{2}$

Consider the integers $1, 2, 3, \ldots, (N-2), (N-1), N$. It is observed that the sum of the first and last integers in the series is equal to $N+1$, the sum of the second and the second from the last is equal to $N+1$, and so on. In any series where N is even, there are $N/2$ such pairs, and the sum of the series is given by $N(N+1)/2$. Where N is odd, there are $(N-1)/2$ such pairs, plus the middle term, which is equal to $(N+1)/2$. The sum of the series is then

$$\frac{(N-1)}{2} (N+1) + \frac{(N+1)}{2} = \frac{N(N+1)}{2}$$

An expression frequently encountered in statistics is $\displaystyle\sum_{i=1}^{N} X_iY_i$. This refers to the sum of the products of two sets of paired numbers. If, for example, 5, 6, 12, 15 are the scores, X, of four people on a test, and 2, 3, 7, 10 are the scores, Y, of the same four people on another test, then $\displaystyle\sum_{i=1}^{N} X_iY_i$ refers to the sum of products and is equal to $5 \times 2 + 6 \times 3 + 12 \times 7 + 15 \times 10$, or 262.

The following are examples which illustrate the application of the above theorems.

Example 1 $\displaystyle\sum_{i=1}^{N} (X + c) = \sum_{i=1}^{N} X + \sum c = \sum_{i=1}^{N} X + Nc$

Example 2 $\displaystyle\sum_{i=1}^{N} (X + Y + c) = \sum_{i=1}^{N} X + \sum_{i=1}^{N} Y + \sum_{i=1}^{N} c = \sum_{i=1}^{N} X + \sum_{i=1}^{N} Y + Nc$

Example 3 $\displaystyle\sum_{i=1}^{N} (X + c)^2 = \sum_{i=1}^{N} (X^2 + 2cX + c^2) = \sum_{i=1}^{N} X^2 + \sum_{i=1}^{N} 2cX + \sum_{i=1}^{N} c^2$

$$= \sum_{i=1}^{N} X^2 + 2c \sum_{i=1}^{N} X + Nc^2$$

Example 4 $\displaystyle\sum_{i=1}^{N} (X + Y)^2 = \sum_{i=1}^{N} (X^2 + 2XY + Y^2) = \sum_{i=1}^{N} X^2 + \sum_{i=1}^{N} 2XY + \sum_{i=1}^{N} Y^2$

$$= \sum_{i=1}^{N} X^2 + 2 \sum_{i=1}^{N} XY + \sum_{i=1}^{N} Y^2$$

Example 5 $\displaystyle\sum_{i=1}^{N} [(X+c)^2 - c^2] = \sum_{i=1}^{N} [X^2 + 2cX + c^2 - c^2]$

$$= \sum_{i=1}^{N} X^2 + \sum_{i=1}^{N} 2cX = \sum_{i=1}^{N} X^2 + 2c \sum_{i=1}^{N} X$$

Example 6 $\displaystyle\sum_{i=1}^{N} [(X+Y)^2 - (X-Y)^2] = \sum_{i=1}^{N} [(X^2 + Y^2 + 2XY) - (X^2 + Y^2 - 2XY)]$

$$= \sum_{i=1}^{N} [X^2 + Y^2 + 2XY - X^2 - Y^2 + 2XY]$$

$$= \sum_{i=1}^{N} 4XY = 4 \sum_{i=1}^{N} XY$$

Example 7 $\displaystyle\sum_{i=1}^{N} (X - 3Y)^2 = \sum_{i=1}^{N} (X^2 + 9Y^2 - 6XY) = \sum_{i=1}^{N} X^2 + \sum_{i=1}^{N} 9Y^2 - \sum_{i=1}^{N} 6XY$

$$= \sum_{i=1}^{N} X^2 + 9 \sum_{i=1}^{N} Y^2 - 6 \sum_{i=1}^{N} XY$$

Example 8   Let X denote the integers 1, 2, 3, . . . , N.

$$\sum_{i=1}^{N} (X + c) = \sum_{i=1}^{N} X + Nc = \frac{N(N+1)}{2} + Nc$$

The notation used in elementary statistics is simple, and skill in its manipulation can be acquired with a little practice. A good understanding of the nature of statistical method and its applications can be acquired with very little in the way of mathematical training at all. A little knowledge of arithmetic and a little elementary algebra go a long way in the study of statistics.

EXERCISES

1   Indicate with examples the differences between (*a*) population and sample, (*b*) finite and infinite populations, (*c*) descriptive and sampling statistics, (*d*) parameters and estimates, (*e*) dependent and independent variables, (*f*) continuous and discrete variables, (*g*) experiments and correlational studies.

2   Classify the following as nominal, ordinal, interval, or ratio variables: (*a*) height, (*b*) weight, (*c*) examination marks, (*d*) sex, (*e*) eye color, (*f*) calendar time, (*g*) age, (*h*) racial origin, (*i*) temperature, (*j*) ratings of scholastic success.

3   Write the following in summation notation:
   a $X_1 + X_2 + \cdots + X_{15}$
   b $Y_1 + Y_2 + \cdots + Y_N$

$\quad$ **c** $\quad (X_1 + Y_1) + (X_2 + Y_2) + \cdots + (X_7 + Y_7)$
$\quad$ **d** $\quad X_1Y_1 + X_2Y_2 + \cdots + X_NY_N$
$\quad$ **e** $\quad X_1{}^3Y_1 + X_2{}^3Y_2 + \cdots + X_N{}^3Y_N$
$\quad$ **f** $\quad (X_1 + c) + (X_2 + c) + \cdots + (X_8 + c)$
$\quad$ **g** $\quad cX_1 + cX_2 + \cdots + cX_{25}$
$\quad$ **h** $\quad X_1/c + X_2/c + \cdots + X_N/c$
$\quad$ **i** $\quad cX_1{}^2Y_1 + cX_2{}^2Y_2 + \cdots + cX_N{}^2Y_N$

4 $\quad$ Write each of the following in full:

$\quad$ **a** $\displaystyle\sum_{i=1}^{2} X_i$ $\qquad\qquad$ **d** $\displaystyle c\sum_{i=1}^{3} X_i{}^2Y_i$

$\quad$ **b** $\displaystyle\sum_{i=1}^{3} X_iY_i$ $\qquad\qquad$ **e** $\displaystyle\sum_{i=1}^{4} X_i + 4c$

$\quad$ **c** $\displaystyle\sum_{i=1}^{5} (X_i + Y_i)$ $\qquad\quad$ **f** $\displaystyle\frac{1}{c}\sum_{i=1}^{5} X_i + c\sum_{i=1}^{5} Y_i$

5 $\quad$ Show that

$$\sum_{i=1}^{N} (X_i + c)^2 = \sum_{i=1}^{N} X_i{}^2 + 2c\sum_{i=1}^{N} X_i + Nc^2$$

6 $\quad$ Consider the following paired observations:

$\quad X_1 = 3 \qquad\qquad Y_1 = 9$
$\quad X_2 = 5 \qquad\qquad Y_2 = 2$
$\quad X_3 = 5 \qquad\qquad Y_3 = 1$
$\quad X_4 = 4 \qquad\qquad Y_4 = 5$
$\quad X_5 = 8 \qquad\qquad Y_5 = 3$

$\quad$ Calculate:

$\quad$ **a** $\sum X_i{}^2$ $\qquad\qquad\qquad$ **f** $\sum (X_i - 5)(Y_i - 4)$
$\quad$ **b** $\sum Y_i{}^2$ $\qquad\qquad\qquad$ **g** $\sum (5X_i - 4Y_i)^2$
$\quad$ **c** $\sum (X_i - 5)^2$ $\qquad\qquad$ **h** $\sum (X_i - 3^2)^2$
$\quad$ **d** $\sum (Y_i - 4)^2$ $\qquad\qquad$ **i** $\sum (X_i/Y_i)$
$\quad$ **e** $\sum X_iY_i$

$\quad$ In all instances the summation is understood to extend over the five paired observations.

7 $\quad$ Which of the following are true and which are false?

$\quad$ **a** $\displaystyle\sum_{i=1}^{N} X_i \sum_{i=1}^{N} Y_i = \sum_{i=1}^{N} X_iY_i$

$\quad$ **b** $\displaystyle\left(\sum_{i=1}^{N} X_i\right)^2 = \sum_{i=1}^{N} X_i{}^2$

$\quad$ **c** $\displaystyle\sum_{i=1}^{N} (X_i + c)(X_i - c) = \sum_{i=1}^{N} X^2 - Nc^2$

$\quad$ **d** $\displaystyle\sum_{i=1}^{N} (X_i + Y_i)^2 = \sum_{i=1}^{N} X_i{}^2 + \sum_{i=1}^{N} Y_i{}^2 + 2\sum_{i=1}^{N} X_iY_i$

8 $\quad$ What is the sum of the first 100 integers?

9 If $X_1 = 4$, $X_2 = 6$, $X_3 = 3$, and $X_4 = 7$, obtain the following:

   a $\sum\limits^{4} (X_i^2 - 4X_i)$ **b** $\sum\limits^{4} (X_i^3 - X_i^2 + X_i)$

10 Obtain

   a $\sum\limits^{4} 6$ **b** $\sum\limits^{N} c$ **c** $\sum\limits^{N} 2$

FREQUENCY DISTRIBUTIONS
AND THEIR
GRAPHIC REPRESENTATION

2.1 INTRODUCTION

The data obtained from the conduct of experiments or correlational studies are frequently collections of numbers. Classification and description of these numbers are required to assist interpretation. Under certain circumstances advantages attach to the classification of the data in the form of frequency distributions. Such classification may help the investigator to understand important features of the data. A frequency distribution is an arrangement of the data that shows the frequency of occurrence of the different values of the variable or the frequency of occurrence of values falling within arbitrarily defined intervals of the variable. The meaning of this latter statement will become clear as we proceed. This chapter discusses the organization of data in the form of frequency distributions, and the graphic representation of frequency distributions. Chaps. 3 and 4, to follow, discuss the descriptive numerical properties of frequency distributions, or the properties of the collections of numbers of which these distributions are comprised.

2.2 CLASSIFICATION OF DATA

Consider the data in Table 2.1. These are the intelligence quotients of 100 children obtained from a psychological test. As a first step in the direction of classification we may rank order the 100 intelligence quotients in order of magnitude, proceeding from the largest to the smallest as shown in

Table 2.1*

Intelligence quotients made by 100 pupils on a mental test

109	111	82	105	134
113	90	79	100	117
80	90	121	75	93
99	90	92	96	82
101	104	80	81	83
104	93	109	72	110
111	91	109	111	81
122	83	92	101	77
99	103	93	91	67
108	93	84	84	100
102	84	96	89	81
107	95	91	107	102
109	93	82	103	116
86	78	73	104	104
103	108	76	94	108
72	87	121	80	127
105	103	106	119	90
93	89	110	103	100
99	79	117	114	117
93	82	98	89	119

* Tables 2.1 to 2.6 are reproduced from R. W. B. Jackson and George A. Ferguson, *Manual of educational statistics*, University of Toronto, Department of Educational Research, Toronto, 1942.

Table 2.2

Rank distribution of intelligence quotients shown in Table 2.1

134	109	102	93	82
127	109	101	92	82
122	108	101	92	82
121	108	100	91	82
121	108	100	91	81
119	107	100	91	81
119	107	99	90	81
117	106	99	90	80
117	105	99	90	80
117	105	98	90	80
116	104	96	89	79
114	104	96	89	79
113	104	95	89	78
111	104	94	87	77
111	103	93	86	76
111	103	93	84	75
110	103	93	84	73
110	103	93	84	72
109	103	93	83	72
109	102	93	83	67

Table 2.2. An arrangement of this kind is called a *rank distribution*. Such an arrangement of data has few advantages. Inspection of the rank data, however, shows that many scores occur more than once; thus there are five 103's, three 100's, and so on. This suggests that the data might be arranged in columns, as shown in Table 2.3, one column listing the possible scores and the other listing the number of times each score occurs. Such an arrangement of data is a *frequency distribution*, and the number of times a particular score value occurs is a *frequency*, represented by the symbol f.

In Table 2.3 the data have been classified in as many classes as there are score values within the total range of the variable. The number of classes is large. Usually it is advisable to reduce the number of classes by arranging the data in arbitrarily defined groupings of the variable; thus all scores within the range 65 to 69, that is, all scores with the values 65, 66, 67, 68,

Table 2.3 Frequency distribution of intelligence quotients of Table 2.1 with as many classes as score values

Score	f	Score	f	Score	f	Score	f
134	1	117	3	100	3	83	2
133	—	116	1	99	3	82	4
132	—	115	—	98	1	81	3
131	—	114	1	97	—	80	3
130	—	113	1	96	2	79	2
129	—	112	—	95	1	78	1
128	—	111	3	94	1	77	1
127	1	110	2	93	7	76	1
126	—	109	4	92	2	75	1
125	—	108	3	91	3	74	—
124	—	107	2	90	4	73	1
123	—	106	1	89	3	72	2
122	1	105	2	88	—	71	—
121	2	104	4	87	1	70	—
120	—	103	5	86	1	69	—
119	2	102	2	85	—	68	—
118	—	101	2	84	3	67	1

and 69, may be grouped together. All scores within the ranges 70 to 74, 75 to 79, and so on, may be similarly grouped. Such groupings of data are usually done by entering a tally mark for each score opposite the range of the variable within which it falls and counting these tally marks to obtain the number of cases within the range. This procedure is shown in Table 2.4.

The range of the variable adopted is called the *class interval*. In the illustration in Table 2.4 the class interval is 5. This arrangement of data is also a frequency distribution, and the number of cases falling within each class interval is a frequency. The only difference between Tables 2.3 and 2.4 is in the class interval, which is 1 in the former case and 5 in the latter.

2.3 CONVENTIONS REGARDING CLASS INTERVALS

In the arrangement of data with a class interval of 1, as shown in Table 2.3, the original observations are retained and may be reconstructed directly from the frequency distribution without loss of information. If the class interval is greater than 1, say, 2, 5, or 10, some loss of information regarding individual observations is incurred; that is, the original observations cannot be reproduced exactly from the frequency distribution. If the class interval is large in relation to the total range of the set of observations, this loss of information may be appreciable. If the class interval is small, the classification of data in the form of a frequency distribution may lead

Table 2.4   Frequency distribution of the intelligence quotients of Table 2.1

Class interval	Tally	Frequency
130–134	/	1
125–129	/	1
120–124	///	3
115–119	丗 /	6
110–114	丗 //	7
105–109	丗 丗 //	12
100–104	丗 丗 丗 /	16
95–99	丗 //	7
90–94	丗 丗 丗 //	17
85–89	丗	5
80–84	丗 丗 丗	15
75–79	丗 /	6
70–74	///	3
65–69	/	1
Total		100

to very little gain in convenience over the utilization of the original observations.

The rules listed below are widely used in the selection of class intervals. These rules lead in most cases to a convenient handling of the data.

1 Select a class interval of such a size that between 10 and 20 such intervals will cover the total range of the observations. For example, if the smallest observation in a set were 7 and the largest 156, a class interval of 10 would be appropriate and would result in an arrangement of the data into 16 intervals. If the smallest observation were 2 and the largest 38, a class interval of 3 would result in an arrangement of 14 intervals. If the observations ranged from 9 to 20, a class interval of 1 would be convenient.

2 Select class intervals with a range of 1, 2, 3, 4, 5, 10, or 20 points. These will meet the requirements of most sets of data.

3 Start the class interval at a value which is a multiple of the size of that interval. For example, with a class interval of 5, the intervals should start with the values 5, 10, 15, 20, etc. With a class interval of 2, the intervals should start with the values 2, 4, 6, 8, 10, etc. This is, of course, highly arbitrary.

4 Arrange the class intervals according to the order of magnitude of the observations they include, the class interval containing the largest observations being placed at the top.

2.4   EXACT LIMITS OF THE CLASS INTERVAL

Where the variable under consideration is continuous, and not discrete, we select a unit of measurement and record our observations as discrete values. When we record an observation in discrete form and the variable is a continuous one, we imply that the value recorded represents a value falling within certain limits. These limits are usually taken as one-half a unit above and below the value reported. Thus when we report a measurement to the nearest inch, say, 16 in., we mean that, if a more accurate form of measurement had been used, the value obtained would fall within the limits 15.5 and 16.5 in. Similarly, a measurement made to the nearest tenth part of an inch, say, 31.7 in., is understood to fall within the limits 31.65 and 31.75 in. In a reaction-time experiment a particular observation measured to the nearest thousandth of a second might be, say, .196 sec. This assumes that had a more accurate timing device been used, the measurement would have been found to fall somewhere within the limits .1955 and .1965 sec.

Class intervals are usually recorded to the nearest unit and thereby reflect the accuracy of measurement. For various reasons it is frequently necessary to think in terms of so-called *exact limits* of the class interval. These are sometimes spoken of as *class boundaries*, or *end values*, and sometimes as *real limits*. Consider the class interval 95 to 99 in Table 2.4. We grouped within this interval all measurements taking the values 95, 96, 97, 98, and 99. The limits of the lower value are 94.5 and 95.5, while those of the upper value are 98.5 and 99.5. The total range, or exact limits, which the interval is presumed to cover is then clearly 94.5 and 99.5, which means all values greater than or equal to 94.5 and less than 99.5.

The above discussion is applicable to continuous variables only. With discrete variables no distinction need be made between the class interval and the exact limits of the interval, the two being identical.

Table 2.5 shows the frequency distribution of the intelligence quotients of Table 2.1. Column 1 shows the class interval as usually written, while col. 2 records the exact limits. In practice, of course, the exact limits are rarely recorded as in Table 2.5.

2.5   DISTRIBUTION OF OBSERVATIONS
WITHIN THE CLASS INTERVAL

The grouping of data in class intervals results in a loss of information regarding the individual observations themselves. Scores may differ one from another within a limited range, and yet all be grouped within the same interval. In the calculation of certain statistics and in the preparation of graphs it becomes necessary to make certain assumptions regarding the values within the intervals. Two separate assumptions may be made, depending on the purposes we have in mind.

The first assumption states that the observations are uniformly distributed over the exact limits of the interval. This assumption is made in the calculation of such statistics as the median, quartiles, and percentiles and in the drawing of histograms. In Table 2.5 it will be observed that 16 cases

Table 2.5 Class intervals, exact limits, and mid-points for frequency distribution of intelligence quotients

1 Class interval	2 Exact limits	3 Mid-point of interval	4 Frequency
130–134	129.5–134.5	132.0	1
125–129	124.5–129.5	127.0	1
120–124	119.5–124.5	122.0	3
115–119	114.5–119.5	117.0	6
110–114	109.5–114.5	112.0	7
105–109	104.5–109.5	107.0	12
100–104	99.5–104.5	102.0	16
95–99	94.5–99.5	97.0	7
90–94	89.5–94.5	92.0	17
85–89	84.5–89.5	87.0	5
80–84	79.5–84.5	82.0	15
75–79	74.5–79.5	77.0	6
70–74	69.5–74.5	72.0	3
65–69	64.5–69.5	67.0	1
Total			100

fall within the interval 100 to 104, which has the exact limits 99.5 to 104.5. The assumption states that these 16 cases are distributed over the interval as follows:

Interval	Frequency
103.5–104.5	3.2
102.5–103.5	3.2
101.5–102.5	3.2
100.5–101.5	3.2
99.5–100.5	3.2
Total	16.0

The second widely used assumption states that all the observations are concentrated at the mid-point of the interval, that is, that all the observations for that interval are the same and equal to the value corresponding to the mid-point of the interval. The mid-point of any class interval is halfway between the exact limits of the interval. In the above example the mid-point of the interval 99.5 to 104.5 is 102. This second assumption is ordinarily made in the calculation of such statistics as means, standard deviations, and in the drawing of frequency polygons.

The determination of the mid-point of a class interval should present no difficulty. The mid-point may be conveniently obtained by adding one-half of the range of the class interval to the lower exact limit of that interval. Thus with the interval 100 to 104 the lower limit is 99.5 and one-half the class interval is 2.5. The mid-point is therefore 99.5 + 2.5, or 102. Consider a 10-point class interval written in the form 100 to 109. Here the lower limit is 99.5 and one-half the class interval is 5. The mid-point is then 99.5 + 5, or 104.5. Table 2.5, col. 3, shows the mid-points of the corresponding class intervals.

2.6 CUMULATIVE FREQUENCY DISTRIBUTIONS

Situations occasionally arise where our concern is not with the frequencies within the class intervals themselves, but rather with the number or percentage of values "greater than" or "less than" a specified value. Such information may be made readily available by the preparation of a cumulative frequency distribution. The cumulative frequencies are obtained by adding successively, starting from the bottom, the individual frequencies. Table 2.6 shows the cumulative frequencies and cumulative percentages for a distribution of intelligence quotients.

Table 2.6 Cumulative frequencies and cumulative percentage frequencies for distribution of intelligence quotients

1 Class interval (IQ's)	2 Frequency	3 Cumulative frequency	4 Cumulative percentage frequency
130–134	1	106	100.0
125–129	3	105	99.1
120–124	4	102	96.2
115–119	10	98	92.5
110–114	8	88	83.0
105–109	15	80	75.5
100–104	20	65	61.3
95–99	14	45	42.5
90–94	11	31	29.2
85–89	8	20	18.9
80–84	6	12	11.3
75–79	5	6	5.7
70–74	0	1	.9
65–69	1	1	.9
Total	106		

2.7 TABULAR REPRESENTATION

Statistical data are frequently arranged and presented in the form of tables. Such tables should be designed to enable the reader to grasp with minimal effort the information which they intend to convey. While very considerable variety in the design of statistical tables is possible, a number of general rules should be observed. Kenney (1954) lists six such rules, and these are as follows:[1]

1   Every table must be self-explanatory. To accomplish this the title should be short, but not at the expense of clearness.

2   Full explanatory notes, when necessary, should be incorporated in the table, either directly under the descriptive title and before the body of the table, or else directly under the table.

3   The columns and rows should be arranged in logical order to facilitate comparisons.

4   In tabulating long columns of figures, space should be left after every five or ten rows. Long unbroken columns are confusing, especially when one is comparing two numbers in a row but in widely separated columns.

5   If the numbers tabulated have more than three significant figures, the digits should be grouped in threes. Thus, one should write 4 685 732, not 4685732.

6   Double lines at the top (or at the top and bottom) may enhance the effectiveness of a table. If the table nicely fills the width of a page, no side lines should be used. In such cases the omission of the side lines will have the tendency to emphasize the other vertical lines and cause the interior columns to stand out better. The columns should not be widely separated, and the form of a narrow, compact table should have its side lines.

Tables presented as part of a manuscript should be appropriately numbered and should be inserted where possible in close proximity to the place where they are referred to in the text, otherwise the reader is put to some inconvenience.

The appropriate design of statistical tables can become a matter of some complexity. This is particularly the case where it is necessary to present data which are cross-classified in a variety of ways.

2.8 GRAPHIC REPRESENTATION OF FREQUENCY DISTRIBUTIONS

Graphic representation is often of great help in enabling us to comprehend the essential features of frequency distributions and in comparing one frequency distribution with another. A graph is the geometrical image of a set of data. It is a mathematical picture. It enables us to think about a problem in visual terms. Graphs are used not only in the practical handling of real sets of data, but also as visual models in thinking about statistical problems. Many problems can be reduced to visual form, and such reduction often facilitates their understanding and solution. Graphs have become a part of our everyday activity. Newspapers, popular magazines,

[1] Reproduced, with permission, from John F. Kenney and E. S. Keeping, *Mathematics of statistics*, part 1, 3d ed., copyright 1954, D. Van Nostrand Company, Inc., Princeton, N.J.

trade publications, business reports, and scientific periodicals use graphic representation extensively. Graphic representation has been carefully studied, and much has been written on the subject. While graphic representation has many ramifications, we shall consider here only those aspects of the subject which are useful in visualizing the important properties of frequency distributions and the ways in which one frequency distribution may differ from another.

2.9   HISTOGRAMS

A histogram is a graph in which the frequencies are represented by areas in the form of bars. Table 2.7 presents measures of auditory reaction time for a sample of 188 subjects.

Figure 2.1 shows the frequencies plotted in the form of a histogram. To prepare such a histogram proceed as follows. Obtain a piece of suitably cross-sectioned graph paper. Paper subdivided into tenths of an inch with heavy lines 1 in. apart is convenient. Draw a horizontal line to represent reaction time in seconds and a vertical line to represent frequencies. Select an appropriate scale, both for reaction time and frequencies. In the present case if we allow $\frac{5}{10}$ in. for each class interval and $\frac{1}{20}$ in. for each unit of frequency, we obtain a graph roughly 6 in. long and 4 in. tall. The scale is arbitrary. The scale suggested in this case, however, results in a graph of convenient size. The mid-points of the interval are written along the horizontal base line, and the frequency scale along the vertical. For each class

Table 2.7   Frequency distribution of auditory reaction times for a sample of 188 University of Chicago undergraduates*

Class interval, sec	Mid-point of interval	Frequency	Cumulative frequency
.34–.35	.345	2	188
.32–.33	.325	2	186
.30–.31	.305	4	184
.28–.29	.285	5	180
.26–.27	.265	11	175
.24–.25	.245	17	164
.22–.23	.225	28	147
.20–.21	.205	69	119
.18–.19	.185	37	50
.16–.17	.165	12	13
.14–.15	.145	1	1
Total		188	

* Adapted from L. L. Thurstone, *A factorial study of perception*, University of Chicago Press, Chicago, 1944.

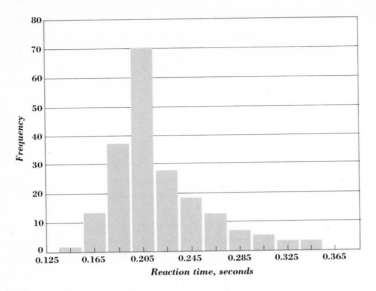

Fig. 2.1 **Histogram for data of Table 2.7. Auditory reaction times for 188 students.**

interval the corresponding frequency is plotted and a horizontal line drawn the full length of the interval. To complete the graph we may join the ends of these lines to the corresponding ends of the intervals on the horizontal axis, although practice in this regard varies. Both the horizontal and vertical axes must be appropriately labeled. A concise statement of what the graph is about should accompany it. Observe that the width of each bar corresponds to the exact limits of the interval. Observe also that in this type of graph the frequencies are represented as equally distributed over the whole range of the interval.

2.10   FREQUENCY POLYGONS

In a histogram we assume that all the cases within a class interval are uniformly distributed over the range of the interval. In a frequency polygon we assume that all cases in each interval are concentrated at the mid-point of the interval. In this fact resides the essential difference between a histogram and a frequency polygon. Instead of drawing a horizontal line the full length of the interval, as in the histogram, we make a dot above the mid-point of each interval at a height proportional to the frequency. It is customary to show an additional interval at each end of the horizontal scale and to join these dots to the dots of the adjacent interval. A frequency distribution based on the same data as the histogram in Fig. 2.1 is shown in Fig. 2.2.

Observe that the frequency distribution in Fig. 2.2 is not a smooth continuous curve, since the lines joining the various points are straight lines. If

Fig. 2.2 Frequency polygon for data of Table 2.7. Auditory reaction times for 188 students.

we subdivide our intervals into smaller intervals, we shall of course obtain irregular frequencies, there being too few members in each interval. Consider, however, a circumstance where our intervals become smaller and smaller and at the same time the total number of cases becomes larger and larger. If we carry this process to the extreme situation where we have an indefinitely small interval and an indefinitely large number of cases, we arrive at the concept of a continuous frequency distribution.

2.11 CUMULATIVE FREQUENCY POLYGONS

The drawing of a cumulative frequency polygon differs from that of a frequency polygon in two respects. *First,* instead of plotting points corresponding to frequencies, we plot points corresponding to cumulative frequencies. *Second,* instead of plotting points above the mid-point of each interval, we plot our points above the top of the exact limits of the interval. This is done because we wish our graph to visually represent the number of cases falling above or below particular values. In plotting the cumulative frequency distribution shown in Table 2.7, we would plot the cumulative frequency 188 against the top of the exact upper limit of the interval, that is, .355, the frequency 186 against .335, and so on. Figure 2.3 shows the cumulative frequency distribution for the data appearing in the last column of Table 2.7.

We may convert our raw frequencies to percentages such that all the frequencies added together add up to 100 instead of to the number of cases. We may then determine the cumulative percentage frequencies. We

Fig. 2.3   Cumulative frequency polygon for data of Table 2.7. Auditory reaction times for 188 students.

may then graph these frequencies and obtain thereby a cumulative percentage polygon, or ogive. The advantage of this type of diagram is that from it we can read off directly the percentage of observations less than any specified value.

2.12   SOME CONVENTIONS FOR THE CONSTRUCTION OF GRAPHS

1   In the graphing of frequency distributions it is customary to let the horizontal axis represent scores and the vertical axis frequencies.

2   The arrangement of the graph should proceed from left to right. The low numbers on the horizontal scale should be on the left, and the low numbers on the vertical scale should be toward the bottom.

3   The distance along either axis selected to serve as a unit is arbitrary and affects the appearance of the graph. Some writers suggest that the units should be selected such that the ratio of height to length is roughly 3:5. This procedure seems to have some aesthetic advantages.

4   Whenever possible the vertical scale should be so selected that a zero point falls at the point of intersection of the axes. With some data this procedure may give rise to a most unusual looking graph. In such cases it is customary to designate the point of intersection as the zero point and make a small break in the vertical axis.

5   Both the horizontal and vertical axes should be appropriately labeled.

6 Every graph should be assigned a descriptive title which states precisely what it is about.

2.13 HOW FREQUENCY DISTRIBUTIONS DIFFER

Comparison of a number of frequency distributions represented in either tabular or graphic form indicates that they differ one from another. An important problem in statistics is the identification and definition of properties or attributes of frequency distributions which describe how they differ. It is customary to designate four important properties of frequency distributions. These are central location, variation, skewness, and kurtosis. These properties may be viewed either as descriptive of the frequency distribution itself or as descriptive of the set of observations of which the distribution is comprised. These alternatives are in effect synonymous. A frequency distribution is a particular kind of arrangement of a set of observations. Central location, variation, skewness, and kurtosis may be discussed either with direct reference to sets of observations or with reference to the observations arranged in frequency-distribution form.

Central location refers to a value of the variable near the center of the frequency distribution. It is a middle point. Measures of central locations are called *averages*. These are discussed in detail in Chap. 3 of this book.

Variation refers to the extent of the clustering about a central value. If all the observations are close to the central value, their variation will be less than if they tend to depart more markedly from the central value. Measures of variation are discussed in Chap. 4.

Skewness refers to the symmetry or asymmetry of the frequency distribution. If a distribution is asymmetrical and the larger frequencies tend to be concentrated toward the low end of the variable and the smaller frequencies toward the high end, it is said to be *positively skewed*. If the opposite holds, the larger frequencies being concentrated toward the high end of the variable and the smaller frequencies toward the low end, the distribution is said to be *negatively skewed*.

Kurtosis refers to the flatness or peakedness of one distribution in relation to another. If one distribution is more peaked than another, it may be spoken of as more *leptokurtic*. If it is less peaked, it is said to be more *platykurtic*. It is conventional to speak of a distribution as leptokurtic if it is more peaked than a particular type of distribution known as the normal distribution, and platykurtic if it is less peaked. The normal distribution is spoken of as *mesokurtic*, which means that it falls between leptokurtic and platykurtic distributions.

Table 2.8 presents hypothetical data illustrating frequency distributions with different properties. The distribution in col. 2 is a *symmetrical binomial*, a type of distribution which is of much importance in statistical work and will be considered in detail in a later chapter. The distribution in col. 3 has central frequencies which are greater than those for the binomial. It is more peaked than the binomial, and as far as kurtosis is concerned, it can be said to be leptokurtic. The distribution in col. 4 has smaller central

Table 2.8 Hypothetical data illustrating frequency distributions of different shapes

1 Class interval	2 Symmetrical binomial	3 Leptokurtic	4 Platykurtic	5 Rectangular	6 Bimodal	7 U-shaped	8 Positively skewed	9 Negatively skewed	10 J-shaped
70–79	1	3	5	16	5	30	2	10	50
60–69	7	8	14	16	10	20	6	25	30
50–59	21	13	20	16	35	10	10	40	20
40–49	35	40	25	16	14	4	15	20	10
30–39	35	40	25	16	14	4	20	15	7
20–29	21	13	20	16	35	10	40	10	5
10–19	7	8	14	16	10	20	25	6	4
0–9	1	3	5	16	5	30	10	2	2
N	128	128	128	128	128	128	128	128	128

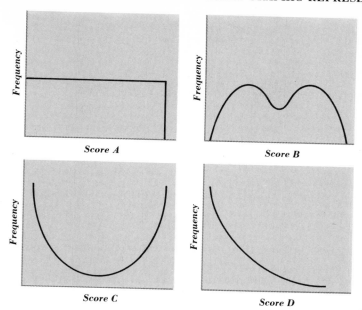

Fig. 2.4 Four types of distributions: *A*, rectangular distribution; *B*, bimodal distribution; *C*, U-shaped distribution; *D*, J-shaped distribution.

frequencies than the binomial and larger frequencies toward the extremities. It can be spoken of as platykurtic. The distribution in col. 5 has uniform frequencies over all class intervals and is described as *rectangular*. The distribution in col. 6 has two humps, or modes. It is said to be *bimodal*. In the distribution in col. 7 the largest frequencies occur at the extremities whereas the central frequencies are the smallest. Such a distribution is said to be *U-shaped*. The distributions in cols. 2 to 7 are all symmetrical and have the same measures of central location although they differ in variation. Column 8 illustrates a *positively skewed* and col. 9 a *negatively skewed* distribution. Extreme skewness leads to the type of distribution shown in col. 10, which is described as *J-shaped*. Figure 2.4 illustrates a rectangular distribution, a bimodal distribution, a U-shaped distribution, and a J-shaped distribution.

2.14 THE PROPERTIES OF FREQUENCY DISTRIBUTIONS REPRESENTED GRAPHICALLY

The differing characteristics of frequency distributions can be readily represented in graphical form. Consider the three distributions in Fig. 2.5. These distributions appear identical in shape. They are markedly different, however, in terms of the central values about which the observations in each distribution appear to concentrate; that is, they have different

Fig. 2.5   Three frequency distributions identical in shape but with different averages.

averages although they may be identical in all other respects. Distribution A has a lower average than B and B than C.

Now consider the distributions in Fig. 2.6. Inspection of these three distributions suggests that while the observations in each case appear to concentrate about the same average, they are nonetheless markedly different one from another. In the case of distribution A the observations appear to be more closely concentrated about the average than in the case of B, and the same applies to B in relation to C. Thus these distributions differ in variation. The observations in A are less variable than the observations in B, and those in B are less variable than those in C.

Examine now the distributions in Fig. 2.7. These three distributions have different averages and possibly different measures of variation. They differ also in skewness. Distribution B is symmetrical about the average; that is, if we were to fold it over about the average, we should find that it had the

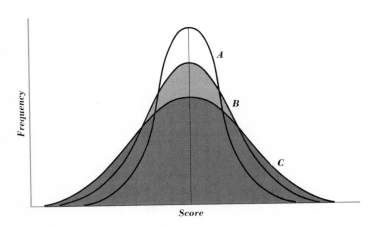

Fig. 2.6   Three frequency distributions with the same average but with different variation.

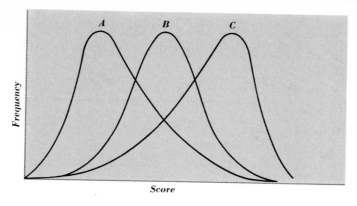

Fig. 2.7 Three frequency distributions differing in skewness.

same shape on both sides. *A* and *C* are asymmetrical, the shape to the left of the average being different from the shape to the right. Distribution *A* is positively skewed, the longer tail extending toward the high end of the scale. Distribution *C* is negatively skewed, the longer tail extending toward the low end of the scale.

Consider now the graphical representation of kurtosis as shown in Fig. 2.8. Distribution *A* is a symmetrical bell-shaped distribution known as the normal distribution. Distribution *B* is observed to be flatter on top than the normal distribution and is referred to as platykurtic, while distribution *C* is more peaked than the normal and is spoken of as leptokurtic.

In the above discussion the meaning which attaches to the descriptive properties of collections of measurements arranged in frequency distributions is largely intuitive and is derived from the inspection of distributions in tabular or graphic form. To proceed with the study of data interpretation we require precisely defined numerical measures of central location, varia-

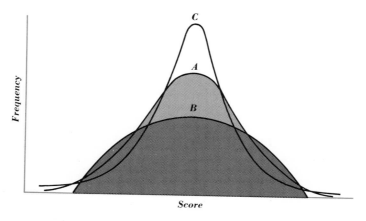

Fig. 2.8 Three frequency distributions differing in kurtosis.

tion, skewness, and kurtosis. Chapters 3 and 4 to follow are concerned with the more precise and formal delineation of these properties, their numerical description, and calculation.

EXERCISES

1   The following are marks obtained by a group of 40 university students on an English examination:

42	88	37	75	98	93	73	62
96	80	52	76	66	54	73	69
83	62	53	79	69	56	81	75
52	65	49	80	67	59	88	80
44	71	72	87	91	82	89	79

Prepare a frequency distribution and a cumulative frequency distribution for these data using a class interval of 5.

2   Write down the exact limits and the mid-points of the class intervals for the frequency distribution obtained by answering Exercise 1 above.

3   Write down an acceptable set of class intervals, the exact limits of the intervals, and their mid-points, in preparing frequency distributions for the following data: (*a*) error scores ranging from 24 to 87 made by a sample of rats in running a maze; (*b*) intelligence quotients ranging from 96 to 137 for a group of school children; (*c*) scholastic aptitude test scores ranging from 227 to 896 obtained by a group of university students; (*d*) response latencies ranging from .276 to .802 sec for a group of experimental subjects.

4   Prepare a frequency distribution for the following test scores:

2	11	6	4	18	1	9	2	2	15
8	16	12	11	17	3	3	5	3	7
11	9	5	16	16	16	4	9	5	7
4	10	4	4	15	15	5	5	11	18
5	10	9	8	7	7	2	6	13	1

Obtain a cumulative percentage frequency distribution for these data.

5   Prepare histograms for the data in Exercises 1 and 4 above.

6   Prepare frequency polygons for the data in Exercises 1 and 4 above.

7   Prepare cumulative frequency polygons for the data in Exercises 1 and 4 above.

8   Frequency distributions of intelligence quotients are available for (*a*) a random sample of the population at large and (*b*) a random sample of university students. In what ways and for what reasons might you expect these two distributions to differ?

9   In what ways might you possibly expect the frequency distribution of marks on a history examination to differ from that for a mathematics examination?

AVERAGES

Chapter 2 discussed the organization of collections of numbers in the form of frequency distributions, and how such frequency distributions could be represented graphically. We now proceed to a consideration of how a collection of numbers, whether organized in the form of a frequency distribution or not, may be described. What are the descriptive properties of collections of numbers or their corresponding frequency distributions? One such property is the *average*.

The average is a measure of central location. The word "average" is commonly used to refer to a value obtained by adding together a set of measurements and then dividing by the number of measurements in the set. This is one type of average only and is called the arithmetic mean. In general, an average is a central reference value which is usually close to the point of greatest concentration of the measurements and may in some sense be thought to typify the whole set. For certain purposes a particular measurement may be viewed as a certain distance above or below the average. Averages in common use in psychology and education are the arithmetic mean and the median. Other measures of central location include the mode, the geometric mean, and the harmonic mean. This chapter contains a discussion of the arithmetic mean, the median, and the mode. Discussion of the geometric mean and the harmonic mean has been omitted because of the infrequency of their use in practice. By far the most important and widely used measure of central location is the arithmetic mean. This statistic is the appropriate measure of central location for interval and ratio variables. The median and mode are sometimes viewed as appropriate measures for ordinal and nominal variables, respectively, although they can also be used with interval and ratio variables.

3.2   THE ARITHMETIC MEAN

By definition the arithmetic mean is the sum of a set of measurements divided by the number of measurements in the set. Consider the following measurements: 7, 13, 22, 9, 11, 4. The sum of these six measurements is 66. The arithmetic mean is therefore 66 divided by 6, or 11.

In general, if N measurements are represented by the symbols X_1, X_2, X_3, . . . , X_N, the arithmetic mean in algebraic language is as follows:

$$[3.1] \quad \bar{X} = \frac{X_1 + X_2 + X_3 + \cdots + X_N}{N} = \frac{\sum\limits_{i=1}^{N} X_i}{N}$$

The symbol $\bar{X}$, spoken of as X bar, is used to denote the arithmetic mean of the values of X. $\sum\limits_{i=1}^{N}$, the Greek letter sigma, describes the operation of summing the N measurements. The summation extends from $i = 1$ to $i = N$.

3.3   CALCULATING THE MEAN FROM FREQUENCY DISTRIBUTIONS

Consider a situation where different values of X occur more than once. The arithmetic mean is then obtained by multiplying each value of X by the frequency of its occurrence, adding together these products, and then dividing by the total number of measurements. Consider the following measurements: 11, 11, 12, 12, 12, 13, 13, 13, 13, 13, 14, 14, 15, 15, 15, 16, 16, 17, 17, 18. The value 11 occurs with a frequency of 2, 12 with a frequency of 3, 13 with a frequency of 5, and so on. These data may be written as follows:

X_i	f_i	f_iX_i
18	1	18
17	2	34
16	2	32
15	3	45
14	2	28
13	5	65
12	3	36
11	2	22
Total	20	280

This is a frequency distribution with a class interval of 1. The symbol f_i is used to denote the frequency of occurrence of the particular value X_i. Multiplying each value X_i by the frequency of its occurrence and adding together the products f_iX_i, we obtain the sum 280. The arithmetic mean is then 280 divided by 20, or 14.0.

In general, where X_1, X_2, X_3, . . . , X_k occur with frequencies f_1, f_2, f_3, . . . , f_k, where k is the number of *different* values of X, the arithmetic mean

$$[3.2] \quad \bar{X} = \frac{f_1X_1 + f_2X_2 + f_3X_3 + \cdots + f_kX_k}{N} = \frac{\sum_{i=1}^{k} f_iX_i}{N}$$

Observe that here the summation is over k terms, the number of *different* values of the variable X. Observe also that $\sum_{i=1}^{N} X_i = \sum_{i=1}^{k} f_iX_i$. The above discussion suggests a simple method for calculating the mean from data grouped in the form of a frequency distribution regardless of the size of the class interval. The mid-point of the interval may be used to represent all values falling within the interval. We assume that the variable X takes values corresponding to the mid-points of the intervals, and these are weighted by the frequencies. We multiply the mid-points of the intervals by the frequencies, sum these products, and divide this sum by N to obtain the mean. More explicitly, the steps involved are as follows. *First*, calculate the

Table 3.1   Calculating the mean for distribution of test scores

1 Class interval	2 Mid-point X_i	3 Frequency f_i	4 Frequency × mid-point f_iX_i
45–49	47	1	47
40–44	42	2	84
35–39	37	3	111
30–34	32	6	192
25–29	27	8	216
20–24	22	17	374
15–19	17	26	442
10–14	12	11	132
5–9	7	2	14
0–4	2	0	0
Total		76	1,612

$$\sum_{i=1}^{k} f_iX_i = 1,612 \qquad \bar{X} = 1,612/76 = 21.21$$

mid-points of all intervals. *Second*, multiply each mid-point by the corresponding frequency. *Third*, sum the products of mid-points by frequencies. *Fourth*, divide this sum by N to obtain the mean. To illustrate, consider Table 3.1.

The mid-points of the intervals X_i appear in col. 2. The frequencies f_i appear in col. 3. The products of the mid-points by the frequencies $f_i X_i$ are shown in col. 4. The sum of these products $\sum_{i=1}^{k} f_i X_i$ is 1,612, N is 76, and the mean X is obtained by dividing 1,612 by 76 and is 21.21.

3.4   THE MEAN OF COMBINED GROUPS

Consider a group of n_1 measurements with mean $\overline{X}_1$ and a set of n_2 measurements with mean $\overline{X}_2$. Denote the ith measurement in the first group by the symbol X_{i1} and the ith measurement in the second group by the symbol X_{i2}. The first subscript identifies the particular measurement. The second subscript identifies the group. Thus X_{72} would in this notation identify the seventh measurement in the second group of measurements. Let $n_1 + n_2 = N$, the total number of measurements in the two groups. The mean of all the measurements in the two groups taken together is

[3.3] $$\overline{X} = \frac{\sum_{i=1}^{n_1} X_{i1} + \sum_{i=1}^{n_2} X_{i2}}{n_1 + n_2} = \frac{n_1 \overline{X}_1 + n_2 \overline{X}_2}{N}$$

To illustrate, the mean of the four measurements 1, 3, 8, and 8 is 5. The mean of the six measurements 4, 4, 5, 6, 8, and 15 is 7. The mean of all ten measurements taken together is then

$$\overline{X} = \frac{4 \times 5 + 6 \times 7}{10} = 6.2$$

The above result may be extended to apply to any number of groups. With more than two groups, say, k, we simply multiply the number of cases in each group by the group mean, sum the k products thus obtained, and divide by N, the total number of measurements in the k groups. Thus with k groups

[3.4] $$\overline{X} = \frac{\sum_{i=1}^{k} n_i \overline{X}_i}{N}$$

3.5   SOME PROPERTIES OF THE ARITHMETIC MEAN

The sum of the deviations of all the measurements in a set from their arithmetic mean is zero. The arithmetic mean of the measurements 7, 13, 22, 9, 11, and 4 is 11. The deviations of these measurements from this mean are -4, 2, 11, -2, 0, and -7. The sum of these deviations is zero.

Proof of this result is as follows:

$$\sum_{i=1}^{N} (X - \bar{X}) = \sum_{i=1}^{N} X - \sum_{i=1}^{N} \bar{X}$$
$$= N\bar{X} - N\bar{X}$$
$$= 0$$

Observe that since $\bar{X} = \left(\sum_{i=1}^{N} X\right)/N$ it follows that $\sum_{i=1}^{N} X = N\bar{X}$. Also adding $\bar{X}$, the mean, N times is the same as multiplying $\bar{X}$ by N; thus if $\bar{X}$ is 11 and N is 6, we observe that $11 + 11 + 11 + 11 + 11 + 11 = 6 \times 11 = 66$.

The sum of squares of deviations about the arithmetic mean is less than the sum of the squares of deviations about any other value. The deviations of the measurements 7, 13, 22, 9, 11, 4 from the mean 11 are $-4, 2, 11, -2, 0, -7$. The squares of these deviations are 16, 4, 121, 4, 0, 49. The sum of squares is 194. Had any other origin been selected, the sum of squares of deviations would be greater than the sum of squares about the mean. Select a different origin, say 13. The deviations are $-6, 0, 9, 4, 2, -9$. Squaring these we have 36, 0, 81, 16, 4, 81. The sum of these squares is 218, which is greater than the sum of squares about the mean. Selection of any other origin will demonstrate the same result.

This property of the mean indicates that it is the centroid, or center of gravity, of the set of measurements. Indeed, the mean is the central value about which the sum of squares of deviations is a minimum. This result may be readily demonstrated. Consider deviations from an origin $\bar{X} + c$, where $c \neq 0$. A deviation of an observation from this origin is

[3.5] $X_i - (\bar{X} + c) = (X_i - \bar{X}) - c$

Squaring and summing over N observations we obtain

[3.6] $\displaystyle\sum_{i=1}^{N} [X_i - (\bar{X} + c)]^2 = \sum_{i=1}^{N} (X_i - \bar{X})^2 + \sum_{i=1}^{N} c^2 - 2c \sum_{i=1}^{N} (X_i - \bar{X})$

Because the sum of deviations about the mean is zero, the third term to the right is zero. Also c^2 summed N times is Nc^2, and we write

[3.7] $\displaystyle\sum_{i=1}^{N} [X_i - (\bar{X} + c)]^2 = \sum_{i=1}^{N} (X_i - \bar{X})^2 + Nc^2$

This expression states that the sum of squares of deviations about an origin $\bar{X} + c$ may be viewed as comprised of two parts, the sum of squares of deviations about the mean $\bar{X}$ and Nc^2. The quantity Nc^2 is always positive. Hence the sum of squares of deviations about an origin $\bar{X} + c$ will always be greater than the sum of squares about $\bar{X}$. Thus the sum of squares of deviations about the arithmetic mean is less than the sum of squares of deviations about any other value.

Any mean calculated on a sample of size N is an estimate of a population mean, which is the value obtained where it is possible to measure all members of the population. The mean has the property that for most distributions it is a better, or more accurate, or more efficient, estimate of the population mean than other measures of central location such as the median and the mode. This is one reason why it is most frequently used. Proof of this result is beyond the scope of this book.

Reference has been made to a number of properties of the arithmetic mean. What importance attaches to these properties, or why should they be discussed? The fact that the sum of deviations about the mean is zero greatly simplifies many forms of algebraic manipulation. Any term involving the sum of deviations about the mean will vanish. The fact that the sum of squares of deviations about the mean is a minimum in effect implies an alternative definition of the mean; namely, *the mean is that measure of central location about which the sum of the squares is a minimum.* In effect, the mean is a measure of central location in the *least-squares* sense. The method of least squares is of considerable importance in statistics and is used, for example, in the fitting of lines and curves. The mean may be regarded as a point located by the method of least squares. The properties pertaining to change of origin and change of unit are of importance in that they lead to simplified methods of computing the mean where a fairly large number of observations is involved. The fact that the sample mean provides a better estimate of a population parameter than other measures of central location is of primary importance. Throughout statistics we are concerned with the problem of making statements about population values from our knowledge of sample values. Obviously, the more accurate these statements are, the better.

3.6 THE MEDIAN

Another commonly used measure of central location is the median. The median is a point on a scale such that half the observations fall above it and half below it. The observations 2, 7, 16, 19, 20, 25, and 27 are arranged in order of magnitude. Here N is an odd number and the median is 19; three observations fall above it and three below it. If another observation, say, 31, is included, the median is then taken as the arithmetic mean of the two middle values 19 and 20; that is, the median is $(19 + 20)/2$, or 19.5. Consider a situation where certain values of the variable occur more than once, as, for instance, with the observations 7, 7, 7, 8, 8, 8, 9, 9, 10, 10. The three 8's are assumed to occupy the interval 7.5 to 8.5. The median is obtained by interpolation. In this instance we must interpolate two-thirds of the way into the interval to obtain a point above and below which half the observations fall. The median is then taken as $7.5 + 0.66 = 8.16$.

With a frequency distribution represented in graphical form, the ordinate at the median divides the total area under the curve into two equal parts.

3.7 CALCULATING THE MEDIAN FROM FREQUENCY DISTRIBUTIONS

In calculating the median from data grouped in the form of a frequency distribution the problem is to determine a value of the variable such that one-half the observations fall above this value and the other half below. The method will be illustrated with reference to the data in Table 3.2

First, record the cumulative frequencies as shown in col. 3. *Second*, determine $N/2$, one-half the number of cases, in this example 38. *Third*, find the class interval in which the 38th case, the middle case, falls. The 38th case falls within the interval 15 to 19, and the exact limits of this interval are 14.5 and 19.5. Clearly, the 38th case falls very close to the top of this interval because we know from an examination of our cumulative frequencies that 39 cases fall below the top of this interval, that is, below 19.5. *Fourth*, interpolate between the exact limits of the interval to find a value above and below which 38 cases fall. To interpolate, observe that 26 cases fall within the limits 14.5 and 19.5, and we assume that these 26 cases are uniformly distributed in rectangular fashion between these exact limits. Now to arrive at the 38th, or middle, case, we require 25 of the 26 cases within this interval, because $2 + 11 + 25 = 38$. This means that we must find a point between 14.5 and 19.5 such that 25 cases fall below and 1 case above this point. The proportion of the interval we require is $\frac{25}{26}$ which is $\frac{25}{26} \times 5$ units of score, or 4.81. We add this to the lower limit of the interval to obtain the median, which is $14.50 + 4.81$, or 19.31.

Let us summarize the steps involved:

1 Compute the cumulative frequencies.

2 Determine $N/2$, one-half the number of cases.

3 Find the class interval in which the middle case falls, and determine the exact limits of this interval.

4 Interpolate to find a value on the scale above and below which one-half the total number of cases falls. This is the median.

For the student who has difficulty in following the above a simple formula may be employed.

[3.8] $$\text{Median} = L + \frac{N/2 - F}{f_m} h$$

where L = exact lower limit of interval containing the median
F = sum of all frequencies below L
f_m = frequency of interval containing median
N = number of cases
h = class interval

In the present example $L = 14.5$, $F = 13$, $f_m = 26$, $N = 76$, and $h = 5$. We

Table 3.2 Frequency distribution of psychological
test scores

1 Class interval	2 Frequency	3 Cumulative frequency
45–49	1	76
40–44	.2	75
35–39	3	73
30–34	6	70
25–29	8	64
20–24	17	56
15–19	26	39
10–14	11	13
5–9	2	2
0–4	0	0
Total	76	

then have

$$\text{Median} = 14.5 + \frac{\frac{76}{2} - 13}{26} \times 5 = 19.31$$

3.8 THE MODE

Another measure of central location is the *mode*. In situations where different values of X occur more than once the mode is the most frequently occurring value. Consider the observations 11, 11, 12, 12, 12, 13, 13, 13, 13, 13, 14, 14, 14, 15, 15, 15, 16, 16, 17, 17, 18. Here the value 13 occurs 5 times, more frequently than any other value; hence the mode is 13.

In situations where all values of X occur with equal frequency, where that frequency may be equal to or greater than 1, no modal value can be calculated. Thus for the set of observations, 2, 7, 16, 19, 20, 25, and 27 no mode can be obtained. Similarly, the observations, 2, 2, 2, 7, 7, 7, 16, 16, 16, 19, 19, 19, 20, 20, 20, 25, 25, 25, 27, 27, 27 do not permit the calculation of a modal value. All values occur with a frequency of 3.

In the case where two adjacent values of X occur with the same frequency, which is larger than the frequency of occurrence of other values of X, the mode may be taken rather arbitrarily as the mean of the two adjacent values of X. Consider the observations 11, 11, 12, 12, 12, 13, 13, 13, 13, 14, 14, 14, 14, 15, 15, 16, 16, 17, 18. Here the values 13 and 14 both occur with a frequency of 4, which is greater than the frequency of occurrence of the remaining values. The mode may be taken as $(13 + 14)/2$, or 13.5.

Where two nonadjacent values of X occur such that the frequencies of both are greater than the frequencies in adjacent intervals, then each value of X may be taken as a mode and the set of observations may be spoken of as bimodal. Consider the observation 11, 11, 12, 12, 12, 13, 13, 13, 13, 13, 14, 14, 14, 15, 15, 15, 15, 16, 16, 16, 17, 17, 18. Here the value 13 occurs five times, and this is greater than the frequency of occurrence of the adjacent values. Also 15 occurs four times, and this is also greater than the frequency of occurrence of the adjacent values. This set of observations may be said to be bimodal.

With data grouped in the form of a frequency distribution the mode is taken as the mid-point of the class interval with the largest frequency.

The mode is a statistic of limited practical value. It does not lend itself readily to algebraic manipulation. It has little meaning unless the number of measurements under consideration is fairly large.

3.9   COMPARISON OF THE MEAN, MEDIAN, AND MODE

The arithmetic mean may be regarded as an appropriate measure of central location for interval and ratio variables. All the particular values of the variable are incorporated in its calculation. The median is an ordinal statistic. Its calculation is based on the ordinal properties of the data. If the observations are arranged in order, the median is the middle value. Its calculation does not incorporate all the particular values of the variable, but merely the fact of their occurrence above or below the middle value. Thus the sets of numbers 5, 7, 20, 24, 25, and 10, 15, 20, 52, 63 have the same median, namely 20, although their means are quite different. The mode, the value or class with the greatest frequency, is a nominal statistic. Its calculation does not depend on particular values of the variable or their order, but merely on their frequency of occurrence.

A comparison of the mean, median, and mode may be made when all three have been calculated for the same frequency distribution. If the frequency distribution is represented graphically, the mean is a point on the horizontal axis which corresponds to the centroid, or center of gravity, of the distribution. If a cutout of the distribution is made from heavy cardboard and balanced on a knife edge, the point of balance will be the mean. The median is a point on the horizontal axis where the ordinate divides the total area under the curve into two equal parts. Half the area falls to the left and half to the right of the ordinate at the median. The mode is a point on the horizontal axis which corresponds to the highest point of the curve.

If the frequency distribution is symmetrical, the mean, median, and mode coincide. If the frequency distribution is skewed, these three measures do not coincide. Figure 3.1 shows the mean, median, and mode for a positively skewed frequency distribution. We note that the mean is greater than the median, which in turn is greater than the mode. If the distribution is negatively skewed the reverse relation holds.

A question may be raised regarding the appropriate choice of a measure

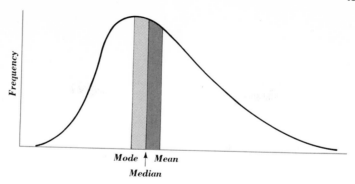

Fig. 3.1 Relation between the mean, median, and mode in a positively skewed frequency distribution.

of central location. In practical situations this question is rarely in doubt. The arithmetic mean is usually to be preferred to either the median or the mode. It is rigorously defined, easily calculated, and readily amenable to algebraic treatment. It provides also a better estimate of the corresponding population parameter than either the median or the mode.

The median is, however, to be preferred in some situations. Observations may occur which appear to be atypical of the remaining observations in the set. Such observations may greatly affect the value of the mean. Consider the observations 2, 3, 3, 4, 7, 9, 10, 11, 86. Observation 86 is quite atypical of the remaining observations, and its presence greatly affects the value of the mean. The mean is 15, a value greater than eight of the nine observations. The median is 7. Under circumstances such as this it may prove advisable in treating the data to use statistical procedures that are based on the ordinal properties of the data in preference to procedures that incorporate all the particular values of the variable and may be grossly affected by atypical values. The median, an ordinal statistic, may under such circumstances be preferred to the mean. In the above example the set of observations is grossly asymmetrical. If the distribution of the variables shows gross asymmetry, the median may be the preferred statistic, because, regardless of the asymmetry of the distribution, it can always be interpreted as the middle value.

For a strictly nominal variable the mode, the most frequently occurring class or value, is the only "most typical" statistic that can be used. It is rarely used with interval, ratio, and ordinal variables where means and medians can be calculated.

EXERCISES 1 In 100 rolls of a die the frequencies of the six possible events are as follows:

X_i	f_i
6	17
5	14
4	20
3	15
2	15
1	19
	$N = 100$

Compute the arithmetic means for this distribution.

2 The following is a frequency distribution of examination marks:

Class interval	f_i
90–94	1
85–89	4
80–84	2
75–79	8
70–74	9
65–69	14
60–64	6
55–59	6
50–54	4
45–49	3
40–44	3
	$N = 60$

Compute the arithmetic mean.

3 How does the addition of a constant and multiplication by a constant affect the arithmetic mean?

4 For the following data determine the mean of the combined groups:
  a $n_1 = 12$ $\bar{X}_1 = 60$ **b** $n_1 = 25$ $\bar{X}_1 = 10$
      $n_2 = 8$ $\bar{X}_2 = 40$ $n_2 = 75$ $\bar{X}_2 = 50$

5 The sum of squares of deviations of 10 observations from a mean of 50 is 225. What is the sum of squares of deviations from an arbitrary origin of 60 (Eq. [3.7])?

6 Compute medians for the following data:
  a 3, 7, 15, 26, 51
  b 3, 9, 22, 25, 31, 46
  c 6, 25, 30, 35, 45, 64

d   12, 19, 24, 24, 29, 42
e   4, 4, 5, 5, 6

7 Compute modes for the following data:
a   2, 2, 5, 5, 5, 6, 6, 6, 7, 8, 12
b   3, 3, 4, 4, 4, 5, 7, 7, 9, 12

8 Compute medians and modes for the data in Exercises 1 and 2 above.

MEASURES OF VARIATION, SKEWNESS, AND KURTOSIS

4.1 INTRODUCTION

Of great concern to the statistician is the variation in the events of nature. The variation of one measurement from another is a persisting characteristic of any sample of measurements. Measurements of intelligence, eye color, reaction time, and skin resistance, for example, exhibit variation in any sample of individuals. Anthropometric measurements such as height, weight, diameter of the skull, length of the forearm, and angular separation of the metatarsals show variation between individuals. Anatomical and physiological measurements vary; also the measurements made by the physicist, chemist, botanist, and agronomist. Statistics has been spoken of as the study of variation. Fisher (1948) has observed, "The conception of statistics as the study of variation is the natural outcoming of viewing the subject as the study of populations; for a population of individuals in all respects identical is completely described by a description of any one individual, together with the number in the group. The populations which are the object of statistical study always display variation in one or more respects." The experimental scientist is frequently concerned with the different circumstances, conditions, or sources which contribute to the variation in the measurements he obtains. The analysis of variance (Chap. 15) developed by Fisher is an important statistical procedure whereby the variation in a set of experimental data can be partitioned into components which may be attributed to different causal circumstances.

How may the variation in any set of measurements be described? Consider the following measurements for two samples:

Sample A	10	12	15	18	20
Sample B	2	8	15	22	28

We note that the two samples have the same mean, namely, 15. Simple inspection indicates, however, that the measurements in sample B are more variable than those in sample A; they differ more one from another. Among the possible measures used to describe this variation are the range, the mean deviation, and the standard deviation. The most important of these is the standard deviation.

4.2 THE RANGE

The *range* is the simplest measure of variation. In any sample of measurements the range is taken as the difference between the largest and smallest measurements. The range for the measurements 10, 12, 15, 18, and 20 is 20 minus 10, or 10. The range for the measurements 2, 8, 15, 22, and 28 is 28 minus 2, or 26. The measurements in the second set quite clearly exhibit greater variation than those in the first set, and this reflects itself in a much greater range. The range has two disadvantages. First, for large samples it is an unstable descriptive measure. The sampling variance of the range for small samples is not much greater than that of the standard deviation but increases rapidly with increase in N. Second, the range is not independent of sample size, except under special circumstances. For distributions that taper to zero at the extremities a better chance exists of obtaining extreme values for large than for small samples. Consequently, ranges calculated on samples composed of different numbers of cases are not directly comparable. Despite these disadvantages the range may be effectively used in the application of tests of significance with small samples.

4.3 THE MEAN DEVIATION

Consider the following measurements:

Sample A	8	8	8	8	8
Sample B	1	4	7	10	13
Sample C	1	5	20	25	29

Intuitively, the measurements in sample A are less variable than those in B, which in turn are less variable than those in C. Indeed, the measurements in A exhibit no variation at all. The means of the three samples are 8, 7, and 16. If we express the measurements as deviations from their sample means, we obtain

Sample A	0	0	0	0	0
Sample B	−6	−3	0	+3	+6
Sample C	−15	−11	+4	+9	+13

Inspection of these numbers suggests that as variation increases, the departure of the observations from their sample mean increases. We may use this characteristic to define a measure of variation. One such measure is the *mean deviation*. The mean deviation is the arithmetic mean of the absolute deviations from the arithmetic mean. An absolute deviation is a deviation without regard to algebraic sign. To obtain the mean deviation we simply calculate the deviations from the arithmetic mean, sum these, disregarding algebraic sign, and divide by N. For sample A above, the mean deviation is zero. For sample B the mean deviation is $(6 + 3 + 0 + 3 + 6)/5 = \frac{18}{5} = 3.6$. For sample C the mean deviation is $(15 + 11 + 4 + 9 + 13)/5 = \frac{52}{5} = 10.4$.

The mean deviation is given in algebraic language by the formula

[4.1] $$\mathrm{MD} = \frac{\Sigma |X - \bar{X}|}{N}$$

Here $X - \bar{X}$ is a deviation from the mean and $|X - \bar{X}|$ is a deviation without regard to algebraic sign. The bars mean that signs are ignored.

Hitherto, symbols above and below the summation sign Σ have been used to indicate the limits of the summation. In the above formula for the mean deviation these symbols have been omitted, the summation being clearly understood to extend over the N members in the sample. In this and subsequent chapters symbols indicating the limits of summation will, for convenience, be omitted where these are understood clearly from the context to extend over N sample members. Where any possibility of doubt could exist, the symbols above and below the summation sign will be inserted.

The mean deviation is infrequently used. It is not readily amenable to algebraic manipulation. This circumstance stems from the use of absolute values. In general, in statistical work the use of absolute values should be avoided, if at all possible. It is of interest to note that the sum of absolute deviations about the median is a minimum. Consider the numbers 1, 5, 20, 25, 29. The median is 20. The sum of absolute deviations is $19 + 15 + 0 + 5 + 9 = 48$. The corresponding sum of deviations about any other origin, say, 19, will be greater than 48. The sum of absolute deviations about the origin 19 is $18 + 14 + 1 + 6 + 10 = 49$. The median could be defined as that value about which the sum of absolute deviations is a minimum.

The mean deviation is discussed here primarily for pedagogic reasons. It illustrates how one particular measure of variation may be defined.

4.4   THE SAMPLE VARIANCE AND STANDARD DEVIATION

Some of the deviations about the mean are positive; others are negative. The sum of deviations is zero. One method for dealing with the presence of negative signs is to use the absolute deviations, as in the calculation of the mean deviation as defined in formula [4.1]. This procedure has little to recommend it. An alternate, and in general preferable, procedure is to square the deviations about the mean, sum these squares, and use this sum of squares in the definition of a measure of variation. For example, the mean of the measurements 1, 4, 7, 10, and 13 is 7. The deviations from the mean are $-6, -3, 0, +3$, and $+6$. The squares of these deviations are 36, 9, 0, 9, and 36. The sum of squares is 90.

The sum of squares of deviations about the mean is used in the definition of a statistic known as the *variance*. Two methods exist for defining the variance. Both are in common use. One method defines the variance by dividing the sum of squares of deviations about the mean by N, the number of cases. Denote this statistic by s^2. Thus

[4.2] $$s^2 = \frac{\Sigma(X - \overline{X})^2}{N}$$

In the illustrative example in the paragraph above, the sum of squares of deviations about the mean was 90, and the variance, according to formula [4.2], is $s^2 = 90/5 = 18$.

An alternate method of defining the variance is to divide the sum of squares by $N - 1$ rather than N. Thus, according to this definition, the variances s^2 is given by

[4.3] $$s^2 = \frac{\Sigma(X - \overline{X})^2}{N - 1}$$

Both formulas, [4.2] and [4.3], provide alternate definitions of the variance. These formulas have no derivation, but are obtained by a process of plausible reasoning. The reader will note that no notational distinction is made here between the variance defined by formula [4.2] and that defined by formula [4.3].

What is the essential difference between the variance as defined by formula [4.2] and the variance as defined by formula [4.3]? To understand the answer to this question the reader should recall that in Chapter 1 a distinction was made between sample values, or estimates, and population values or parameters. Both formulas, [4.2] and [4.3], provide estimates of a population variance σ^2. For certain algebraic reasons when we divide $\Sigma(X - \overline{X})^2$ by N we obtain a biased estimate of σ^2. This estimate will show a systematic tendency to be less than σ^2. It is biased. When we divide $\Sigma(X - \overline{X})^2$ by $N - 1$, however, we obtain an unbiased estimate of σ^2. Such an estimate will show no systematic tendency to be either greater than or less than σ^2.

In all situations where an estimate of a population variance σ^2 is required, the statistic s^2 which divides the sum of squares by $N - 1$ and not N should be used. In the great majority of situations discussed in this book an unbiased estimate is required. In some situations, involving descriptive statistics, convenience and simplicity dictate the use of an estimate which divides the sum of squares by N and not $N - 1$. *In general in this book the symbol s^2 will be used to refer to the unbiased estimate obtained by dividing* *the sum of squares by $N - 1$, as in formula [4.3]. In a few situations the definition of formula [4.2] will be used. All such instances will be clearly specified in the text.*

The reader should note that a population variance is defined as $\sigma^2 = \Sigma(X - \mu)^2/N_p$, where μ is the population mean and N_p is the number of members in the population. Thus in any situation where the variance of a complete population is required we divide the sum of squares by the number of members in the population.

While N is the number of measurements or observations, the quantity $N - 1$ is the number of deviations about the mean that are free to vary. To illustrate, consider the measurements 7, 8, and 15. The mean is 10, and the deviations about the mean are $-3, -2, +5$. The sum of deviations about the mean is zero; that is, $(-3) + (-2) + (5) = 0$. Because this is so, if any two of the deviations are known, the third deviation is fixed. It cannot vary. In this example, the sum of squares of deviations about the mean is $9 + 4 + 25 = 38$. Although this sum of squares is obtained by adding together three squared deviations, only two of these squared deviations can exhibit freedom of variation. The number of values that are free to vary is called the number of *degrees of freedom*. A quantity of the kind $\Sigma(X - \overline{X})^2$ is said to have associated with it $N - 1$ degrees of freedom, because $N - 1$ of the N squared deviations of which it is composed can vary. Some intuitive plausibility attaches to the idea that in the definition of a measure of variation we should divide the sum of squares by the number of values that can exhibit freedom of variation. The concept of degrees of freedom is a very useful and general concept in statistics, and is elaborated in more detail elsewhere in this book.

In the above discussion the definition of the variance evolves initially from a consideration of deviations about the arithmetic mean. An alternative, and perhaps more elemental, approach is to begin by considering the differences between each value and every other value. With two measurements only, X_1 and X_2, we may consider the difference between them, $X_1 - X_2$. With three measurements, X_1, X_2, and X_3, we may consider the differences $X_1 - X_2$, $X_1 - X_3$, and $X_2 - X_3$. In general, for N measurements the number of such differences is $N(N - 1)/2$. To illustrate, for the measurements 1, 4, 7, 10, and 13, the differences between each measurement and every other measurement are $-3, -6, -9, -12, -3, -6, -9, -3, -6, -3$. Note that the sign of the difference depends on the order of the measurements. If we obtain the sum of squares of the differences between each measurement and every other measurement and divide by the number of such differences, the result is closely related to s^2; in fact it is simply twice s^2. In our example the sum of squares of differences is 450. We divide this by 10 to obtain 45.0, which is seen to be twice the variance, 22.5, as calculated by formula [4.3]. In general, in algebraic notation it may be shown that

[4.4] $$\frac{\Sigma(X_i - X_j)^2}{N(N - 1)/2} = 2s^2$$

where the summation is understood to extend over $N(N-1)/2$ differences. This result means that s^2 is a descriptive index of how different each value is from every other value; in fact it is an average of the squared differences divided by 2.

The variance is a statistic in squared units. If $X - \bar{X}$ is a deviation in feet, then $(X - \bar{X})^2$ is a deviation in feet squared. For many purposes it is desirable to use a measure of variation which is not in squared units, but in units of the original measurements themselves. We obtain this result by taking the square root of either formula 4.2 or formula 4.3. This statistic is called the *standard deviation*. Thus

[4.5] $$s = \sqrt{\frac{\Sigma(X - \bar{X})^2}{N}}$$

or

[4.6] $$s = \sqrt{\frac{\Sigma(X - \bar{X})^2}{N - 1}}$$

Here again, as with the variance, the standard deviation obtained using N is biased, and that obtained using $N - 1$ is unbiased.

4.5 AN ILLUSTRATIVE APPLICATION

Our understanding of the nature of the variance and the standard deviation will be enriched by considering illustrative situations where these statistics are of interest. Consider a simple experiment designed to investigate the effect of a drug on a cognitive task such as coding. An experimental group of subjects, who receive the drug, and a control group, who do not receive the drug, are used. Each group contains 10 subjects. Let us assume the scores on the coding task for the two groups are as follows:

Experimental	5	7	17	31	45	47	68	85	96	99
Control	29	36	37	42	49	58	62	63	69	70

The mean score for the experimental group is 50.0, and that for the control, 51.5. The investigator might be led to conclude from inspecting these means that the drug had little or no effect on the performance of the subjects. The standard deviations for the two groups are, respectively 35.63 and 14.86, the experimental group being much more variable in performance than the control group. Quite clearly the treatment is exerting a substantial influence on the variation in performance, although its influence on level of performance is negligible. In the analysis of experimental data the investigator must attend to, and if possible interpret, differences in the standard deviation, or variance, as well as differences in the arithmetic mean.

4.6 CALCULATING THE SAMPLE VARIANCE AND THE STANDARD DEVIATION FROM UNGROUPED DATA

For purposes of calculation, it is convenient to write the variance and the standard deviation in a different form. The variance may be written as

$$s^2 = \frac{\Sigma(X - \bar{X})^2}{N - 1}$$
$$= \frac{\Sigma(X^2 + \bar{X}^2 - 2X\bar{X})}{N - 1}$$
$$= \frac{\Sigma X^2 + N\bar{X}^2 - 2N\bar{X}^2}{N - 1}$$
$$= \frac{\Sigma X^2 - N\bar{X}^2}{N - 1}$$

In this derivation note that the summation of $\bar{X}^2$ over N is simply $N\bar{X}^2$; also the summation of $2X\bar{X}$ is $2\bar{X}\Sigma X = 2N\bar{X}^2$. The standard deviation is given by

$$s = \sqrt{\frac{\Sigma X^2 - N\bar{X}^2}{N - 1}}$$

Thus to calculate the standard deviation using this formula, we sum the squares of the original observations, subtract from this N times the square of the arithmetic mean, divide by $N - 1$, and then take the square root. For example, the five observations 1, 4, 7, 10, and 13 have a mean of 7. The squares of these observations are 1, 16, 49, 100, and 169. The sum of these squared observations is 335. The variance is then

$$s^2 = \frac{\Sigma X^2 - N\bar{X}^2}{N - 1} = \frac{335 - 5 \times 7^2}{5 - 1} = 22.50$$

and the standard deviation is $\sqrt{22.50} = 4.74$.

An alternative formula for the standard deviation which avoids the calculation of the arithmetic mean and may, therefore, be useful for certain computational purposes is

[4.7] $$s = \sqrt{\frac{N\Sigma X^2 - (\Sigma X)^2}{N(N - 1)}}$$

This formula requires one operation of division only.

In computing a standard deviation on a calculating machine, the measurements are entered on the machine and the sums and sums of squares of measurements are obtained in a single operation. This yields the information required to calculate the standard deviation. In most cases it is advisable to repeat the operation as a check.

4.7 THE EFFECT ON THE STANDARD DEVIATION OF ADDING OR MULTIPLYING BY A CONSTANT

If a constant is added to all the observations in a sample, the standard deviation remains unchanged. An examiner may conclude, for example, that an examination is too difficult. He may decide to add 10 points to all the marks assigned. The standard deviation of the original marks will be the same as the standard deviation of marks with the 10 points added. This result follows directly from the fact that if X is an observation, the corresponding observation with the constant c added is $X + c$. If $\overline{X}$ is the mean of the original observations, the mean with the constant added is $\overline{X} + c$. A deviation from the mean of the observations with the constant added is then $(X + c) - (\overline{X} + c)$, which is readily observed to be equal to $X - \overline{X}$. Since the deviations about the mean are unchanged by the addition of a constant, the standard deviation will remain unchanged. To illustrate, by adding a constant, say 5, to the measurements 1, 4, 7, 10, and 13, we obtain 6, 9, 12, 15, and 18. The mean of the original measurements is 7, and the mean of the measurements with the constant added is $7 + 5$, or 12. The deviations from the mean are in both instances the same, namely, $-6, -3,$ 0, $+3$, and $+6$. The standard deviation in both instances is 4.74.

If all measurements in a sample are multiplied by a constant, the standard deviation is also multiplied by the absolute value of that constant. If the standard deviation of examination marks is 4 and all marks are multiplied by the constant 3, then the standard deviation of the resulting marks is $3 \times 4 = 12$. To demonstrate this result, we observe that if $\overline{X}$ is the mean of a sample of measurements, the mean of the measurements multiplied by c is $c\overline{X}$. A deviation from the mean is then $cX - c\overline{X} = c(X - \overline{X})$. By squaring, summing over N observations, and dividing by $N - 1$, we obtain

$$\frac{\Sigma(cX - c\overline{X})^2}{N - 1} = \frac{c^2\Sigma(X - \overline{X})^2}{N - 1} = c^2 s^2$$

Thus if all measurements are multiplied by a constant c, the variance is multiplied by c^2 and the standard deviation by the absolute value of c. If c is a negative number, say -3, s is multiplied by the absolute value 3. By way of illustration, the measurements 1, 4, 7, 10, 13 have a mean of 7, a variance of 22.50, and a standard deviation of 4.74. If the measurements are multiplied by the constant 5, we obtain 5, 20, 35, 50, 65. The mean is now 5×7, or 35. The deviations from the mean are $-30, -15, 0, +15, +30$. Squaring these we obtain 900, 225, 0, 225, 900. The sum of squares is 2,250, the variance is 562.50, and the standard deviation is 23.72, whereas 5 times the original standard deviation of 4.74 is 23.70. The slight discrepancy results from the rounding of decimals.

4.8 STANDARD DEVIATION OF THE FIRST N INTEGERS

We state without proof that the sum of squares of the first N integers is

[4.8] $$\frac{N(N + 1)(2N + 1)}{6}$$

and the standard deviation of the first N integers is

[4.9] $$s = \sqrt{\frac{N(N + 1)}{12}}$$

Consider the integers 1, 2, 3, 4, 5, 6, 7, 8, 9, 10. Applying the above formulas, the sum of squares is 385 and the standard deviation is 3.03. These results may be readily checked by direct calculation.

Formula [4.9] is obtained directly from the definition of the standard deviation as $s = \sqrt{\Sigma(X - \bar{X})^2/(N-1)}$. If the standard deviation is defined as $s = \sqrt{\Sigma(X - \bar{X})^2/N}$, the standard deviation of the first N integers is $s = \sqrt{(N^2 - 1)/12}$.

These formulas are of particular use in relation to problems involving ranks (Chap. 21). Where ranks are used, the observations are represented by the first N integers.

4.9 THE VARIANCE OF COMBINED GROUPS

On occasion we know the means and variances of two samples of measurements and may wish to obtain the variance of the two samples combined. We may, for example, have means and variances of examination marks for two classes of university students and may wish to find the variance of marks for all the individuals in the two classes. Let the number of cases, mean, and variance for one group be n_1, $\bar{X}_1$, and s_1^2, and for the other n_2, $\bar{X}_2$, and s_2^2. Let $\bar{X}$ and s^2 be the mean and variance for the combined groups. Also, let $n_1 + n_2 = N$, $\bar{X}_1 - \bar{X} = d_1$, and $\bar{X}_2 - \bar{X} = d_2$. We state without proof that the variance of the combined group is

[4.10] $$s^2 = \frac{1}{N-1}[(n_1 - 1)s_1^2 + (n_2 - 1)s_2^2 + n_1 d_1^2 + n_2 d_2^2]$$

This formula can be extended from two to any number of groups, say, k.

To illustrate, seven measurements 1, 6, 8, 10, 13, 18, and 21 have a mean of 11 and a variance of 48.00; that is, $n_1 = 7$, $\bar{X}_1 = 11$, and $s_1^2 = 48.00$. The five measurements 1, 4, 7, 10, 13 have a mean of 7 and a variance of 22.50; that is, $n_2 = 5$, $\bar{X}_2 = 7$, and $s_2^2 = 22.50$. The mean of all 12 measurements taken together, the combined group, is 9.33. The quantity $d_1 = 11 - 9.33 = 1.67$ and $d_2 = 7 - 9.33 = -2.33$. The variance of the combined group is

$$s^2 = \tfrac{1}{11}\ [(7-1) \times 48 + (5-1) \times 22.50 + 7(1.67)^2 + 5(-2.33)^2] = 38.61$$

The standard deviation $s = \sqrt{38.61} = 6.21$. The above result may be checked by direct calculation.

4.10 STANDARD SCORES

A deviation from the mean divided by the standard deviation is called a *standard score* and is represented by the symbol z. Thus

[4.11] $$z = \frac{X - \bar{X}}{s} = \frac{x}{s}$$

Deviation scores $X - \bar{X}$, or x, have a mean of zero and a standard deviation

s. The subtraction of the mean from all measurements in a sample does not change the standard deviation. Standard scores have zero mean and unit standard deviation. As previously shown, if all measurements in a sample are multiplied by a constant, the standard deviation is also multiplied by that constant. The deviations from the mean $X - \bar{X}$ have a standard deviation s. If all deviations are divided by s, which amounts to multiplying by the constant $1/s$, the standard deviation of the scores thus obtained is $s/s = 1$.

To illustrate, the following observations have been expressed in raw-score, deviation-score, and standard-score form.

Individual	X	x	z
A	3	−7	−1.11
B	6	−4	−.63
C	7	−3	−.47
D	9	−1	−.16
E	15	5	.79
F	20	10	1.58
Sum	60	.00	.00
Mean	10	.00	.00
s	6.32	6.32	1.00

Because standard scores have zero mean and unit standard deviation, they are readily amenable to certain forms of algebraic manipulation. Many formulations can be derived more conveniently using standard scores than using raw or deviation scores.

The use of standard scores means, in effect, that we are using the standard deviation as the unit of measurement. In the above example individual A is 1.11 standard deviations, or standard deviation units, below the mean, while individual F is 1.58 standard deviation units above the mean.

Standard scores are frequently used to obtain comparability of observations obtained by different procedures. Consider examinations in English and mathematics applied to the same group of individuals and assume the means and standard deviations to be as follows:

	$\bar{X}$	s
English	65	8
Mathematics	52	12

In effect, in relation to the performance of the individuals in the group, a score of 65 on the English examination is the equivalent of a score of 52 on the mathematics examination. To illustrate, a score one standard deviation above the mean, that is, $65 + 8$, or 73, on the English examination can be considered to be the equivalent of a score one standard deviation above the mean, that is, $52 + 12$, or 64, on the mathematics examination. If an individual makes a score of 57 on the English examination and a score of 58 on the mathematics examination, we may compare his relative performance on the two subjects by comparing his standard scores. On English his standard score is $(57 - 65)/8 = -1.0$, and on mathematics his standard score is $(58 - 52)/12 = .5$. Thus on English his performance is one standard deviation unit below the average, while on mathematics his performance is .5 standard deviation unit above the average. Quite clearly, this individual did much more poorly in English than in mathematics relative to the performance of the group of individuals taking the examinations, although this is not reflected in the original marks assigned. To attain rigorous comparability of scores, the distributions of scores on the two tests should be identical in shape. The meaning of this statement will become clear as we proceed.

The reader should note that the sum of squares of standard scores, Σz^2, is equal to $N - 1$. We observe that $z^2 = (X - \bar{X})^2/s^2$; hence

$$\Sigma z^2 = \frac{\Sigma(X - \bar{X})^2}{s^2} = \frac{\Sigma(X - \bar{X})^2}{\Sigma(X - \bar{X})^2/(N - 1)} = N - 1$$

The reader should note here that if s^2 is defined as $\Sigma(X - \bar{X})^2/N$, the sum of squares of standard scores is N and not $N - 1$.

4.11   ADVANTAGES OF THE VARIANCE AND STANDARD DEVIATION AS MEASURES OF VARIATION

The variance and standard deviation have many advantages over other measures of variation. Much statistical work involves their use. The variance has certain additive properties, and may on occasion be partitioned into additive components, each of which may be related to some causal circumstance. The sample standard deviation is a more stable or accurate estimate of the population parameter than other measures of variation. It provides a more stable estimate of the standard deviation in the population than the sample mean deviation, for example, does of the mean deviation in the population. The variance and standard deviation are more amenable to mathematical manipulation than other measures. They enter into formulas for the computation of many types of statistics. They are widely used as measures of error. In later discussion on sampling statistics the reader will observe that the standard error is in effect the standard deviation of errors made in estimating population parameters from sample values. These errors result from the operation of chance factors in random sampling. A full appreciation of the importance and meaning of the variance and standard deviation in their many ramifications requires considerable familiarity with statistical ideas.

4.12 MOMENTS ABOUT THE MEAN

The mean and the standard deviation are closely related to a family of descriptive statistics known as *moments*. The first four moments about the arithmetic mean are as follows:

[4.12]
$$m_1 = \frac{\Sigma(X - \overline{X})}{N} = 0$$
$$m_2 = \frac{\Sigma(X - \overline{X})^2}{N} = \frac{N-1}{N} s^2$$
$$m_3 = \frac{\Sigma(X - \overline{X})^3}{N}$$
$$m_4 = \frac{\Sigma(X - \overline{X})^4}{N}$$

In general, the rth moment about the mean is given by

[4.13]
$$m_r = \frac{\Sigma(X - \overline{X})^r}{N}$$

The term "moment" originates in mechanics. Consider a lever supported by a fulcrum. If a force f_1 is applied to the lever at a distance x_1 from the origin, then $f_1 x_1$ is called the moment of the force. Further, if a second force f_2 is applied at a distance x_2, the total moment is $f_1 x_1 + f_2 x_2$. If we square the distances x, we obtain the second moment; if we cube them, we obtain the third moment; and so on. When we come to consider frequency distributions, the origin is the analogue of the fulcrum and the frequencies in the various class intervals are analogous to forces operating at various distances from the origin. Observe that the first moment about the mean is zero and the second moment is $(N-1)/N$ times the sample variance. The third moment is used to obtain a measure of skewness, and the fourth moment, a measure of kurtosis.

4.13 MEASURES OF SKEWNESS AND KURTOSIS

A commonly used measure of skewness may be obtained from the second and third moments and is defined as

[4.14]
$$g_1 = \frac{m_3}{m_2 \sqrt{m_2}}$$

The rationale of this statistic is based on the observation that when a distribution is symmetrical, the sum of cubes of deviations above the mean will balance the sum of cubes of deviations below the mean. Thus when the distribution is symmetrical, $m_3 = 0$ and $g_1 = 0$. If the distribution has a long tail to the right, the sum of cubes of deviations above the mean will be greater than the corresponding sum below the mean. Under these circumstances, the distribution is positively skewed and g_1 is positive. Conversely, if the distribution is negatively skewed, g_1 is negative. The second moment is used in the denominator of the expression for g_1 in order to make the measure independent of the scale of measurement.

The most acceptable measure of kurtosis is obtained from the second and fourth moments and is defined as

[4.15] $$g_2 = \frac{m_4}{m_2{}^2} - 3$$

When g_2 is zero, the distribution is a particular type of symmetrical distribution known as a *normal distribution*. When g_2 is less than zero, the distribution is flatter on top than the normal distribution. When g_2 is greater than zero, the distribution is more peaked than the normal distribution. The rationale underlying g_2 as a measure of kurtosis is based on the fact that if two frequency distributions have the same standard deviation and one is more peaked than the other, the more peaked must, of necessity, have thicker tails. As a consequence, the sum of the deviations about the mean raised to the fourth power will be greater for the more peaked distribution than for the less peaked.

EXERCISES

1   For the measurements 2, 5, 9, 10, 15, 19, compute (*a*) the range, (*b*) the mean deviation, (*c*) the sample variance, (*d*) the standard deviation.

2   The variance calculated on a sample of 100 cases is 15. What is the sum of squares of deviations about the arithmetic mean?

3   A biased variance estimate calculated on a sample of 5 cases is 10. What is the corresponding unbiased variance estimate?

4   The variance for a sample of N measurements is 20. What will the variance be if all measurements are (*a*) multiplied by a constant 5 and (*b*) divided by a constant 5?

5   Show that

$$\Sigma(X - \bar{X})^2 = \Sigma X^2 - N\bar{X}^2$$

6   Obtain the variance and the standard deviation of the combined groups for the following data:

     a $n_1 = 6$ $\bar{X}_1 = 20$ $s_1{}^2 = 50$
            $n_2 = 4$ $\bar{X}_2 = 10$ $s_2{}^2 = 40$
     b $n_1 = 14$ $\bar{X}_1 = 50$ $s_1{}^2 = 400$
            $n_2 = 6$ $\bar{X}_2 = 75$ $s_2{}^2 = 250$

7   Express the measurements 2, 4, 5, 6, 8 in standard-score form. What is the sum of squares of standard scores?

8   The mean and standard deviation of marks in a statistics examination for a class of 26 students are, respectively, 62 and 10. Three students make scores of 50, 75, and 90. What are their corresponding standard scores?

9   Calculate the second, third, and fourth moments for the observations 4, 6, 10, 14, 16. Compute also measures of skewness and kurtosis.

PROBABILITY AND THE
BINOMIAL DISTRIBUTION

5.1 INTRODUCTION

In Chaps. 2, 3, and 4 the reader was introduced to certain statistics which are used to describe the properties of frequency distributions, or the collections of numbers of which frequency distributions are comprised. In the present chapter the nature of probability is considered.

Probability is used to introduce the reader to one particular type of theoretical distribution known as the *binomial distribution*. The binomial distribution is only one of a number of theoretical distributions which are used in practical situations to estimate probabilities. Why are experimentalists concerned with questions of probability?

In experimental work a line of theoretical speculation may lead to the formulation of a particular hypothesis. An experiment is conducted, and data obtained. How are the data interpreted? Do the data support the acceptance or rejection of the hypothesis? What rules of evidence apply? Questions of this type involve considerations of probability. The answers are in probabilistic terms. The assertions of the investigator are not made with certainty but have associated with them some degree of doubt, however small.

Consider a hypothetical illustration. Two methods for the treatment of a disease are under consideration. Two groups of 20 patients suffering from the disease are selected. Method A is applied to one group, method B to the other. Following a period of treatment, 16 patients in group A and 10 patients in group B show marked improvement. How may this difference be evaluated? May it be argued from the data that treatment A is in general superior to treatment B? Here the investigator proceeds by adopting a trial hypothesis that no difference exists between the two treatments, that one treatment is no better than the other. He then estimates the probability of obtaining by random sampling under this trial hypothesis a difference equal to or greater than the one observed. If this probability is small, say the chances are less than 5 in 100, he may consider this sufficient evidence for the rejection of the trial hypothesis and may be prepared to assert that one method of treatment is better than the other. If the probability is not small and the observed difference may be expected to occur quite frequently under the trial hypothesis, say the chances are 20 in 100, then the evidence does not warrant the conclusion that one treatment is better than the other.

In general, the interpretation of the data of experiments is in probabilistic terms. The theory of probability is of the greatest importance in scientific work where questions about the correspondences between the deductive consequences of theory and observed data are raised. Probability theory had its origins in games of chance. It has become basic to the thinking of the scientist.

5.2 THE NATURE OF PROBABILITY

Diverse views of the nature of probability may be entertained. The topic is controversial. No inclusive summary of these different views will be attempted here. We shall discuss three approaches to probability: (1) the subjective, or personalistic, (2) the formal mathematical, and (3) the empirical relative-frequency approach. These different ways of regarding probability are not incompatible.

The term *probability* may be used subjectively to refer to an attitude of doubt with respect to some future event. For example, the assertions may be made that "It will probably rain tomorrow," or "The probability is small that I shall live to be 90 years old," or "There is a high probability that a particular horse will win the Kentucky Derby." Frequently, numerical terms are used in making assertions of this kind, such as, "The odds are even that it will rain tomorrow," or "I estimate that the chances are about 95 in 100 that I shall die before I am 90 years old," or "The chances are three to one that a particular horse will win the Kentucky Derby." All such assertions, whether numerical terms are used or not, refer to feelings of degrees of doubt or confidence with regard to future outcomes. This subjective usage is sometimes spoken of as psychological, or personalistic, probability.

A second usage defines the probability of an event as the ratio of the number of favorable cases to the total number of *equally likely* cases. This usage stems from a consideration of games of chance involving cards, dice, and coins. For example, on examining the structure of a die the assertion may be made that no basis exists for choosing one of the six alternatives in preference to another; consequently all six alternatives may be considered equally likely. The probability of throwing a particular result, say, a 3, in a single toss is then $\frac{1}{6}$, there being one favorable case among six equally likely alternatives. This approach to probability involves a concept of equally likely cases, which has a degree of intuitive plausibility in relation to cards, dice, and coins. Difficulties present themselves, however, when we attempt to apply this approach in situations where it is impossible to delineate cases which can be construed to be equally likely. These difficulties have led to the argument that equally likely means the same as equally probable; therefore the definition is circular because it defines probability in terms of itself. Arguments have been advanced to escape this circularity. These need not detain us. The difficulty, however, is readily resolved by observing that the concept of equally likely in this definition of probability is a formal postulate and is not empirical. In effect we say, "Let us postulate that certain events are equally likely, and given this postulate let us deduce certain consequences." This means that a theory of probability employing this postulate is a formal mathematical model. It may or may not correspond to empirical events. It may be demonstrated, however, that this model does approximate closely certain empirical events and consequently is of value in dealing with practical problems.

The situation here is somewhat analogous to that in ordinary Euclidian geometry. Euclidian geometry is a formal system comprised of a set of axioms, or primitive postulates, and their deductive consequences, called theorems. The proofs of the theorems hold regardless of questions of correspondence with the empirical world. We know, however, on the basis of lengthy experience, that these theorems can be shown to correspond closely to the world around us. In consequence, Euclidian geometry provides a valuable model for dealing with problems in engineering, surveying, building construction, and many other fields. Both with Euclidian geometry and probability it is useful to draw a clear distinction between the formal mathematical system and the empirical events for which the formal system may serve as a model.

A third approach to probability is through a consideration of relative frequencies. If a series of trials is made, say, N, and a given event occurs r times, then r/N is the relative frequency. This relative frequency may be considered an estimate of a value p. If a longer series of trials is made, the relative frequency will usually be closer to p. The difference between r/N and p may be made as small as we like by increasing the value of N. The probability p is defined as the limit approached by the relative frequency as the number of trials is increased. This approach to probability requires that a population of events be defined. Probability is the relative frequency in the population. It is a population parameter. The relative frequency in a sample of observations is an estimate of that parameter. To illustrate, consider a coin. The population of events may be regarded as an indefinitely large number of tosses which theoretically could be made. The proportion of heads p in this population is the probability of a head. This is often assumed to be $\frac{1}{2}$. If the coin is tossed 100 times and a proportion .47 of heads is obtained, this may be taken as an estimate of p.

The ways of regarding probability described here, the subjective or personalistic, the formal mathematical, and the empirical through the study of relative frequencies, are not incompatible, and indeed it may be argued that all three must of necessity coexist. While subjective, or personalistic, probability may be an interesting topic of psychological inquiry, in practical statistical work use is made mainly of the formal mathematical and relative-frequency approaches, the latter being the operational complement of the former.

5.3 POSSIBLE OUTCOMES

Questions of probability frequently involve a consideration of the number of possible outcomes, sometimes spoken of as a set. For example, in tossing a single coin, there are two possible outcomes—either a head or a tail will occur. In tossing two coins, the four possible outcomes are HH, HT, TH, TT. This means that both coins may be heads, the first a head and the second a tail, the first a tail and the second a head, or both tails. In tossing three coins, there are eight possible outcomes, which may be represented as HHH, HHT, HTH, THH, HTT, THT, TTH, TTT. In tossing four coins, the number of possible outcomes is 16. In throwing a single die, the number of possible outcomes is six; that is, either 1, 2, 3, 4, 5, or 6 may appear. In throwing two dice, the number of possible outcomes may be listed as follows:

11	21	31	41	51	61
12	22	32	42	52	62
13	23	33	43	53	63
14	24	34	44	54	64
15	25	35	45	55	65
16	26	36	46	56	66

These are the numbers which may come up in throwing two dice. Both dice may come up 1, the first may come up 2 and the second 1, the first 3 and the second 1, and so on. Thus there are 36 possible outcomes to this experiment. In drawing a single card from a deck of 52 cards, the number of possible outcomes is 52. In drawing one card from a deck of 52 cards and another from a different deck, the number of possible outcomes is $52 \times 52 = 2,704$.

In the illustrative situations described above we may be willing to assume that the possible outcomes are equally likely. With most coins, dice, and cards this is a justifiable assumption, the validity of which may be readily checked by experiment. Under this assumption the probability of obtaining two heads, HH, in tossing two coins is $\frac{1}{4}$. Two heads may occur once in four equally probable outcomes. We denote this probability by $p(\text{HH}) = \frac{1}{4}$. The probability of obtaining three tails in tossing three coins is $p(\text{TTT}) = \frac{1}{8}$, there being eight equally likely outcomes. The probability of obtaining two 6's in throwing two dice is $p(6,6) = \frac{1}{36}$. This event is seen to occur once in the set of 36 possible, and equally likely, outcomes.

The illustrative applications above regard a probability as the ratio of the number of favorable cases to the total number of equally likely cases. This approach may be extended to situations where the number of possible outcomes are not equally likely, provided a method exists for either deducing or estimating the expected relative frequencies of the possible outcomes. In rolling two dice, we may decide simply to sum the two numbers which appear and to regard this sum as an outcome. The set of all possible outcomes, thus defined, consists of the 11 numbers 2, 3, 4, . . . , 11, 12. These outcomes are not, of course, equally likely. Their expected relative frequencies may be obtained, however, from a consideration of the 36 possible, equally likely, outcomes involved in rolling two dice. A 2 can occur only once in 36 equally likely outcomes; that is $p(2) = \frac{1}{36}$. A 3 can occur twice; that is, $p(3) = \frac{2}{36}$. Likewise $p(4) = \frac{3}{36}$, $p(5) = \frac{4}{36}$, and so on. Thus the possible outcomes are not equally likely, but have different relative frequencies or probabilities. In this illustrative example we were able to deduce the probabilities associated with the possible outcomes from more elementary considerations involving equally likely cases. In many practical situations this is not possible, and the required probabilities must be estimated empirically. For example, what is the probability that a person chosen at random from the population of the city of Boston is over 60 years of age? The estimation of this probability requires an empirical method. Presumably in this case the required probability can be estimated using census data.

5.4 JOINT AND CONDITIONAL PROBABILITIES

Consider a population of members classified with regard to two characteristics A and B, there being three classes, or strata, of A and two classes, or strata, of B, as follows:

	A_1	A_2	A_3	
B_1	.40	.15	.05	.60
B_2	.00	.10	.30	.40
	.40	.25	.35	1.00

The probability that a person is B_1 is .60; that is, $p(B_1) = .60$. The probability that he is A_1 is .40; that is, $p(A_1) = .40$. The probability that he is both A_1 and B_1 is .40; that is, $p(A_1B_1) = .40$. This latter probability is a *joint* probability. It is the probability that a member will fall simultaneously within two classes. All the probabilities in the above table are joint probabilities.

Now given the fact that a member is B_1, what is the probability that he is A_1, A_2, or A_3? To answer this question we divide the joint probabilities $p(A_1B_1) = .40$, $p(A_2B_1) = .15$, and $p(A_3B_1) = .05$ by the probability $p(B_1) = .60$ to obtain $p(A_1/B_1) = .67$, $p(A_2/B_1) = .25$, and $p(A_3/B_1) = .08$. These are *conditional* probabilities. They are the probabilities of A_1, A_2, and A_3 given B_1. Note that the sum of these three probabilities is 1.00. Likewise, if a member is B_2 what is the probability that he is A_1, A_2, or A_3? Here $p(A_1/B_2) = .00$, $p(A_2/B_2) = .25$, and $p(A_3/B_2) = .75$. These conditional probabilities may be written in tabular form as follows:

	A_1	A_2	A_3	
B_1	.67	.25	.08	1.00
B_2	.00	.25	.75	1.00

Likewise, the conditional probabilities of B given A are

	A_1	A_2	A_3
B_1	1.00	.60	.14
B_2	.00	.40	.86
	1.00	1.00	1.00

5.5 THE ADDITION AND MULTIPLICATION OF PROBABILITIES

In throwing a die six possible events may occur. If we are prepared to assume, as in the formal mathematical approach to probability, that these six events are equally likely, then the probability of obtaining a 1, 2, 3, 4, 5, or 6 in a single throw is $\frac{1}{6}$, the ratio of the number of favorable cases to the number of equally likely cases. Consider now the probability of obtaining either a 1, 2, or 3 in a single throw. Since there are now three favorable cases among six equally likely cases, this probability is readily observed to be $\frac{1}{6} + \frac{1}{6} + \frac{1}{6} = \frac{1}{2}$. This is an application of the addition theorem of probability. This theorem states that *the probability that any one of a number of mutually exclusive events will occur is the sum of the probabilities of the separate events.* "Mutually exclusive" means that if one event occurs, the others cannot. To illustrate further, in tossing two coins four possible events may occur. Both coins may be heads, both may be tails, the first may be a head and the second a tail, or the first may be a tail and the second a head. These events exhaust the possible outcomes. They may be represented as HH, TT, HT, TH. Again, if we assume these four events to be equally likely, the probability of any one of the four events is $\frac{1}{4}$. By the addition theorem the probability of either two heads or two tails, that is, HH or TT, is $\frac{1}{4} + \frac{1}{4} = \frac{1}{2}$.

In throwing two dice the number of possible outcomes is 36, and the probability of any particular outcome, assuming these to be equally likely, is $\frac{1}{36}$, which is the product of the two independent probabilities, or $\frac{1}{6} \times \frac{1}{6}$. This is an application of the multiplication theorem of probability. This theorem states that *the probability of the joint occurrence of two or more mutually independent events is the product of their separate probabilities.*

By mutually independent is meant that the occurrence of one event does not affect the occurrence of the other events. To illustrate, the probability of obtaining four heads in four tosses of a coin is $\frac{1}{2} \times \frac{1}{2} \times \frac{1}{2} \times \frac{1}{2} = \frac{1}{16}$. The probability of drawing the ace, king, and queen of spades in that order in drawing one card from each of three well-shuffled decks of 52 cards is $\frac{1}{52} \times \frac{1}{52} \times \frac{1}{52} = 1/140,608$. The probability of drawing the ace, king, and queen of spades in that order, and without replacement, from a single deck of 52 cards is $\frac{1}{52} \times \frac{1}{51} \times \frac{1}{50}$, or $1/132,600$. The probability that the first card is the ace of spades is $\frac{1}{52}$. Having drawn one card; 51 cards remain, and the probability that the second card is the king of spades is $\frac{1}{51}$. Similarly, the probability that the third card is the queen of spades is $\frac{1}{50}$. The probability of the combined event is the product of the separate probabilities.

5.6   PROBABILITY DISTRIBUTIONS

In Chap. 2 of this book the concept of a frequency distribution was discussed. A frequency distribution was defined as an arrangement of the data that shows the frequency of occurrence of particular values of the variable, or the frequency of occurrence of values falling within defined ranges of the variable. For example, the following is a frequency distribution for a collection of 50 measurements.

Class interval	f
40–49	5
30–39	10
20–29	20
10–19	10
0–9	5
Total	50

This distribution specifies the frequency of events falling within any class interval. Instead of specifying the frequency, however, a probability could be specified, the probabilities being obtained by dividing the frequency falling within any class interval by the total frequency or the total number of cases, as follows:

Class interval	p
40–49	.10
30–39	.20
20–29	.40
10–19	.20
0–9	.10
Total	1.00

The probability for the interval 40 to 49 is .10. This means that if a member were selected at random from the original group of 50 members the probability is .10 that he would fall within the interval 40 to 49. Similarly the probability that the member would fall within the interval 30 to 39 is .20, and so on. Such an arrangement is called a *probability distribution*. A probability distribution specifies the probabilities associated with particular values of the variable or the probabilities associated, where class intervals are used, with defined ranges of the variable.

Probability distributions can be represented graphically by plotting probabilities along the vertical axes, instead of frequencies, as in the case of frequency distributions.

In sections to follow, a particular type of theoretical probability distribution, known as the *binomial distribution*, is discussed. Before proceeding with a discussion of the binomial distribution it will prove useful to review the reader's knowledge of permutations and combinations.

5.7   PERMUTATIONS AND COMBINATIONS

A knowledge of permutations and combinations is useful in dealing with many problems involving probabilities.

Consider two objects labeled A and B. Two arrangements are possible, AB and BA. With three objects labeled A, B, and C, six arrangements are possible. These are ABC, ACB, BAC, BCA, CAB, and CBA. These arrangements are called *permutations*. In general, if there are n distinguishable objects, the number of permutations of these objects taken n at a time is given by $n!$, or n factorial, which is the product of all integers from n to 1, or $n(n-1)(n-2) \cdots 3 \times 2 \times 1$. For $n=3$, $n! = 3 \times 2 \times 1 = 6$. For $n=5$, $n! = 5 \times 4 \times 3 \times 2 \times 1 = 120$. Consider the number of seating arrangements of eight guests in eight chairs at a dinner table. The first guest may sit in any one of eight chairs. When the first guest is seated, the second guest may sit in any one of the remaining seven chairs. Thus the number of possible arrangements for the first two guests is $8 \times 7 = 56$. When the first two guests are seated, the third guest may occupy any one of the remaining six chairs, and so on for the remaining guests. The number of possible seating arrangements for the eight guests is $8!$, or 40,320, a number which may help explain the indecision of many hostesses.

Instead of considering the number of ways of arranging n things n at a time, we may consider the number of ways of arranging n things r at a time, where r is less than n. Thus the possible arrangements of the objects A, B, and C taken two at a time are AB, AC, BA, BC, CA, and CB. Here we observe that there are three ways of selecting the first object and two ways of selecting the second. The number of arrangements is then $3 \times 2 = 6$. Similarly, on considering the number of arrangements of 10 objects taken three at a time, we observe that there are 10 ways of selecting the first, 9 ways of selecting the second, 8 ways of selecting the third. The number of arrangements is then $10 \times 9 \times 8 = 720$. In general, the number of permutations of n things taken r at a time is

[5.1] $P_r{}^n = n(n-1) \; \cdots \; (n-r+1) = \dfrac{n!}{(n-r)!}$

The number of *different* ways of selecting objects from a set, ignoring the order in which they are arranged, is the number of combinations. Given the objects A, B, C, and D, the number of permutations of two from this set is $4 \times 3 = 12$. The arrangements are AB, BA, AC, CA, AD, DA, BC, CB, BD, DB, CD, and DC. Note that each arrangement occurs in two different orders. If we ignore the order in which each pair of objects is arranged, we have the number of combinations. In this example each pair occurs in two different orders. The number of combinations is then $4 \times \frac{3}{2} = 6$. In general, the number of different combinations of n things taken r at a time is

[5.2] $C_r{}^n = \dfrac{n!}{r!(n-r)!}$

The number of combinations of 10 things taken three at a time is $10!/3!7! = 120$.

The number of combinations of n things taken n at a time is clearly 1, because there is only one way of picking all n objects if we ignore the order of their arrangement.

5.8   THE BINOMIAL DISTRIBUTION

In tossing 10 coins what is the probability of obtaining 0, 1, 2, . . . , 10 heads? We are required to determine the probability of obtaining 0 heads and 10 tails, 1 head and 9 tails, 2 heads and 8 tails, and so on. Let us designate the 10 coins by the letters A, B, C, D, E, F, G, H, I, and J. Let us assume that all 10 coins are unbiased and that the probability of throwing a head or a tail on a single toss of any coin is $\frac{1}{2}$.

Let us attend first to the probability of throwing 0 heads and 10 tails in tossing all 10 coins. The probability that coin A is not a head is $\frac{1}{2}$, that B is not a head is $\frac{1}{2}$, that C is not a head is $\frac{1}{2}$, and so on. Therefore, from the multiplication theorem of probability, the probability that all 10 coins are not heads, or that they are tails, is obtained by multiplying $\frac{1}{2}$ ten times; that is, $(\frac{1}{2})^{10}$, or $1/1{,}024$. Thus in tossing 10 coins there is 1 chance in 1,024 of obtaining 0 heads or 10 tails.

Now consider the problem of obtaining one head and nine tails. The probability that coin A is a head is $\frac{1}{2}$. The probability that all the remaining nine coins are tails is $(\frac{1}{2})^9$. Therefore the probability that A is a head and all other nine coins are tails is $(\frac{1}{2})^{10}$. It is readily observed, however, that one head can occur in 10 different ways. A may be a head and all other coins tails, B may be a head and all the others tails, and so on. Since one head can occur in 10 different ways, the probability of obtaining one head and nine tails is $10(\frac{1}{2})^{10} = 10/1{,}024$. Thus in tossing 10 coins there are 10 chances in 1,024 of obtaining one head and nine tails.

Determining the probability of obtaining two heads and eight tails may be similarly approached. The probability that coins A and B are heads is $(\frac{1}{2})^2$. The probability that all the remaining coins are tails is $(\frac{1}{2})^8$. The probability that A and B are heads and all the remaining coins are tails is $(\frac{1}{2})^{10}$. We readily observe, however, that two heads can occur in quite a number of different ways. This number is the number of combinations of ten things taken two at a time, C_2^{10}, which is $10 \times 9/2 = 45$. Therefore the probability of obtaining two heads and eight tails is $45(\frac{1}{2})^{10}$, or $45/1{,}024$. Similarly, the probability of obtaining three heads and seven tails is $C_3^{10}(\frac{1}{2})^{10} = 120/1{,}024$. Likewise, the probability of obtaining four heads and six tails is $C_4^{10}(\frac{1}{2})^{10} = 210/1{,}024$; and so on. The probabilities of obtaining different numbers of heads in tossing 10 coins are then as follows:

No. of heads	Probability
10	1/1,024
9	10/1,024
8	45/1,024
7	120/1,024
6	210/1,024
5	252/1,024
4	210/1,024
3	120/1,024
2	45/1,024
1	10/1,024
0	1/1,024

The above probabilities are the successive terms in the expansion of the symmetrical binomial $(\frac{1}{2} + \frac{1}{2})^{10}$. This expansion is

$$(\tfrac{1}{2} + \tfrac{1}{2})^{10} = (\tfrac{1}{2})^{10} + 10(\tfrac{1}{2})^{10} + \frac{10 \times 9}{1 \times 2} (\tfrac{1}{2})^{10}$$

$$+ \frac{10 \times 9 \times 8}{1 \times 2 \times 3} (\tfrac{1}{2})^{10} + \cdots + (\tfrac{1}{2})^{10}$$

This is a particular case of the symmetrical binomial $(\frac{1}{2} + \frac{1}{2})^n$ whose terms are

[5.3] $$(\tfrac{1}{2} + \tfrac{1}{2})^n = (\tfrac{1}{2})^n + n(\tfrac{1}{2})^n + \frac{n(n-1)}{1 \times 2} (\tfrac{1}{2})^n$$

$$+ \frac{n(n-1)(n-2)}{1 \times 2 \times 3} (\tfrac{1}{2})^n + \cdots + (\tfrac{1}{2})^n$$

The symmetrical binomial is a particular case of the general form of the binomial $(p + q)^n$, where p is the probability that an event will occur and q is the probability that it will not occur; that is, $p + q = 1$. This binomial may be written as follows:

[5.4] $(p + q)^n = p^n + np^{n-1}q + \dfrac{n(n-1)}{1 \times 2} \, p^{n-2}q^2$

$$+ \dfrac{n(n-1)(n-2)}{1 \times 2 \times 3} p^{n-3}q^3 + \cdots + q^n$$

The terms of this expansion for $n = 2$, $n = 3$, and $n = 4$ are as follows:

$(p + q)^2 = p^2 + 2pq + q^2$
$(p + q)^3 = p^3 + 3p^2q + 3pq^2 + q^3$
$(p + q)^4 = p^4 + 4p^3q + 6p^2q^2 + 4pq^3 + q^4$

The binomial $(p + q)^n$ can be readily illustrated by considering a problem involving the rolling of dice. What are the probabilities of obtaining five, four, three, two, one, and zero 6's in rolling five dice presumed to be unbiased. The probability of obtaining a 6 in rolling a single die is $\frac{1}{6}$, whereas the probability of not obtaining a 6 is $\frac{5}{6}$. The required probabilities are given by the six terms of the binomial $(\frac{1}{6} + \frac{5}{6})^5$. These terms are

$(\frac{1}{6} + \frac{5}{6})^5 = (\frac{1}{6})^5 + 5(\frac{1}{6})^4(\frac{5}{6}) + \dfrac{5 \times 4}{1 \times 2} \, (\frac{1}{6})^3(\frac{5}{6})^2$

$$+ \dfrac{5 \times 4}{1 \times 2} \, (\tfrac{1}{6})^2(\tfrac{5}{6})^3 + 5(\tfrac{1}{6})\,(\tfrac{5}{6})^4 + (\tfrac{5}{6})^5$$

Thus the probabilities of obtaining five, four, three, two, one, and zero 6's in rolling five dice are calculated as

No. of 6's	Probability
5	1/7,776
4	25/7,776
3	250/7,776
2	1,250/7,776
1	3,125/7,776
0	3,125/7,776

There is one chance in 7,776 of obtaining five 6's, 25 chances in 7,776 of obtaining four 6's, and so on. This distribution is seen to be asymmetrical.

Any term in the binomial expansion may be written as

[5.5] $C_r{}^n p^r q^{n-r} = \dfrac{n!}{r!(n-r)!} \, p^r q^{n-r}$

where $C_r{}^n$ is the number of combinations of n things taken r at a time. Thus the probability of obtaining three heads in 10 tosses of a coin is

$$\dfrac{10!}{3!(10-3)!} \, (\tfrac{1}{2})^3(\tfrac{1}{2})^7 = \dfrac{120}{1,024}$$

The coefficients C_r^n in any expansion are

$$1, \; n, \; \frac{n(n-1)}{1 \times 2}, \; \frac{n(n-1)(n-2)}{1 \times 2 \times 3}, \; \ldots$$

These coefficients may be rapidly obtained for different values of N from what is known as Pascal's triangle. The coefficients for different values of N are written in rows in the form of a triangle as shown in Table 5.1. The

Table 5.1 Pascal's triangle

n											
0						1					
1					1		1				
2				1		2		1			
3			1		3		3		1		
4		1		4		6		4		1	
5	1		5		10		10		5		1
6	1	6		15		20		15		6	1
7	1	7	21		35		35		21	7	1
8	1	8	28	56		70		56	28	8	1
9	1	9	36	84	126		126	84	36	9	1
10	1	10	45	120	210	252	210	120	45	10	1

number in any row is the sum of the two numbers to the left and right on the row above. This device is very useful in generating expected frequencies and probabilities. For example, for $N = 10$, the entries in the triangle are the expected frequencies of heads, or tails, in tossing 10 coins 1,024 times. The required probabilities in this case are obtained by dividing the frequencies by 1,024.

5.9 PROPERTIES OF THE BINOMIAL

For the symmetrical binomial, where $p = q = \frac{1}{2}$, the mean, variance, skewness, and kurtosis are

[5.6]
$$\mu = n/2$$
$$\sigma^2 = n/4$$
$$\gamma_1 = 0$$
$$\gamma_2 = -2/n$$

Here we use the symbols μ, σ^2, γ_1, and γ_2 instead of $\overline{X}$, s^2, g_1, and g_2, because the binomial is a theoretical distribution. μ, σ^2, γ_1, and γ_2 may be viewed as population parameters rather than sample estimates. The above formulas may be illustrated by considering the tossing of five coins 32 times. The expected frequencies of zero, one, two, three, four, and five heads are 1, 5, 10, 10, 5, 1. These frequencies are the coefficients of the expansion $(\frac{1}{2} + \frac{1}{2})^5$. Using the formula $\mu = n/2$, the mean is $\mu = 2.5$. This is the expected mean number of heads in tossing five unbiased coins 32 times. The variance of the distribution is $\sigma^2 = n/4 = \frac{5}{4} = 1.25$. Because the distribution is symmetrical, the measure of skewness $\gamma_1 = 0$. The measure of kurtosis $\gamma_2 = -2/n = -\frac{2}{5} = -.40$. Note that as n increases in size, γ_2 becomes smaller. As n increases in size, γ_2 approaches zero as a limit.

The above values may be obtained by direct calculation. Denote the number of heads by X, the frequencies by f, and a deviation from the mean of X by x.

X	f	fX	x	fx^2	fx^3	fx^4
5	1	5	2.50	6.25	15.625	39.0625
4	5	20	1.50	11.25	16.875	25.3125
3	10	30	.50	2.50	1.250	.6250
2	10	20	$-.50$	2.50	-1.250	.6250
1	5	5	-1.50	11.25	-16.875	25.3125
0	1	0	-2.50	6.25	-15.625	39.0625
Total	32	80		40.00	.000	130.0000

The arithmetic mean, μ, of this distribution is $\Sigma fX/N = \frac{80}{32} = 2.50$. The variance, σ^2, is $\Sigma fx^2/N = \frac{40}{32} = 1.25$. σ^2 is here defined as $\Sigma fx^2/N$, and not $\Sigma fx^2/(N-1)$. Here we are dealing with a theoretical model, and not a sample value. The skewness γ_1 is readily seen to be zero because the third moment $\Sigma fx^3/N$ is zero. The fourth moment is $\Sigma fx^4/N = \frac{130}{32} = 4.0625$ and the kurtosis is $\gamma_2 = 4.0625/1.25^2 - 3 = -.40$. Here N denotes the number of tosses, or observations, whereas n denotes the number of coins.

In general the mean, variance, skewness, and kurtosis of any binomial distribution are given by

$$\mu = np$$
$$\sigma^2 = npq$$
$$[5.7] \quad \gamma_1 = \frac{p-q}{\sqrt{npq}}$$
$$\gamma_2 = \frac{1-6pq}{npq}$$

In tossing an unbiased die, the probability p of throwing a 6 is $\frac{1}{6}$ and the probability q of not throwing a 6 is $\frac{5}{6}$. The expected probability distribution of 6's in tossing 10 dice is given by the terms of the expansion $(\frac{1}{6} + \frac{5}{6})^{10}$. The mean of this distribution is $\mu = np = \frac{10}{6} = 1.667$. The variance is $\sigma^2 = npq = 10 \times \frac{1}{6} \times \frac{5}{6} = 1.389$. The skewness $\gamma_1 = .566$ and the kurtosis $\gamma_2 = .12$.

5.10 A HYPOTHETICAL EXPERIMENT

The binomial distribution is frequently used as a model in evaluating experimental results. Such uses of the binomial may be illustrated with reference to a hypothetical experiment.

An individual asserts that he has certain psychic powers which enable him to predict the outcome of future events. An experiment is arranged involving the tossing of a coin. The individual is required to predict the outcome in 10 tosses. If we operate on the working hypothesis that the individual possesses no powers of the type claimed, the probability of a correct prediction by chance alone in a single toss of the coin is $\frac{1}{2}$. From the binomial expansion $(\frac{1}{2} + \frac{1}{2})^{10}$ we can ascertain the probabilities of different numbers of correct predictions. Thus the probability of the individual successfully predicting the outcome in all 10 trials by chance alone is 1/1,024, or .00098. The probability of nine successful predictions and one failure is 10/1,024, or .00977. The probability of eight successful predictions and two failures is 45/1,024, or .04395, and so on. The probability of nine or more successful predictions is .00977 + .00098 = .01075, and the probability of eight or more successful predictions is .04395 + .00977 + .00098 = .05470. Now clearly, before undertaking the experiment, some agreement must be reached regarding the number of correct predictions we are prepared to accept as evidence for the rejection of the hypothesis that the individual possesses no powers of the type claimed.

We may agree arbitrarily that if the results obtained in the experiment could have occurred by chance with a small probability only, say, equal to or less than .05, then these results would be accepted as at least not incompatible with the claims for psychic powers. We observe that the probability of eight or more correct predictions by chance alone is .05470. This is greater than the .05 probability we have agreed to accept; consequently eight correct predictions would in this case not be considered sufficient evidence. The only possibilities here which would prove acceptable within the criterion adopted are nine or ten correct predictions.

The experiment is conducted; seven correct predictions and three failures are obtained. The probability of seven or more correct predictions occurring by chance alone in ten trials may be calculated from the binomial distribution and is 176/1,024, or .17189. Thus there are about 17 chances in 100 of obtaining a result by ordinary guessing equal to or better than the one observed. In consequence, the experimental results provide no acceptable basis for rejecting the working hypothesis that the individual possesses no powers of the type claimed.

Let us suppose that the individual had made 10 correct predictions. Could we reasonably argue from this result that the individual in question did in fact possess psychic power? Quite clearly, such a result is not incompatible with the assertion of psychic power and provides no basis for rejecting that assertion. We observe, however, that circumstances other than the possession of psychic power may possibly have led to the result obtained; that is, alternative explanations of the results may be possible.

In experimental situations of the type described we would ordinarily require more than 10 trials. Let us suppose that 1,000 trials had been made and 550 correct predictions obtained. The probabilities required to evaluate this result would then be generated by the expansion $(\frac{1}{2} + \frac{1}{2})^{1,000}$. Quite clearly, the calculation of the required probabilities directly from the binomial would involve almost prohibitive arithmetical labor. Fortunately, a very close approximation to the required probabilities can be readily obtained from the normal probability distribution, which we shall now consider.

EXERCISES

1   In rolling a die, what is the probability of obtaining either a 5 or a 6?

2   In rolling two dice on one occasion, what is the probability of obtaining either a 7 or an 11? Note that there are 36 possible outcomes.

3   In rolling two dice, what is the probability that neither a 2 nor a 9 will appear?

4   In rolling two dice, what is the probability of obtaining a value less than 6?

5   On four consecutive rolls of a die a 6 is obtained. What is the probability of obtaining a 6 on the fifth roll?

6   In dealing four cards without replacement from a well-shuffled deck, what is the probability of obtaining four aces?

7   An urn contains four black and three white balls. If they are drawn without replacement, what is the probability of the order BWBWBWB?

8   In seating eight people at a table with eight chairs, what is the number of possible seating arrangements?

9   In how many ways can two people seat themselves at a table with four chairs?

10   In tossing five coins, what is the probability of obtaining fewer than three heads?

11   A multiple choice tests contains 100 questions. Each question has five alternatives. If a student guesses all questions, what score might he expect to obtain?

12   In how many ways can a committee of three be chosen from a group of five men?

13 Assume that intelligence and honesty are independent. If 10 per cent of a population is intelligent and 60 per cent is honest, what is the probability that an individual selected at random is both intelligent and honest?

14 A husband engages in random verbal behavior for 20 min in an hour. His wife is similarly engaged for 30 min in that hour. Neither listens to the other. Estimate the number of minutes of silence in the hour.

15 A coin is tossed 10 times. What is the probability that the third head will appear on the tenth toss?

16 The chances are three out of four that the weather tomorrow will be like the weather today. Today is Sunday and is a sunny day. What is the probability that it will be sunny all week? What is the probability that it will be sunny until Wednesday and not sunny on Thursday, Friday, and Saturday?

17 What is the expected distribution of heads in tossing six coins 64 times?

18 What is the expected distribution of 6's in rolling six dice 64 times?

19 What is the probability of obtaining either nine or more heads or three or fewer heads in tossing 12 coins?

20 A man tosses six coins and rolls six dice simultaneously. What is the probability of five or more heads and five or more 6's?

THE NORMAL CURVE

6.1 INTRODUCTION

In Chap. 5 the binomial distribution was discussed in detail. The symmetrical binomial $(\frac{1}{2} + \frac{1}{2})^{10}$ was used to illustrate the binomial expansion. Instead of considering $(\frac{1}{2} + \frac{1}{2})^{10}$ we might consider the more general form $(\frac{1}{2} + \frac{1}{2})^{n}$. As n increases in size this distribution will approach a continuous frequency curve. This frequency curve, which is bell shaped, is the *normal curve* or normal distribution. The frequency distributions of many events in nature are found in practice to be approximated closely by the normal curve, and they are said to be normally distributed. Errors of measurement and errors made in estimating population values from sample values are often assumed to be normally distributed. The frequency distributions of many physical, biological, and psychological measurements are observed to approximate the normal form. Because the frequency of occurrence of many events in nature can be shown empirically to conform fairly closely to the normal curve, this curve can be used as a model in dealing with problems involving these events. The present chapter discusses the normal curve in detail. Before proceeding, however, with a detailed discussion of the normal curve it may be helpful to the reader to consider briefly the nature of functions and frequency curves in general.

6.2 FUNCTIONS AND FREQUENCY CURVES

When two variables are so related that the values of one depend on the values of the other they are said to be functions of each other. A function is descriptive of change in one variable with change in another. The area of a circle is a function of the radius, and the volume of a cube is a function of the length of the edge. Consider the equation $Y = bX + a$. This is a linear function. It is the equation for a straight line; Y and X are variables; b and a are constants. If b and a are known, different values of X can be substituted in the equation and the corresponding values of Y obtained. If the paired values of Y and X are plotted on graph paper, Y on the vertical and X on the horizontal axis, a straight line results. Y and X bear a functional relation to each other, and this relation is linear. Y is sometimes spoken of as the dependent and X the independent variable. A functional relation may be written in the general form $Y = f(X)$. This simply states that Y is some function of X. Here the nature of the function is not specified.

Consider now the binomial $(p + q)^n$. The terms in the binomial expansion are the expected relative frequencies or probabilities associated with particular events. Inspection of formula [5.4] indicates that any term in the binomial expansion is given by

[6.1] $p_r = C_r^n p^r q^{n-r}$

where p_r is the probability of the rth event. This expression is a function. For fixed values of n, p, and q, different values of r may be substituted on the right and the corresponding values of p_r obtained. Here p_r is the dependent and r the independent variable. The variable r is restricted to the $n + 1$ values $0, 1, 2, \ldots, n$; consequently p_r is also restricted to a fixed number of possible values. The paired values of p_r and r may be plotted on graph paper, p_r on the vertical and r on the horizontal axis. The resulting graph is a visual description of the functional relation between the event r and its relative frequency or probability p_r.

In the binomial the variable r is discrete and not continuous. In tossing 50 coins, for example, the number of heads or tails obtained is a discrete number. The value of p_r changes from r to $r + 1$ by discrete steps. We observe, however, that as n increases in size we obtain a larger and larger number of graduations of the distribution and by increasing the size of n we can make the graduations as fine as we like. By considering the situation where n becomes indefinitely large, that is, n approaches infinity, we arrive at the conception of a continuous frequency curve or function. This curve is the limiting form of the binomial.

Frequency curves are in certain instances conceptualized as extending along the X axis from minus infinity to plus infinity; that is, the curves taper off to zero at the two extremities. Although this is so, the area between the curve and the horizontal axis is always finite. For convenience this area is often taken as unity.

On occasion it becomes necessary to find the proportion of the total area of the curve between ordinates erected at particular values of X, that is, between $X = a$ and $X = b$ as shown in Fig. 6.1. This proportion is the probability that a particular value of X drawn at random from the population which the curve describes falls between a and b. Because of this, frequency curves are often referred to as probability curves or probability distributions. Statisticians use a variety of theoretical frequency curves as models. The normal curve is one of the more important of these.

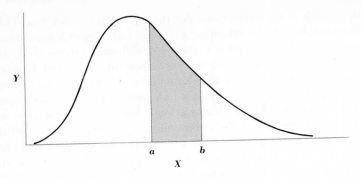

Fig. 6.1 Frequency curve showing area between $X = a$ and $X = b$.

6.3 THE NORMAL CURVE

In tossing n coins the frequency distribution of heads or tails is approximated more closely by the normal distribution as n increases in size. The normal curve is the limiting form of the symmetrical binomial. The equation for the normal curve is

[6.2] $$Y = \frac{N}{\sigma\sqrt{2\pi}}\, e^{-(X-\mu)^2/2\sigma^2}$$

where Y = height of curve for particular values of X
  π = a constant = 3.1416
  e = base of Napierian logarithms = 2.7183
  N = number of cases, which means that the total area under the
     curve is N
 μ and σ = mean and standard deviation of the distribution, respectively

We have used the notation μ and σ in this formula to represent the mean and standard deviation, instead of X and s, because the formula is a theoretical model. Presumably μ and σ may be regarded as population parameters. If N, μ, and σ are known, different values of X may be substituted in the equation and the corresponding values of Y obtained. If paired values of X and Y are plotted graphically, they will form a normal curve with mean μ, standard deviation σ, and area N.

The normal curve is usually written in standard-score form. Standard scores have a mean of zero and a standard deviation of 1. Thus $\mu = 0$ and $\sigma = 1$. The area under the curve is taken as unity; that is, $N = 1$. With these substitutions we may write

[6.3] $$y = \frac{1}{\sqrt{2\pi}}\, e^{-z^2/2}$$

Here z is a standard score on X and is equal to $(X - \mu)/\sigma$. The score z is a deviation in standard deviation units measured along the base line of the curve from a mean of zero, deviations to the right of the mean being positive and those to the left negative. The curve has unit area and unit stan-

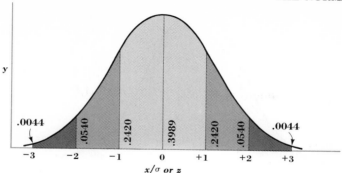

Fig. 6.2 Normal curve showing height of the ordinate at different values of x/σ, or z.

dard deviation. By substituting different values of z in the above formula, different values of y may be calculated. When $z = 0$, $y = 1/\sqrt{2\pi} = .3989$. This follows from the fact that $e^0 = 1$. Any term raised to the zero power is equal to 1. Thus the height of the ordinate at the mean of the normal curve in standard-score form is given by the number .3989. For $z = +1$, $y = .2420$, and for $z = +2$, $y = .0540$. Similarly, the height of the curve may be calculated for any value of z. In practice the student is not required to substitute different values of z in the normal-curve formula and solve for y to obtain the height of the required ordinate. These values may be obtained from Table A of the Appendix. This table shows different values of y corresponding to different values of z. It also shows the area of the curve falling between the ordinates at the mean and different values of z.

The general shape of the normal curve can be observed by inspection of Fig. 6.2. The curve is symmetrical. It is asymptotic at the extremities; that is, it approaches but never reaches the horizontal axis. It can be said to extend from minus infinity to plus infinity. The area under the curve is finite.

6.4 AREAS UNDER THE NORMAL CURVE

For many purposes it is necessary to ascertain the proportion of the area under the normal curve between ordinates at different points on the base line. We may wish to know (1) the proportion of the area under the curve between an ordinate at the mean and an ordinate at any specified point either above or below the mean, (2) the proportion of the total area above or below an ordinate at any point on the base line, (3) the proportion of the area falling between ordinates at any two points on the base line.

Table A of the Appendix shows the proportion of the area between the mean of the unit normal curve and ordinates extending from $z = 0$ to $z = 3$. Let us suppose that we wish to find the area under the curve between the ordinates at $z = 0$ and $z = +1$. We note from Table A that this area is .3413

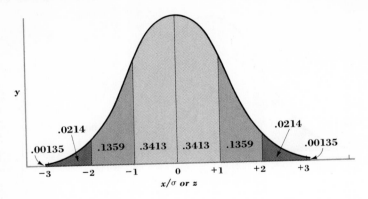

Fig. 6.3 Normal curve showing areas between ordinates at different values of x/σ, or z.

of the total. Thus approximately 34 per cent of the total area falls between the mean and one standard deviation unit above the mean. The proportion of the area of the curve between $z = 0$ and $z = 2$ is .4772. Thus about 47.7 per cent of the area of the curve falls between the mean and two standard deviation units above the mean. The proportion of the area between $z = 0$ and $z = 3$ is .49865, or a little less than 49.9 per cent.

The proportion of the area falling between $z = 0$ and $z = +1$ is .3413. Since the curve is symmetrical the proportion of the area falling between $z = 0$ and $z = -1$ is also .3413. The proportion of the area falling between the limits $z = \pm 1$ is therefore $.3413 + .3413 = .6826$, or roughly 68 per cent. The proportion of the area falling between $z = \pm 2$ is $.4772 + .4772 = .9544$, or about 95 per cent. The proportion between $z = \pm 3$ is $.49865 + .49865 = .99730$ or 99.73 per cent. The area outside these latter limits is very small and is only .27 per cent. For rough practical purposes the curve is sometimes taken as extending from $z = \pm 3$.

Consider the determination of the proportion of the total area above or

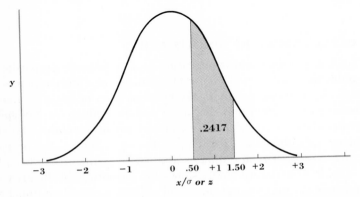

Fig. 6.4 Normal curve showing area between ordinates at $z = .50$ and $z = 1.50$.

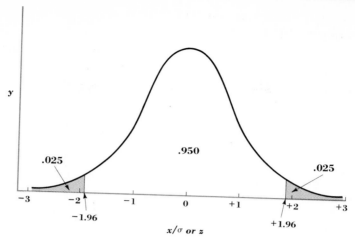

Fig. 6.5 Normal curve showing values of z which include a proportion which is .95 of the total area.

below any point on the base line of the curve. For illustration, let the point be $z = 1$. The proportion of the area between the mean and $z = 1$ is .3413. The proportion of the area below the mean is .5000. The proportion of the total area below $z = 1$ is .5000 + .3413 = .8413. The proportion above this point is 1.0000 − .8413 = .1587. Similarly, the proportion of the area above or below any point on the base line can be readily ascertained.

Consider the problem of finding the area between ordinates at any two points on the base line. Let us assume that we require the area between $z = .5$ and $z = 1.5$. From Table A of the Appendix we note that the proportion of the area between the mean and $z = .50$ is .1915. We note also that the area between the mean and $z = 1.5$ is .4332. The area between $z = .50$ and $z = 1.5$ is obtained by subtracting one area from the other and is .4332 − .1915 = .2417. The area for any other segment of the curve may be similarly obtained.

On occasion we wish to find values of z which include some specified proportion of the total area. For example, the values of z above and below the mean which include a proportion .95 of the area may be required. We select a value of z above the mean which includes a proportion .475 of the total area and a value of z below the mean which also includes a proportion .475 of the total area. From Table A of the Appendix we observe that the proportion .475 of the area falls between $z = 0$ and $z = 1.96$. Since the curve is symmetrical the proportion .475 of the area falls between $z = 0$ and $z = -1.96$. Thus a proportion .95, or 95 per cent, of the total area falls within the limits $z = \pm 1.96$. Also a proportion .05, or 5 per cent, falls outside these limits. Similarly, it may be shown that 99 per cent of the area of the curve falls within, and 1 per cent outside, the limits $z = \pm 2.58$. Fig. 6.5 is a normal curve showing values of z which include a proportion .95 of the total area.

6.5 AREAS UNDER THE NORMAL CURVE— ILLUSTRATIVE EXAMPLE

The distribution of intelligence quotients obtained by the application of a particular test is approximately normal with a mean of 100 and a standard deviation of 15. We are required to estimate what per cent of individuals in the population have intelligence quotients of 120 and above. The intelligence quotient of 120 in standard-score form is $z = (120 - 100)/15 = 1.33$. Thus an intelligence quotient of 120 is 1.33 standard deviation units above the mean. Reference to a table of areas under the normal curve shows that the proportion of the area above a standard score of 1.33 is .092. Thus we estimate that on this particular test about 9.2 per cent of the population have intelligence quotients equal to or greater than 120.

We are required to estimate for the same test the middle range of intelligence quotients which includes 50 per cent of the population. A table of areas under the normal curve shows that 25 per cent of the area under the curve falls between the mean and a standard score of $-.675$. Also 25 per cent of the area falls between the mean and a standard score of $+.675$. Thus 50 per cent of the area falls between the limits of $z = \pm.675$. The standard-score scale has a mean of zero and a standard deviation of unity. Here we must transform standard scores to the original scale of intelligence quotients with a mean of 100 and a standard deviation of 15. To transform standard scores to intelligence quotients we multiply the standard score by 15 and add 100. Thus the standard score $-.675$ is transformed to $15 \times (-.675) + 100 = 89.88$ and $+.675$ to $15 \times .675 + 100 = 110.12$. Thus we estimate that about 50 per cent of the population have intelligence quotients within a range of roughly 90 and 110.

6.6 THE NORMAL APPROXIMATION TO THE BINOMIAL

The observation has been made that as n increases in size the symmetrical binomial is more closely approximated by the normal distribution. This means that the normal distribution may be used to estimate binomial probabilities. Consider a situation where ten coins are tossed a large number of times. What is the probability of obtaining either seven or more heads? Here the mean of the binomial is $\mu = 10 \times \frac{1}{2} = 5.0$ and the standard deviation is $\sigma = \sqrt{\frac{10}{4}} = 1.58$. Because the normal distribution is continuous, and not discrete, we consider the value 7 as covering the exact limits 6.5 to 7.5. Thus we must ascertain the proportion of the area of the normal curve falling above an ordinate at 6.5, the mean of the curve being 5.0 and the standard deviation 1.58. In standard-score form the value 6.5 is equivalent to $z = (6.5 - 5.0)/1.58 = .949$. The proportion of the area of the normal curve falling above an ordinate at $z = .949$ can be readily ascertained from Table A of the Appendix and is .171. Thus using the normal-curve approximation to the binomial we estimate the probability of obtaining seven or more heads in tossing ten coins as .171. We may compare this with the exact probabilities obtained directly from the binomial expansion shown in

Table 6.1 Comparison of binomial probabilities with corresponding normal approximations for $n = 10$ and $p = 1/2$

No. of heads	Exact binomial probability	Normal approximation
10	.001	.002
9	.010	.011
8	.044	.044
7	.117	.114
6	.205	.205
5	.246	.248
4	.205	.205
3	.117	.114
2	.044	.044
1	.010	.011
0	.001	.002
Total	1.000	1.000

Table 6.1. This probability is .172. Here we note that the discrepancy between the estimate obtained from the normal curve and the exact binomial probability is trivial.

Table 6.1 compares the binomial and normal probabilities for $n = 10$ and $p = \frac{1}{2}$. We note that in this instance the differences between the exact binomial probabilities and the corresponding normal approximations are small.

The accuracy of the approximation depends both on n and p; as n increases in size the accuracy of the approximation is improved. For any n as p departs from $\frac{1}{2}$ the approximation becomes less accurate.

6.7 SUMMARY OF PROPERTIES OF THE NORMAL CURVE

The following is a summary of properties of the normal curve.

1 The curve is symmetrical. The mean, median, and mode coincide.

2 The maximum ordinate of the curve occurs at the mean, that is, where $z = 0$, and in the unit normal curve is equal to .3989.

3 The curve is asymptotic. It approaches but does not meet the horizontal axis and extends from minus infinity to plus infinity.

4 The points of inflection of the curve occur at points ± 1 standard deviation unit above and below the mean. Thus the curve changes from convex to concave in relation to the horizontal axis at these points.

5 Roughly 68 per cent of the area of the curve falls within the limits ±1 standard deviation unit from the mean.

6 In the unit normal curve the limits $z = ±1.96$ include 95 per cent and the limits $z = ±2.58$ include 99 per cent of the total area of the curve, 5 per cent and 1 per cent of the area, respectively, falling beyond these limits.

EXERCISES 1 Find the height of the ordinate of the normal curve at the following z values: $-2.15, -1.53, +.07, +.99, +2.76$.

2 Consider a normally distributed variable with $\bar{X} = 50$ and $s = 10$. For $N = 200$ find the height of the ordinates at the following values of X: 25, 35, 49, 57, and 63.

3 Find the proportion of the area of the normal curve (a) between the mean and $z = 1.49$, (b) between the mean and $z = 1.26$, (c) to the right of $z = .25$, (d) to the right of $z = 1.50$, (e) to the left of $z = -1.26$, (f) to the left of $z = .95$, (g) between $z = ±.50$, (h) between $z = -.75$ and $z = 1.50$, (i) between $z = 1.00$ and $z = 1.96$, (j) between $z = 1.00$ and $z = 1.01$.

4 Find a value of z such that the proportion of the area (a) to the right of z is .25, (b) to the left of z is .90, (c) between the mean and z is .40, (d) between $±z$ is .80.

5 On the assumption that IQ's are normally distributed in the population with a mean of 100 and a standard deviation of 15, find the proportion of people with IQ's (a) above 135, (b) above 120, (c) below 90, (d) between 75 and 125.

6 A teacher decides to fail 25 per cent of the class. Examination marks are roughly normally distributed, with a mean of 72 and a standard deviation of 6. What mark must a student make to pass?

7 Error scores on a maze test for a particular strain of rats are known through prolonged experimentation to have an approximately normal distribution with a mean of 32 and a standard deviation of 8. In one experiment a control sample of six animals contains one animal with an error score of 66. What arguments may be advanced for discarding the results for this animal?

8 Scores on a particular psychological test are normally distributed, with a mean of 50 and a standard deviation of 10. The decision is made to use a letter-grade system A, B, C, D, and E with the proportions .10, .20, .40, .20, and .10 in the five grades, respectively. Find the score intervals for the five letter grades.

9 The following are data for test scores for two age groups:

	11-year group	14-year group
$\overline{X}$	48	56
s	8	12
N	500	800

Assuming normality, estimate how many of the 11-year-olds do better than the average 14-year-old and how many of the 14-year-olds do worse than the average 11-year-old.

CORRELATION

7.1 INTRODUCTION

Hitherto we have considered the description of a single variable. We now approach the problem of describing the degree of simultaneous or concomitant variation of two variables. The data under consideration, sometimes called bivariate data, consist of pairs of measurements. The data, for example, may be measures both of height and weight for a group of school children, or measures both of intelligence and scholastic performance for a group of university students, or error scores for a group of experimental animals in running two different mazes. The essential feature of the data is that one observation can be paired with another observation for each member of the group. The study of this type of data has two closely related aspects, correlation and prediction. *Correlation* is concerned with describing the degree of relation between variables. *Prediction* is concerned with estimating one variable from a knowledge of another. We shall use for illustration the record of scores obtained on a psychological test administered to students entering university and examination marks obtained by these same students at the end of the first year of university work. The investigator may concern himself with obtaining a simple summary description of the degree of relation or correlation between test scores and examination marks. On the other hand, he may focus attention on the prediction of examination marks from a knowledge of psychological test scores, his purpose being to use psychological test scores to provide estimates, on university entrance, of subsequent scholastic performance.

Historically, the study of the prediction of one variable from a knowledge of another preceded the development of measures of correlation. In the year 1885 Francis Galton published a paper called *Regression towards*

Mediocrity in Hereditary Stature. Galton was interested in predicting the physical characteristics of offspring from a knowledge of the physical characteristics of their parents. He observed, for example, that the offspring of tall parents tended on the average to be shorter than their parents, whereas the offspring of short parents tended on the average to be taller than their parents. He used the word "regression" to refer to this effect. In modern statistics the term regression no longer has the biological implication assigned to it by Galton. In general, regression has to do with the prediction of one variable from a knowledge of another. Karl Pearson extended Galton's ideas of regression and developed the methods of correlation extensively used today.

The most widely used measure of correlation is the Pearson *product-moment correlation coefficient.* This measure is used where the variables are quantitative, that is, of the interval or ratio type. Other varieties of correlation have been developed for use with nominal and ordinal variables. One measure commonly used to describe the relationship between two nominal variables is the *contingency coefficient.* Methods used with ordinal variables are called *rank-order correlation methods.* These special types of correlation will be discussed in later chapters.

In this chapter we shall present a discussion of correlation and proceed in Chap. 8 to a discussion of prediction and its relation to correlation. The reader will bear in mind that correlation and prediction are two closely related topics. Certain topics pertaining to the interpretation of the correlation coefficient and assumptions underlying its use can only be discussed following a consideration of prediction.

7.2 RELATIONS BETWEEN PAIRED OBSERVATIONS

Consider a group comprised of N members. Denote these by A_1, A_2, A_3, . . . , A_N. Measurements are available on each member on two variables, X and Y. The data may be represented symbolically as follows:

	Measurement	
Members	X	Y
A_1	X_1	Y_1
A_2	X_2	Y_2
A_3	X_3	Y_3
.	.	.
.	.	.
.	.	.
A_N	X_N	Y_N

Let us assume that measurements have been arranged in order of magnitude on X extending from X_1, the highest, to X_N, the lowest, measurement. Given this arrangement on X, we may consider the possible arrangements of Y with respect to X. Consider an arrangement where the values of Y are in order of magnitude extending from the highest to the lowest. Thus the member who is highest on X is also highest on Y, the member who is next highest on X is next highest on Y, and so on, until the member who is lowest on X is also lowest on Y. This situation represents the maximum positive relation between the two variables. Consider now an arrangement where the values of Y are reversed so that Y_1 is the lowest and Y_N is the highest. The member who is highest on X is lowest on Y, the member who is next highest on X is next lowest on Y, and so on, until the member who is lowest on X is highest on Y. This situation represents the maximum negative relation between the variables. Consider a situation where the arrangement of Y is strictly random in relation to X. Values of Y may be inserted in a hat, shuffled, drawn at random, and paired with values of X. This is a situation of independence. The two sets of variate values bear a random relation to each other. Under this arrangement we may state that no relation exists between X and Y. Between the two extreme arrangements, representing the maximum positive and negative relation, we may consider arrangements which represent varying degrees of relation in either a positive or negative direction. To illustrate, let us assume that the values of X for the members A_1, A_2, A_3, A_4, and A_5 are the integers 5, 4, 3, 2, and 1. If the values of Y are the same integers and are also arranged in the order 5, 4, 3, 2, and 1, we have a maximum positive relation. If values of Y are arranged in the order 4, 5, 3, 2, and 1, we have clearly a high positive relation, although not the highest possible. If values of Y are arranged in the order 1, 2, 3, 4, and 5, we have a maximum negative relation. Again an arrangement on Y of the kind 1, 2, 4, 3, 5 would be high negative, although not the highest possible.

Relations of the kind described above may be examined by plotting the paired measurements on graph paper, each pair of observations being represented by a point. Such a plotting of measurements is sometimes called a *scatter diagram*. Inspection of a scatter diagram yields an intuitive appreciation of the degree of relation between the two variables. Figure 7.1 shows four such diagrams.

Figure 7.1a is a graphical representation of a high positive relation. Note that the points fall very close to a straight line. If the points fall exactly on a straight line, a perfect positive relation exists between the variables. Figure 7.1b shows a low positive relation. Figure 7.1c shows a relation which is more or less random. No systematic tendency is observed for high values of X to be associated with high values of Y and low values of X to be associated with low values of Y, or vice versa. Figure 7.1d shows a fairly high negative relation. Again, if all the points fall exactly along a straight line, a perfect negative relation exists. It is obvious that between the two extremes of a perfect positive and a perfect negative relation an indefinitely large number of possible arrangements of points may occur repre-

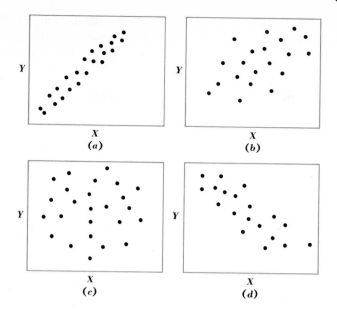

Fig. 7.1 (*a*) **High positive correlation.** (*b*) **Low positive correlation.**
(*c*) **Zero correlation.** (*d*) **Negative correlation.**

senting an indefinitely large number of possible relations between the two variables.

7.3 THE CORRELATION COEFFICIENT

Measures of correlation are conventionally defined to take values ranging from -1 to $+1$. A value of -1 describes a perfect negative relation. All points lie on a straight line, and X decreases as Y increases. A value of $+1$ describes a perfect positive relation. All points lie on a straight line, and X increases as Y increases. A value of 0 describes the absence of a relation. The variable X is independent of Y or bears a random relation to Y. Measures of correlation take positive values where the relation is positive and negative values where the relation is negative.

The most commonly used measure of correlation is the Pearson product-moment correlation coefficient. Many forms of correlation are particular cases of this coefficient. Let X and Y be two sets of paired observations with standard deviations s_x and s_y. We may represent the paired observations in standard-score form by taking deviations from the mean and dividing by the standard deviation. Thus

$$z_x = \frac{X - \bar{X}}{s_x}$$

$$z_y = \frac{Y - \bar{Y}}{s_y}$$

The standard scores have a mean of zero and a standard deviation of unity. The product-moment correlation coefficient, denoted by the letter r, is the sum of products of standard scores, divided by $N - 1$. The formula for r in standard-score form is

[7.1] $$r = \frac{\Sigma z_x z_y}{N - 1}$$

Thus the correlation coefficient may be obtained by converting the two variables to standard-score form, summing their product, and dividing by $N - 1$.

A brief and rather incomplete digression on the rationale underlying the above coefficient is appropriate here. Consider a set of paired observations in standard-score form. The sum of products of standard scores $\Sigma z_x z_y$ is readily observed to be a measure of the degree of relationship between the two variables. Let us consider the *maximum* and *minimum* values of $\Sigma z_x z_y$. This sum of products is observed to take its maximum possible value when (1) the values of z_x and z_y are in the same order and (2) every value of z_x is equal to the value of z_y with which it is paired, the two sets of paired standard scores being identical. If the paired standard scores are plotted on graph paper, all points will fall exactly along a straight line with positive slope. Since all pairs of observations are such that $z_x = z_y$, we may write $z_x z_y = z_x^2 = z_y^2$ and $\Sigma z_x z_y = \Sigma z_x^2 = \Sigma z_y^2$. The quantity $\Sigma z_x^2 = \Sigma z_y^2 = N - 1$, as shown in Sec. 4.12. Thus we observe that the maximum possible value of $\Sigma z_x z_y$ is equal to $N - 1$. Similarly, $\Sigma z_x z_y$ will take its minimum possible value when (1) the values of z_x and z_y are in inverse order and (2) every value of z_x has the same absolute numerical value as the z_y with which it is paired, but differs in sign. This minimum value of $\Sigma z_x z_y$ is readily shown to be equal to $-(N - 1)$. Graphically, all points will fall exactly along a straight line with negative slope. When z_x and z_y bear no systematic relation to each other, the expected value of $\Sigma z_x z_y$ will be zero. We may define a coefficient of correlation as the ratio of the observed value of $\Sigma z_x z_y$ to the maximum possible value of this quantity; that is, r is defined as $\Sigma z_x z_y / (N - 1)$. Since $\Sigma z_x z_y$ has a range extending from $N - 1$ to $-(N - 1)$, the coefficient r will extend from $+1$ to -1. The reader will note that for any particular set of paired standard scores the maximum and minimum values of $\Sigma z_x z_y$ obtained by arranging the paired scores in direct and inverse order are not necessarily $N - 1$ and $-(N - 1)$. A maximum value equal to $N - 1$ will occur only when the paired observations have the characteristic that every value of z_x is equal to the value of z_y with which it is paired. A minimum value of $-(N - 1)$ will occur only when every value of z_x is equal to z_y in absolute value, but differs in sign. When the data do not have these characteristics, the limits of the range of r, for the particular set of paired observations under consideration, will be less than $+1$ and greater than -1.

7.4 CALCULATION OF THE CORRELATION COEFFICIENT FROM UNGROUPED DATA

The formula for the correlation coefficient in standard-score form is $r = \Sigma z_x z_y / (N - 1)$. The calculation of a correlation coefficient using this formula is somewhat laborious, as it requires the conversion of all values to standard scores. Since $z_x = (X - \bar{X})/s_x$ and $z_y = (Y - \bar{Y})/s_y$ by substitution we may write the formula for the correlation coefficient in deviation-score form. Thus

$$[7.2] \quad r = \frac{\Sigma(X - \bar{X})(Y - \bar{Y})}{\sqrt{\Sigma(X - \bar{X})^2 \Sigma(Y - \bar{Y})^2}} = \frac{\Sigma xy}{\sqrt{\Sigma x^2 \Sigma y^2}}$$

where x and y are deviations from the means $\bar{X}$ and $\bar{Y}$, respectively.

Table 7.1 Calculation of the correlation coefficient from ungrouped data using deviation scores

1 X	2 Y	3 x	4 y	5 x^2	6 y^2	7 xy
5	1	−1	−3	1	9	+3
10	6	+4	+2	16	4	+8
5	2	−1	−2	1	4	+2
11	8	+5	+4	25	16	+20
12	5	+6	+1	36	1	+6
4	1	−2	−3	4	9	+6
3	4	−3	0	9	0	0
2	6	−4	+2	16	4	−8
7	5	+1	+1	1	1	+1
1	2	−5	−2	25	4	+10
60	40	0	0	134	52	48
$\bar{X} = 6.0$	$\bar{Y} = 4.0$			Σx^2	Σy^2	Σxy

$$r = \frac{\Sigma xy}{\sqrt{\Sigma x^2 \Sigma y^2}} = \frac{48}{\sqrt{134 \times 52}} = +.58$$

The above formula for the correlation coefficient may be used for computational purposes. The calculation is illustrated in Table 7.1. The first two columns contain the paired observations on X and Y. These columns are summed and divided by N to obtain the means $\bar{X}$ and $\bar{Y}$. Column 3 contains the deviations from the mean of X, and col. 4 the deviations from the mean of Y. Columns 5 and 6 contain the squares of these deviations. These columns are summed to obtain Σx^2 and Σy^2. Column 7 contains the products of x and y, and this column is summed to obtain Σxy. The correlation coefficient in this example is $+.58$.

For certain purposes it is desirable to express the formula for the correlation in terms of the raw scores or the original observations. This formula is as follows:

$$[7.3] \quad r = \frac{N\Sigma XY - \Sigma X \Sigma Y}{\sqrt{[N\Sigma X^2 - (\Sigma X)^2][N\Sigma Y^2 - (\Sigma Y)^2]}}$$

This is one of the more convenient formulas to use where a calculating machine is available. Some modern calculating machines are so designed that pairs of observations may be entered successively on the machine and the terms ΣXY, ΣX^2, ΣY^2, ΣX, and ΣY obtained in a single operation. Where a calculating machine is not available, this formula usually involves rather large and unwieldly numbers and the formula in deviation form may be preferred.

The application of the formula for computing the correlation coefficient

Table 7.2   Calculation of the correlation coefficient from ungrouped data using raw scores

1 X	2 Y	3 X^2	4 Y^2	5 XY
5	1	25	1	5
10	6	100	36	60
5	2	25	4	10
11	8	121	64	88
12	5	144	25	60
4	1	16	1	4
3	4	9	16	12
2	6	4	36	12
7	5	49	25	35
1	2	1	4	2
60 ΣX	40 ΣY	494 ΣX^2	212 ΣY^2	288 ΣXY

$$
\begin{aligned}
r &= \frac{N\Sigma XY - \Sigma X \Sigma Y}{\sqrt{[N\Sigma X^2 - (\Sigma X)^2][N\Sigma Y^2 - (\Sigma Y)^2]}} \\
&= \frac{10 \times 288 - 60 \times 40}{\sqrt{(10 \times 494 - 60^2)(10 \times 212 - 40^2)}} \\
&= \frac{480}{\sqrt{1{,}340 \times 520}} = +.58
\end{aligned}
$$

from raw scores is illustrated in Table 7.2. The first two columns contain the paired observations on X and Y. These columns are summed to obtain ΣX and ΣY. Columns 3 and 4 contain the squares of the observations, and these are summed to obtain ΣX^2 and ΣY^2. Column 5 contains the product terms XY, and the sum of this column is ΣXY. The correlation is $+.58$, which checks with the value obtained by the previous method using deviation scores.

7.5 BIVARIATE FREQUENCY DISTRIBUTIONS

In Chap. 2 we discussed the construction of frequency distributions for a single variable. A frequency distribution was defined as an arrangement of the data showing the frequency of occurrence of the observations within defined ranges of the values of the variable, the defined ranges being the class intervals. Where one variable only is involved, the distribution may be spoken of as *univariate*. The frequency-distribution idea may be readily extended to two-variable situations. A frequency distribution involving two variables is known as a *bivariate* frequency distribution.

A bivariate frequency distribution is a table comprised of a number of rows and columns. The columns correspond to class intervals of the X variable and the rows to class intervals of the Y variable. Each pair of observations is entered as a tally in its appropriate cell. To illustrate, Table 7.3 shows a bivariate frequency distribution for a set of paired observations, these being scores on two forms of a French reading test. In constructing such a distribution a person who makes a score of 27 on Form A and a score of 31 on Form B is entered as a tally in the cell that is common to the row corresponding to the class interval 25 to 29 on Form A and the column corresponding to the class interval 30 to 34 on Form B. Similarly, every pair of observations is entered as a tally in its appropriate cell. The tallies in each cell are then counted, and their number recorded. These numbers are the bivariate frequencies. By summing the bivariate frequencies in the rows we obtain, as shown in Table 7.3, the frequency distribution for the Y variable, and by summing the columns we obtain the frequency distribution for the X variable. The separate frequency distributions of X and Y are usually written at the bottom and to the right of the table. In the selection of class intervals for X and Y the usual conventions regarding class intervals apply.

Methods exist for the calculation of correlation coefficients from bivariate frequency distributions. These methods are now infrequently used and will not be described here. The idea of a bivariate frequency table is, however, of some importance.

7.6 THE VARIANCE OF SUMS AND DIFFERENCES

Let X and Y be two sets of measurements for the same group of individuals. These, for example, may be marks on mathematics and history examinations for a group of university students. What is the variance of $X + Y$? If

Table 7.3*   Bivariate frequency distribution for two forms of French reading test

X-Form B

Y-Form A	0-4	5-9	10-14	15-19	20-24	25-29	30-34	35-39	f_y
35-39							2	1	3
30-34						1	1		2
25-29						3	2	1	6
20-24				2	3	3	1		9
15-19			1	9	8				18
10-14	1	1	3	2	1				8
5-9	1	3	3						7
0-4	1	1							2
f_x	3	5	7	13	12	7	6	2	

*   Table 7.3 is reproduced from R. W. B. Jackson and George A. Ferguson, *Manual of educational statistics*, University of Toronto, Department of Educational Research, Toronto, 1942.

mathematics and history marks are added together, what is the variance of the sums?

The sum of X and Y is $X + Y$. The mean of the sum of X and Y is $\bar{X} + \bar{Y}$, or the sum of the two means. We may then write the variance of sums as follows:

$$[7.4] \quad s_{x+y}^2 = \frac{\Sigma[(X+Y) - (\bar{X} + \bar{Y})]^2}{N-1}$$

$$= \frac{\Sigma[(X - \bar{X}) + (Y - \bar{Y})]^2}{N-1}$$

$$= \frac{\Sigma(X - \bar{X})^2}{N-1} + \frac{\Sigma(Y - \bar{Y})^2}{N-1} + \frac{2\Sigma(X - \bar{X})(Y - \bar{Y})}{N-1}$$

$$= s_x^2 + s_y^2 + 2rs_x s_y$$

The variance of the sum of X and Y is the sum of the two variances plus $2rs_x s_y$. If the correlation between the two variables is zero, then $2rs_x s_y = 0$ and the variance of sums is simply the sum of the two variances. Terms of the kind rs_xs_y are sometimes called *covariance* terms, or *covariances*.

Similarly, the variance of the differences between X and Y, the variance of $X - Y$, is readily shown to be

[7.5] $\quad s_{x-y}^2 = s_x^2 + s_y^2 - 2rs_xs_y$

The variance of differences is the sum of the two variances minus the covariance term $2rs_xs_y$.

The sum of more than two variables may be usefully considered. Consider three variables X_1, X_2, and X_3. The variances and covariances may for convenience be written in the form of a covariance table, or matrix, as follows:

	1	2	3
1	s_1^2	$r_{12}s_1s_2$	$r_{13}s_1s_3$
2	$r_{12}s_1s_2$	s_2^2	$r_{23}s_2s_3$
3	$r_{13}s_1s_3$	$r_{23}s_2s_3$	s_3^2

The variances appear along the main diagonal and the covariances on either side of the main diagonal. The variance of sums is simply the sum of all the elements in the covariance table or matrix. Thus, for three variables the variance of sums is:

$$s_{1+2+3}^2 = s_1^2 + s_2^2 + s_3^2 + 2r_{12}s_1s_2 + 2r_{13}s_1s_3 + 2r_{23}s_2s_3$$

The argument may be generalized to any number of variables. The variance of the sum of k variables is very simply the sum of all the elements of the covariance table or matrix with k rows and columns.

EXERCISES 1 Would you expect the correlation between the following to be positive, negative, or about zero? (*a*) The intelligence of parents and their offspring, (*b*) scholastic success and annual income 10 years after graduation, (*c*) age and mental ability, (*d*) marks on examinations in physics and mathematics, (*e*) wages and the cost of living, (*f*) birth rate and the numerosity of storks, (*g*) scores on a dominance-submission test for husbands and scores for their wives.

2 The following are paired measurements:

X	5	8	9	7	6	1
Y	3	7	8	8	5	9

Compute the correlation between X and Y.

3 Show that

$$\frac{\Sigma xy}{(N-1)s_x s_y} = \frac{\Sigma xy}{\sqrt{\Sigma x^2 \Sigma y^2}}$$

4 When $N = 2$, what are the possible values of the correlation coefficient?

5 The correlation coefficient is not necessarily equal to $+1$ when the paired measurements are in exactly the same rank order. Discuss.

6 Calculate the correlation coefficient for the following data using formula [7.3].

X	Y	X	Y	X	Y
22	18	19	25	11	17
15	16	7	36	5	6
9	31	6	27	26	45
7	8	46	45	19	30
4	2	11	18	8	18
45	36	27	18	1	3
19	12	19	37	9	7
26	16	36	42	18	28
35	47	25	20	46	21
49	22	10	12	9	25

7 Show that $s_{x-y}^2 = s_x^2 + s_y^2 - 2rs_x s_y$.

8 Under what conditions will the variance of the sum of two variables equal the variance of the difference between two variables?

9 Is the correlation between X and Y changed by adding a constant to X or by multiplying X by a constant?

10 The formulas in this chapter assume that the variance is defined as $s^2 = \Sigma(X - \bar{X})^2/(N - 1)$. What is the formula for r in standard-score form if the variance is defined as $s^2 = \Sigma(X - \bar{X})^2/N$?

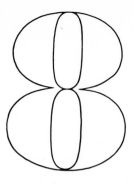

PREDICTION IN RELATION TO CORRELATION

8.1 INTRODUCTION

Psychologists and educationists are frequently concerned with problems of prediction. The educational psychologist is interested in predicting the scholastic performance of a child from a knowledge of intelligence test scores. The industrial psychologist in the selection of an individual for a particular type of employment makes a prediction about the subsequent job performance of that individual from information available at the time of selection. The clinical psychologist may direct his attention to predicting the patient's receptivity to treatment from information obtained prior to treatment. In many areas of human endeavor predictions about the subsequent behavior of individuals are required. A somewhat elaborate statistical technology has evolved for dealing with the prediction problem. In this chapter we shall restrict attention to the simplest aspect of prediction, the prediction of one variable from a knowledge of another.

Prediction and correlation are closely related topics, and an understanding of one requires an understanding of the other. The presence of a zero correlation between two variables X and Y may usually be interpreted to mean that they bear no systematic relation to each other. A knowledge of X tells us nothing about Y, and a knowledge of Y tells us nothing about X. In predicting X from Y or Y from X no prediction better than a random guess is possible. The presence of a nonzero correlation between X and Y, implies that if we know something about X we know something about Y, and vice versa. If knowing X implies some knowledge of Y, a prediction of Y from X is possible which is better than a random guess about Y made in the absence of a knowledge of X. The greater the absolute value of the correlation between X and Y, the more accurate the prediction of one variable from the other. If the correlation between X and Y is either -1 or $+1$, perfect prediction is possible.

8.2   THE LINEAR REGRESSION OF Y ON X

Any set of paired observations may be plotted on graph paper, each pair of observations being represented by a point. Consider the data shown in Table 8.1, cols. 2 and 3. These columns contain intelligence quotients and reading-test scores for a group of 18 school children. These data are plotted in graphical form in Fig. 8.1. While the arrangement of points when plotted graphically shows considerable irregularity, we observe a tendency for reading-test scores to increase as intelligence quotients increase.

Table 8.1   Calculations for regression line of Y on X for ungrouped data*

1 Pupil no.	2 IQ X	3 Reading score Y	4 X²	5 XY	6 Expected reading score Y′
1	118	66	13,924	7,788	68
2	99	50	9,801	4,950	55
3	118	73	13,924	8,614	68
4	121	69	14,641	8,349	70
5	123	72	15,129	8,856	71
6	98	54	9,604	5,292	54
7	131	74	17,161	9,694	77
8	121	70	14,641	8,470	70
9	108	65	11,664	7,020	61
10	111	62	12,321	6,882	63
11	118	65	13,924	7,670	68
12	112	63	12,544	7,056	64
13	113	67	12,769	7,571	65
14	111	59	12,321	6,549	63
15	106	60	11,236	6,360	60
16	102	59	10,404	6,018	57
17	113	70	12,769	7,910	65
18	101	57	10,201	5,757	57
Sum	2,024	1,155	228,978	130,806	

*Reproduced from R. W. B. Jackson and George A. Ferguson, *Manual of educational statistics*, University of Toronto, Department of Educational Research, Toronto, 1942.

Fig. 8.1   Scatter diagram for data of Table 8.1.

Let us suppose that we are given a child's intelligence quotient only and are required to predict his reading-test scores. How shall we proceed? Clearly, the data show considerable irregularity. An exact correspondence between the two sets of scores does not exist. In this situation we may proceed by fitting a straight line to the data. This straight line provides an average statement about the change in one variable with change in the other. It describes the trend in the data and is based on all the observations. If, then, we are given a child's intelligence quotient and are required to predict his reading-test score, we use the properties of the line. The method used in fitting a line to a set of points in a situation of this kind is the *method of least squares*. If our interest resides in predicting Y from X, the method of least squares locates the line in a position such that the sum of squares of distances from the points to the line taken *parallel to the Y axis* is a minimum. This line is known as the regression line of Y on X.

The general equation of any straight line is given by

[8.1] $Y = bX + a$

The quantity a is a constant. It is the distance on the Y axis from the origin to the point where the line cuts the Y axis. It is the value of Y corresponding to $X = 0$. If we substitute $X = 0$ in the equation for a straight line, we observe that $Y = a$. The quantity b is the slope of the line. The slope of any line is simply the ratio of the distance in a vertical direction to the distance in a horizontal direction, as illustrated in Fig. 8.2. The slope describes the rate of increase in Y with increase in X. If a and b are known, the location of the line is uniquely fixed and for any given value of X we can compute a corresponding value of Y.

Where the regression line of Y on X is fitted by the method of least squares, the slope of the line b_{yx} and the point where the line cuts the Y axis a_{yx} may be calculated by the following formulas:

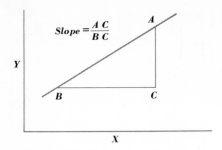

Fig. 8.2   The slope of a line.

[8.2] $b_{yx} = \dfrac{\Sigma XY - (\Sigma X \Sigma Y/N)}{\Sigma X^2 - [(\Sigma X)^2/N]}$

[8.3] $a_{yx} = \dfrac{\Sigma Y - b_{yx}\Sigma X}{N}$

The quantity ΣX is the sum of X, ΣY is the sum of Y, ΣXY is the sum of products of X and Y, ΣX^2 is the sum of squares of X, and N is the number of cases.

To illustrate, consider the data of Table 8.1. Columns 2 and 3 provide intelligence quotients and reading scores for the 18 school children. Column 4 provides the values X^2, and col. 5 the products XY. Summing the columns, we obtain

$$\Sigma XY = 130,806$$
$$\Sigma X = 2,024$$
$$\Sigma Y = 1,155$$
$$\Sigma X^2 = 228,978$$
$$N = 18$$

Applying formulas [8.2] and [8.3], we have

$$b_{yx} = \frac{130,806 - 2,024 \times 1,155/18}{228,978 - (2,024)^2/18} = .6708$$

$$a_{yx} = \frac{1,155 - .6708 \times 2,024}{18} = -11.25$$

The regression line of Y on X is then described by the equation $Y' = .6708X - 11.25$. The symbol Y' has been introduced to refer to the estimated value of Y, that is, the value of Y estimated from a knowledge of X. Y' is a distance from the X axis to the line corresponding to any value of X. By substituting any value of X in the formula we obtain Y', the estimated value of Y. Column 6 of Table 8.1 shows the estimated reading-test scores obtained by applying this regression equation.

8.3   THE LINEAR REGRESSION OF X ON Y

Above we have considered the regression of Y on X. The regression line has been located in order to minimize the sum of squares of the distances from the points to the line taken *parallel to the Y axis.* Given reading-test scores and intelligence quotients, we concerned ourselves with predicting reading-test scores from intelligence quotients. If, however, we wish to predict intelligence quotients from reading-test scores, a different regression line is used. This is the regression line of X on Y. This line is located in a position such as to minimize the sum of squares of the distances from the points to the line *parallel to the X axis.* We see, therefore, that two regression lines may be fitted to any set of paired observations, the regression line of Y on X and the regression line of X on Y. The regression of Y on X is used in predicting Y from X. The regression of X on Y is used in predicting X from Y. These two lines will differ except in the particular case where all the points fall exactly on a straight line. Under this circumstance the two regression lines coincide.

The formula for the regression line of X on Y is given by

[8.4] $\quad X' = b_{xy}Y + a_{xy}$

The symbol X' is used to refer to the predicted value of X, the value estimated from a knowledge of Y; b_{xy} is the slope of the regression line; and a_{xy} is the point where the line intercepts the X axis. The values of b_{xy} and a_{xy} may be calculated from the formulas

[8.5] $\quad b_{xy} = \dfrac{\Sigma XY - (\Sigma X \Sigma Y / N)}{\Sigma Y^2 - [(\Sigma Y)^2 / N]}$

and

[8.6] $\quad a_{xy} = \dfrac{\Sigma X - b_{xy} \Sigma Y}{N}$

8.4 REGRESSION LINES FOR A BIVARIATE FREQUENCY DISTRIBUTION

Where data are grouped in the form of a bivariate frequency table the frequencies in each row or each column of the table constitute a frequency distribution. Table 8.2 is a bivariate frequency distribution for scores on a verbal intelligence test and the Binet intelligence test. We note, for example, that 104 individuals make scores between 100 and 109 on the Binet. The frequencies in the 100–109 column comprise the frequency distribution of scores on the verbal test for all individuals with IQ's between 100 and 109 on the Binet. The mean score on the verbal test for these 104 individuals can be readily calculated from the distribution in the 100–109 column. If we know only that an individual's IQ falls between 100 and 109, the best estimate we can make of his verbal-test score is that he is at the mean of those individuals with IQ's between 100 and 109. The means for all column arrays may be calculated. These are the mean verbal-test scores of the individuals falling within particular class intervals on the Binet scale. A straight line may be fitted to this set of means by the method of least squares. This line is the regression of Y on X, the regression line used in

Table 8.2 Bivariate frequency distribution of scores obtained by 500 Scottish school boys on a verbal intelligence test and the Binet intelligence test showing regression lines of Y on X and X on Y*

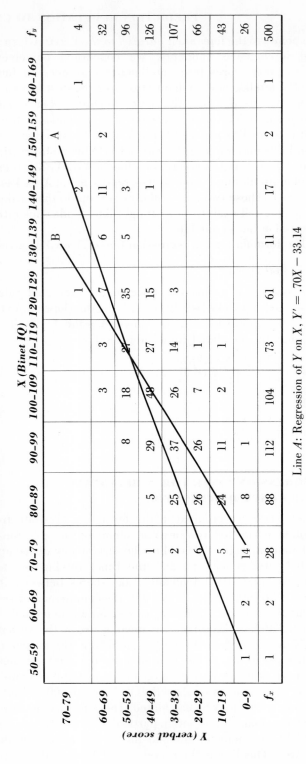

Y (verbal score)	50–59	60–69	70–79	80–89	90–99	100–109	110–119	120–129	130–139	140–149	150–159	160–169	f_y
70–79								1		2		1	4
60–69						3	3	7	6	11	2		32
50–59					8	18	27	35	5	3			96
40–49			1	5	29	48	27	15		1			126
30–39			2	25	37	26	14	3					107
20–29			6	26	26	7	1						66
10–19			5	24	11	2	1						43
0–9	1	2	14	8	1								26
f_x	1	2	28	88	112	104	73	61	11	17	2	1	500

X (Binet IQ)

Line *A*: Regression of Y on X, $Y' = .70X - 33.14$
Line *B*: Regression of X on Y, $X' = .85Y + 70.25$

* Data from A. M. Macmeeken, *The intelligence of a representative group of Scottish children*, University of London Press, London, 1940. Reproduced with permission of The Scottish Council for Research in Education, Edinburgh, Scotland.

predicting verbal-test scores from Binet IQ's. Similarly, the means for the row arrays may be calculated and a line fitted to these means by the method of least squares. This line is the regression of X on Y, the regression line used in predicting Binet IQ's from verbal-test scores. The two regression lines are shown in Table 8.2.

8.5   RELATION OF REGRESSION TO CORRELATION

If all points in a scatter diagram fall exactly along a straight line, the two regression lines coincide. Perfect prediction is possible. The correlation coefficient in this case is either -1 or $+1$. Where the correlation departs from either -1 or $+1$, the two regression lines have an angular separation. In general, as the degree of relationship between two variables decreases, the angular separation between the two regression lines increases. Where no systematic relationship exists at all, the two variables being independent, the two regression lines are at right angles to each other.

A simple relationship exists between the correlation coefficient and the slopes of the two regression lines. The slopes of the regression lines when expressed in deviation-score form are given by

[8.7]
$$b_{yx} = \frac{\Sigma(X - \overline{X})(Y - \overline{Y})}{(N-1)s_x{}^2}$$
$$b_{xy} = \frac{\Sigma(X - \overline{X})(Y - \overline{Y})}{(N-1)s_y{}^2}$$

Since $r = \Sigma(X - \overline{X})(Y - \overline{Y})/(N-1)s_x s_y$,

[8.8]
$$b_{yx} = r\frac{s_y}{s_x}$$
$$b_{xy} = r\frac{s_x}{s_y}$$

Multiplying these two expressions, we obtain

[8.9] $b_{yx}b_{xy} = r^2$

Thus the product of the slopes of the two regression lines is the square of the correlation coefficient. The geometric mean of the two slopes is the correlation coefficient.

Because of the above relation between correlation and regression, we may write equations for the two regression lines, using the correlation coefficient. The two equations are as follows:

[8.10]
$$Y' = r\frac{s_y}{s_x}(X - \overline{X}) + \overline{Y}$$
$$X' = r\frac{s_x}{s_y}(Y - \overline{Y}) + \overline{X}$$

These are commonly used equations for predicting a raw score on one variable from a knowledge of a raw score on another.

If measurements are represented in standard-score form, the correlation coefficient may be written as $r = \Sigma z_x z_y / (N - 1)$. If the pairs of standard scores are plotted graphically and two regression lines fitted to the data, the equation of these lines may readily be shown to be

[8.11]
$$z'_y = rz_x$$
$$z'_x = rz_y$$

where z'_y and z'_x are the predicted or estimated standard scores. Both regression lines have the same slope, which is equal to the correlation coefficient. In this case the slope of the regression of Y on X relative to the X axis is the same as the slope of the regression of X on Y relative to the Y axis.

8.6 ERRORS OF ESTIMATE

In predicting one variable from a knowledge of another, distances from either the X or Y axis to the regression line are used as the predicted values. A difference between an observed value and a predicted value is an error of estimate. Thus in predicting Y from X, the predicted value Y' is a distance from the X axis to the regression line, and the difference between the observed value of Y and the predicted value, or $Y - Y'$, is an error of estimate. If the pairs of observations, when plotted graphically as points, all fall exactly along a straight line, all values of $Y - Y' = 0$ and perfect prediction is possible. If the points appear to be arranged at random when plotted graphically, many values of $Y - Y'$ will be large. The more accurate the predictions possible, the smaller the values of $Y - Y'$ will tend to be. The variance of the errors of estimate, that is, of $Y - Y'$, is taken as a measure of the accuracy of estimate and is given by

[8.12]
$$s_{y.x}^2 = \frac{\Sigma (Y - Y')^2}{N - 1}$$

The square root of this quantity is the *standard error of estimate* in predicting Y from X.

The reader should note that $s_{y.x}^2$, as defined above, is not an unbiased estimate of $\sigma_{y.x}^2$. The number of degrees of freedom associated with the sum of squares $\Sigma (Y - Y')^2$ is not $N - 1$ but $N - 2$, there being $N - 2$ deviations about the regression line which are free to vary. Here we have rather arbitrarily defined the standard error of estimate using $N - 1$, and not $N - 2$, to simplify subsequent exposition. For purposes of descriptive statistics it is algebraically more convenient to use $N - 1$ and not $N - 2$. When, however, a problem of estimation is involved, a definition using $N - 2$ should be used.

In predicting X from a knowledge of Y, the variance of the errors of estimate is

[8.13]
$$s_{x.y}^2 = \frac{\Sigma (X - X')^2}{N - 1}$$

where X is an observed value and X' is a value of X estimated from a knowledge of Y. The square root of this quantity is the standard error of estimate in predicting X from Y.

The standard error of estimate is related to the correlation coefficient by the simple relation

[8.14] $s_{y.x} = s_y \sqrt{1 - r^2}$

and similarly

[8.15] $s_{x.y} = s_x \sqrt{1 - r^2}$

By transposing these formulas we obtain relations as follows:

[8.16] $r = \sqrt{1 - \dfrac{s_{y.x}^2}{s_y^2}} = \sqrt{1 - \dfrac{s_{x.y}^2}{s_x^2}}$

The above constitutes, in effect, an alternative definition of the correlation coefficient. If all pairs of points when plotted graphically fall exactly on a straight line, both $s_{y.x}^2 = 0$ and $s_{x.y}^2 = 0$. In consequence, r will be either $+1$ or -1, depending on whether we take the positive or the negative square root. If the points are arranged at random when plotted graphically, X and Y being independent of each other, $s_{x.y}^2 = s_x^2$, $s_{y.x}^2 = s_y^2$, and $r = 0$. The value of the correlation is seen, therefore, to depend on the ratio of two variances, $s_{y.x}^2/s_y^2$ or $s_{x.y}^2/s_x^2$. These two ratios are equal.

8.7 THE VARIANCE INTERPRETATION OF THE CORRELATION COEFFICIENT

A correlation coefficient is not a proportion. A coefficient of .60 does not represent a degree of relationship twice as great as a coefficient of .30. The difference between coefficients of .40 and .50 is not equal to the difference between coefficients of .50 and .60. The question arises as to how correlation coefficients of different sizes may be interpreted. One of the more informative ways of interpreting the correlation coefficient is in terms of variance.

A score on Y may be viewed as comprised of two parts, an estimated value Y' and an error of estimation $(Y - Y')$. Hence $Y = Y' + (Y - Y')$. These two parts are independent of each other; that is, they are uncorrelated. The variances of the two parts are directly additive, and we may write

[8.17] $s_y^2 = s_{y'}^2 + s_{y.x}^2$

where s_y^2 = variance of Y
$\qquad s_{y'}^2$ = variance of values of Y predicted from X, that is, values on regression line
$\qquad s_{y.x}^2$ = variance of errors of estimation

The variance $s_{y.x}^2 = s_y^2(1 - r^2)$. By substitution we obtain

$$s_y{}^2 = s_{y'}{}^2 + s_y{}^2(1 - r^2)$$

Dividing this equation by $s_y{}^2$ and writing it explicit for r^2, we have

[8.18] $r^2 = \dfrac{s_{y'}{}^2}{s_y{}^2}$

Similarly, it may be shown that

[8.19] $r^2 = \dfrac{s_{x'}{}^2}{s_x{}^2}$

These expressions state that r^2 is the ratio of two variances, the variance of the predicted values of Y or X divided by the variance of the observed values of Y or X.

The variance $s_{y'}{}^2$ is that part of the variance of Y which can be predicted from, explained by, or attributed to the variance of X. It is a measure of the amount of information we have about Y from our information about X. If $r = .80$, $r^2 = .64$, and we can state that 64 per cent of the variance of the one variable is predictable from the variance of the other variable. We know 64 per cent of what we would have to know to make a perfect prediction of the one variable from the other. Thus r^2 can quite meaningfully be interpreted as a proportion and $r^2 \times 100$ as a per cent. In general, in attempting to conceptualize the degree of relationship represented by a correlation coefficient it is more meaningful to think in terms of the square of the correlation coefficient instead of the correlation coefficient itself. The values of $r^2 \times 100$ for values of r from .10 to 1.00 are as follows:

r	$r^2 \times 100$
.10	1
.20	4
.30	9
.40	16
.50	25
.60	36
.70	49
.80	64
.90	81
1.00	100

Thus a correlation of .10 represents a 1 per cent association, a correlation of .50 represents a 25 per cent association, and the like. A correlation of .7071 is required before we can state that 50 per cent of the variance of the one variable is predictable from the variance of the other. With a correlation as high as .90 the unexplained variance is 19 per cent.

The existence of a correlation between two variables is indicative of a functional relationship, but does not necessarily imply a causal relationship. Whether a functional relationship can be regarded as a causal relationship is a matter of interpretation. The correlation between the intelligence of parents and their offspring has been frequently reported to be of the order of .50. This may be interpreted as indicative of a causal relationship. Frequently two variables may correlate because both are correlated with some other variable or variables. For example, given a group of children with a substantial range of ages, a correlation may be found between a measure of intelligence and a measure of motor ability. Such a correlation may come about because the measures of intelligence and motor ability are both correlated with age. If the effects of age are removed, the correlation may vanish.

8.8 ASSUMPTIONS UNDERLYING THE CORRELATION COEFFICIENT

In interpreting the correlation coefficient it is assumed that the fitting of two straight regression lines to the data does not distort or conceal the functional relation between the two variables. If the relation is curvilinear, a coefficient of zero may be obtained and yet a close relation may exist between the two variables. Figure 8.3 shows a curvilinear relation between X and Y. If X is known, a fairly accurate prediction can be made of Y. If however, two straight regression lines are fitted to the data, these lines will be about at right angles to each other and r will be about zero. If a strictly random relation exists between X and Y, the correlation will be zero. The above example demonstrates that the converse does not hold. If the correlation is zero, it does not necessarily follow that X and Y bear a random relation to each other. This may mean that the linear-regression model is a poor fit to the data. In interpreting the correlation coefficient it is ordinarily assumed that the linear-regression model is a good fit to the data and that a correlation of zero means a random relation. Consider a situation where $r = .80$. This means that 64 per cent of the variance of the one variable is

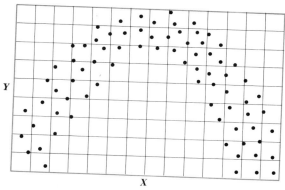

Fig. 8.3   Scatter diagram showing curvilinear relation.

predictable from the other and the residual 36 per cent is due to other factors. The assumption is that these factors do *not* include, at least to any appreciable extent, a lack of goodness of fit of the linear-regression lines to the data. If a large proportion of the residual 36 per cent did result because of nonlinearity, this would affect the interpretation of the data. In interpreting a correlation coefficient the investigator should satisfy himself that the linear-regression lines are a good fit to the data. Any gross departure from linearity can readily be detected by inspection of the bivariate frequency table. For small values of N, curvilinear relations may be difficult to detect. In practice, for many of the variables used in psychology and education the assumption of linearity of regression is in most instances reasonably well satisfied.

In calculating a correlation coefficient it need not be assumed that the distributions of the two variables are normal. Correlations can be computed for rectangular and other types of distributions. If the two variables have different shapes, however, this circumstance will impose constraints upon the correlation coefficient. If a positively skewed distribution is correlated with a negatively skewed distribution, the differences in the shapes of the distributions will influence the correlation coefficient. Some part of the departure of the correlation coefficient from unity will result because of the different shapes of the two distributions. In such a situation as this the differences in shapes of the distributions will in effect ensure that one or the other or both regression lines are nonlinear. In psychological research substantial differences in the shapes of the distributions under study occasionally are found. Under these circumstances it is common practice to transform the variables to a binomial or to an approximately normal form. Such transformations will frequently tend to eliminate curvilinearity of regression.

Many other circumstances affect the correlation coefficient. Among these may be mentioned sampling error and errors of measurement. The effects of these on the correlation coefficient are discussed in later chapters.

EXERCISES 1 The following are paired measurements:

X	1	5	6	6	2
Y	2	4	5	3	1

Compute (*a*) the correlation between X and Y, (*b*) the slope of the regression line for predicting X from a knowledge of Y, (*c*) the slope of the regression line for predicting Y from a knowledge of X, (*d*) the regression equation for predicting a standard score on X from a standard score on Y, (*e*) the regression equation for predicting a raw score on X from a raw score on Y, (*f*) the variance of the errors of estimation in predicting Y from X.

2 The following are marks on a college entrance examination X and first-year averages Y for a sample of 20 students.

X	Y	X	Y	X	Y	X	Y
55	61	70	75	63	85	77	84
79	72	80	61	64	87	62	72
59	69	89	79	69	70	85	70
81	89	92	90	75	90	55	60
62	52	60	55	84	67	66	67

Compute (a) the correlation between entrance examination marks and first-year averages, (b) the regression equation for predicting first-year averages from examination marks, (c) the predicted first-year averages for the 20 students, (d) the variance of the errors of estimation.

3 Standard scores on variable X for four individuals are $-2.0, -1.68, .18,$ 1.16. The correlation between X and Y is $.50$. What are the estimated standard scores on Y? What is the standard error in estimating standard scores on Y from standard scores on X?

4 From the data $\bar{X} = 40.3$, $\bar{Y} = 12.5$, $s_x = 12.6$, $s_y = 3.6$, and $r_{xy} = .60$, write the regression equations for predicting Y from X and X from Y.

5 Show that $r_{xy} = \sqrt{b_{yx} b_{xy}}$.

6 A correlation of $.7071$ may be interpreted to mean that 50 per cent of the variance of one variable is predictable from the other variable. Is this statement correct if the regression lines are not linear?

7 A variance $s_y{}^2 = 400$ and the correlation between X and Y is $.50$. What is the variance of the errors of estimation in predicting Y from X? What is the variance of the predicted values?

8 What correlation between X and Y is required in order to assert that 75 per cent of the variance of X depends on the variance of Y?

9 In predicting Y from a knowledge of X, the standard error of estimate is 5, and the mean of the errors of estimation is zero. Assuming the errors to be normally distributed, indicate the limits above and below the mean that include 95 per cent of the errors of estimation.

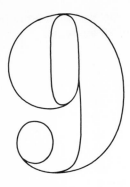

SAMPLING

9.1 INTRODUCTION

In Chap. 1 the concepts of population and sample were discussed. A *population* is any defined aggregate of objects, persons, or events, the variables used as the basis for classification or measurement being specified. A *sample* is any subaggregate drawn from the population. Any statistic calculated on a sample of observations is an *estimate* of a corresponding population value or *parameter*. The symbol $\overline{X}$ is used to refer to the arithmetic mean of X calculated on a sample of size N. The symbol μ is used to refer to the mean of the population. Similarly, s^2 is used to refer to the variance in the sample and σ^2 is the corresponding population parameter. $\overline{X}$ is an estimate of μ, and s^2 is an estimate of σ^2. Likewise, any other statistic calculated on a sample is an estimate of a corresponding population parameter. In most situations the parameters are unknown and must be estimated in some manner from the sample data.

Much statistical work in practice is concerned with the use of sample statistics as estimates of population parameters, and more particularly with describing the magnitude of error which attaches to such statistics. The body of statistical method concerned with the making of statements about population parameters from sample statistics is called *sampling statistics*, and the logical process involved is called *statistical inference*, this being a rigorous form of inductive inference. If inferences about population parameters are to be drawn from sample statistics, certain conditions must attach to the methods of sampling used.

9.2 METHODS OF SAMPLING

In drawing inferences about the characteristics of populations from sample statistics, the assertion is frequently made that the sample should be drawn *at random* from the population. The sample is spoken of as a *random sample*. The word "random" is used in at least three ways. It may refer to our subjective experience that certain events are haphazard or completely lacking in order. It may be used in a theoretical sense to refer to an assumption about the equiprobability of events. Thus a random sample is one such that every member of the population has an equal probability of being included in it. When the word is used in this way the meaning is assigned to it within the framework of probability theory. The word "random" is also used in an operational sense to describe certain operations or methods. Thus the drawing of numbers from a hat after they have been thoroughly mixed, or the drawing of cards from a deck after they have been well shuffled, or certain techniques used in sweepstakes, lotteries, and other games of chance are examples of random operations or methods. Sampling theory in statistics is based on the theoretical use of the word "random," that is, on the idea of the equiprobability of each population member being included in the sample. This equiprobability assumption underlies the derivation of many formulas used in sampling statistics. Practical operational methods of sampling are frequently such as to ensure that the theoretical assumption of equiprobability is closely approximated in practice. If methods of sampling ensure approximate equiprobability, then clearly the deductive consequences of a theory based on the idea of equiprobability can be used in dealing with practical sampling problems. The correspondence between the consequences of theory and what occurs in practice can, of course, always be checked by experiment.

For a finite population, if the members are listed or catalogued, a random sampling procedure may readily be applied. Consider, for example, a population containing 2,000 members from which a sample of 200 members is required. All names, or some identifying code number, may be entered on slips of paper. These may be placed in a container from which 200 slips are drawn. A more convenient procedure is to use a table of random numbers. The members of the population may be labeled from 1 to 2,000, perhaps according to the order they appear on the list. Numbers may then be read directly from a table of random numbers. These numbers will identify which of the 200 members from the population of 2,000 should be included in the sample. Tables of random numbers are found in many texts on statistics or in Fisher and Yates, *Statistical Tables for Biological, Agricultural, and Medical Research*. If a list is arranged alphabetically, or in some other systematic fashion, every *n*th name may be chosen in the construction of the sample. This is sometimes spoken of as *systematic sampling*. Although in most practical situations such a sample may be viewed as random, the possibility exists that the variable under investigation may not be independent of the basis of ordering, and a biased sample may result.

On occasion, samples are drawn from lists which do not provide a complete record of all members of the population, but are viewed, perhaps erroneously, as representative of the population. A telephone directory is such a list. Names chosen from a telephone directory may yield a biased sample of the population at large, because ownership or nonownership of a telephone may not be independent of the variable under investigation, e.g., how a person intends to vote in an election.

Forms of modified random sampling are sometimes used. One example is *stratified random sampling*. This procedure requires prior knowledge, perhaps obtained from census data, about the number or proportion of members in the population of various strata. Thus we may know the number of males and females, the number in various age groups, and the like. In constructing the sample, members are drawn at random from the various strata. If the members are drawn such that the proportions in the various strata in the sample are the same as the proportions in those strata in the population, the sample is a *proportional stratified sample*. For example, a university may have 10,000 students, of which 7,000 are males and 3,000 are females. A sample of 100 students is required. We may draw 70 males by a random method from the subpopulation, or stratum, of males, and 30 females from the subpopulation, or stratum, of females. Such a sample is a proportional stratified sample.

In much experimentation using human or animal subjects, the populations from which the samples are drawn may not be amenable to precise definition, and the methods of sampling described above may have little relevance. For example, a laboratory experiment may use a sample of experimental animals. Can such a sample be viewed meaningfully as a random sample from a known population of animals? In a study of the therapeutic effects of different operative procedures applied to a certain class of brain tumor, all cases admitted to a particular hospital during a specified time period may be used. This number may be small. In what sense may this group of cases be viewed as a random sample drawn from a larger group or population? Again, in an educational experiment the pupils in two or three classes in a particular school and grade may be used as subjects. Can such a group of subjects be viewed as a random sample or its equivalent? Such samples as these are clearly not random. The method by which they are selected is not a random method. Despite this the investigator may wish to draw inferences that transcend the particular samples under study. He may wish to argue that his findings are probably true for some large, although perhaps ill-defined, population of experimental animals; or that the therapeutic effects of his operative procedures may be extended to all patients suffering from the particular class of brain tumor; or that the results of the educational experiment may be generalized to a much larger group of school pupils in the same grade.

In situations of the above type, where strict random sampling procedures have not been used, does any basis exist for valid inference? A common practice is to investigate a posteriori a variety of characteristics of the sample. It may be possible to show that the sample does not differ appreciably in these characteristics from a larger group or population. Thus in the educational experiment the sample of students may be studied with respect to age, sex, IQ, socioeconomic level of the parents, and other characteristics. The sample may not differ much in these respects from a larger group or population in the same grade. Because the sample shows no bias on a number of known characteristics, that is it may not differ from a random sample as far as these characteristics are concerned, the investigator may be prepared to regard it as representative of the larger group or population and treat it as if it were a random sample. Frequently precise knowledge is lacking about a larger reference group or population, and the investigator must rely on accumulated past experience and intuition in the attempt to detect possible bias. Clearly, where possible, random sampling is to be preferred to methods such as these. It must be recognized, however, that were we to insist on rigorous random sampling methods, much experimentation would not be possible.

In practice, experiments are clearly not always conducted in the way our statistical preconceptions suggest they should be conducted. Experienced experimentalists are frequently aware of this. Much of the art of the experimentalist is concerned with reaching conclusions from data which do not satisfy some of the conditions necessary for rigorous inference.

9.3 SAMPLING ERRORS

Let us now consider the nature of the errors associated with particular sample values. What precisely is a sampling error? A sampling error is a difference between a population value, or parameter, and a particular sample value. Thus, if μ is the population value of the mean and $\bar{X}_i$ is an estimate based on a random sample of size N, then the difference $\mu - \bar{X}_i = e_i$, where e_i is a sampling error. Let us suppose that we know that the mean scholastic aptitude test score for a population of 5,000 university students is $\mu = 562$. A sample of 100 provides an estimated mean of $\bar{X}_i = 566$. The sampling error in this case is $\mu - \bar{X}_i = 562 - 566 = -4$. Ordinarily, μ is not known, and we are unable to specify e_i exactly for any particular sample. Despite this, meaningful statements can be made about the magnitude of error which attaches to $\bar{X}_i$ as an estimate of the parameter μ. The reader should note that the concept of error in any context always implies a parametric, true, fixed, or standard value from which a given observed value may depart in greater or less degree. The idea that something in the nature of a parametric or true value can meaningfully be defined is essential to the concept of error. Without some appropriate definition of such a value, the concept of error has no meaning, and no theory of error is possible. Also no science is possible.

How may the magnitude of error be estimated and described? Common sense suggests that in the measurement of any quantity some appreciation of the magnitude of error may be obtained by repeating the measurements a number of times, presumably under constant conditions, and observing how these repeated measurements vary from each other. Thus in the measurement of the length of a bar of metal a series of separate measurements may be made under constant conditions. Let us suppose that five such measurements are 55.95, 56.23, 56.25, 56.41, and 56.54 in. In this case each measurement is an estimate of the same "true" length; hence the variation observed with repeated measurement is due to error. Let us suppose that five additional measurements are made using another measuring operation or procedure, these measurements being 54.80, 55.31, 56.44, 56.52, 57.29 in. This latter set shows greater variation with repetition than the former. We may conclude that the magnitude of error associated with this latter set of measurements is greater.

The above example is concerned with errors associated with particular observations, namely, measurements of a bar of metal. In considering the magnitude of error associated with, say, a sample mean $\bar{X}_i$, as an estimate of a population mean μ, the situation is similar. The problem may be approached experimentally by considering how values of $\bar{X}_i$ vary in repeated samples of size N. Thus the mean scores on the scholastic aptitude test for five different samples of 100 students, drawn from a population of 5,000, may be 552, 558, 562, 568, and 569. These five sample means may be viewed as estimates of the same population mean μ; that is, the value that would have been obtained were information available on all 5,000 members of the population. The variation of these five means one from another may be attributed to sampling error.

In general, a number, say k, of samples of size N may be drawn at random from the same population and a mean calculated for each sample. These means may be represented by the symbols $\bar{X}_1, \bar{X}_2, \bar{X}_3, \ldots, \bar{X}_k$. We may write

$$\mu - \bar{X}_1 = e_1$$
$$\mu - \bar{X}_2 = e_2$$
$$\mu - \bar{X}_3 = e_3$$
$$\cdots \cdots \cdots$$
$$\mu - \bar{X}_k = e_k$$

The variance and the standard deviation are ordinarily used to describe the magnitude of variation in any set of observations or values. In describing the magnitude of variation in sample means with repeated sampling, the variance and standard deviation are also used. These statistics describe the magnitude of sampling error, that is, the magnitude of error associated with $\bar{X}_i$ as an estimate of μ. Note that the variance and standard deviation of sample means $\bar{X}_i$ are the same as the variance and standard deviation of the sampling error e_i, because μ is a constant.

9.4 SAMPLING DISTRIBUTIONS

In the above discussion the problem of estimating error has been approached *experimentally*; that is, we considered the actual drawing of a number of samples and approached the experimental study of error through observed sample-to-sample fluctuation. Consider for illustrative purposes a small finite population of eight members. Let the members of the population be cards numbered from 1 to 8. These cards may be shuffled, a sample of four cards drawn without replacement, and a mean calculated for the sample. This procedure may be repeated 100 times, and a frequency distribution made of the 100 sample means. This distribution is an *experimental sampling distribution*, and its standard deviation is a measure of the fluctuation in means from sample to sample.

A *theoretical*, as distinct from an experimental, approach may be used. Given a finite population of eight members, a limited number of different samples of four cards exist. The number of such samples is the number of combinations of eight things taken four at a time, or $C_4^8 = 70$. Each of these 70 samples may be considered equiprobable. The means for the 70 possible samples may be ascertained and a frequency distribution prepared. This frequency distribution is a *theoretical sampling distribution*. It is obtained by direct reference to probability considerations. No drawing of actual samples is involved. The standard deviation of the theoretical sampling distribution is a measure of fluctuation in means from sample to sample.

In the above example the population is small and finite. In practice, most of the populations with which we deal are indefinitely large, or if finite, they are so large that for all practical purposes they can be considered indefinitely large. In the study of sampling error the approach used in dealing with an indefinitely large population is a simple extension of that used with a small finite population. The distinction between an experimental and theoretical sampling distribution still applies. When the population is indefinitely large, the theoretical sampling distribution of, for example, the mean is the frequency distribution of means of the indefinitely large number of samples of size N which theoretically could be drawn.

The theoretical sampling distributions are known for all commonly used statistics. The standard deviation of the sampling distribution is called the *standard error*. Thus a standard error is always a standard deviation which describes the variability of a statistic over repeated sampling. The standard deviation of a theoretical sampling distribution is, in effect, a population parameter. It is descriptive of the variation of a statistic in a complete population of sample values. The standard deviation of the theoretical sampling distribution of the mean is represented by the symbol $\sigma_{\bar{x}}$. In practice this standard deviation must be estimated from sample data. This estimate in the case of the mean may be represented by the symbol $s_{\bar{x}}$. For most statistics fairly simple formulas are available for estimating the standard deviation of the theoretical sampling distribution.

The theoretical sampling distributions of some statistics are normal, or approximately so; others are not. For example, the theoretical sampling distribution of the mean $\bar{X}$ is normally distributed in sampling from an indefinitely large normally distributed population. The sampling distribution of the correlation coefficient presents a complicated problem. It is not normally distributed except under certain special circumstances. When the shape of the sampling distribution is known, certain kinds of statements can be made about a population value from a sample estimate. For example, it is possible to fix limits above and below a sample value and assert with a known degree of confidence that the population parameter falls within those limits. The fixing of such limits requires a knowledge of the shape of the sampling distribution.

9.5 SAMPLING DISTRIBUTION OF MEANS FROM A FINITE POPULATION

In practice, most samples are viewed as drawn from indefinitely large populations. The essential ideas of sampling may, however, be conveniently illustrated with reference to a small finite population. Suppose, as mentioned above, that we have a population of eight cards numbered from 1 to 8. These cards may be shuffled, and a sample of four cards drawn at random. After each card is drawn it is not returned; that is, the sampling is without replacement. A mean $\bar{X}$ may be calculated for this sample. The four cards may now be returned, the eight cards shuffled, another sample of four cards drawn, and another mean calculated. Let us continue this procedure until 100 samples of four cards have been drawn and their means calculated. Table 9.1, col. 3, shows the frequency distribution of 100 such sample means. This distribution is an experimental sampling distribution of means. It shows experimentally how the means of samples of four drawn at random without replacement from a population of eight vary from sample to sample. The mean of the experimental sampling distribution $\bar{X}_{\bar{x}}$, that is, the mean of the 100 means based on samples of four, is found to be 4.56. The mean of the population from which the samples have been drawn is the mean of the integers from 1 to 8 and is 4.50. The standard deviation of the 100 means $s_{\bar{x}}$ is found to be .834.

The investigation of the fluctuation in sample means may be approached theoretically. The number of different samples of four in sampling without replacement from a population of eight members is the number of combinations of eight things taken four at a time, or $C_4^8 = 70$. These 70 samples may be considered equiprobable. A listing of the 70 samples may readily be made, and the means calculated. The sample with the smallest mean will be 1, 2, 3, 4; the mean $\bar{X} = 2.50$. The sample with the largest mean will be 5, 6, 7, 8; here $\bar{X} = 6.50$. Thus $\bar{X}$ will range from 2.50 to 6.50. Table 9.1, col. 5, shows the frequency distribution of the 70 sample means. This distribution is a theoretical sampling distribution of the mean of samples of four from a small finite population of eight members. It is based on the idea that there are 70 possible combinations of eight things taken four at a time, all combinations being equiprobable.

The mean of the theoretical sampling distribution may be calculated. This mean $\mu_{\bar{x}}$ is found to be 4.50. The standard deviation $\sigma_{\bar{x}}$ is found to be .866. These values do not differ markedly from the mean and standard deviation of the experimental sampling distribution, these being $\bar{X}_{\bar{x}} = 4.56$ and $s_{\bar{x}} = .834$. Presumably, had a larger number of samples been drawn, say, 200 or 1,000, the experimental sampling distribution would be observed to approximate more closely to the theoretical distribution.

The mean and standard deviation of the theoretical sampling distribution of Table 9.1 were calculated directly from the 70 possible sample means. These values may, however, be readily obtained without using this time-consuming method. It can be shown that the mean of the theoretical sampling distribution is equal to the population mean; that is, $\mu_{\bar{x}} = \mu$. In our example the mean of the sampling distribution of samples of four from a population of eight members is observed to be 4.50. Likewise, the population mean, that is, the mean of the integers from 1 to 8, is also 4.50. The

Table 9.1 Experimental and theoretical sampling distributions of means of samples of four drawn from a population of eight members

ΣX	$\bar{X}$	Experimental distribution		Theoretical distribution	
1	**2**	f **3**	p **4**	f **5**	p **6**
10	2.50	1	.010	1	.014
11	2.75	2	.020	1	.014
12	3.00	0	.000	2	.029
13	3.25	5	.050	3	.043
14	3.50	7	.070	5	.071
15	3.75	7	.070	5	.071
16	4.00	8	.080	7	.100
17	4.25	11	.110	7	.100
18	4.50	13	.130	8	.114
19	4.75	10	.100	7	.100
20	5.00	10	.100	7	.100
21	5.25	9	.090	5	.071
22	5.50	7	.070	5	.071
23	5.75	4	.040	3	.043
24	6.00	3	.030	2	.029
25	6.25	1	.010	1	.014
26	6.50	2	.020	1	.014
Total	....	100	1.000	70	.998

standard deviation of the theoretical sampling distribution is given by the formula

[9.1] $$\sigma_{\bar{x}} = \frac{\sigma}{\sqrt{N}} \sqrt{\frac{N_p - N}{N_p - 1}}$$

where σ = standard deviation in population
N_p = number of members in population
N = sample size

In our example σ is the standard deviation of the integers from 1 to 8 and is equal to 2.29. Population and sample size are, respectively, 8 and 4. Hence

$$\sigma_{\bar{x}} = \frac{2.29}{\sqrt{4}} \sqrt{\frac{8 - 4}{8 - 1}} = .866$$

If, then, the standard deviation σ of the population is known, we can readily obtain from the above formula the standard deviation of the theoretical sampling distribution and use this as a measure of fluctuation in means from sample to sample.

A knowledge of the standard deviation of a theoretical sampling distribution is of limited usefulness unless additional information is available on the shape of the distribution. In certain instances sampling distributions are normal, or approximately normal in form. The theoretical sampling distribution of Table 9.1 departs appreciably from the normal form. If, however, both sample and population size were increased, the distribution would approximate more closely to the normal form. For $N = 30$ and $N_p = 100$, the normal distribution would be a good approximate fit. If the sampling distribution is approximately normal, we can, given its standard deviation, readily estimate the probability of obtaining values equal to or greater than any given size in random sampling from the population.

9.6 SAMPLING DISTRIBUTION OF MEANS FROM AN INDEFINITELY LARGE POPULATION

Many populations may be conceptualized as comprised of an indefinitely large number of members. Most applications of sampling theory encountered in psychology and education assume such populations. Sampling from an indefinitely large population is essentially the same as sampling from a finite population with replacement, that is, where each sample member is returned to the population prior to the drawing of the next member. In sampling without replacement from an indefinitely large population the probabilities remain unchanged regardless of the size of the sample drawn. Similarly, in sampling from a finite population with replacement, the population is not depleted and the probabilities are unchanged by the number of prior draws. It follows that problems of sampling from an indefinitely large population can be approached through the study of finite populations where samples are drawn with replacement.

To illustrate sampling from an indefinitely large population, an artificial

Table 9.2* Population from which samples were drawn: frequency distribution of numbers

Number	Frequency	Number	Frequency
1	1	14	174
2	2	15	154
3	4	16	127
4	7	17	96
5	14	18	67
6	26	19	43
7	43	20	26
8	67	21	14
9	96	22	7
10	127	23	4
11	154	24	2
12	174	25	1
13	181		——
		Total	1,611

* Tables 9.2 to 9.4 are reproduced from R. W. B. Jackson and George A. Ferguson, *Manual of educational statistics*, University of Toronto, Department of Educational Research, Toronto, 1942.

population was constructed. This population was comprised of 1,611 cards containing the numbers from 1 to 25. The distribution of numbers is approximately normally distributed. The distribution of this population is shown in Table 9.2. The mean μ of the population is 13, and the standard deviation σ is 3.56. The cards were inserted in a box, and samples of 10 cards drawn with replacement; that is, a card was drawn, its number noted, and the card then returned to the box before the next draw. Altogether 100 samples of 10 cards were drawn, and the 100 means calculated. Table 9.3 shows the means of the samples. Table 9.4, col. 2, shows a frequency distribution of these means. This distribution is an experimental sampling distribution of means based on samples of size 10. The mean of the sampling distribution $\overline{X}_{\bar{x}}$ is 13.205, and the standard deviation $s_{\bar{x}}$ is 1.139. This standard deviation is a description based on experimental data of the sample-to-sample fluctuation of means of samples of size 10 drawn at random from this population.

The mean and standard deviation of the sampling distribution need not be estimated by the rather laborious experimental approach described above. It can be shown that the mean of the theoretical sampling distribution of the mean in sampling from an indefinitely large population is equal to the population mean; that is, $\mu_{\bar{x}} = \mu$. It can also be shown that the stan-

Table 9.3   **Means of samples of 10 drawn from the population in table 9.2**

10.9	13.5	11.7	13.3	13.8	12.5	15.0	12.7	14.3	12.7
13.0	13.2	14.0	13.1	13.2	12.7	12.6	11.5	13.2	12.9
12.4	13.9	14.1	12.2	13.1	11.7	11.5	14.6	12.6	12.9
13.9	14.0	11.7	12.1	13.2	13.6	14.4	14.0	12.2	13.7
12.6	11.6	11.8	12.1	13.1	13.2	12.5	14.0	16.4	12.2
12.6	13.7	13.6	14.0	12.1	13.2	14.8	13.6	12.5	14.5
14.4	13.9	13.8	15.1	14.2	14.4	13.5	12.7	14.5	14.4
12.9	11.3	14.5	13.0	12.0	13.3	12.7	14.8	11.3	11.0
12.7	14.6	15.2	14.1	16.1	14.7	12.3	11.2	14.3	14.7
12.9	12.3	11.9	14.0	14.5	12.4	11.9	12.3	12.4	12.6

dard deviation of the sampling distribution is given by

[9.2] $\sigma_{\bar{x}} = \dfrac{\sigma}{\sqrt{N}}$

where σ = standard deviation in population
N = size of sample

The reader will observe that the difference between this formula and the formula previously given for the standard deviation of the sampling distribution for the means of samples from a finite population resides in the absence here of the term $\sqrt{(N_p - N)/(N_p - 1)}$. As N_p increases, this term approaches 1 as a limit. It is equal to 1 when the population is indefinitely

Table 9.4   **Experimental and theoretical sampling distribution of 100 sample means for samples of size 10 drawn from the population of table 9.2**

	Frequency	
Class interval	*Experimental*	*Theoretical*
16.5–17.4	—	.1
15.5–16.4	2	1.4
14.5–15.4	13	8.4
13.5–14.4	27	24.6
12.5–13.4	31	34.3
11.5–12.4	22	22.8
10.5–11.4	5	7.2
9.5–10.4	—	1.1
Total	100	99.9

large. The standard deviation of the theoretical sampling distribution is $\sigma_{\bar{x}} = 3.56/\sqrt{10} = 1.126$. This is very close to the standard deviation of the experimental sampling distribution $s_{\bar{x}}$, which was found to be 1.139.

The theoretical sampling distribution of the means of samples drawn from a normal population is normal. Thus if we know that the population distribution is normal, we know that the sampling distribution of means is normal. Regardless of the shape of the population distribution, the sampling distribution of means will approximate the normal form as N increases in size. For practical purposes the distribution may be taken as approximately normal for samples of reasonable size, except in the case of fairly gross departures of the population from normality. The theoretical normal frequencies have been calculated for our illustrative example. These theoretical frequencies are shown in Table 9.4, col. 3. These are the expected normal frequencies for a normal curve with a mean of 13.00 and a standard deviation of 1.139. The differences between the experimental and the theoretical normal distribution are not very great.

Examination of the formula $\sigma_{\bar{x}} = \sigma/\sqrt{N}$ indicates that the standard error of the mean is directly related to the standard deviation of the population and inversely related to the size of sample. Thus the greater the variation of the variable in the population, the greater the standard error; also the larger the size of N, the smaller the standard error. The standard error of means of samples of $N = 1$, the smallest sample size possible, is equal to the population standard deviation. For any fixed value of σ the standard error can be made as small as we like by increasing the size of the sample.

9.7 SAMPLING DISTRIBUTION OF PROPORTIONS

Many problems require the use of proportions. A study of the sampling distribution of a proportion may be approached either experimentally or theoretically. To illustrate, consider an urn containing a finite number, N_p, of black and white chips. Denote the proportion of black and white chips by θ and $1 - \theta$, respectively. Let us draw a large number of samples of size N at random, without replacement, from the urn, observe the proportion of black chips in each sample, and make a frequency distribution of these proportions. This frequency distribution is an experimental sampling distribution of proportions for samples of size N. As with the arithmetic mean, we may use a theoretical, as distinct from an experimental, approach. In drawing samples of N from a finite population of N_p members, the number of different equiprobable samples is $C_N{}^{N_p}$. The proportion of black chips may be calculated from each of these samples, and a frequency distribution made of the proportions. This distribution is a theoretical sampling distribution of proportions.

For illustrative purposes consider a hypothetical population of three black and three white chips. Denote the members of this population by B_1, B_2, B_3, W_4, W_5, W_6, the subscript identifying the particular population member. We may consider the set of equiprobable samples of three members which may be drawn without replacement from this population. The number of such samples is $C_3{}^6 = 20$. The first sample is $B_1B_2B_3$, and the proportion of black chips is $p = 1.00$. The second sample is $B_1B_2W_4$ with $p = .67$; the third, $B_1B_2W_5$ with $p = .67$, and so on. A frequency distribution may be made of these proportions, and is as follows:

p	f	f/N
0	1	.05
.33	9	.45
.67	9	.45
1.00	1	.05
Total	20	1.00

The standard deviation, or standard error, of this theoretical sampling distribution may be readily calculated and is $\sigma_p = .224$. This standard error may be obtained directly by the formula

[9.3] $$\sigma_p = \sqrt{\frac{\theta(1 - \theta)}{N} \times \frac{N_p - N}{N_p - 1}}$$

In the above example $\theta = .50$ since three of the six chips are black. Population and sample size respectively are $N_p = 6$ and $N = 3$. Hence

$$\sigma_p = \sqrt{\frac{.5 \times .5}{3} \times \frac{6 - 3}{6 - 1}} = .224$$

which agrees with the value obtained by direct calculation.

The discussion above relates to sampling without replacement from a finite population. As with the arithmetic mean, the term $(N_p - N)/(N_p - 1)$ in formula [9.3] approaches unity for any finite value of N as N_p approaches infinity. Thus this term may be considered equal to unity for an indefinitely large population, and we obtain

[9.4] $$\sigma_p = \sqrt{\frac{\theta(1 - \theta)}{N}}$$

as the standard error of a proportion in sampling from an indefinitely large population, or in sampling from a finite population with replacement.

Formula [9.4] may be obtained by reference to the binomial. If the proportion of black chips in a population is θ, the expected, or theoretical, sampling distribution of the *number* of black chips in samples of size N, as distinct from the proportion of black chips, is given by the terms of the binomial $[\theta + (1 - \theta)]^N$. The mean and standard deviation of this distribution are $N\theta$ and $\sqrt{N\theta(1 - \theta)}$, respectively. We are interested in the distribution of the proportion, instead of the number, of black chips in the samples. To obtain the standard deviation of the distribution of the proportion, as distinct from the number, of black chips in samples of size N, we multiply $\sqrt{N\theta(1-\theta)}$ by $1/N$ to obtain $\sigma_p = \sqrt{\theta(1-\theta)/N}$ which is the same as formula [9.4] above. To illustrate, let $\theta = .25$ and $1 - \theta = .75$. The expected distribution of the *number* of black chips in samples of size 10 is given by expanding the binomial $(.25 + .75)^{10}$. The mean in this example is $10 \times .25 = 2.5$, and the standard deviation is $\sqrt{10 \times .25 \times .75} = 1.37$. The standard deviation of the distribution of the *proportion* of black chips in samples of size 10 is obtained by dividing 1.37 by 10 and is .137.

Formulas [9.3] and [9.4] assume that θ is known. In practice, θ is usually not known and the sample value p is used as an estimate of θ.

9.8 SAMPLING DISTRIBUTION OF DIFFERENCES

For certain purposes a knowledge of the sampling distribution of the difference between two statistics, such as the difference between two arithmetic means or two proportions, is required. To conceptualize the sampling distribution of the difference between, say, two arithmetic means, let us consider two indefinitely large populations whose means are equal; that is, $\mu_1 = \mu_2$. Let $\bar{X}_1$ be the mean of a sample of N_1 cases drawn at random from the first population and $\bar{X}_2$ be the mean of a sample of N_2 cases drawn from the second population. The difference between means is $\bar{X}_1 - \bar{X}_2$. Since $\mu_1 = \mu_2$ this difference results from sampling error. A large number of pairs of samples may be drawn, and a frequency distribution made of the differences. It describes how the differences between means chosen at random from two populations, where $\mu_1 = \mu_2$, will vary with repeated sampling. From this distribution we may estimate the probability of obtaining a difference of any specified size in drawing samples at random from populations where $\mu_1 = \mu_2$. By considering an indefinitely large number of pairs of samples we arrive at the concept of a theoretical sampling distribution of differences between sample means. In this situation the individual measurements in the two populations are not paired with one another. The samples are independent. The means may be viewed as paired at random. No correlation exists between the pairs of means.

The variance of the sampling distribution of differences describes how the differences vary with repeated sampling. Consider the case of independent samples. If $\sigma_{\bar{x}_1}^2 = \sigma_1^2/N_1$ is the variance of the sampling distribution of means drawn from one population and $\sigma_{\bar{x}_2}^2 = \sigma_2^2/N_2$ is the corresponding variance from the other population, then the variance of the sampling distribution of differences between means is the sum of the two variances. Thus

[9.5] $\qquad \sigma_{\bar{x}_1-\bar{x}_2}^2 = \sigma_{\bar{x}_1}^2 + \sigma_{\bar{x}_2}^2 = \dfrac{\sigma_1^2}{N_1} + \dfrac{\sigma_2^2}{N_2}$

When $\sigma_1^2 = \sigma_2^2 = \sigma^2$, the variances in the two populations being equal, we may write

[9.6] $\qquad \sigma_{\bar{x}_1-\bar{x}_2}^2 = \sigma^2 \left(\dfrac{1}{N_1} + \dfrac{1}{N_2} \right)$

Consider now a situation where measurements are paired with one another. Such data arise, for example, where measurements are made on the same group of subjects under both control and experimental conditions. The paired measurements may be correlated. In this instance, in approaching the sampling distribution of differences between means, we conceptualize two populations of paired measurements with equal means; thus $\mu_1 = \mu_2$. Denote the correlation between the paired measurements by the symbol ρ_{12}. Samples of size N are drawn at random, and the differences between means obtained. The distribution of differences between means for an indefinitely large number of samples is the sampling distribution of differences for correlated populations.

For correlated populations the variance of the sampling distribution of differences may be shown to be

[9.7] $$\sigma_{\bar{x}_1 - \bar{x}_2}^2 = \sigma_{\bar{x}_1}^2 + \sigma_{\bar{x}_2}^2 - 2\rho_{12}\sigma_{\bar{x}_1}\sigma_{\bar{x}}$$

where ρ_{12} is the correlation in the population. Note that the formula for independent samples is a particular case of the more general formula for correlated samples. It is the particular case which arises when $\rho_{12} = 0$. In the correlated case $N_1 = N_2 = N$.

Formulas [9.5] to [9.7] are simple applications of the formula for the variance of differences.

EXERCISES

1 Indicate the difference between (a) a random sample and a stratified random sample, (b) a stratified random sample and a proportional stratified sample, (c) an experimental and a theoretical sampling distribution.

2 How would you proceed to draw a random sample of 100 university students?

3 How would you proceed to draw a systematic stratified sample of 100 students from the students in a university?

4 Would a random sample of names selected from a telephone book be considered appropriate for the study of voting behavior?

5 Samples of three cards are drawn at random from a population of eight cards numbered from 1 to 8. Obtain (a) the theoretical sampling distribution of means, (b) the standard deviation of the sampling distribution, (c) the probability of obtaining a mean equal to or greater than 7.

6 The standard deviation of the sampling distribution of $\bar{X}$ is $\sigma/\sqrt{N}$. What is the standard deviation of the sampling distribution of $N\bar{X}$ or ΣX?

7   A university has a population of 1,000 students. The standard deviation of scholastic aptitude test scores in this population is 80. Calculate the standard errors of mean scholastic aptitude test scores for a sample of 100 students drawn with and without replacement.

8   The pages of this book, excluding the Appendix, may be defined as a population. Some pages contain no formulas, some one, some two, and so on. (*a*) Obtain the population distribution of the number of formulas per page; that is, ascertain the number of pages with no formulas, the number with one, with two, and so on. Count only numbered formulas. (*b*) Calculate the mean and standard deviation for the distribution of (*a*) above. (*c*) Draw without replacement five random samples of 30 pages each, using an appropriate random sampling method. Calculate for each sample the mean number of formulas per page and the standard deviation of means for the five samples. How does this standard deviation compare with that obtained from formula [9.1]? Repeat (*c*) using a random sample drawn with replacement, and compare the resulting standard deviation with that obtained from formula [9.2].

9   The variance of the sampling distribution of the mean for samples of 100 cases drawn from an indefinitely large population is 20. How large should the samples be to reduce this variance by one-half? How large should the samples be to reduce the standard deviation of the distribution by one-half?

10   A population consists of six black and two white chips. Obtain the frequency distribution of proportions of black chips in samples of four drawn from this population. What is the standard deviation of this distribution?

11   Will a negative correlation between paired observations increase or decrease the standard error of the difference between two means?

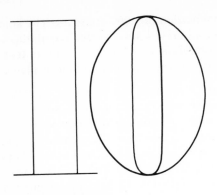

ESTIMATION

10.1 INTRODUCTION

This chapter considers some aspects of the problem of estimating population parameters from sample values. A distinction is commonly made between two types of estimates, *point estimates* and *interval estimates*. A point estimate is the value obtained by direct calculation on the sample values. If in a particular sample the mean $\overline{X} = 26.88$, this is a point estimate of the parameter μ. Another approach is to specify an interval within which we may assert with some known degree of confidence that the population mean lies. Thus, for example, instead of the point estimate $\overline{X}$ we may perform a simple calculation, which will shortly be described, and assert with 95 per cent confidence that the population mean falls within the limits 24.92 and 28.84. These values are called *confidence limits*, and the interval they contain is called a *confidence interval*. Such an interval is an interval estimate.

10.2 PROPERTIES OF ESTIMATES

Methods of estimation are sometimes said to yield *unbiased, consistent, efficient*, and *sufficient* estimates. These are desirable characteristics and serve as criteria for preferring one method of estimation to another.

A method of estimation provides an unbiased estimate when the mean of a large number of sample values, obtained by repeated sampling, approaches the population value in the limit as the number of samples increases. This simply means that a statistic is unbiased when it displays no systematic tendency to be either greater than or less than the population parameter; that is, it is not subject to a constant error. The arithmetic mean is an unbiased estimate. The sample mean $\overline{X}$ exhibits no systematic tendency to be either greater than or less than the parameter μ. Stated in somewhat different language, an estimate is unbiased when its *expected* value is equal to the parameter it purports to estimate.

The expected value of a statistic is the value we should expect to obtain upon averaging the values of the statistic over an indefinitely large number of repeated random samples. It is the value we should expect to obtain in the long run. The expected value of a variable is obtained by multiplying each value of the variable by its associated probability and summing the products. For example, consider the following frequency distribution with corresponding probabilities:

X	f	p
2	40	.40
3	50	.50
4	10	.10
	100	1.00

The expected value of X, that is, $E(X)$, is $2 \times .40 + 3 \times .50 + 4 \times .10 = 2.70$. Note that $E(X)$ is also the mean of X obtained by multiplying the values of X by the frequency of X and dividing by the total frequency. In general the expected value of the mean, $E(\overline{X})$, is the population mean μ. A statistic is a *biased* estimate when its expected value does not tend toward the population value but departs in systematic fashion from it. For example, it may be shown that the expected value of the variance defined as $s^2 = \Sigma(X - \overline{X})^2/N$ is not σ^2 but is $\sigma^2(N-1)/N$. Thus the value of the variance, so defined, which we would expect to get in the long run is a biased estimate of σ^2.

A method of estimation is said to yield a consistent estimate if that estimate approaches the population parameter more closely as sample size increases. The arithmetic mean is a consistent estimate in that it tends to draw closer to the population parameter with increase in sample size.

The efficiency of a method of estimation is related to its sampling variance. The relative efficiency of two methods of estimation is the ratio of the two sampling variances, the larger variance being placed in the denominator. To illustrate, when the distribution of a variable in the population is symmetrical and unimodal, both the mean and the median are estimates of the same population parameter, μ. The sampling variance of the mean is $\sigma_{\overline{x}}^2 = \sigma^2/N$, and of the median, $\sigma_{\text{mdn}}^2 = 1.57\sigma^2/N$. The relative efficiency is then

[10.1]   Relative efficiency $= \dfrac{\sigma^2/N}{1.57\sigma^2/N} = .64$

Thus for this type of population the mean is more efficient than the median. Relative efficiency has meaning in terms of sample size. A median calculated on a sample of 100 cases has a sampling variance equal to that of a mean calculated on 64 cases. The mean is a more economical estimate and the saving achieved by its use in preference to the median is 36 cases in 100.

A method of estimation is sufficient if it is more efficient than any other possible method of estimation, that is, if its sampling variance is less. A sufficient method of estimation uses all the information in the sample. In this context, the concept of information is assigned a precise mathematical meaning.

10.3 CONFIDENCE INTERVALS FOR MEANS OF LARGE SAMPLES

The calculation of confidence intervals for a mean based on a large sample is a relatively simple procedure. The sampling variance of the sampling distribution of the mean, as previously stated, is $\sigma_{\bar{x}}^2 = \sigma^2/N$. The population variance σ^2 is unknown. If we use an unbiased variance s^2 as an estimate of σ^2, our estimate of the sampling variance is $s_{\bar{x}}^2 = s^2/N$ and the estimated standard error is given by

$$[10.2] \quad s_{\bar{x}} = \frac{s}{\sqrt{N}}$$

This is the commonly used formula for estimating the standard error of the arithmetic mean.

Consider now the ratio $z = (\bar{X} - \mu)/s_{\bar{x}}$. This ratio is a deviation of a sample mean from its population mean, divided by an estimate of the standard deviation of the sampling distribution. It is a standard score. Assuming the normality of z, it is correct to state that the probability is .95 that the following statement is true:

$$[10.3] \quad -1.96 \leqslant \frac{\bar{X} - \mu}{s_{\bar{x}}} \leqslant 1.96$$

This inequality specifies the confidence interval in standard-score form. In effect it states that the chances are 95 in 100 that $(\bar{X} - \mu)/s_{\bar{x}}$ falls between ± 1.96. Ordinarily, however, we are not interested in the confidence interval in standard-score form, but in raw-score form. To convert the inequality to raw-score form, we multiply by $s_{\bar{x}}$ and add $\bar{X}$ to each term to obtain

$$[10.4] \quad \bar{X} - 1.96 s_{\bar{x}} \leqslant \mu \leqslant \bar{X} + 1.96 s_{\bar{x}}$$

This states that the chances are 95 in 100 that μ falls between $\bar{X} \pm 1.96 s_{\bar{x}}$; thus the upper limit is 1.96 standard error units above the sample mean, and the lower limit is 1.96 standard error units below the mean. The figure 1.96 derives, as the reader will recall, from the fact that 95 per cent of the area of the normal curve falls within the limits ± 1.96 standard deviation units from the mean. To illustrate the fixing of confidence intervals, let the mean IQ of a random sample of 100 secondary school children be 114 and the standard deviation 17. The standard deviation here is the square root of the unbiased variance estimate. Our estimate of the standard error of the mean is $s_{\bar{x}} = 17/\sqrt{100} = 1.70$. The 95 per cent confidence interval is then given by $114 \pm 1.96 \times 1.70$. The upper limit is 117.33 and the lower limit is 110.67. Thus we may assert with 95 per cent confidence that the population mean falls within these limits. The 99 per cent confidence limits are given by $\bar{X} \pm 2.58 s_{\bar{x}}$. The figure 2.58 derives from the fact that 99 per cent of the area of the normal curve falls within the limits ± 2.58 standard deviation units above and below the mean. In the above example the 99 per cent confidence limits are given by $114 \pm 2.58 \times 1.70$. These limits are 109.61 and 118.39.

What meaning attaches to the statement that we are 95 per cent confident that the actual population mean falls within certain specific limits? A particular sample may have a mean $\bar{X} = 26.88$ with 95 per cent confidence intervals 24.92 and 28.84. Another sample of the same size may have a mean $\bar{X} = 25.68$ with 95 per cent confidence intervals 23.72 and 27.64. Presumably we could draw a large number of samples, obtain a large number of upper and lower limits, and prepare frequency distributions of these upper and lower limits. These two distributions would be experimental sampling distributions for the 95 per cent confidence limits. Without elaborating the details of this situation, we state that about 95 per cent of the intervals so obtained would include the population mean and about 5 per cent of the intervals would not include the population mean. Thus the statement that we are 95 per cent confident implies that we expect about 95 per cent of our assertions to be correct and the remaining 5 per cent to be incorrect, or that the odds are 19:1 that the confidence interval includes the population value. The use of a 95 per cent confidence interval is fairly common. If a greater degree of confidence is desired, a 99 per cent interval may be used. This interval is, very roughly, 1.3 times as great as the 95 per cent interval. Thus as we increase our level of confidence, the interval is increased. Likewise, of course, as we decrease the level of confidence, the interval is decreased. Any desired level of confidence can be obtained by varying the size of the confidence interval. As the confidence level is decreased and approaches zero, the confidence interval approaches zero as a limit. As the confidence level is increased and approaches 100, the confidence interval approaches infinity as a limit. In practice, 95 and 99 per cent confidence intervals are widely used.

Implicit in the above discussion is the assumption that the ratio $(\bar{X} - \mu)/s_{\bar{x}}$ is normally distributed. This ratio is not normally distributed when N is small, but approaches the normal form as N increases in size. It is a not uncommon statistical convention to consider a sample of 30 or more observations as large and a sample of less than 30 as small. This, of course, is highly arbitrary.

10.4   THE DISTRIBUTION OF t

In drawing samples from a normal population with mean μ and variance σ^2, the distribution of the ratio

$$\frac{\overline{X} - \mu}{\sigma_{\overline{x}}}$$

is normal. This ratio is in standard-score form with zero mean and unit standard deviation. It is a deviation of a sample mean from a population mean, divided by the standard deviation of the sampling distribution of means. Where σ^2 is unknown, we estimate it from the sample data, using in this instance an unbiased estimate. We obtain thereby an estimate of $\sigma_{\overline{x}}$. Denote this by $s_{\overline{x}}$. We may now consider the ratio

[10.5] $$t = \frac{\overline{X} - \mu}{s_{\overline{x}}} = \frac{\overline{X} - \mu}{\sqrt{\dfrac{\Sigma(X - X)^2}{N(N-1)}}}$$

This ratio contains the variable sample values $\overline{X}$ and $s_{\overline{x}}$ in the numerator and denominator, respectively. This is a t ratio. It departs appreciably from the normal form for small N. Its theoretical sampling distribution is called the distribution of t. If samples of, say, 5 or 10 members are drawn from a normal population, a value of t calculated for each sample, and a frequency distribution of the different values of t prepared, the resulting distribution will not be normally distributed. It will be symmetrical but leptokurtic. The theoretical sampling distribution of t for small N is also symmetrical and leptokurtic. It tapers off to infinity at the two extremities. It is, however, thicker at the extremities than the corresponding normal curve. A different t distribution exists for each number of degrees of freedom. As the number of degrees of freedom increases, the t distribution approaches the normal form. Figure 10.1 compares the normal distribution with the distribution of t for various degrees of freedom.

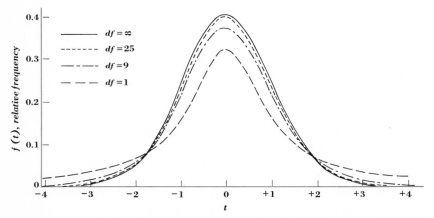

Fig. 10.1   Distribution of t for various degrees of freedom. (*From D. Lewis, Quantitative methods in psychology, McGraw-Hill Book Company, New York, 1960.*)

Hitherto we have considered two theoretical model frequency distributions, the binomial distribution and the normal distribution. The *t distribution* is a third theoretical model distribution with wide application to many sampling problems. It was developed originally in 1908 by W. S. Gosset who wrote under the pen name "Student."

In sampling problems the *t* distribution is used in a manner directly analogous to the normal distribution. In the normal distribution 95 per cent of the total area under the curve falls within plus and minus 1.96 standard deviation units from the mean and 5 per cent of the area falls outside these limits. Likewise, 99 per cent of the area under the normal curve falls within plus and minus 2.58 standard deviation units from the mean and 1 per cent of the area falls outside these limits. In the *t* distribution, the distances along the base line of the curve that include 95 per cent and 99 per cent of the total area are different for different numbers of degrees of freedom. It is customary in tabulating areas under the *t* curve to use degrees of freedom, *df*, instead of *N*. While the *df* associated with the sample variance is $N - 1$, the *df* associated with other statistics may be $N - 2$, $N - 3$, and the like. Consequently, tables of *t* by degrees of freedom instead of *N* are more generally applicable. The distances from the mean, measured along the base line of the *t* distribution, that include 95 per cent and 99 per cent of the total area (analogous to the 1.96 and 2.58 of the normal distribution) for selected degrees of freedom are as follows:

df	95%	99%
1	12.71	63.66
2	4.30	9.93
3	3.18	5.84
4	2.78	4.60
5	2.57	4.03
10	2.23	3.17
15	2.13	2.95
20	2.09	2.85
30	2.04	2.75
120	1.98	2.62
∞	1.96	2.58

Note that as the number of degrees of freedom approaches infinity, *t* approaches the values 1.96 and 2.58. The difference between *t* for about 30 degrees of freedom and *t* for an indefinitely large number of degrees of freedom is sometimes interpreted for practical purposes as trivial. A more complete tabulation of t is given in Table B of the Appendix. A distinction is often made between *large* and *small* sample statistics. This distinction resides in the fact that the normal distribution is frequently found to be an appropriate model for use with sampling problems involving large samples. With small samples the distribution of t provides for many statistics a more appropriate model.

10.5 DEGREES OF FREEDOM

In the above discussion on the distribution of t, mention is made of the number of degrees of freedom. This concept was discussed also in Chap. 4, where the sample variance was defined as the sum of squares of deviations about the arithmetic mean, divided by the number of degrees of freedom. The degree of freedom concept requires further elaboration.

As stated, and illustrated in Chap. 4, the number of degrees of freedom is the number of values of the variable that are free to vary. The measurements 10, 14, 6, 5, and 5, when represented as deviations from a mean of 8, become $+2, +6, -2, -3, -3$. The sum of these deviations is zero. In consequence, if any four deviations are known, the remaining deviation is determined. The number of degrees of freedom is 4.

This type of situation may be represented in symbolic form. Let X_1, X_2, X_3 be three measurements with mean $\bar{X}$. The sum of deviations is $(X_1 - \bar{X}) + (X_2 - \bar{X}) + (X_3 - \bar{X}) = 0$. If $\bar{X}$ and any two of the values of X are known, the third value of X is determined. The number of degrees of freedom here is 2. The calculation of the variance and standard deviation requires the sum of squares of deviations about the mean, $\Sigma(X - \bar{X})^2$. $N - 1$ of the values of which this sum of squares is comprised are free to vary independently. The number of degrees of freedom associated with the sum of squares is $N - 1$. Dividing this sum of squares by the number of degrees of freedom associated with it, as distinct from the number of observations, yields an unbiased estimate of the population variance σ^2. The symbol df is frequently used to represent degrees of freedom.

The number of degrees of freedom depends on the nature of the problem. In fitting a line to a series of points by the method of least squares the number of degrees of freedom associated with the sum of squares of deviations about the line is $N - 2$. If there are two points only, a straight line will fit the points exactly and the sum of squares of deviations about the line will, of course, be zero. No freedom of variation is possible. With three points $df = 1$; with 15 points, $df = 13$. The equation of a straight line is $Y = bX + a$, where b is the slope of the line and a is the point where it cuts the Y axis. Both b and a are estimated from the data. It may be said that 2 degrees of freedom are lost in estimating b and a from the data. If b, a, and any $N - 2$ deviations from the line are known, the remaining two deviations are determined.

The concept of degrees of freedom has a geometric interpretation. A point on a line is free to move in one dimension only and has 1 degree of freedom. A point on a plane has freedom of movement in two dimensions and has 2 degrees of freedom. A point in a space of three dimensions has 3 degrees of freedom. Likewise, a point in a space of k dimensions has k degrees of freedom. It has freedom of movement in k dimensions.

The concept of degrees of freedom is widely used in statistical work and will be discussed subsequently in connection with contingency tables and the analysis of variance. The essence of the idea is simple. The number of degrees of freedom is always the number of values that are free to vary, given the number of restrictions imposed upon the data. It seems intuitively obvious that in the study of variation we should concern ourselves with the number of values that enjoy freedom to vary within the restrictions of the problem situation.

10.6 CONFIDENCE INTERVALS OF MEANS FOR SMALL SAMPLES

The line of reasoning used in determining confidence intervals for small samples is similar to that for large samples. With small samples, however, the distribution of t is used instead of the normal distribution in fixing the limits of the interval. For large samples the 95 and 99 per cent confidence intervals for the mean are given, respectively, by $\overline{X} \pm 1.96s_{\bar{x}}$ and $\overline{X} \pm 2.58s_{\bar{x}}$. For small samples an unbiased estimate of σ^2 is used in estimating the standard error. The value of t used in fixing the limits of the 95 and 99 per cent intervals will vary, depending on the number of degrees of freedom. Consider an example where $\overline{X} = 24.26$, $s^2 = 64$, $N = 16$, and $df = 16 - 1$. On reference to Table B of the Appendix we observe that for 15 degrees of freedom 95 per cent of the area of the distribution falls within a t of ± 2.13 from the mean. The standard error using the unbiased variance estimate is $8/\sqrt{16}$. The 95 per cent confidence limits are given by $24.26 \pm 2.13 \times 8/\sqrt{16}$. These limits are 19.88 and 28.64. We may assert with 95 per cent confidence that the population mean falls within these limits. The 99 per cent limits are given by $24.26 \pm 2.95 \times 8/\sqrt{15}$. These limits are 18.16 and 30.36.

10.7 STANDARD ERRORS AND CONFIDENCE INTERVALS OF PROPORTIONS

The estimate of the standard error of a proportion is given by

$$[10.6] \quad s_p = \sqrt{\frac{p(1-p)}{N}} = \sqrt{\frac{pq}{N}}$$

where $1 - p = q$. Also, it may be readily shown that the standard error of a per cent is given by

$$[10.7] \quad s_p = 100 \sqrt{\frac{pq}{N}}$$

If it can be assumed that the sampling distribution of a proportion can be approximately represented by a normal distribution, then the 95 and 99 per cent confidence limits for a proportion are given by $p \pm 1.96 s_p$ and $p \pm 2.58 s_p$, respectively. Whether or not the sampling distribution can be represented by a normal distribution depends both on the size of the sample and on the value of p. For any given value of N the sampling distribution of a proportion becomes increasingly skewed as p and q depart from .50. Quite clearly, the formula for the standard error of a proportion should not be used with reference to a normal curve for extreme values of p and q. It has been suggested that the formula for the standard error of a proportion should be used only when Np or Nq, whichever is the smaller, is equal to or greater than 5. Thus when $p = .10$ and $N = 20$, $Np = 2$. The use of the formula $s_p = \sqrt{pq/N}$ would be considered inappropriate here. When $p = .10$ and $N = 100$, $Np = 10$. Presumably, here the differences between the binomial and the normal distribution are quite small and can safely be ignored.

10.8 STANDARD ERRORS AND CONFIDENCE INTERVALS OF OTHER STATISTICS

The standard error of the median may be estimated by

$$[10.8] \quad s_{\text{mdn}} = \frac{1.253 s}{\sqrt{N}}$$

where s is obtained from the unbiased estimate of σ^2. Confidence limits at the 95 and 99 per cent levels may be located by taking $\pm 1.96 s_{\text{mdn}}$ and $\pm 2.58 s_{\text{mdn}}$ about the sample median. The above formulation assumes normality of the parent population and a large N. In many situations where the median is used, the distribution of the variable is not normal. This, indeed, is one of the reasons for using the median instead of the mean. In consequence the above formulation is of limited use. Confidence intervals for the median involving no assumptions about the shape of the distribution of the variable in the population, other than its continuity, have been worked out by Nair (1940). His method is described by Kenney and Keeping (1954) and Johnson (1949). Given N observations arranged in ascending order, $X_1, X_2, \ldots, X_N$, the median is the middle value. The problem of fixing, say, 95 per cent limits is approached by locating two values of X_i in this ascending series such that the probability that these values will include the population median is not less than .95.

The standard error of the standard deviation for large samples from a normal population is estimated by

$$[10.9] \quad s_s = \frac{s}{\sqrt{2N}}$$

The 95 and 99 per cent confidence limits can readily be obtained by taking

$s \pm 1.96s_s$ and $s \pm 2.58s_s$, respectively. In using this formula a sample substantially greater than 30 should be regarded as large. The method of determining confidence limits for s based on small samples, and indeed the method which is perhaps most appropriate in all cases regardless of size of N involves a knowledge of the distribution of chi square, or χ^2. For a simple discussion of this method see Freund (1967) or Johnson (1949). The application of χ^2 to a variety of statistical problems will be discussed in Chap. 13.

EXERCISES 1 Using a large-sample procedure, obtain the 95 and 99 per cent confidence intervals for a mean of 105, where $N = 100$ and $s = 10$. Obtain also the 75 and 85 per cent confidence intervals.

2 A random sample of 400 observations has a mean of 50 and a standard deviation of 18. Estimate the 95 and 99 per cent confidence limits for the mean.

3 How is the standard error of the mean affected by tripling sample size?

4 Using a large-sample procedure, estimate for the following data the 95 and 99 per cent confidence intervals for means:

	$\bar{X}$	N	$\Sigma(X - \bar{X})^2$
a	26.2	7	77.0
b	58.3	11	249.0
c	46.3	25	1,525.0
d	8.4	16	444.7

5 Find the value of t for $df = 20$ such that the proportion of the area (a) to the right of t is .025, (b) to the left of t is .0005, (c) between the mean and t is .45, (d) between $\pm t$ is .90.

6 Obtain the values required in (a), (b), (c), and (d) of Exercise 5 above for $df = 5$.

7 What proportion of the area of the t distribution falls (a) above $t = 3.169$ where $df = 10$, (b) below $t = -1.725$ where $df = 20$, (c) between $t = \pm 3.659$ where $df = 29$, (d) between $t = 2.131$ and $t = 2.602$ where $df = 15$, (e) between $t = -4.541$ and $t = 3.182$ where $df = 3$?

8 Estimate the 95 and 99 per cent confidence limits for $p = .75$ where $N = 169$.

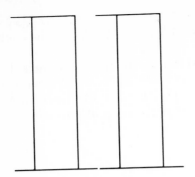

TESTS OF SIGNIFICANCE: MEANS

INTRODUCTION

In Chaps. 9 and 10 we considered the sampling error associated with single sample values. Sampling distributions, standard errors of single sample values, and confidence intervals were discussed. In practical statistical work in psychology we are infrequently concerned with simply describing the magnitude of error associated with single sample values. Experimental data very often require a comparison and evaluation of two or more means, proportions, standard deviations, or other statistics obtained from separate samples or from the same sample for measurements obtained under two or more experimental conditions. To illustrate, an investigator may wish to explore the effects of a tranquilizing drug on the estimation of time intervals as part of a study on time perception. He may administer a drug to an experimental group of subjects and a placebo, an inactive simulation of the drug, to a control group and measure the errors in time estimation made by the two groups. The mean error for the two groups may be calculated. The experiment requires an evaluation of the difference between these two means. Both means are subject to sampling error. May the difference between the two means be probably ascribed to sampling error, or may it be argued with confidence that the drug affects time perception? A decision is required between these alternatives. Statistical procedures which lead to decisions of this kind are known as *tests of significance.*

Tests of significance may be applied to the difference between statistics calculated on independent samples or between statistics obtained under different conditions for the same sample. Sometimes a test of significance is applied to test the difference between a single sample statistic and a fixed value. An example is the procedure used to test whether a correlation coefficient is significantly different from zero. In this case the fixed value is zero. While many tests of significance involve a comparison of two sample statistics, or a single sample statistic and a fixed value, such tests can readily be extended to cover situations where more than two sample statistics are involved. For example, in the experiment mentioned above on the effects of a drug on time perception, the experiment could be designed to include the administration of the drug in different dosages to different groups of subjects. Three or four or five different dosages might be used, resulting in three or four or five different means. The means could be compared two at a time to ascertain whether or not the differences between them could be attributed to sampling error. A more efficient form of analysis, the analysis of variance, provides a procedure for making an over-all test in this type of situation.

11.2 THE NULL HYPOTHESIS

Consider an experiment using an experimental and a control group. A treatment is applied to the experimental group. The treatment is absent for the control group. Measurements are made on both groups. Presumably any significant difference between the two groups can be ascribed with confidence to the treatment and to no other cause. Let $\overline{X}_1$ and $\overline{X}_2$ be the means for the experimental and the control group, respectively. Both means are subject to sampling error. The means $\overline{X}_1$ and $\overline{X}_2$ are estimates of the population means μ_1 and μ_2. The trial hypothesis may be formulated that no difference exists between μ_1 and μ_2. This hypothesis is a null hypothesis and may be written

[11.1] $H_0{:}\mu_1 - \mu_2 = 0$

The symbol H_0 represents the null hypothesis. Very simply, this hypothesis asserts that no difference exists between the two population means. Note that the statement $\mu_1 - \mu_2 = 0$ is the same as $\mu_1 = \mu_2$. Thus an alternative formulation of the hypothesis is to assert that the two samples are drawn from populations having the same mean. In general, regardless of the particular statistics used, the null hypothesis is a trial hypothesis asserting that no difference exists between population parameters. Thus a null hypothesis about two variances would take the form $H_0{:}\sigma_1^2 - \sigma_2^2 = 0$, or $H_0{:}\sigma_1^2 = \sigma_2^2$.

The logical steps used by an investigator in applying a test of significance are these. *First*, he assumes the null hypothesis; that is, he operates on the trial hypothesis that the treatment applied will have no effect. *Second*, he examines the empirical data. Where the hypothesis pertains to two means he examines the difference between the two means, $\overline{X}_1 - \overline{X}_2$. *Third*, the question is asked, what is the probability of obtaining a difference equal to or greater than the one observed in drawing samples at random from populations where the null hypothesis is assumed to be true? In the case of two means, what is the probability of obtaining a difference equal to or greater than $\bar{X}_1 - \bar{X}_2$ in drawing random samples from populations where $\mu_1 - \mu_2 = 0$? *Fourth*, if this probability is *small*, the observed result being highly improbable on the basis of the null hypothesis, the investigator may be prepared to reject the null hypothesis. This means that the observed difference cannot reasonably be explained by sampling error and presumably may be attributed to the treatment applied. Thus the result may be said to be significant. If this probability cannot be considered small and the observed result is not highly improbable, then sampling error may account for the difference observed. Hence we cannot with confidence infer that the difference results from the treatment applied.

In the testing of any statistical hypothesis, it is necessary to specify an alternative hypothesis. This alternative is accepted if the initial hypothesis is rejected. Thus in the testing of the hypothesis $H_0 : \mu_1 - \mu_2 = 0$, the alternative may be $H_1 : \mu_1 - \mu_2 \neq 0$. Under certain circumstances, as described in detail in Sec. 11.5, some advantages may attach to a test of a null hypothesis against the alternative $\mu_1 - \mu_2 > 0$, or the alternative $\mu_1 - \mu_2 < 0$. The alternative hypothesis under consideration should be clearly recognized.

11.3 TWO TYPES OF ERROR

In reaching a decision about the null hypothesis H_0, two types of error may arise. An alternative H_1 may be accepted when the null hypothesis H_0 is true. This is called a Type I error. The null H_0 may be accepted when an alternative hypothesis H_1 is true. This is called a Type II error. The probabilities of committing Type I and Type II errors are represented, respectively, by α and β.

The situation may be represented as follows:

	H_0 *is true*	H_1 *is true*
Accept H_1	α	correct decision
Accept H_0	correct decision	β

When we accept H_1, and H_1 is true, this is a correct decision about nature. When we accept H_0, and H_0 is true, this is also a correct decision about nature. The acceptance of H_1 when H_0 is true and the acceptance of H_0 when H_1 is true are both errors. Both are incorrect decisions about nature.

The situation above is somewhat analogous to that which arises in the acceptance or rejection of applicants for employment or admission to a university. For example, a university admissions officer may accept applicants who subsequently are successful in the university, or he may reject applicants who would have failed if they had been accepted. In either case a correct decision may be said to have been made. On the other hand, the admissions officer may reject an applicant who would have been successful if he had been accepted, or he may accept an applicant who subsequently fails. In both instances an error has been made.

11.4 LEVELS OF SIGNIFICANCE

The probability of Type I, or α, error is called the level of significance of a test. Ordinarily the investigator adopts, perhaps rather arbitrarily, a particular level of significance. It is a common convention to adopt levels of significance of either .05 or .01. If the probability is equal to or less than .05 of asserting that there is a difference between two means, for example, when no such difference exists, then the difference is said to be significant at the .05, or 5 per cent, level or less. Here the chances are 5 in 100, or less, that the difference could result when the treatment applied is having no effect. If the probability is .01, or less, the difference is said to be significant at the .01, or 1 per cent, level. The .05 and .01 probability levels are descriptive of our degree of confidence that a real difference exists, or that the observed difference is not due to the caprice of sampling. Usually in evaluating an experimental result, it is unnecessary to determine the probabilities with a high degree of accuracy. For most practical purposes it is sufficient to designate the probability as $p \leqslant .05$, or $p \leqslant .01$, or possibly $p \leqslant .001$ if the result is highly significant.

Decisions regarding the rejection, or otherwise, of the null hypothesis at a level of significance α are commonly made without reference to Type II, or β, error. A detailed description of the functioning of β error is beyond the scope of this book. A few comments, however, may be appropriate. Consider, for example, the difference between two means, μ_1 and μ_2. For any particular value of α, say .05, the value of β is a function of sample size N and the actual difference between μ_1 and μ_2. For specified α and N, the value of β, which is the probability of failing to recognize the existence of a difference when such a difference exists, will decrease with increase in the difference between μ_1 and μ_2. This means that the larger the difference between μ_1 and μ_2 the less likely we are to accept H_0. For a specified difference between μ_1 and μ_2 and a specified N the value of β will increase as α decreases. Accordingly, if too strict a level of significance is adopted, we may fail to reject the null hypothesis when in fact a fairly large difference between μ_1 and μ_2 exists. For any specified difference between μ_1 and μ_2, and any α, the Type II error β is a function of sample size N. The smaller the sample the greater the value of β. This means that although a large difference may exist between μ_1 and μ_2, it may be difficult to prove for small samples.

Although, quite clearly, failure to reject the null hypothesis does not imply that the null hypothesis is true, many investigators exhibit an inclination to conclude, even for quite small samples, that no difference, or a trivial difference, exists when a required level of significance is not achieved. Our discussion of Type II error clearly indicates that such conclusions are unwarranted. It would, of course, be possible to establish dual criteria such that if α were greater than, say, .05 and β were less than, say, .05, the investigator, in practice, might be allowed to conclude that the difference was nonexistent or inconsequential.

11.5   DIRECTIONAL AND NONDIRECTIONAL TESTS

An investigator may wish to test the null hypothesis, $H_0:\mu_1 - \mu_2 = 0$, against the alternative, $H_1:\mu_1 - \mu_2 \neq 0$. This means that if H_0 is rejected, the decision is that a difference exists between the two means. No assertion is made about the direction of the difference. Such a test is a *nondirectional* test. A test of this kind is sometimes called a *two-tailed* or *two-sided* test, because if the normal distribution, or the distribution of t, is used, the two tails, or the two sides, of the distribution are employed in the estimation of probabilities. Consider a 5 per cent significance level. If the sampling distribution is normal, 2.5 per cent of the area of the curve falls to the right of 1.96 standard deviation units above the mean, and 2.5 per cent falls to the left of 1.96 standard deviation units below the mean. The area outside these limits is 5 per cent of the total area under the curve. Under the null hypothesis the chances are 2.5 in 100 of getting a difference of 1.96 standard deviation units in one direction because of chance factors alone, and 2.5 in 100 in the other direction. Hence the chances in either direction are 5 in 100. Thus for significance at the 5 per cent level for a nondirectional test, when the sampling distribution is normal, the observed difference must be equal to or greater than 1.96 times the standard deviation of the sampling distribution of differences. For significance at the 1 per cent level a value of 2.58 is required for a nondirectional test. A nondirectional test is appropriate if concern is with the absolute magnitude of the difference, that is, with the difference regardless of sign.

Under certain circumstances we may wish to make a decision about the direction of the difference. It has been argued that few instances exist in scientific work where the traditional nondirectional test is of interest. If concern is with the direction of the difference we may test the hypothesis $H_0:\mu_1 - \mu_2 \leq 0$ against the alternative $H_1:\mu_1 - \mu_2 > 0$, or the hypothesis $H_0:\mu_1 - \mu_2 \geq 0$ against the alternative $H_2:\mu_1 - \mu_2 < 0$. Note that the symbol H_0 has been used to denote three different hypotheses: an hypothesis of no difference, an hypothesis of equal to or less than, and an hypothesis of equal to or greater than. Conventionally the term null hypothesis has been restricted to an hypothesis of no difference. It is not inappropriate, as pointed out by Kaiser (1960), to extend the meaning of the null hypothesis to include hypotheses of equal to or less than and equal to or greater than. Such tests are directional one-sided, or one-tailed, tests. If the normal, or t, distribution is used, one side or one tail only is employed to estimate the required probabilities. To reject $H_0:\mu_1 - \mu_2 \leq 0$ and accept $H_1:\mu_1 - \mu_2 > 0$, using the normal distribution, a normal deviate $> +1.64$ is required for significance at the .05 level. Likewise to reject $H_0:\mu_1 - \mu_2 \geq 0$ and accept $H_2:\mu_1 - \mu_2 < 0$, the corresponding normal deviate is < -1.64. The figure 1.64 derives from the fact that for a normal distribution 5 per cent of the area of the curve falls beyond $+1.64$ standard deviation units above the mean, and 5 per cent beyond -1.64 standard deviation units below the mean. For significance at the .01 level for a directional test, normal deviates $> +2.33$ or < -2.33 are required.

Kaiser (1960) discusses a directional two-sided test, which in the end result uses much the same procedure as the one described above. His procedure requires a decision between three hypotheses, $H_0:\mu_1 - \mu_2 = 0$, $H_1:\mu_1 - \mu_2 > 0$, and $H_2:\mu_1 - \mu_2 < 0$. The rules here, at the 5 per cent significance level, are that we decide upon H_0 when the normal deviate falls between -1.64 and $+1.64$. We decide upon H_1 when the normal deviate is greater than $+1.64$ and upon H_2 when the normal deviate is less than -1.64. This test amounts in practice to the same thing as making two directional one-sided tests. Involved in its development are errors of the third kind. These relate to a decision about a difference in the wrong direction.

When is it appropriate to use a directional as distinct from a nondirectional test? This question is open to some controversy. Clearly there are many instances in research where the direction of the differences is of substantial interest; indeed, it has been argued that there are few, if any, instances where the direction is not of interest. At any rate it is the opinion of this writer that directional tests should be used more frequently.

11.6 SIGNIFICANCE OF THE DIFFERENCE BETWEEN TWO MEANS FOR INDEPENDENT SAMPLES

Let $\overline{X}_1$ and $\overline{X}_2$ be two sample means based on N_1 and N_2 cases, respectively. We proceed by combining the data for the two samples to obtain the best unbiased estimate of the population variance. This estimate is obtained by adding together the two sums of squares of deviations about the two sample means and dividing this by the total number of degrees of freedom. This unbiased estimate of the population variance may be written as

[11.2] $$s^2 = \frac{\sum\limits^{N_1} (X - \overline{X}_1)^2 + \sum\limits^{N_2} (X - \overline{X}_2)^2}{N_1 + N_2 - 2}$$

The two terms in the numerator are sums of squares of deviations about the means of the two samples of N_1 and N_2 cases, respectively. The total number of degrees of freedom on which s^2 is based is $N_1 + N_2 - 2$. We lose two degrees of freedom because deviations are taken separately about the means of the two samples. The unbiased variance estimate s^2 is used to obtain an estimate of the standard error of the difference between the two means. This standard error is given by

[11.3] $$s_{\bar{x}_1-\bar{x}_2} = \sqrt{\frac{s^2}{N_1} + \frac{s^2}{N_2}}$$

The difference between means, $\bar{X}_1 - \bar{X}_2$, is then divided by this estimate of the standard error to obtain the ratio

[11.4] $$t = \frac{\bar{X}_1 - \bar{X}_2}{s_{\bar{x}_1-\bar{x}_2}} = \frac{\bar{X}_1 - \bar{X}_2}{\sqrt{s^2/N_1 + s^2/N_2}}$$

This ratio has a distribution of t with $N_1 + N_2 - 2$ degrees of freedom. The values of t required for significance at the .05 and .01 levels will vary, depending on the number of degrees of freedom, and may be obtained by consulting Table B of the Appendix.

The formula given above for s^2 is not very convenient from a computational viewpoint. A more convenient formula is

[11.5] $$s^2 = \frac{\sum^{N_1} X^2 - \left(\sum^{N_1} X\right)^2 / N_1 + \sum^{N_2} X^2 - \left(\sum^{N_2} X\right)^2 / N_2}{N_1 + N_2 - 2}$$

This, if desired, may be further modified by writing

$$\frac{\left(\sum^{N_1} X\right)^2}{N_1} = N_1 \bar{X}_1^2 \qquad \text{and} \qquad \frac{\left(\sum^{N_2} X\right)^2}{N_2} = N_2 \bar{X}_2^2$$

Let the following be error scores obtained for two groups of experimental animals in running a maze under different experimental conditions.

Group A	16	9	4	23	19	10	5	2
Group B	20	5	1	16	2	4		

The following statistics are calculated from these data:

	Group A	Group B
N	8	6
$\sum X$	88	48
$\bar{X}$	11	8
$\sum X^2$	1,372	702

The unbiased estimate of variance is

$$s^2 = \frac{1{,}372 - 88^2/8 + 702 - 48^2/6}{8 + 6 - 2} = 60.17$$

The t ratio is then

$$t = \frac{11 - 8}{\sqrt{60.17/8 + 60.17/6}} = .72$$

The number of degrees of freedom in this example is $8 + 6 - 2 = 12$. For 12 degrees of freedom a t equal to 2.179 is required for significance at the .05 level. In this example the difference between means is not significant. No adequate grounds exist for rejecting the null hypothesis. We are not justified in drawing the inference from these data that the two experimental conditions are exerting a differential effect on the behavior of the animals.

The t test described here assumes that the distributions of the variables in the populations from which the samples are drawn are normal. It assumes also that these populations have equal variances. This latter condition is referred to as homogeneity of variance. The t test should be used only when there is reason to believe that the population distributions do not depart too grossly from the normal form and the population variances do not differ markedly from equality. Tests of normality and homogeneity of variance may be applied, but these tests are not very sensitive for small samples.

11.7 SIGNIFICANCE OF THE DIFFERENCE BETWEEN TWO MEANS FOR CORRELATED SAMPLES

Consider a situation where a single group of subjects is studied under two separate experimental conditions. The data may, for example, be autonomic response measures under stress and nonstress or measures of motor performance in the presence or absence of a drug. The data are composed of pairs of measurements. These may be correlated. This circumstance leads to a test of significance between means different from that for independent samples. A procedure for testing significance may be applied without actually computing the correlation coefficient between the paired observations. This method is sometimes called the difference method. Its nature is simply described. Given a set of N paired observations, the difference between each pair may be obtained. Denote any pair of observations by X_1 and X_2 and the difference between any pair $X_1 - X_2$ by D. The mean difference over all pairs is $(\Sigma D)/N = \bar{D}$. It is readily observed that the difference between the means of the two groups of observations is equal to the mean difference. The difference between any pair of observations is $X_1 - X_2 = D$. Summing over N pairs yields $\Sigma X_1 - \Sigma X_2 = \Sigma D$. Dividing by N, we obtain $\bar{X}_1 - \bar{X}_2 = \bar{D}$. Since the mean difference is the difference between the two means, we may test for the significance of the differences between means by testing whether or not $\bar{D}$ is significantly different from zero. Here in effect we treat the D's as a variable and test the difference between the mean of this variable and zero.

An unbiased estimate of the variance of the D's is given by

[11.6] $s_D{}^2 = \dfrac{\Sigma(D - \bar{D})^2}{N - 1}$

where N is the number of paired observations. Using this unbiased estimate, the sampling variance of $\bar{D}$ is given by

[11.7] $s_{\bar{D}}{}^2 = \dfrac{s_D{}^2}{N}$

To test whether $\bar{D}$ is significantly different from zero, we divide $\bar{D}$ by its standard error to obtain

[11.8] $t = \dfrac{\bar{D}}{s_{\bar{D}}}$

The number of degrees of freedom used in evaluating t is one less than the number of pairs of observations, or $N - 1$. The reader should note that the $\bar{D}$ in the numerator of the above formula is in effect $\bar{D} - 0$, which is of course $\bar{D}$. This test is concerned with the significance of $\bar{D}$ from zero.

The above formula for t is not convenient computationally. A more convenient computational formula is

[11.9] $t = \dfrac{\Sigma D}{\sqrt{[N\Sigma D^2 - (\Sigma D)^2]/(N - 1)}}$

The data below are those obtained for a group of 10 subjects on a choice–reaction-time experiment under stress and nonstress conditions, the stress agent being electric shock. The figures are the number of false reactions over a series of trials. The problem here is to test whether the

Subject	Stress X_1	Nonstress X_2	D	D^2
1	7	5	2	4
2	9	15	−6	36
3	4	7	−3	9
4	15	11	4	16
5	6	4	2	4
6	3	7	−4	16
7	9	8	1	1
8	5	10	−5	25
9	6	6	0	0
10	12	16	−4	16
Sum	76	89	−13	127
Mean	7.60	8.90	−1.30	

means under the two conditions are significantly different. These means are 7.60 and 8.90. The difference between them is equal to the mean of the differences, or -1.30. The sum and sum of squares of D are, respectively, -13 and 127. Hence

$$t = \frac{-13}{\sqrt{[10 \times 127 - (-13)^2]/(10-1)}} = -1.18$$

We may ignore the negative sign of t and consider only its absolute magnitude. The number of degrees of freedom associated with this value of t is 9. For 9 degrees of freedom we require a t of 2.262 for significance at the 5 per cent level. The observed value of t is well below this, and the difference between means is not significant. We cannot justifiably argue from these data that the mean number of false reactions under the two conditions is different.

The method described above takes into account the correlation between the paired measurements. This results because the variance of differences is related to the correlation between the paired measurements by the formula

[11.10] $s_D{}^2 = s_1{}^2 + s_2{}^2 - 2r_{12}s_1s_2$

When s_1, s_2, and r_{12} have been computed, as will frequently be the case, the variance of differences $s_D{}^2$ can be readily obtained from the above formula and need not be obtained by direct calculation on the differences themselves. A positive correlation between the paired measurements will reduce the size of $s_D{}^2$ and $s_{\bar{D}}{}^2$.

11.8 SIGNIFICANCE OF THE DIFFERENCE BETWEEN MEANS WHERE POPULATION VARIANCES ARE UNEQUAL

The t test for the significance of the difference between means assumes equality of the population variances. Where the assumption of equality of variance is untenable, the ordinary t test should not be applied. Approximate methods for use where the variances are unequal have been suggested by Cochran and Cox (1950) and by Welch (1938). The method of Cochran and Cox makes an adjustment in the value of t required for significance at the 5 or 1 per cent level, or other critical level as may be required. The method proposed by Welch makes an adjustment in the number of degrees of freedom.

To use the Cochran and Cox method we proceed by calculating the standard error of the differences between the two means, using the formula

[11.11] $s_{\bar{X}_1 - \bar{X}_2} = \sqrt{\dfrac{\Sigma(X - \bar{X}_1)^2}{N_1(N_1 - 1)} + \dfrac{\Sigma(X - \bar{X}_2)^2}{N_2(N_2 - 1)}} = \sqrt{s_{\bar{X}_1}{}^2 + s_{\bar{X}_2}{}^2}$

The difference between the sample means is then divided by the standard error of the difference to obtain

$$t = \frac{\overline{X}_1 - \overline{X}_2}{s_{\overline{x}_1 - \overline{x}_2}}$$

One sample is based on N_1 cases with $N_1 - 1$ degrees of freedom, the other on N_2 cases with $N_2 - 1$ degrees of freedom. Assume that a two-tailed test at the 5 per cent level is appropriate. Refer to a table of t and obtain the critical value of t required for significance at the 5 per cent level with $N_1 - 1$ degrees of freedom. Obtain also the value of t required with $N_2 - 1$ degrees of freedom. Denote these two values of t as t_1 and t_2. The approximate value of t required for significance at the 5 per cent level is given by the formula

[11.12] $$t_{.05} = \frac{s_{\overline{x}_1}{}^2 t_1 + s_{\overline{x}_2}{}^2 t_2}{s_{\overline{x}_1}{}^2 + s_{\overline{x}_2}{}^2}$$

The value of t obtained by dividing the difference between means by the standard error of their difference must be equal to or greater than $t_{.05}$ before significance at the 5 per cent level can be claimed.

Consider the following data:

Sample A	Sample B
$N_1 = 13$	$N_2 = 9$
$\overline{X}_1 = 26.99$	$\overline{X}_2 = 15.10$
$\Sigma(X - \overline{X}_1)^2 = 1{,}128$	$\Sigma(X - \overline{X}_2)^2 = 1{,}269$
$s_{\overline{x}_1}{}^2 = 7.23$	$s_{\overline{x}_2}{}^2 = 17.62$

The standard error of the difference between means is

$$s_{\overline{x}_1 - \overline{x}_2} = \sqrt{\frac{1{,}128}{13(13 - 1)} + \frac{1{,}269}{9(9 - 1)}} = \sqrt{7.23 + 17.62} = 4.98$$

Divide this into the difference between means to obtain

$$t = \frac{26.99 - 15.10}{4.98} = 2.39$$

For $13 - 1 = 12$ and $9 - 1 = 8$ degrees of freedom, the values of t required for significance at the 5 per cent level are, respectively, 2.179 and 2.306. The value required for significance at the 5 per cent level in testing the significance of the difference between means is then

$$t_{.05} = \frac{7.23 \times 2.179 + 17.62 \times 2.306}{7.23 + 17.62} = 2.27$$

This value 2.27 is less than the obtained value 2.39. Consequently we may conclude that the difference between means is significant at the 5 per cent level.

Another approximate method proposed by Welch (1947) requires the calculation of a t value as above by dividing the difference between means by their standard error. We then refer this value to the table of t using the following formula for the number of degrees of freedom:

$$df = \frac{(s_{\bar{x}_1}{}^2 + s_{\bar{x}_2}{}^2)^2}{(s_{\bar{x}_1}{}^2)^2/(N_1 + 1) + (s_{\bar{x}_2}{}^2)^2/(N_2 + 1)} - 2$$

Applying this formula to the previous data we obtain

[11.13] $$df = \frac{(7.23 + 17.62)^2}{7.23^2/14 + 17.62^2/10} - 2 = 15.76$$

The value of df will seldom be a whole number. If df is taken as 16, the value of t required for significance at the 5 per cent level is 2.12. If df is taken as 15, the value is 2.13. In either case the observed value of t, 2.39, exceeds the value required for significance at the 5 per cent level and we may conclude that the difference between means is significant. This result is in agreement with that obtained using the Cochran and Cox procedure.

The above procedures are approximate. For a more accurate method the reader is referred to Welch (1947) and Aspen (1949). The latter author has prepared tables which assist the comparison of means involving two variances, separately estimated. The problem has also been discussed by Gronow (1951).

11.9 SIGNIFICANCE OF THE DIFFERENCE BETWEEN MEANS WHEN THE POPULATION DISTRIBUTIONS ARE NOT NORMAL

The t test for the significance of the difference between means assumes normality of the distributions of the variables in the populations from which the samples are drawn. Where the variables are not normally distributed, what effect will this have on the probabilities, and significance levels, as estimated from the distribution of t?

Under certain conditions the sampling distribution of means of size N, where N is large, is closely approximated by the normal distribution. This result holds regardless of the shape of the distribution in the population from which the samples are drawn. The closeness of the approximation improves as N becomes increasingly large. The implication of this is that for large samples the nonnormality of the populations will not seriously affect the estimation of probabilities, except perhaps in cases of very extreme skewness.

A number of investigators have studied the effect of nonnormal populations on the t test for small samples. The empirical evidence suggests that even for quite small samples, say, of the order of 5 or 10, reasonably large departures from normality will not seriously affect the estimation of probabilities for a two-tailed t test. A one-tailed t test is, however, apparently more seriously affected by nonnormality.

Where the data show fairly gross departures from normality it is probably advisable to use *nonparametric*, or *distribution-free*, methods. These methods provide tests which are independent of the shapes of the distributions in the populations from which the samples are drawn. They deal with the ordinal or sign properties of the data. A number of such tests are described in Chaps. 21 and 22 of this book. Nonparametric methods are being used with increasing frequency in psychological research.

EXERCISES **1** The following are data for two samples of subjects under two experimental conditions:

Sample A	2	5	7	9	6	7
Sample B	4	16	11	9	8	

Test the significance of the difference between means using a nondirectional test.

2 The following are data for two independent samples:

	Sample A	Sample B
$\overline{X}$	124	120
N	50	36
$\Sigma(X - \overline{X})^2$	5,512	5,184

Test whether the mean for sample A is equal to or greater than that for sample B.

3 The following are paired measurements obtained for a sample of eight subjects under two conditions:

Condition A	8	17	12	19	5	6	20	3
Condition B	12	31	17	17	8	14	25	4

Test the significance of the difference between means using a nondirectional test.

4 Calculate t for the following data:

	Sample A	Sample B
$\overline{X}$	20	25
N	25	10
ΣX^2	12,500	7,900

5　For a sample of 26 paired measurements $\Sigma D = 52$ and $\Sigma D^2 = 400$. Calculate t.

6　What advantages attach to matched groups or paired observations in experimentation?

7　The means for two independent samples of 10 and 17 cases are 9.63 and 14.16, respectively. The unbiased variance estimates are 64.02 and 220.30. Compare the methods proposed by Cochran and Cox with those proposed by Welch to test the significance of the difference between the two means.

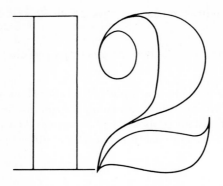

TESTS OF SIGNIFICANCE:
OTHER STATISTICS

12.1 INTRODUCTION

In Chap. 11 problems associated with the application of tests of significance to arithmetic means were discussed. Not infrequently, tests of significance for proportions, variances, correlation coefficients, or other statistics are required. The general rationale underlying the application of such tests of significance is precisely the same as that for arithmetic means, although the technical procedures used in estimating the required probabilities are different. The present chapter discusses procedures for applying tests of significance to proportions, variances, and correlation coefficients, for independent and correlated samples.

12.2 SIGNIFICANCE OF THE DIFFERENCE
BETWEEN TWO INDEPENDENT PROPORTIONS

Questions arise in the interpretation of experimental results which require a test of significance of the difference between two independent proportions. The data are comprised of two samples drawn independently. Of the N_1 members in the first sample, f_1 have the attribute A. Of the N_2 members in the second sample, f_2 have the attribute A. The proportions having the attributes in the two samples are $f_1/N_1 = p_1$ and $f_2/N_2 = p_2$. Can the two samples be regarded as random samples drawn from the same population? Is p_1 significantly different from p_2? To illustrate, in a public opinion poll the proportion .65 in a sample of urban residents may express a favorable attitude toward a particular issue as against a proportion .55 in a sample of rural residents. May the difference between the proportions be interpreted as indicative of an actual urban-rural difference in opinion? To illustrate further, the proportion of failures in air-crew training in two training periods may be .42 and .50. Does this represent a significant change in the proportion of failures, or may the difference be attributed to sampling considerations?

The standard error of a single proportion is estimated by the formula

$$s_p = \sqrt{\frac{pq}{N}}$$

where p = sample value of a proportion
$\quad\quad q = 1 - p$

The standard error of the difference between two proportions based on independent samples is estimated by

[12.1] $$s_{p_1-p_2} = \sqrt{pq\left(\frac{1}{N_1} + \frac{1}{N_2}\right)}$$

where p is an estimate based on the two samples combined. The value p is obtained by adding together the frequencies of occurrence of the attribute in the two samples and then dividing this by the total number in the two samples. Thus

$$p = \frac{f_1 + f_2}{N_1 + N_2}$$

where f_1 and f_2 are the two frequencies.

The justification for combining data from the two samples to obtain a single estimate of p resides in the fact that in all cases where the difference between two proportions is tested, the null hypothesis is assumed. This hypothesis states that no difference exists in the population proportions. Because the null hypothesis is assumed, we may use an estimate of p based on the data combined for the two samples. This procedure is analogous to that used in the t test for the difference between means for independent samples where the sums of squares for the two samples are combined to obtain a single variance estimate.

To test the difference between two proportions we divide the observed difference between the proportions by the estimate of the standard error of the difference to obtain

[12.2] $$z = \frac{p_1 - p_2}{s_{p_1-p_2}} = \frac{p_1 - p_2}{\sqrt{pq[(1/N_1) + (1/N_2)]}}$$

The value z may be interpreted as a deviate of the unit normal curve, provided N_1 and N_2 are reasonably large and p is neither very small nor very large. As usual for a two-tailed test, values of 1.96 and 2.58 are required for significance at the 5 and 1 per cent levels.

How large should the N's be and how far should p depart from extreme values before this ratio can be interpreted as a deviate of the unit normal curve? An arbitrary rule may be used here. If the smaller value of p or q multiplied by the smaller value of N exceeds 5, then the ratio may be interpreted with reference to the normal curve. Thus if $p = .60$, $q = .40$, $N_1 = 20$, and $N_2 = 30$, the product $.40 \times 20 = 8$ and the normal curve may be used.

To illustrate, we refer to data obtained in a study of the attitudes of Canadians to immigrants and immigration policy. Independent samples of French- and English-speaking Canadians were used. Subjects were asked whether they agreed or disagreed with present government immigration practices. In the French-speaking sample of 300 subjects, 176 subjects indicated agreement. The proportion p_1 is $176/300 = .587$. In the English-speaking sample of 500 subjects, 384 indicated agreement. The proportion p_2 is $384/500 = .768$. By combining data for the two samples we obtain a value

$$p = \frac{176 + 384}{300 + 500} = .700$$

The value of q is $1 - .700 = .300$. The estimate of the standard error of the difference is $s_{p_1-p_2} = \sqrt{.700 \times .300 \left(\frac{1}{300} + \frac{1}{500}\right)} = .033$. The required z value is $z = (.768 - .587)/.033 = 5.48$.

Interpreting the value 5.48 as a unit-normal-curve deviate we observe immediately that the difference is highly significant. The chances are one in a great many millions that the observed difference could result from sampling. We may very safely conclude from these data that a real difference exists between French- and English-speaking Canadians on the question asked.

An alternative, but closely related, method exists for testing the significance of the difference between proportions for independent samples. This method uses chi square and is described in Chap. 13.

12.3 SIGNIFICANCE OF THE DIFFERENCE BETWEEN TWO CORRELATED PROPORTIONS

Frequently in psychological work we wish to test the significance of the difference between two proportions based on the same sample of individuals or on matched samples. The data consist of pairs of observations and are usually nominal in type. The paired observations may exhibit a correlation, which must be taken into consideration in testing the difference between proportions. To illustrate, a psychological test may be administered to a sample of N individuals. The proportions passing items 1 and 2 are p_1 and p_2. Paired observations are available for each individual. One individual may "pass" item 1 and also "pass" item 2. A second individual may "pass" item 1 and "fail" item 2. A third individual may "fail" both items. The paired observations may be tabulated in a 2×2 table. A tendency may exist for individuals who pass item 1 to also pass item 2 and for those who fail item 1 to also fail item 2. Thus the items are correlated. A further illustration arises where attitudes are measured with an attitude scale before and after a program designed to induce attitude change. On any particular attitude item a before response and an after response are available. Thus the data are comprised of a set of paired observations. To apply a test of significance to the difference between before and after proportions on any particular item requires that the correlation between responses be taken into account.

We proceed by tabulating the data in the form of a fourfold, or 2×2, table. A table with four cell frequencies is obtained. By way of illustration, assume that the data are "pass" and "fail" on two test items. The data may be represented schematically as follows:

		Frequencies Item 2					*Proportions* Item 2		
		Fail	*Pass*				*Fail*	*Pass*	
Item 1	**Pass**	A	B	$A + B$	*Item 1*	**Pass**	a	b	p_1
	Fail	C	D	$C + D$		**Fail**	c	d	q_1
		$A + C$	$B + D$	N			q_2	p_2	1.00

The capital letters represent frequencies. The small letters are proportions obtained by dividing the frequencies by N. The proportions passing the two items are p_1 and p_2. We wish to test the significance of the difference between p_1 and p_2.

An estimate of the standard error of the difference between two correlated proportions is given by the formula

[12.3] $\quad s_{p_1 - p_2} = \sqrt{\dfrac{a + d}{N}}$

This formula is due to McNemar (1947). It takes into account the correlation between the paired observations. A normal deviate z is obtained by dividing the difference between the two proportions by the standard error of the difference. Thus

[12.4] $\quad z = \dfrac{p_1 - p_2}{\sqrt{\dfrac{a + d}{N}}}$

When the sum of the two cell frequencies, $A + D$, is reasonably large, this ratio can be interpreted as a unit-normal-curve deviate, values of 1.96 and 2.58 being required for significance at the 1 and 5 per cent levels for a two-tailed test. In this context a reasonably large value of $A + D$ may be taken as about 20 or above.

It may be shown that the formula for the value of z given above reduces to

[12.5] $\quad z = \dfrac{D - A}{\sqrt{A + D}}$

where A and D are the cell frequencies. For computational purposes this is the more useful formulation.

To illustrate, consider the following fictitious data relating to attitude change. Let us assume an initial testing followed by a program intended to produce a change in attitude, and then a second testing with the same attitude scale. On a particular question let the data for the two testings be as follows:

		Frequencies 2d					*Proportions* 2d		
		Disagree	*Agree*				*Disagree*	*Agree*	
1st	*Agree*	10	50	60	*1st*	*Agree*	.05	.25	.30
	Disagree	110	30	140		*Disagree*	.55	.15	.70
		120	80	200			.60	.40	1.00

Inspection of the above tables indicates a high correlation in response between the first and second testings. We wish to test the significance of the difference between .40 and .30. The standard error of the difference between the two proportions is

$$s_{p_1-p_2} = \sqrt{\frac{.05 + .15}{200}} = .0316$$

The value of z is

$$z = \frac{.40 - .30}{.0316} = 3.16$$

In this case the difference is significant. It exceeds the value of 2.58 required for significance at the 1 per cent level for a two-tailed test. Arguments may be advanced for the use of a one-tailed test with the above data. It may be assumed that knowledge of a program intended to induce attitude change may warrant a hypothesis about the direction of the change. In either case the result is significant.

12.4 SIGNIFICANCE OF THE DIFFERENCE BETWEEN VARIANCES FOR INDEPENDENT SAMPLES

Occasions arise where a test of the significance of the difference between the variances of measurements for two independent samples is required. In the conduct of a simple experiment using control and experimental groups, the effect of the experimental condition may reflect itself not only in a mean difference between the two groups but also in a variance differ-

ence. For example, in an experiment
distracting agent, such as noise, on n
distraction may be to greatly increa'
addition possibly to exerting some
obtained in any experiment should
comparison. A common situation,
difference between variances is r(
significance of the difference be'
equality of variances in the popu
that is, it assumes that $\sigma_1^2 = \sigma_1$
as homogeneity of variance.

Let s_1^2 and s_2^2 be two variances ~~
may consider the difference $s_1^2 - s_2^2$. An alterna~
the ratio s_1^2/s_2^2 or s_2^2/s_1^2. If the two variances are equal, ~
unity. If they differ and $s_1^2 > s_2^2$, then $s_1^2/s_2^2 > 1$ and $s_2^2/s_1^2 < 1$. ~~
ture of the variance ratio from unity is indicative of a difference between
variances, the greater the departure the greater the difference. Quite
clearly, a test of the significance of the departure of the ratio of two variances from unity will serve as a test of the significance of the difference
between the two variances.

To apply such a test the sampling distribution of the ratio of two variances is required. To conceptualize such a sampling distribution, consider
two normal populations A and B with the same variance σ^2. Draw samples
of N_1 cases from A and N_2 cases from B, calculate unbiased variance
estimates s_1^2 and s_2^2, and compute the ratio s_1^2/s_2^2. Continue this procedure
until a large number of variance ratios is obtained. Always place the
variance of the sample drawn from A in the numerator and the variance
of the sample drawn from B in the denominator. Some of the variance ratios
will be greater than unity; others will be less than unity. The frequency
distribution of the variance ratios for a large number of pairs of variances
is an experimental sampling distribution. The corresponding theoretical
sampling distribution of variance ratios is known as the distribution of F.
The variance ratio is known as an F ratio; that is, $F = s_1^2/s_2^2$, or $F = s_2^2/s_1^2$.

In the above illustration samples of N_1 are drawn from one population
and samples of N_2 from another. $N_1 - 1$ and $N_2 - 1$ degrees of freedom are
associated with the two variance estimates. A separate sampling distribution of F exists for every combination of degrees of freedom. Table D of the
Appendix shows values of F required for significance at the 5 and 1 per
cent levels for varying combinations of degrees of freedom. This table
shows values of F equal to or greater than unity. It does not show values of
F less than unity. The number of degrees of freedom associated with the
variance estimates in the numerator and denominator are shown along the
top and to the left, respectively, of Table D. The numbers in lightface type
are the values for significance at the 5 per cent level, and those in boldface
type the values at the 1 per cent level. These values cut off 5 and 1 per cent
of one tail of the distribution of F.

In testing the significance of the difference between two variances, the

$H_0: \sigma_1^2 = \sigma_2^2 = \sigma^2$ is assumed. We then find the ratio of the variance estimates. These are

$$\frac{(X - \bar{X}_1)^2}{N_1 - 1}$$

$$s_2^2 = \frac{\Sigma(X - \bar{X}_2)^2}{N_2 - 1}$$

No prior grounds exist for deciding which variance estimate should be placed in the numerator and which in the denominator of the F ratio. In practice the larger of the two variance estimates is always placed in the numerator and the smaller in the denominator. In consequence the F ratio in this situation is always greater than unity. The F ratio is calculated, referred to Table D of the Appendix, and a significance level determined. At this point a slight complication arises. The obtained significance level must be doubled. Table D shows values required for significance at the 5 and 1 per cent levels. In comparing the variances for two independent groups these become the 10 and 2 per cent levels. The reason for this complication resides in the fact that the larger of the two variances has been placed in the numerator of the F ratio. This means that we have considered one tail only of the F distribution. Not only must we consider the probability of obtaining s_1^2/s_2^2 but also the probability of s_2^2/s_1^2. Where interest is in the significance of the difference, regardless of direction, the required per cent or probability levels are simply obtained by doubling those shown in Table D.

Table D has been prepared for use with the analysis of variance (Chap. 15) which makes extensive use of the F ratio. In the analysis of variance the decision as to which variance estimate should be put in the numerator and which in the denominator is made on grounds other than their relative size. Consequently, in the analysis of variance, F ratios less than unity can occur and Table D provides the appropriate probabilities without any doubling procedure.

To illustrate, a psychological test is administered to a sample of 31 boys and 26 girls. The sum of squares of deviations, $\Sigma(X - \bar{X})^2$, is 1,926 for boys and 2,875 for girls. Unbiased variance estimates are obtained by dividing the sum of squares by the number of degrees of freedom. The df for boys is $31 - 1 = 30$ and for girls $26 - 1 = 25$. The variance estimate for boys is $1{,}926/30 = 64.20$ and for girls $2{,}875/25 = 115.00$.

Are boys significantly different from girls in the variability of their performance on this test? The F ratio is $115.00/64.20 = 1.79$. The df for the numerator is 25 and for the denominator 30. Referring this F to Table D we see that a value of F of about 1.88 is required for significance at the 5 per cent level, and doubling this we obtain the 10 per cent level. It is clear, therefore, that the difference between the variances for boys and girls cannot be considered statistically significant. The evidence is insufficient to warrant rejection of the null hypothesis.

12.5 SIGNIFICANCE OF THE DIFFERENCE BETWEEN CORRELATED VARIANCES

Given a set of paired observations, the two variances are not independent estimates. Data of this kind arise when the same subjects are tested under two experimental conditions, or matched samples are used. For example, in an experiment designed to study the effects of an educational program on attitude change, attitudes may be measured, an educational program applied, and attitudes remeasured. It may be hypothesized that some change in variance of attitude-test scores may result. An increase in variance may mean that the effect of the program is to reinforce existing attitudes, producing more extreme attitudes among individuals at both ends of the attitude continuum. A decrease in variance may mean that the effect of the program is to produce an attitudinal regression to greater uniformity.

If s_1^2 and s_2^2 are the two unbiased variance estimates and r_{12} is the correlation between the paired observations, the quantity

[12.6] $$t = \frac{(s_1^2 - s_2^2)\ \sqrt{N-2}}{\sqrt{4s_1^2 s_2^2 (1 - r_{12}^2)}}$$

has a t distribution with $N - 2$ degrees of freedom.

By way of illustration let s_1^2 and s_2^2 be unbiased variance estimates of attitude scale scores before and after the administration of an educational program. Let $s_1^2 = 153.20$ and $s_2^2 = 102.51$ where $N = 38$. The correlation between the before-and-after attitude measures is .60. Are the two variances significantly different from each other? We obtain

$$t = \frac{(153.20 - 102.51)\ \sqrt{38-2}}{\sqrt{4 \times 153.20 \times 102.51(1 - .36)}} = 1.52$$

The number of degrees of freedom is $38 - 2 = 36$. For significance at the 5 per cent level, a value of t equal to or greater than about 2.03 is required. The evidence is insufficient to warrant rejection of the null hypothesis. We cannot argue that the intervening educational program has changed the variability of attitudes.

12.6 SAMPLING DISTRIBUTION OF THE CORRELATION COEFFICIENT

We may draw a large number of samples from a population, compute a correlation coefficient for each sample, and prepare a frequency distribution of correlation coefficients. Such a frequency distribution is an experimental sampling distribution of the correlation coefficient. To illustrate, casual observation suggests that a positive correlation exists between height and weight. A number of samples of 25 cases may be drawn at random from a population of adult males, and a correlation coefficient between height and weight computed for each sample. These coefficients will display variation one from another. By arranging them in the form of a frequency distribution an experimental sampling distribution of the correlation coefficient is obtained. The mean of this distribution will tend to approach the population value of the correlation coefficient with increase in the number of samples. Its standard deviation will describe the variability of the coefficients from sample to sample. A further illustration may prove helpful. By throwing a pair of dice a number of times, say, a white one and a red one, a set of paired observations is obtained. A correlation coefficient may be calculated for the paired observations. Since the two dice are independent, the expected value of this correlation coefficient is zero. However, for any particular sample of N throws, a positive or a negative correlation may result. A large number of samples of N throws may be obtained, a correlation coefficient computed for each sample, and a frequency distribution of the coefficients prepared. The mean of this experimental sampling distribution will tend to approach zero, the population value of the correlation coefficient, and its standard deviation will be descriptive of the variability of the correlation in drawing samples of size N from this particular kind of population. Note that here, as in all sampling problems, a distinction is drawn between a population value and an estimate of that value based on a sample. The symbol ρ is used to refer to the population value of the correlation coefficient, and r is the sample value.

The shape of the sampling distribution of the correlation coefficient depends on the population value ρ. As ρ departs from zero, the sampling distribution becomes increasingly skewed. When ρ is high positive, say, $\rho = .80$, the sampling distribution has extreme negative skewness. Similarly, when ρ is high negative, say, $\rho = -.80$, the distribution has extreme positive skewness. When $\rho = 0$, the sampling distribution is symmetrical and for large values of N, say, 30 or above, is approximately normal. The reason for the increase in skewness in the sampling distribution as ρ departs from zero is intuitively plausible. In sampling, for example, from a population where $\rho = .90$, values greater than 1.00 cannot occur, whereas values extending from .90 to -1.00 are theoretically possible. The range of possible variation below .90 is far greater than the range above .90. This suggests that the sample values may exhibit greater variability below than above .90, a circumstance which leads to negative skewness.

The standard deviation of the theoretical sampling distribution of ρ, the standard error, is given by the formula

$$[12.7] \quad \sigma_r = \frac{1 - \rho^2}{\sqrt{N - 1}}$$

When ρ departs appreciably from zero, this formula is of little use, because the departures of the sampling distribution from normality make interpretation difficult.

Difficulties resulting from the skewness of the sampling distribution of the correlation coefficient are resolved by a method developed by R. A. Fisher. Values of r are converted to values of z_r using the transformation

$$[12.8] \quad z_r = \tfrac{1}{2} \log_e (1 + r) - \tfrac{1}{2} \log_e (1 - r)$$

Values of z_r corresponding to particular values of r need not be computed directly from the above formula, but may be simply obtained from Table E in the Appendix. For $r = .50$ the corresponding $z_r = .549$, for $r = .90$ $z_r = 1.472$, and so on. For negative values of r the corresponding z_r values may be given a negative sign. In a number of sampling problems involving correlation, r's are converted to z_r's, and a test of significance is applied to the z_r's instead of to the original r's.

One advantage of this transformation resides in the fact that the sampling distribution of z_r is for all practical purposes independent of ρ. The distribution has the same variability for a given N regardless of the size of ρ. Another advantage is that the sampling distribution of z_r is approximately normal. Values of z_r can be interpreted in relation to the normal curve. The standard error of z_r is given by

[12.9] $$s_{z_r} = \frac{1}{\sqrt{N - 3}}$$

The standard error is seen to depend entirely on the sample size.

The z_r transformation may be used to obtain confidence limits for r. Let $r = .82$ for $N = 147$. The corresponding $z_r = 1.157$. The standard error of z_r, given by $1/\sqrt{N - 3}$, is .083. The 95 per cent confidence limits are obtained by taking 1.96 times the standard error above and below the observed value of z_r, or $z_r \pm 1.96 s_{z_r}$. These are $1.157 + 1.96 \times .083 = 1.320$ and $1.157 - 1.96 \times .083 = .994$. These two z_r's may now be converted back to r's, where $z_r = 1.320$, $r = .867$ and where $z_r = .994$, $r = .759$. Thus we may assert with 95 per cent confidence that the population value of the correlation coefficient falls within the limits .759 and .867. In practice we are infrequently concerned with fixing confidence intervals for correlation coefficients. The most frequently occurring problems are testing the significance of a correlation coefficient from zero and testing the significance of the difference between two correlation coefficients.

12.7 SIGNIFICANCE OF A CORRELATION COEFFICIENT

Testing the significance of the correlation between a set of paired observations is a frequent problem in psychological research. We begin by assuming the null hypothesis that the value of the correlation coefficient is equal to zero, or $H_0{:}\rho = 0$. A test of significance may then be applied using the distribution of t. The t value required is given by the formula

[12.10] $$t = r \sqrt{\frac{N - 2}{1 - r^2}}$$

The number of degrees of freedom associated with this value of t is $N - 2$. The loss of 2 degrees of freedom results because testing the significance of r from zero is equivalent to testing the significance of the slope of a regression line from zero. The reader will recall that the correlation coefficient is the slope of a regression line in standard-score form. The number of degrees of freedom associated with the variability about a straight line fitted to a set of points is two less than the number of observations. A straight line will always fit two points exactly, and no freedom to vary is possible. With three points there is 1 degree of freedom, with four points 2 degrees of freedom, and so on.

Consider an example where $r = .50$ and $N = 20$. We obtain

$$t = .50 \sqrt{\frac{20 - 2}{1 - .50^2}} = 2.45$$

The $df = 20 - 2 = 18$. Referring to the table of t, Table B in the Appendix, we find that for this df a t of 2.10 is required for significance at the 5 per cent level and a t of 2.88 at the 1 per cent level. The sample value of r falls between these two values. It may be said to be significant at the 5 per cent level.

Table F of the Appendix presents a tabulation of the values of r required for significance at different levels. We note that where the number of degrees of freedom is small, a large value of r is required for significance. For example, where $df = 5$, a value of $r > .754$ is required before we can argue at the 5 per cent level that the r is significant. Even for $df = 20$, a value of $r > .423$ is required for significance at the 5 per cent level. This means that little importance can be attached to correlation coefficients calculated on small samples unless these coefficients are fairly substantial in size.

12.8 SIGNIFICANCE OF THE DIFFERENCE BETWEEN TWO CORRELATION COEFFICIENTS FOR INDEPENDENT SAMPLES

Consider a situation where two correlation coefficients, r_1 and r_2, are obtained on two independent samples. The correlation coefficients may, for example, be correlations between intelligence-test scores and mathematics-examination marks for two different freshman classes. We wish to test whether r_1 is significantly different from r_2, that is, whether the two samples can be considered random samples from a common population. The null hypothesis is $H_0: \rho_1 = \rho_2$ or $H_0: \rho_1 - \rho_2 = 0$.

The significance of the difference between r_1 and r_2 can be readily tested using Fisher's z_r transformation. Convert r_1 and r_2 to z_r's using Table E of the Appendix. As stated previously, the sampling distribution of z_r is approximately normal with a standard error given by $s_{z_r} = 1/\sqrt{N - 3}$. The standard error of the difference between two values of z_r is given by

$$[12.11] \quad s_{z_{r1} - z_{r2}} = \sqrt{s_{z_{r1}}^2 + s_{z_{r2}}^2} = \sqrt{\frac{1}{N_1 - 3} + \frac{1}{N_2 - 3}}$$

By dividing the difference between the two values of z_r by the standard error of the difference, we obtain the ratio

$$[12.12] \quad z = \frac{z_{r1} - z_{r2}}{\sqrt{1/(N_1 - 3) + 1/(N_2 - 3)}}$$

This is a unit-normal-curve deviate and may be so interpreted. Values of 1.96 and 2.58 are required for significance at the 1 per cent and 5 per cent levels.

To illustrate, let the correlations between intelligence scores and mathematics-examination marks for two freshman classes be .320 and .720. Let the number of students in the first class be 53 and in the second 23. Are the two coefficients significantly different? The corresponding z_r values obtained from Table E of the Appendix are .332 and .908. The required normal deviate is

$$z = \frac{.908 - .332}{\sqrt{1/(53 - 3) + 1/(23 - 3)}} = 2.18$$

The difference between the two correlations is significant at the 5 per cent level.

The application of a test of significance in a situation of this kind is simple. The interpretation of what the difference in correlation means may be difficult.

12.9 SIGNIFICANCE OF THE DIFFERENCE BETWEEN TWO CORRELATION COEFFICIENTS FOR CORRELATED SAMPLES

Consider a situation where three measurements have been made on the same sample of individuals. Three correlation coefficients result, r_{12}, r_{13}, and r_{23}. If we wish to compare r_{12} and r_{13}, or r_{12} and r_{23}, or r_{13} and r_{23}, the method described in the preceding section does not apply. Here the two coefficients under comparison are not based on independent samples but are based on the same sample and are correlated.

To test the difference between r_{12} and r_{13} under these conditions, we may calculate a value t by the formula

$$[12.13] \quad t = \frac{(r_{12} - r_{13}) \sqrt{(N - 3)(1 + r_{23})}}{\sqrt{2(1 - r_{12}{}^2 - r_{13}{}^2 - r_{23}{}^2 + 2r_{12}r_{13}r_{23})}}$$

This expression follows the distribution of t with $N - 3$ degrees of freedom. Note that to apply this test the correlation r_{23} is required.

Let X_2 and X_3 be two psychological tests used to predict a criterion measure of scholastic success X_1. The three correlation coefficients based on a sample of 100 cases are $r_{12} = .60$, $r_{13} = .50$, and $r_{23} = .50$. Are X_2 and X_3 significantly different as predictors of scholastic success? Is there a reasonable probability that the difference between the two correlations r_{12} and r_{13} can be explained in terms of sampling error? The value of t is

$$t = \frac{(.60 - .50) \sqrt{(100 - 3)(1 + .50)}}{\sqrt{2(1 - .60^2 - .50^2 - .50^2 + 2 \times .60 \times .50 \times .50)}} = 1.29$$

For $df = .97$, a t of about 1.99 is required for significance at the 5 per cent level. In consequence, the difference between the two correlation coefficients cannot be said to be significant.

The above test has certain restrictive assumptions underlying its development and because of these is perhaps not entirely satisfactory.

EXERCISES

1 Given two random samples of size 100 with sample values $p_1 = .80$ and $p_2 = .60$, test the significance of the difference between p_1 and p_2.

2 Consider two test items A and B. In a sample of 100 people, 30 pass item A and fail item B, whereas 20 fail item A and pass item B. Are the proportions passing the two items significantly different from each other?

3 In a market survey 24 out of 96 males and 63 out of 180 females indicate a preference for a particular brand of cigarettes. Do the data warrant the conclusion that a sex difference exists in brand preference?

4 On an attitude scale, 63 and 39 individuals from a sample of 140 indicate agreement to items A and B, respectively, and 29 individuals indicate agreement to both items. Is there a significant difference in the response elicited by the two items?

5 Given two independent samples of size 20 with $s_1^2 = 400$ and $s_2^2 = 625$, test the hypothesis that the variances are significantly different from each other.

6 Given two correlated samples of size 20 with $s_1^2 = 400$, $s_2^2 = 625$, and $r_{12} = .7071$, test the hypothesis that the variances are significantly different from each other.

7 Calculate values of z_r for $r = .70$, $r = .05$, $r = -.60$, and $r = -.99$.

8 Calculate, using formula [12.10], values of t for the following values of r and N:

r	.20	.30	.40	.50
N	50	40	30	20

9 The correlation between psychological-test scores and academic achievement for a sample of 147 freshmen is .40. The corresponding correlation for a sample of 125 sophomores is .59. Do these correlations differ significantly?

10 Three psychological tests are administered to a sample of 50 students. The correlations are $r_{12} = .70$, $r_{13} = .40$, and $r_{23} = .60$. Is r_{12} significantly different from r_{13}?

THE ANALYSIS OF FREQUENCIES
USING CHI SQUARE

13.1 INTRODUCTION

We have previously discussed the application of the binomial, normal, t, and F distributions. Another distribution of considerable theoretical and practical importance is the distribution of chi square, or χ^2. In many experimental situations we wish to compare *observed* with *theoretical* frequencies. The observed frequencies are those obtained empirically by direct observation or experiment. The theoretical frequencies are generated on the basis of some hypothesis, or line of theoretical speculation, which is independent of the data at hand. The question arises as to whether the differences between the observed and theoretical frequencies are significant. If they are, this constitutes evidence for the rejection of the hypothesis or theory that gave rise to the theoretical frequencies.

Consider, for example, a die. We may formulate the hypothesis that the die is unbiased, in which case the probability of throwing any of the six possible values in a single toss is $\frac{1}{6}$. The frequencies expected on the basis of this hypothesis are the theoretical frequencies. In a series of 300 throws the expected or theoretical frequencies of 1, 2, 3, 4, 5, and 6 are 50, 50, 50, 50, 50, and 50. Let us now experiment by throwing the die 300 times. The observed frequencies of the values from 1 to 6 are 43, 55, 39, 56, 63, and 44. May the differences between the observed and theoretical frequencies be considered to result from sampling error? Are the differences highly improbable on the basis of the null hypothesis, thereby providing evidence for the rejection of the hypothesis that the die is unbiased?

As a further illustration, let us formulate the hypothesis that in litters of rabbits the probability of any birth being either male or female is $\frac{1}{2}$. Using the binomial distribution we ascertain that the expected or theoretical frequencies of 0, 1, 2, 3, 4, 5, and 6 males in 64 litters of 6 rabbits are 1, 6, 15, 20, 15, 6, and 1. By counting the number of males in 64 litters of six rabbits, the corresponding observed frequencies are 0, 3, 14, 19, 20, 6, and 2. Do the observed and theoretical frequencies differ?

Consider another example. In a market research project two varieties of soap, *A* and *B*, are distributed to a random sample of 200 housewives. After a period of use the housewives are asked which they prefer. The results indicate that 115 prefer *A* and 85 prefer *B*. The hypothesis may be formulated that no difference exists in consumer preference for the two varieties of soap; that a 50:50 split exists. Do the observed frequencies constitute evidence for the rejection of this hypothesis?

The statistic χ^2 is used in situations of the type described above where a comparison of observed and theoretical frequencies is required. It has extensive application in statistical work. For our purposes here χ^2 is defined by

[13.1] $$\chi^2 = \sum \frac{(O - E)^2}{E}$$

where O = an observed frequency
E = an expected or theoretical frequency

Thus to calculate a value of χ^2 we find the differences between the observed and expected values, square these, divide each difference by the appropriate expected value, and sum over all frequencies.

Table 13.1 illustrates the calculation of χ^2 in comparing the observed and expected frequencies for 300 throws of a die. Note that the sum of both the observed and expected frequencies is equal to N; that is, $\Sigma O = \Sigma E = N$. The value of χ^2 obtained in Table 13.1 is 8.72. This is a measure of the dis-

Table 13.1 Calculation of χ^2 in comparing observed and expected frequencies for 300 throws of a die

Value of die	Observed frequency	Expected frequency			$\dfrac{(O-E)^2}{E}$
X	O	E	$O - E$	$(O - E)^2$	E
1	43	50	−7	49	.98
2	55	50	5	25	.50
3	39	50	−11	121	2.42
4	56	50	6	36	.72
5	63	50	13	169	3.38
6	44	50	−6	36	.72
Total	300	300	0	. . .	$\chi^2 = 8.72$

crepancy between the observed and theoretical frequencies. If the discrepancy is large, χ^2 is large. If the discrepancy is small, χ^2 is small. Does a value of $\chi^2 = 8.72$ consitute evidence at an accepted level of significance for rejecting the null hypothesis? The answer to this question demands a consideration of the sampling distribution of χ^2.

13.2 THE SAMPLING DISTRIBUTION OF CHI SQUARE

The sampling distribution of χ^2 may be illustrated with reference to the tossing of coins. Let us assume that in tossing 100 *unbiased* coins 46 heads and 54 tails result. The expected frequencies are 50 heads and 50 tails. A value of χ^2 may be calculated as follows:

	O	E	$O - E$	$(O - E)^2$	$\dfrac{(O - E)^2}{E}$
H	46	50	-4	16	.32
T	54	50	$+4$	16	.32
					$\chi^2 = .64$

In the tossing of 100 coins two frequencies are obtained, one for heads and one for tails. These frequencies are not independent. If the frequency of heads is 46, the frequency of tails is $100 - 46 = 54$. If the frequency of heads is 62, the frequency of tails is $100 - 62 = 38$. Quite clearly, given either frequency, the other is determined. One frequency only is free to vary. In this situation 1 degree of freedom is associated with the value of χ^2.

Let us toss the 100 coins a second time, a third time, and so on, to obtain different values of χ^2. A large number of trials may be made, and a large number of values of χ^2 obtained. The frequency distribution of these values is an experimental sampling distribution of χ^2 for 1 degree of freedom. It describes the variation in χ^2 with repeated sampling. By inspecting this experimental sampling distribution estimates may be made of the proportion of times, or the probability, that values of χ^2 equal to or greater than any given value will occur due to sampling fluctuation for 1 degree of freedom. In the present illustration this assumes, of course, that the coins are unbiased.

Instead of tossing 100 coins, let us throw 100 unbiased dice, obtain observed and expected frequencies, and calculate a value of χ^2. In this situation if any five frequencies are known, the sixth is determined. Five degrees of freedom are associated with the value of χ^2 obtained. The 100 dice may be tossed a great many times, a value of χ^2 calculated for each trial, and a frequency distribution made. This frequency distribution is an experimental sampling distribution of χ^2 for 5 degrees of freedom.

The theoretical sampling distribution of χ^2 is known, and probabilities may be estimated from it without using the elaborate experimental

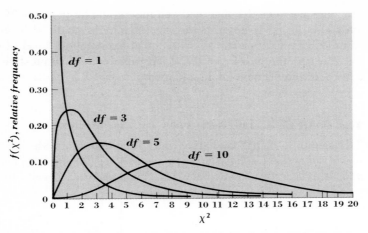

Fig. 13.1   Chi-square distribution and 5 per cent critical regions for various degrees of freedom. (*From Francis G. Cornell, The essentials of educational statistics, John Wiley & Sons, Inc., New York, 1956.*)

approach described for illustrative purposes above. The equation for χ^2 is complex and is not given here. It contains the number of degrees of freedom as a variable. This means that a different sampling distribution of χ^2 exists for each value of df. Figure 13.1 shows different chi-square distributions for different values of df. χ^2 is always positive, a circumstance which results from squaring the difference between the observed and expected values. Values of χ^2 range from zero to infinity. The right-hand tail of the curve is asymptotic to the abscissa. For 1 degree of freedom the curve is asymptotic to the ordinate as well as to the abscissa.

The χ^2 distribution is used in tests of significance in much the same way that the normal, t, or the F distributions are used. The null hypothesis is assumed. This hypothesis states that no actual differences exist between the observed and expected frequencies. A value of χ^2 is calculated. If this value is equal to or greater than the critical value required for significance at an accepted significance level for the appropriate df, the null hypothesis is rejected. We may state that the differences between the observed and expected frequencies are significant and cannot reasonably be explained by sampling fluctuation. Table C in the Appendix shows values of χ^2 required for significance at various probability levels for different values of df. The critical values at the 5 and 1 per cent levels for $df = 1$ are, respectively, 3.84 and 6.64. This means that 5 and 1 per cent of the area of the curve fall to the right of ordinates erected at distances 3.84 and 6.64 measured along the base line from a zero origin. For $df = 5$, the corresponding 5 and 1 per cent critical values are 11.07 and 15.09.

Table C of the Appendix provides the 5 and 1 per cent critical values for $df = 1$ to $df = 30$. This covers the great majority of situations ordinarily encountered in practice. Situations where a χ^2 is calculated based on a df greater than 30 are infrequent. Where df is greater than 30 the expression $\sqrt{2\chi^2} - \sqrt{2df - 1}$ has a sampling distribution which is approximately

Proportion of tails	O	E	O − E	$\frac{(O-E)^2}{E}$
1.0	2 ⎫ 7	0.5 ⎫ 5.5	1.5	0.409
0.9	5 ⎭	5.0 ⎭		
	15	22.5	−7.5	2.500
	68	60.0	8.0	1.067
		105.0	0.0	0.000
	< 0	8.0	8.0	0.508
			−10.0	0.952
			−5.0	0.417
				0.011
		5		3.682
			$\chi^2 = 9.546$	

n research, Prentice-Hall,

normal. Values of this expression required for significance at the 5 and 1 per cent levels are 1.64 and 2.33.

The above discussion of χ^2 has undoubtedly given the reader the impression that χ^2 is a statistic which is exclusively concerned with the comparison of observed and expected frequencies. The distribution of χ^2 is, in fact, a general statistical distribution and its use in the study of frequencies is one approximate, particular application. The more general approach defines χ^2 as the sum of squared normal deviates. Consider a population with mean μ, variance σ^2, and a normal distribution of scores Y. A squared standard score drawn from this population is $z^2 = (Y - \mu)^2/\sigma^2$. If members are drawn from their population one at a time, the frequency distribution of z^2 will be a χ^2 distribution with 1 degree of freedom. If members are drawn two at a time, the distribution of $z_1^2 + z_2^2$ will be a χ^2 distribution with 2 degrees of freedom. In general for samples of size N the quantity Σz_i^2 has a χ^2 distribution with N degrees of freedom.

13.3 GOODNESS OF FIT

Numerous examples may be found to illustrate the goodness of fit of a theoretical to an observed frequency distribution. In one experiment Abbé Mendel observed the shape and color of peas in a sample of plants. The distribution he obtained is shown in Table 13.2. According to his genetic theory the expected frequencies should follow the ratio $9:3:3:1$. The correspondence between observed and expected frequencies is close. The value of $\chi^2 = .470$, and no grounds exist for rejecting the null hypothesis. The data lend confirmation to the theory. The value of χ^2 is smaller than we should ordinarily expect, the probability associated with it being between .90 and .95. Assuming the null hypothesis, a fit as good or better than the one observed may be expected to occur in between 5 and 10 per cent of samples of the same size.

In testing goodness of fit the hypothesis may be entertained that the distribution of a variable conforms to some widely known distribution such as the binomial or normal distribution. Johnson (1949), in order to illustrate the goodness of fit of the theoretical binomial distribution to an observed

Table 13.2 Comparison of observed and expected frequencies in shape and color of peas in experiment by Mendel

	O	E	$O - E$	$(O - E)^2$	$\dfrac{(O - E)^2}{E}$
Round yellow	315	312.75	2.25	5.06	.016
Round green	108	104.25	3.75	14.06	.135
Angular yellow	101	104.25	−3.25	10.56	.101
Angular green	32	34.75	−2.75	7.56	.218
Total	556	556	0.00	. . . .	$\chi^2 = .470$

distribution, tossed 10 coins 512 times and recorded the proportion of tails. His data are shown in Table 13.3, together with the corresponding theoretical binomial frequencies. The mean and standard deviation of the observed distribution are $\bar{X} = 0.5$ and $s = .162$. The mean and standard deviation of the theoretical binomial are $\bar{X} = 0.5$ and $s = .156$.

Note that in the calculation of χ^2 for these data, the small frequencies at the tails of the distributions are combined, a procedure that is generally advisable with data of this type. Problems in the application of chi square resulting from the presence of small frequencies are discussed later in this chapter. With the present data, combining small frequencies reduces the number of frequencies from 11 to 9 and the number of degrees of freedom from 10 to 8. The value of χ^2 for these data is 9.55. The value required for significance at the 5 per cent level for 8 degrees of freedom is 15.51. The conclusion is that the evidence is insufficient to justify rejection of the null hypothesis. Reference to a table of χ^2 shows that a value of χ^2 equal to or greater than the one observed might be expected to occur in about 30 per cent of samples due to sampling fluctuation alone.

Where the theoretical frequency distribution is continuous, we require a method for the estimation of the theoretical frequencies. In fitting a continuous curve we calculate the proportion of the area under the theoretical curve corresponding to each class interval. This proportion multiplied by N is taken as the theoretical frequency within the class interval. This procedure is illustrated in Table 13.4. The data are adapted from McNemar

Table 13.3   **Goodness of fit of binomial distribution to observed distribution of proportion of tails from 512 tosses of 10 coins***

Proportion of tails	O	E	O − E	$\frac{(O - E)^2}{E}$
1.0	2 ⎫ 7	0.5 ⎫ 5.5	1.5	0.409
0.9	5 ⎭	5.0 ⎭		
0.8	15	22.5	−7.5	2.500
0.7	68	60.0	8.0	1.067
0.6	105	105.0	0.0	0.000
0.5	134	126.0	8.0	0.508
0.4	95	105.0	−10.0	0.952
0.3	55	60.0	−5.0	0.417
0.2	23	22.5	0.5	0.011
0.1	8 ⎫ 10	5.0 ⎫ 5.5	4.5	3.682
0.0	2 ⎭	0.5 ⎭		
Total	512	512		$\chi^2 = 9.546$

* Adapted from Palmer Johnson, *Statistical methods in research*, Prentice-Hall, Inc., Englewood Cliffs, N.J., 1949.

Table 13.4 Calculation of normal distribution frequencies for Stanford-Binet IQ's, form M*

	1	2	3	4	5	6	7	8
	Class interval	O	Upper limit	Deviation from mean x	x/s	Proportion below	Proportion within	Expected frequency
	160–	3 ⎱ 16						
	150–159	13 ⎰					.0041	12
	140–149	55	149.5	44.94	2.645	.9959	.0158	47
	130–139	120	139.5	34.94	2.057	.9801	.0512	152
	120–129	330	129.5	24.94	1.468	.9289	.1186	352
	110–119	610	119.5	14.94	.879	.8103	.1958	582
	100–109	719	109.5	4.94	.291	.6145	.2316	688
	90–99	592	99.5	−5.06	−.298	.3829	.1950	579
	80–89	338	89.5	−15.06	−.886	.1879	.1177	350
	70–79	130	79.5	−25.06	−1.475	.0702	.0506	150
	60–69	48	69.5	−35.06	−2.064	.0196	.0155	46
	50–59	7 ⎱	59.5	−45.06	−2.652	.0041	.0041	12
	40–49	4 ⎰ 12						
	30–39	1 ⎰						
	Total	2,970	. . . .	. . . .	. . . .	. . . .	1.0000	2,970

* Adapted from Quinn McNemar, *Psychological statistics*, 4th ed., John Wiley & Sons, Inc., New York, 1969.

(1969) and are Stanford-Binet IQ's, Form M, for a sample of 2,970 individuals.

We are required to calculate the theoretical normal frequencies for the class intervals and test the goodness of fit between the theoretical and the observed. The mean and standard deviation are $\overline{X} = 104.56$ and $s = 16.99$. A normal frequency distribution is required with the same N, $\overline{X}$, and s as the observed distribution. We proceed by combining the small frequencies at the tails of the distribution, as shown in col. 2. This reduces the number of frequencies from 14 to 11. The frequency of 16 at the top of the distribution contains all cases above the exact limit 149.5. The frequency of 12 at the bottom of the distribution contains all cases below the exact limit 59.5. We next record the exact upper limits of the class intervals (col. 3), convert these to deviations from the mean of 104.56 (col. 4), and divide by the standard deviation 16.99 to obtain the standard score x/s (col. 5). Thus the exact upper limits of the class intervals are expressed in standard measure. For example, the exact upper limit of the interval 140 to 149 is 149.5. This as a deviation from the mean is $149.5 - 104.56 = 44.94$. Dividing this by the standard deviation 16.99, we obtain $44.94/16.99 = 2.645$. We then consult a table of areas under the normal curve and ascertain the proportions of the area under the normal curve falling below the standard-score values x/s of col. 5. These proportions are shown in col. 6. We observe that a proportion .9959 of the area of the normal curve falls below 2.645 standard deviation units above the mean, a proportion .9801 falls below 2.057 units above the mean, and so on. By subtraction we obtain the proportions of the area of the normal curve falling within the class intervals (col. 7). The proportion above the exact limit 149.5 is $1.0000 - .9959 = .0041$. The proportion between 139.5 and 149.5 is $.9959 - .9801 = .0158$. The proportion between 129.5 and 139.5 is $.9801 - .9289 = .0512$, and so on. By multiplying these proportions by N we obtain the expected frequencies (col. 8).

The above method simply involves converting the exact limits of the class intervals to standard deviation units, using a table of areas under the normal curve to find the proportion of the area within these limits and multiplying these proportions by N to obtain the expected frequencies.

Table 13.5 shows the calculation of χ^2 in comparing the observed with expected frequencies. A value of $\chi^2 = 17.02$ is obtained. In this case the number of df is $11 - 3 = 8$. Although there are 11 frequencies, 8 only are free to vary. The loss of 3 degrees of freedom results because the observed and expected distributions are made to agree on N, $\overline{X}$, and s. For $df = 8$, the value of χ^2 required for significance at the 5 per cent level is 15.51 and at the 1 per cent level 20.09. The obtained χ^2 falls between these two values at about the 3 per cent level. Thus the chances are 3 in 100 that a fit as good or worse than the one observed would result in random sampling from a normal population. This establishes grounds for rejecting the hypothesis that the distribution of Stanford-Binet IQ's, Form M, is normal. The departures from normality are, however, not gross.

Chi square may be used to test the representativeness of a sample where certain population values are known. This in effect is a test of goodness of

Table 13.5 Goodness of fit of normal frequencies to frequency distribution of Stanford-Binet IQ's, form M*

Class interval	O	E	$O - E$	$\dfrac{(O - E)^2}{E}$
160–	3 ⎫ 16			
150–159	13 ⎭	12	4	1.33
140–149	55	47	8	1.36
130–139	120	152	−32	6.74
120–129	330	352	−22	1.38
110–119	610	582	28	1.35
100–109	719	688	31	1.40
90–99	592	579	13	.29
80–89	338	350	−12	.41
70–79	130	150	−20	2.67
60–69	48	46	2	.09
50–59	7 ⎫	12	0	.00
40–49	4 ⎬ 12			
30–39	1 ⎭			
Total	2,970	2,970	0	$\chi^2 = 17.02$

* Adapted from Quinn McNemar, *Psychological statistics*, 4th ed., John Wiley & Sons, Inc., New York, 1969.

fit. To illustrate, in a study of attitudes toward immigrants, a sample of 200 cases is drawn from the city of Montreal. The observed frequencies and percentages by racial origin are shown in Table 13.6.

The population percentages obtained from census returns are also shown. These population percentages are used to obtain the expected, or

Table 13.6 Application of χ^2 in comparing sample frequencies of racial origin with population frequencies

Racial origin	O	Sample, per cent	Population, per cent	E	$O - E$	$\dfrac{(O - E)^2}{E}$
French	95	47.5	62.5	125	−30	7.20
English	67	33.5	19.4	39	28	20.10
Other	38	19.0	18.1	36	2	.11
Total	200	100.0	100.0	200	0	$\chi^2 = 27.41$

theoretical, frequencies. The value of χ^2 is 27.41. For $df = 2$ this is highly significant, the value required for significance at the 1 per cent level being 9.21. We may conclude that the sample is biased and cannot be considered a random sample with respect to racial origin. Since attitudes toward immigrants may be linked to racial origin, results obtained on this sample may be highly questionable unless a correction is applied to adjust for the sample bias.

13.4 TESTS OF INDEPENDENCE

A frequent application of chi square occurs where the data are comprised of paired observations on two nominal variables. We wish to know whether the variables are independent of each other or associated. To illustrate, Table 13.7 presents data collected by Woo (1928) on the relationship between eyedness and handedness in a sample of 413 subjects. Subjects were tested for eyedness and handedness and grouped in one of three categories on both variables. Paired observations were available for each subject. One subject was left-handed and ambiocular, another right-handed and right-eyed, and so on. The paired observations were entered in a bivariate frequency table as shown in Table 13.7. Such tables are analogous to correlation tables. They are used to study the independence or association of the two variables. Tables of this kind are spoken of as *contingency tables*. With such tables chi square provides an appropriate test of independence.

Table 13.7   Contingency table showing relationship between eye and hand laterality for 413 subjects and calculation of expected values

	Left-eyed	*Ambiocular*	*Right-eyed*	*Total*
Left-handed	34	62	28	124
	(35.4)	(58.5)	(30.0)	
Ambidextrous	27	28	20	75
	(21.4)	(35.4)	(18.2)	
Right-handed	57	105	52	214
	(61.1)	(101.0)	(51.8)	
Total	118	195	100	413

Calculation of expected values:

$$\frac{124 \times 118}{413} = 35.4 \qquad \frac{124 \times 195}{413} = 58.5 \qquad \frac{124 \times 100}{413} = 30.0$$

$$\frac{75 \times 118}{413} = 21.4 \qquad \frac{75 \times 195}{413} = 35.4 \qquad \frac{75 \times 100}{413} = 18.2$$

$$\frac{214 \times 118}{413} = 61.1 \qquad \frac{214 \times 195}{413} = 101.0 \qquad \frac{214 \times 100}{413} = 51.8$$

In applying chi square to a contingency table to test independence, the expected cell frequencies are derived from the data. The expected cell frequencies are those we should expect to obtain if the two variables were independent of each other, given the marginal totals of the rows and columns. Chi square provides a measure of the discrepancy between the observed cell frequencies and those expected on the basis of independence. If the value of chi square is considered significant at some accepted level, usually either the 5 or the 1 per cent level, we reject the null hypothesis that no difference exists between the observed and expected values. We then accept the alternative hypothesis that the two variables are associated.

How are the expected cell frequencies calculated? The marginal totals to the right of Table 13.7 show that 124 subjects were left-handed, 75 ambidextrous, and 214 right-handed. The proportions in these three categories are $\frac{124}{413}$, $\frac{75}{413}$, and $\frac{214}{413}$. These proportions are the probabilities that an individual, selected at random from the sample of 413 individuals, is left-handed, ambidextrous, or right-handed. The marginal totals at the bottom of Table 13.7 show that 118 subjects are left-eyed, 195 ambiocular, and 100 right-eyed. The proportions in these three categories are $\frac{118}{413}$, $\frac{195}{413}$, and $\frac{100}{413}$. These are the probabilities that an individual is left-eyed, ambiocular, or right-eyed. Assuming the independence of the two variables, what are the expected probabilities associated with the joint events, or what is the expected proportion of left-eyed people who are left-handed, of left-eyed people who are ambiocular, and so on? The multiplication theorem of probability states that the joint occurrence of two or more *mutually independent* events is the product of their separate probabilities. The joint probabilities are obtained, therefore, by multiplying the probabilities obtained from the marginal totals. The probability that any individual, selected at random from the 413 individuals, is left-handed is $\frac{124}{413}$. The probability that any individual is left-eyed is $\frac{118}{413}$. If handedness and eyedness are independent, the probability that any individual is both left-eyed and left-handed is the product of the separate probabilities, or $\frac{124}{413}$ $\times$ $\frac{118}{413}$. This is the expected proportion in the top left-hand cell in Table 13.7. We require, however, not the expected proportion, but the expected frequency. This is obtained by multiplying the expected proportion by N, in this case 413. Thus the expected frequency is $\left(\frac{124}{413} \times \frac{118}{413}\right) 413 = (124 \times 118)/413 = 35.4$. We observe that for computational purposes the expected cell frequency is obtained by multiplying together the first row and column totals and dividing by N. Similarly, the other expected cell frequencies may be obtained. The expected frequency of left-handed ambiocular individuals is $(124 \times 195)/413 = 58.5$, that of left-handed right-eyed individuals is $(124 \times 100)/413 = 30.0$, and so on. The expected cell frequencies are shown in parentheses in Table 13.7.

If eye and hand laterality are independent of each other, the 124 observations in the first row of Table 13.7 will be distributed in the three cells in that row in a manner proportional to the column sums. The expected values 35.4, 58.5, and 30.0 are proportional to the column sums 118, 195, and 100. Likewise, the 118 cases in the first column will be distributed in the three cells in that column in a manner proportional to the row sums. The expected values 35.4, 21.4, and 61.1 are proportional to the row sums 124, 75, and 214. A similar proportionality exists throughout the table. The expected cell frequencies in the rows and columns of any contingency table are proportional to the marginal totals.

The calculation of χ^2 for a contingency table is similar to that for tests of goodness of fit. The difference between each observed and expected value is squared and divided by the expected value, obtaining $(O - E)^2/E$. These values are summed over all cells to obtain χ^2. The calculation is perhaps most readily accomplished by writing the data in columnar fashion as shown in Table 13.8. The value of χ^2 obtained is 4.021. The number of degrees of freedom associated with this value of χ^2 is 4. The value of χ^2 required for significance at the 5 per cent level is 9.488. We have therefore no grounds for rejecting the hypothesis of independence between eye and hand laterality. Apparently there is no relationship between these two variables.

Table 13.8   Calculation of χ^2 for data of table 13.7

O	E	$O - E$	$(O - E)^2$	$\dfrac{(O - E)^2}{E}$
34	35.4	−1.4	1.96	.055
62	58.5	3.5	12.25	.209
28	30.0	−2.0	4.00	.133
27	21.4	5.6	31.36	1.465
28	35.4	−7.4	54.76	1.547
20	18.2	1.8	3.24	.178
57	61.1	−4.1	16.81	.275
105	101.0	4.0	16.00	.158
52	51.8	.2	.04	.001
413	412.8	....	....	$\chi^2 = 4.021$

How is the number of degrees of freedom calculated? In testing independence in any contingency table comprised of R rows and C columns the number of degrees of freedom is given by $(R - 1)(C - 1)$. For Table 13.7 $R = 3$ and $C = 3$. The number of degrees of freedom is $(3 - 1)(3 - 1) = 4$. For a table comprised of two rows and two columns, referred to as a 2×2, or fourfold, table, the number of degrees of freedom is $(2 - 1)(2 - 1) = 1$. Consider for explanatory purposes the 2×2 table:

	A_1	A_2	
B_1	25	35	60
B_2	5	35	40
	30	70	100

Given the restrictions of the marginal totals, if one cell value is known, the remaining three values are determined. Thus if we know that the value in the top left cell is 25, the top right cell must be $60 - 25 = 35$, the bottom left $30 - 25 = 5$, and the bottom right $40 - 5 = 35$. If one cell value is known, no freedom of variation remains. One degree of freedom only is associated with the variation in the data. Similarly, in Table 13.7 only four cell values are free to vary. Given the marginal totals and four cell values, the remaining cell values are determined.

A frequently occurring type of contingency table is the 2×2, or fourfold, contingency table. A χ^2 test for independence can be readily obtained for such a table without calculating the expected values. Let us represent the cell and marginal frequencies by the following notation:

A	B	$A + B$
C	D	$C + D$

$A + C$ $B + D$ N

Chi square may then be calculated by the formula

$$[13.2] \quad \chi^2 = \frac{N(AD - BC)^2}{(A + B)(C + D)(A + C)(B + D)}$$

Note that the term in the numerator, $AD - BC$, is simply the difference between the two cross products and the term in the denominator is the product of the four marginal totals.

Consider the following 2×2 table showing the relationship between ratings of successful or unsuccessful on a job and pass or fail on an ability-test item.

Test item

		Fail	Pass	
Job rating	**Successful**	20	40	60
	Unsuccessful	25	15	40
		45	55	100

Is there an association between performance on the job and performance on the test item? Does the item differentiate significantly between the successful and unsuccessful individuals? Chi square is as follows:

$$\chi^2 = \frac{100(20 \times 15 - 40 \times 25)^2}{60 \times 40 \times 45 \times 55} = 8.25$$

For $df = 1$, a $\chi^2 = 8.25$ is significant at better than the 1 per cent level. The data provide fairly conclusive evidence that the test item differentiates between individuals on the basis of their job performance.

13.5 THE APPLICATION OF χ^2 IN TESTING THE SIGNIFICANCE OF THE DIFFERENCE BETWEEN PROPORTIONS

In Chap. 12 procedures were described for testing the significance of the difference between both *independent* and *correlated* proportions. These procedures involved dividing the difference between two proportions by the standard error of the difference to obtain a normal deviate which could be referred to a table of areas under the normal curve. Because of a simple relationship for 1 degree of freedom between χ^2 and the normal deviate, χ^2 provides an alternative but equivalent procedure for testing the significance of the difference between proportions. For 1 degree of freedom it may be shown that χ^2 is equal to the normal deviate squared. Thus $\chi^2 = (x/s)^2 = z^2$ or $\sqrt{\chi^2} = z$.

We shall now consider the use of χ^2 in testing the significance of the difference between proportions for *independent* samples. Let the following be data obtained in response to an attitude-test statement for a group of males and females:

	Frequency					*Proportion*		
	Agree	**Disagree**				**Agree**	**Disagree**	
Males	70	70	140		**Males**	.500	.500	1.00
Females	20	40	60		**Females**	.333	.667	1.00
	90	110	200			.450	.550	1.00

The number of males and females are $N_1 = 140$ and $N_2 = 60$, respectively. The proportions of males and females indicating agreement to the attitude statement are $p_1 = \frac{70}{140} = .500$ and $p_2 = \frac{20}{60} = .333$. Is there a significant difference in the attitudes of males and females? To apply the method previously described we calculate a proportion p based on a combination of data for the two samples. With the above data

$$p = \frac{70 + 20}{140 + 60} = \frac{90}{200} = .450$$
$$q = 1 - p = 1 - .450 = .550$$

The required normal deviate is then

$$z = \frac{p_1 - p_2}{\sqrt{pq[(1/N_1) + (1/N_2)]}}$$
$$= \frac{.500 - .333}{\sqrt{.450 \times .550(\frac{1}{140} + \frac{1}{60})}} = 2.172$$

The difference between the two proportions falls between the 5 and 1 per cent levels. Reference to a table of areas under the normal curve shows that the proportion of the area falling beyond plus and minus 2.172 standard deviation units from the mean is close to .03. The difference may be taken as significant at about the 3 per cent level. Let us now apply the formula for calculating χ^2 for a 2×2 contingency table to the same data. We have

$$\chi^2 = \frac{N(AD - BC)^2}{(A + B)(C + D)(A + C)(B + D)}$$
$$= \frac{200(70 \times 40 - 70 \times 20)^2}{140 \times 60 \times 90 \times 110} = 4.717$$

Consulting a table of χ^2 with 1 degree of freedom, we observe that the proportion of the area in the tail of the distribution of χ^2 is about .03 and the difference between proportions may be said to be significant at about the 3 per cent level. We observe also that $\chi^2 = (2.172)^2 = 4.717$. The two procedures for testing the significance of the difference between proportions for independent samples lead to identical results. From a computational viewpoint the χ^2 test is the more convenient. Considerations pertaining to small frequencies apply also to the application of χ^2 in testing the significance of the difference between proportions (Sec. 13.6).

Where the data are correlated and are composed of paired observations the normal deviate for testing the significance of the differences between proportions is given (Sec. 12.3) by the formula

$$z = \frac{D - A}{\sqrt{A + D}}$$

where D and A are cell frequencies in the bottom right and top left cells, respectively, of a 2×2 table. Instead of calculating a critical ratio and referring this to the normal curve, we may calculate χ^2 by the formula

[13.3] $\chi^2 = \dfrac{(D-A)^2}{A+D}$

For the data shown in Sec. 12.3, where we wish to test the significance of the differences between proportions of agreements to an attitude question for the same individuals tested on two occasions, we obtain a $z = 3.16$. The difference is significant at better than the 1 per cent level. The value of the probability is .0016. The value of χ^2 calculated on the same data is $(3.16)^2$, or 9.986. The probability is the same as before.

13.6 SMALL EXPECTED FREQUENCIES

The distribution of χ^2 used in determining critical significance values is a continuous theoretical frequency curve. Where the expected frequencies are small, the actual sampling distribution of χ^2 may exhibit marked discontinuity. The continuous curve may provide a poor fit to the data, and appreciable error may occur in the estimation of probabilities, these being areas under the continuous χ^2 curve. The situation here is analogous to that found in using the normal curve as a fit to the binomial. For small values of N the continuous normal curve is a poor fit to the discrete binomial.

A correction may be applied known as *Yates's correction for continuity*. To apply this correction we reduce by .5 the obtained frequencies that are greater than expectation and increase by .5 the obtained frequencies that are less than expectation. This brings the observed and expected values closer together and decreases the value of χ^2. This correction should be used where any of the expected frequencies is less than 5, and some writers suggest 10. Indeed, it may be argued that the correction should always be used. For large expected frequencies the correction will be negligible.

The formula used in computing χ^2 from a 2 × 2 table can be written to incorporate Yates's correction for continuity. This formula becomes

[13.4] $\chi^2 = \dfrac{N(|AD - BC| - N/2)^2}{(A+B)(C+D)(A+C)(B+D)}$

The term $|AD - BC|$ is the absolute difference, that is, the difference taken regardless of sign. The correction amounts to subtracting $N/2$ from this absolute difference.

The following data show the relationship between sociometric choices for a group of 20 Protestant and Jewish school children.

		Chosen		
		Protestant	Jewish	
Chooser	Jewish	3	5	8
	Protestant	10	2	12
		13	7	20

The value of χ^2 using Yates's correction is

$$\chi^2 = \frac{20(|6 - 50| - \frac{20}{2})^2}{8 \times 12 \times 13 \times 7} = 2.65$$

This value falls at about the 10 per cent level. The evidence does not justify the rejection of the hypothesis that sociometric choice is independent of whether the child is Jewish or Protestant. We note that if χ^2 is calculated on these data without Yates's correction, a value $\chi^2 = 4.43$ is obtained. This value falls at about the 3.5 per cent level. If Yates's correction had not been used, the result would be considered significant at better than the 5 per cent level.

With 2 or more degrees of freedom the error introduced by small expected frequencies is of less consequence than with 1 degree of freedom. An expectation of not less than 2 in each cell will permit the estimation of roughly approximate probabilities. If the frequencies are 5 or more, good approximations to the exact probabilities are obtained. With certain types of data it is a common practice to combine frequencies. In testing the goodness of fit of a theoretical to an observed frequency distribution, small frequencies at the tails may be combined. On occasion it may be possible without serious distortion of the data to combine rows and columns of a contingency table to increase the expected cell frequencies.

With 1 degree of freedom where the expected frequencies are small, an exact test of significance may be applied. This involves the determination of exact probabilities, as distinct from those estimated from the continuous χ^2 curve. An exact test of significance for a 2×2 table is described in Chap. 22 of this book.

13.7 MISCELLANEOUS OBSERVATIONS ON CHI SQUARE

In this section we shall consider a number of miscellaneous points about χ^2 not hitherto discussed.

One-tailed and two-tailed tests Tables of χ^2 used for tests of significance are based on one tail only, the tail to the right, of the sampling distribution of χ^2. Table C of the Appendix shows that for 1 degree of freedom 5 per cent of the area of the distribution falls to the right of $\chi^2 = 3.84$ and 1 per cent to the right of $\chi^2 = 6.64$. These are *not* critical values for directional, or one-tailed, tests as described in Chap. 11. Although one tail only of the sampling distribution of χ^2 is used, the tabled values are those required for testing the significance of a difference regardless of direction, that is, for two-tailed tests. The critical ratio or normal deviate required for significance at the 5 per cent level for a two-tailed test is 1.96. If this value is squared, we obtain 3.84, the χ^2 value at the 5 per cent level for 1 degree of freedom. For 1 degree of freedom the square root of χ^2 is a normal deviate and may be used with reference to the normal curve in applying two-tailed tests. In effect, because χ^2 is the square of the normal deviate for 1 degree of freedom, both tails of the normal curve are incorporated in the right tail of the χ^2 curve. In many situations where χ^2 is applied, the idea of a directional, or one-tailed, test has little meaning. In tests of goodness of fit and in most tests of independence we are usually not concerned with the direction of the difference observed. If a one-tailed test is required, the proportionate areas in the chi-square tables should be halved. The value of χ^2 required for significance at the 5 per cent level for a one-tailed test is 2.71 for $df = 1$. The corresponding value at the 1 per cent level is 5.41. These are the squares of the normal deviates 1.64 and 2.33 required for significance for a one-tailed test at the 5 and 1 per cent levels, respectively.

Chi square and sample size   The value of χ^2 is related to the size of the sample. If an actual difference exists between observed and expected values, this difference will tend to increase as sample size increases. χ^2 will also increase, and the associated probability value will decrease. Consider the following tables:

1		
6	4	10
4	11	15
10	15	25

$\chi^2 = 2.78$

2		
12	8	20
8	22	30
20	30	50

$\chi^2 = 5.56$

3		
24	16	40
16	44	60
40	60	100

$\chi^2 = 11.12$

As the samples are doubled in size from 25 to 50 to 100, the differences between the observed and expected values, $O - E$, are doubled and the χ^2 values are doubled. If no actual difference exists between observed and expected values, χ^2 will tend to remain unchanged as sample size increases. For a constant difference between observed and expected values χ^2 will decrease as sample size increases. If we double sample size and hold the difference between observed and expected values fixed, the value of χ^2 will be reduced by one-half.

Alternative formula for chi square   We can readily demonstrate that

[13.5] $$\chi^2 = \sum \frac{(O - E)^2}{E} = \sum \frac{O^2}{E} - N$$

This alternative way of writing χ^2 is sometimes useful for computational purposes.

2 × 2 tables with more than 1 degree of freedom   For most 2 × 2 tables the row and column totals are considered fixed and 1 degree of freedom is associated with the variation in the data. Situations arise where either the row or column totals, or both, are free to vary. In a sociometric study on a class of 8 Jewish and 12 Protestant children, each child may be asked to choose one other child with whom he would prefer to play. He cannot choose himself. If choices are independent of whether a child is Jewish or Protestant, what are the expected frequencies of choices? On a strictly random basis, how many Jewish choosers will make Protestant choices, and so on? Since a child cannot choose himself, a Jewish child chooses from among seven Jewish and 12 Protestant children. The probability of a Jewish child choosing a Jewish child in making a choice at random is $\frac{7}{19}$ and of choosing a Protestant child $\frac{12}{19}$. Since eight choices are made, the expected frequency of Jewish choices is $\frac{7}{19} \times 8 = 2.95$. The expected frequency of Protestant choices is $\frac{12}{19} \times 8 = 5.05$. Likewise, we find that the expected frequency of Jewish choices by Protestant children is $\frac{8}{19} \times 12 = 5.05$, and the expected frequency of Protestant choices is $\frac{11}{19} \times 12 = 6.95$. The expected frequencies are tabulated below together with the observed frequencies.

| | Expected chosen | | | | Observed chosen | | |
	P	J			P	J	
Chooser J	5.05	2.95	8	Chooser J	6	2	8
Chooser P	6.95	5.05	12	Chooser P	8	4	12
	12	8	20		14	6	20

In this example the row totals are fixed. The column totals are free to vary from expectation. In these data we observe a tendency for both Jewish and Protestant children to choose Protestant children more frequently than expectation. In this case a χ^2 based on a comparison of the expected and observed cell frequencies has 2 degrees of freedom. χ^2 may of course be applied to the observed frequencies in the usual way with 1 degree of freedom. This is a test of association, within the restrictions of the marginal totals, of the religious affiliation of the choosers and the chosen. It is not a test of randomness of choice. Fourfold tables may occur where both row and column totals are free to vary. Such tables arise where all expected frequencies are derived in a manner entirely independent of the data. χ^2 here has 3 degrees of freedom.

Reduction of an $R \times C$ table to a 2×2 table   A table with R rows and C columns may be reduced to a 2×2 table in order to facilitate a rapid test of association with χ^2. This procedure is legitimate enough provided the points of dichotomy of the two variables are made without reference to the cell frequencies. The investigator may decide a priori to dichotomize about the two medians, or something of the sort. Data are found where the points of dichotomy have been located in order to maximize the association in the data and obtain thereby a significant χ^2. This practice is spurious and should be enthusiastically discouraged.

EXERCISES **1** In 180 throws of a die the observed frequencies of the values from 1 to 6 are 34, 27, 41, 25, 18, and 35. Test the hypothesis that the die is unbiased.

2 A psychological test yields a distribution of scores as shown below. The theoretical normal frequencies are also shown:

Class interval	Observed frequencies	Theoretical frequencies
90–99	1 }	8
80–89	5 }	
70–79	17	17
60–69	30	31
50–59	50	39
40–49	35	34
30–39	10	20
20–29	6	8
10–19	4 }	3
0–9	2 }	
	160	160

Test the goodness of fit between the theoretical and the observed frequencies.

3 How many cell frequencies are free to vary in tables with (*a*) two rows and two columns, (*b*) two rows and three columns, (*c*) three rows and five columns? Assume fixed marginal totals.

4 The following data relate to patients in a mental hospital:

	Rating		
	Improvement	No improvement	
Therapy A	16	28	44
Therapy B	9	37	46
	25	65	90

Test the hypothesis that method of therapy is independent of rating assigned.

5 The following contingency table describes the relation between scores above and below the median on an examination and ratings of job performance for 100 employees.

	Rating			
	Below average	Average	Above average	
Above median	11	25	35	71
Below median	15	7	7	29
	26	32	42	100

Test the hypothesis that job performance is independent of examination results.

6 A sample used in a market survey contains 100 males and 100 females. Of the males 33 and of the females 18 state a preference for brand A. Use χ^2 to test the hypothesis that no sex differences exist in consumer preference.

7 Calculate χ^2 for the following tables, using Yates's correction for continuity:

(a)

	Weight		
	Increase	No increase	
Gentled animals	5	2	7
Ungentled animals	3	5	8
	8	7	15

(b)

	Locus of lesion			
	A	B	C	
Impairment in performance	2	5	4	11
No impairment in performance	3	2	2	7
	5	7	6	18

THE DESIGN OF EXPERIMENTS

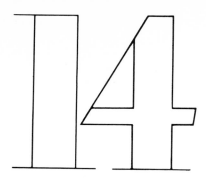

THE STRUCTURE AND PLANNING
OF EXPERIMENTS

14.1 INTRODUCTION

Several subsequent chapters of this book are concerned with the analysis of variance and covariance. These procedures are used in the analysis of the data of experiments. A very brief, and elementary, discussion of the structure and planning of experiments should serve as a useful preliminary to a detailed study of these methods of analysis. The study of the structure and planning of experiments is a field of investigation commonly called the design of experiments. This subject has many aspects, some quite complex. The present discussion will deal in a nonmathematical way with only a few of the simplest aspects of experimental design.

All experiments are concerned with the relations between variables. In the simplest type of experiment two variables only are involved, an independent variable and a dependent variable. To illustrate, an experiment may be initiated to compare three different methods of teaching French, designated A, B, and C. Each method may be applied to a different group of experimental subjects. Following a period of instruction, performance may be measured using an achievement test. In this experiment the different methods of teaching French constitute the levels or categories of the independent variable. The investigator decides which methods will be used and the size of the groups to which they will be applied; that is, he controls the values or categories of the independent variable and the frequency of occurrence of those values or categories. The measures of achievement constitute the dependent variable. The experiment is concerned with the way in which achievement in French depends on the method of instruction used. The essence of the idea of an experiment lies in the simple fact that the investigator selects the values or categories of the treatment variable and the frequency of their occurrence. This enables him to study an indefinitely large number of relations which are not amenable to study by observational or correlational methods, and may not in fact have any existence in nature at all prior to the conduct of the experiment. New knowledge is thereby produced. The tremendous proliferation of human knowledge in the past 20 years is in large measure a result of experimentation. It is clearly desirable, therefore, that some of the principles underlying such experimentation should be understood.

In developing the design for an experiment, the investigator must (1) select the values or categories of the independent variable, or variables, to be compared; (2) select the subjects for the experiment; (3) apply rules or procedures whereby subjects are assigned to the particular values or categories of the independent variable; (4) specify the observations or measurements to be made on each subject. In the experiment mentioned above, on the comparison of different methods of teaching French, the investigator must select the different methods of teaching to be compared. He must choose the experimental subjects to which these methods are applied. He must allocate subjects to methods. Also decisions must be made regarding appropriate measures of achievement which will yield valid comparisons of the methods used.

14.2 TERMINOLOGY

Comment on terminology is appropriate here. An independent variable used in an experiment may be either a *treatment* variable or a *classification* variable. A treatment variable involves a modification in the experimental subjects, a modification which is controlled by the experimenter. Different dosages of a drug or different methods of learning are administered to different groups of subjects. In effect, the subjects are treated in some way by the experimenter. Experimental subjects may, however, be classified on a characteristic which was present prior to, and quite apart from, the experiment, and does not result from the manipulations of the experimenter. Such a variable is a classification variable. Examples are sex, age, disease entity, IQ level, socioeconomic status, and so on. Although the values of a classification variable are not, as it were, created by the experimenter, as is the case with a treatment variable, the investigator nonetheless selects the classification variables which are included in the experiment or are the objects of attention.

Any independent variable, whether of the treatment or classification type, is spoken of as a *factor*. Experiments which investigate simultaneously the effects of two independent variables are spoken of as two-factor, or two-way classification, experiments. When three factors are involved, the experiment is said to be a three-factor, or three-way classification, experiment, and so on. The different values or categories of the independent variable are spoken of as *levels;* thus we may have two, or three, or more levels of a factor.

In the literature on experimental design the unit to which a treatment is applied is frequently spoken of as a *plot*, a term which derives from agricultural experimentation. In experimental work in psychology and education the plot is usually a human subject or an experimental animal. The term "plot" will be used infrequently in this text. Measurements obtained from a plot are sometimes spoken of as the *yield*, a term which stems also from agricultural usage. In psychology and education the measurements or observations made on a group of subjects or animals correspond to the yield for a number of plots. A grouping of experimental units which is homogeneous with regard to some basis of classification is spoken of as a *block*. An experimenter may select 50 subjects, of which 25 are male and 25 female. The two groupings by sex are blocks.

14.3 THE CLASSIFICATION OF VARIABLES IN RELATION TO EXPERIMENTS

As indicated above, experiments are concerned with the relations between variables. In Chap. 1 of this book variables were classified as nominal, ordinal, or interval-ratio types. In a simple two-variable experiment the treatment variables may be nominal, ordinal, or interval-ratio; the dependent variable may also be nominal, ordinal, or interval-ratio. The methods to be applied in the analysis of the data of experiments and the type of questions which experiments can answer are determined by the nature of the variables. Thus, in effect, the nature of the information which an experiment can yield, and the analytic methods which communicate that information to our understanding, depend on the nature of the variables.

To illustrate, consider an experiment in which three types of therapy are applied to three groups of depressed patients. After a time period, observations are made indicating those subjects that show evidence of recovery and those that do not. In this experiment both variables are nominal. The investigator may calculate the proportions in the two groups that show recovery and compare these proportions one with another. The magnitude of the differences between proportions is a measure of the difference between treatments. No further analysis of these data is possible. Consider, on the other hand, an experiment in which both the treatment variable and the dependent variable are of the interval or ratio type. The treatment variable may consist of five equally spaced dosages of a drug, and the dependent variable may be reaction time. Here the investigator may not only compare the mean reaction times for each dosage with every other dosage, but he may explore the nature of the functional relation between the two variables. Reaction time may increase or decrease in a linear fashion with change in the treatment variable; it may increase and then decrease; or some other type of relation may exist. Here we have considered two experiments. In one experiment both variables were nominal; in the other, both were either interval or ratio. In many experiments the treatment variables may be nominal, and the dependent variable may be of the interval or ratio type; or the treatment variable may be ordinal, and the dependent variable may be nominal, and so on. The nature of the variables determines the method of analysis employed and the nature of the conclusions drawn.

14.4 SINGLE-FACTOR EXPERIMENTS

Many experiments involve a single-treatment, or classification, variable with two or more levels. Let us initially consider experiments in which the single factor is a treatment variable with k levels or categories, and not a classification variable. Such experiments are of a variety of types, some of which are discussed here. *First*, a group of experimental subjects may be divided into k independent groups, using a random method. A different treatment may then be applied to each group. One group may be a control group, that is, a group to which no treatment is applied. A meaningful interpretation of the experiment may require a comparison of results obtained under treatment with results obtained in the absence of treatment. Comparisons may be made between treatments and a control, between treatments, or both. *Second*, some single-factor experiments involve a single group of subjects. Each subject receives all k treatments. Repeated observations or measurements are made under k conditions, one of which may be a control condition, on the same subjects. In such an experiment as this the measurements made under the k treatments will not be independent. Positive correlation will usually exist between the paired measurements obtained under any two treatments. These correlations will reduce the magnitude of the error used in the comparisons of the separate treatment means. *Third*, a single-factor experiment may consist of groups that are matched on one or more variables which are known to be correlated with the dependent variable. In an experiment on three different methods of teaching French IQ may be known to be correlated with achievement in French. Three groups of subjects, paired or matched subject by subject by IQ, may be used. The rationale here is that because IQ is correlated with French achievement, a correlation between French achievement scores for paired subjects may result. The error term would be reduced thereby. In some practical experimental situations the gains made by matching are trivial in relation to the work involved. An interesting variant of the matched-group experiment is one in which subjects are not matched subject by subject but are matched with regard to distribution on one or more control variables. This results in the same error reduction as the individual matching of subjects (see McNemar, 1969).

14.5 RANDOMIZATION

In the design of experiments frequent use is made of randomization. In general the purpose of randomization is to ensure that extraneous variables which are concomitant with the dependent variable, and may be correlated with it, will not introduce systematic bias in the experimental results. Consider again the experiment in which three methods of teaching French are compared. If subjects are assigned to the three groups using a random method, IQ, which may be correlated with French achievement, will not vary in any systematic way from group to group. Also, an indefinitely large number of other variables, the specifications of many of which may not have entered the purview of the investigator, will be rendered powerless to introduce systematic bias. Although methods of randomization differ from one type of experiment to another, the general purpose of randomization is to protect the validity of the experiment by controlling the biasing influence of extraneous variables. A quotation from Cochran and Cox (1957) is relevant here.

Randomization is somewhat analogous to insurance, in that it is a precaution against disturbances which may or may not occur and that may or not be serious if they do occur. It is generally advisable to take the trouble to randomize even when it is not expected that there will be any serious bias from failure to randomize. The experimenter is thus protected against unusual events that upset his expectations.

In experimental design the term "randomization" refers not to any subjective impression of haphazard arrangement but to clearly stated operational procedures. These procedures may involve the tossing of coins, the drawing of numbered cards from a well-shuffled deck, or the use of tables of random numbers. Such tables are composed of series of digits from zero to nine. Each digit occurs with approximately equal frequency, and adjacent digits are independent of each other. Tables of random numbers are available in Fisher and Yates (1963), and elsewhere. To illustrate, suppose we wish to select a sample of 12 experimental subjects from a sample of 40 subjects. Identify the subjects by the numbers 1 to 40. Enter a row, column, or diagonal of two-digit numbers in a table of random numbers, and select the first 12 different numbers between 1 and 40 as they occur. This procedure will yield a random set of 12 subjects. Other procedures may, of course, be used.

In a single-factor experiment with k independent groups of n_1, n_2, . . . , n_k subjects, an appropriate procedure is to choose n_1 subjects at random for the first group, n_2 at random for the second, and so on. This procedure is continued until n_k subjects remain for the kth group. In an experiment in which k experimental treatments are administered to the same group of n subjects and repeated measurements are made on the same subjects, an order effect may exist; that is, the result observed may not be independent of the order of treatment for reason of fatigue, practice, or some other cause. For example, measures of reaction time may be made on the same group of subjects under four different dosages of a drug. Here quite clearly an order effect may exist. In such an experiment as this the order of treatments may be randomized for each subject, thereby eliminating any systematic influence of order on the treatment means. When there are two treatments only, A and B, a common practice is to use the order AB for half the subjects and the order BA for the other half, assigning subjects to orders at random. In experiments involving matched groups, the members of each matched pair, triplet, or quadruplet may be allocated to treatments by a random method.

14.6 FACTORIAL EXPERIMENTS

The experiments discussed hitherto in this chapter have involved a single independent variable or factor. Experiments may, however, be designed to study simultaneously the effects of two or more independent variables. For example, the experiment in which three methods of teaching French are compared may be extended to include a comparison of spaced versus massed learning. Under the spaced conditions, subjects receive short periods of intensive instruction separated by time intervals. Under the massed learning conditions, subjects receive intensive instruction for a prolonged time period. The effect on achievement of the six possible combinations of three methods of teaching and two learning conditions may be investigated, each combination being applied to a different group of experimental subjects. Such an experiment is called a *factorial experiment*. Experiments in which the treatments are combinations of levels of two or more factors are said to be factorial. If all possible treatment combinations are studied, the experiment is said to be a *complete factorial experiment*. In some experiments the factors have two levels only. We may speak of a 2×2 experiment, a $2 \times 2 \times 2$ experiment, or a 2^n experiment, where n is the number of factors.

It is informative to examine in more detail the structural features of a factorial experiment. Let us suppose that in our 3×2 experiment on teaching French in relation to spaced versus massed learning, four subjects were used in each of the six groups and the following achievement test scores obtained.

| Learning condition | *Teaching method* | | |
	A	*B*	*C*
Spaced	72 34 96 55	78 29 25 20	16 29 24 41
Massed	81 85 99 62	75 45 36 26	19 19 41 46

Three variables are involved in this experiment. Two of these, method and learning condition, are independent variables, and one, achievement-test score, is the dependent variable. Three variate values are available for each subject—the method used, the learning condition, and achievement-test scores. Method and learning condition are nominal variables. These data could readily be written in columnar fashion as follows:

Learning condition	Teaching method	Test score	Learning condition	Teaching method	Test score
S	A	72	M	A	81
S	A	96	M	A	99
S	A	34	M	A	85
S	A	55	M	A	62
S	B	78	M	B	75
S	B	25	M	B	36
S	B	29	M	B	45
S	B	20	M	B	26
S	C	16	M	C	19
S	C	24	M	C	41
S	C	29	M	C	19
S	C	41	M	C	46

The purpose of this experiment is to explore the relation between the learning condition and test scores and the teaching method and test scores. The relations between the six combinations of learning condition and teaching method may also be studied. An important feature of the structure of this experiment is that the two nominal variables, learning condition and teaching method, are independent of each other. By choosing six groups of equal size, the independence of these two variables is assured. If the numbers of cases in the six groups, when written in the form of a 2×3 contingency table, are proportional to the marginal totals, and χ^2 for the table is zero, the independence of the two treatment variables will also be assured. In the design of factorial experiments the groups should be either of equal size or proportional. Departures from equality or proportionality should be avoided.

One advantage of the factorial experiment is that information is obtained about the interaction between factors. For example, in the experiment on teaching French one method of teaching may interact with a condition of learning and render that combination either better or worse than any other combination. The concept of interaction is discussed more explicitly, and in detail, in Sec. 16.5. One disadvantage of the factorial experiment is that the number of combinations may become quite unwieldy, and, from a practical point of view, the experiment may be very difficult to conduct. Also, the meaningful interpretation of the interactions may prove difficult. In general, in psychology and education, it is usually advisable to avoid factorial experiments with more than a few factors.

The reader should note that factorial experiments may involve repeated measurement on the same subjects. For example, in a 3×2 design, repeated measurements may be made on the same subjects under each of the six treatment combinations. The result is an $N \times 3 \times 2$ arrangement of numbers.

14.7 OTHER EXPERIMENTAL DESIGNS

Situations occur where a basis exists for the classification of experimental subjects into r subgroups or blocks. An experiment may be conducted involving k treatments, where the number of subjects used is such that $N = rk$. Subjects in each block may be assigned at random to treatments, under the restriction that each treatment occur once only in each block. Such an experiment is a *randomized block experiment*. To illustrate, an experiment involves a comparison of four treatments. A sample of 24 experimental animals is used, which consists of six sets of four litter mates. Each set of litter mates is a block. Treatments are allocated to the animals in each block, or vice versa, on a random basis. If we designate the four treatments as A, B, C, and D, the allocation of treatments to subjects may be as follows:

	Animal within block			
Block	1	2	3	4
1	A	D	C	B
2	C	A	B	D
3	A	B	D	C
4	B	C	D	A
5	D	B	A	C
6	D	A	C	B

In previous discussion of order effects in single-factor experiments involving repeated measurements on the same subjects, it was suggested that the order of treatment be randomized for each subject. This is, in fact, a randomized block design, where the subject becomes the analogue of the block, and the order of presentation of treatment conditions A, B, C, D is represented by column headings.

Experiments may be considered in which the number of blocks is equal to the number of treatments or a multiple thereof; thus $N = k^2$ or some multiple of k^2. In such experiments each treatment occurs once only in each block and once only over blocks. Thus a particular arrangement may be:

A	*B*	*D*	*C*
D	*A*	*C*	*B*
C	*D*	*B*	*A*
B	*C*	*A*	*D*

This is a Latin square design. A Latin square is simply an arrangement in which each treatment occurs once only in each row and column. Tables of Latin squares appear in Fisher and Yates (1963). These authors suggest methods of choosing a Latin square at random from sets of possible Latin squares. When an experiment requires more than one Latin square, they must be chosen independently.

14.8 CLASSIFICATION VARIABLES

Hitherto we have discussed experiments in which the independent variables are treatment variables. In many investigations in psychology and education the independent variables are not treatment variables but are, in fact, classification variables; that is, subjects are classified according to a characteristic which was present prior to the conduct of the experiment and did not result from the manipulations of the investigator. For example, normals, neurotics, and psychotics may be compared on flicker-fusion rates, or some other variable. Here the independent variable is a classification, and not a treatment, variable. No treatment by the investigator is involved. Quite clearly in such an experiment it is not possible to assign subjects to experimental groups at random. Randomization is not possible, because the attribute which determines membership in a particular group is not under the control of the investigator.

Three types of situations in which classification variables are involved may be recognized. *First*, the sample of subjects used may be a random sample drawn from a defined population, and the proportions in the various classes, or strata, may, within the limits of sampling error, correspond to the population proportions. The object of the investigation is to describe relations between the classification variable and some other variable. An investigation of this type is, in effect, not an experiment at all, but a correlational study. *Second*, samples may be drawn for comparison at random from two or more subpopulations, but the proportions in these samples may not correspond to the population proportions. For example, an investigator may choose to compare 50 normals with 50 psychotics on a specified variable or variables. Clearly the proportions in the two samples do not correspond to the proportions of normals and psychotics in the population. This, in effect, is a form of disproportional stratified sampling. Also we note that certain classes which exist in the population at large may be excluded. Thus, for example, not all nonpsychotics are necessarily normal. This again is, in effect, a correlational study, the object of which is simply to identify differences between groups. Frequently there is a presupposition that a knowledge of such differences may ultimately prove useful in the construction of causal arguments. *Third*, investigations may be conducted involving classification variables in which the attempt is made to control the influence of certain variables which in a straightforward correlational study would not be controlled. In comparing hospitalized normals with hospitalized psychotics, the two groups may be equated for age, IQ, sex, length of hospitalization, socioeconomic status, and other variables which might be construed to be correlated with the dependent variable. In studies such as this the investigator excludes certain variables from the group of causal influences affecting the results observed. Although no direct causal argument linking the classification variable and the dependent variable can be advanced, the influence of certain variables on the results observed can be excluded. Such investigations narrow the range of possible causal influences. Factorial investigations involving classification variables may perhaps be construed legitimately to be experiments in that by design they involve the experimental control of certain variables which otherwise might be uncontrolled.

14.9 CONCLUDING OBSERVATION

Discussion in this chapter has touched briefly on a few aspects only of the structure and planning of experiments. This subject can be elaborated at great length and complexity. For a more comprehensive discussion the reader is referred to Cochran and Cox (1957), Cox (1961), Finney (1960), and Winer (1962).

EXERCISES 1 Distinguish, with examples, between (*a*) an independent and a dependent variable and (*b*) a treatment and a classification variable.

2 Discuss the rationale underlying the use of matched samples in the planning of experiments.

3 Why does randomization eliminate systematic bias in experimental results?

4 Outline a randomization procedure for allocating 100 experimental subjects to five experimental groups.

5 A group of 10 experimental subjects receives four treatments. Describe a procedure to eliminate order effects.

6 What is the rationale underlying the use of equal or proportional groups in factorial experiments?

7   What is meant by a randomized block experiment?

8   Give an example of a Latin square containing five rows and five columns.

9   How would you distinguish between an experiment and a correlational study?

ANALYSIS OF VARIANCE:
ONE-WAY CLASSIFICATION

15.1 ITS NATURE AND PURPOSE

The analysis of variance is a method for dividing the variation observed in experimental data into different parts, each part assignable to a known source, cause, or factor. We may assess the relative magnitude of variation resulting from different sources and ascertain whether a particular part of the variation is greater than expectation under the null hypothesis. The analysis of variance is inextricably associated with the design of experiments. Obviously, if we are to relate different parts of the variation to particular causal circumstances, experiments must be designed to permit this to occur in a logically rigorous fashion.

The partitioning of variance is a common occurrence in statistics. The particular body of technology known as the analysis of variance was developed by R. A. Fisher and reported by him in 1923. Since that time it has found wide application in many areas of experimentation. Its early applications were in the field of agriculture. If the variance is understood as the square of the standard deviation of a variable X, s_x^2, the analysis of variance does not in fact divide this variance into additive parts. The method divides the sum of squares $\Sigma(X - \bar{X})^2$ into additive parts. These are used in the application of tests of significance to the data.

In its simplest form the analysis of variance is used to test the significance of the differences between the means of a number of different populations. We may wish to test the effects of k treatments. These may be different methods of memorizing nonsense syllables, different methods of instruction, or different dosages of a drug. A different treatment is applied to each of the k samples, each sample being comprised of n members. Members are assigned to treatments at random. The means of the k samples are calculated. The null hypothesis is formulated that the samples are drawn from populations having the same mean. Assuming that the treatments applied are having no effect, some variation due to sampling fluctuation is expected between means. If the variation cannot reasonably be attributed to sampling error, we reject the null hypothesis and accept the alternative hypothesis that the treatments applied are having an effect. With only two means, $k = 2$, this approach leads to the same result as that obtained from the t test for the significance of the difference between means for independent samples.

Consider an agricultural experiment undertaken to compare yields of four varieties of wheat. Thirty-two experimental plots are prepared, and each of the four varieties grown in eight plots. Thus $k = 4$ and $n = 8$. Assume that appropriate precautions have been exercised to randomize uncontrolled factors such as variation in soil fertility from plot to plot. The yield for each plot is obtained, and the mean yield for each variety on the eight plots calculated. Differences in yield reflect themselves in the variation in the four means. If this variation is small and can be explained by sampling error, the investigator has no grounds for rejecting the null hypothesis that no difference exists between the yields of the four varieties. If the variation between means is not small and of such magnitude that it could arise in random sampling in less than 1 or 5 per cent of cases, then the evidence is sufficient to warrant rejection of the null hypothesis and acceptance of the alternative hypothesis that the varieties differ in yield.

In the above agricultural experiment the sampling unit is the plot. In psychological experimentation the analogue of the plot is usually either a human subject or an experimental animal. In an experiment on the relative efficacy of four different methods of memorizing nonsense syllables, four groups of subjects may be selected, a different method used on each group, and means on a measure of recall obtained for the four groups. A comparison of these means provides information on the relative efficacy of the different methods, and the analysis of variance may be used to decide whether the variation between means is greater than that expected from random sampling fluctuation.

The problem of testing the significance of the differences between a number of means results from experiments designed to study the variation in a dependent variable with variation in an independent variable. The independent variable may be varieties of wheat, methods of memorizing nonsense syllables, or different environmental conditions. The dependent variable may be crop yield, number of nonsense syllables recalled, or number of errors made by an animal in running a maze. Experiments which employ one independent variable are said to involve one basis of classification. The analysis of variance may be used in the analysis of data resulting from experiments which involve more than one basis of classification. For example, an experiment may be designed to permit the study both of varieties of wheat and types of fertilizer on crop yield. This experiment employs two independent variables. We wish to discover how crop yield depends on these two variables. The analysis of variance may be used to extract a part of the total variation resulting from the differences in varieties of wheat and another part resulting from differences in fertilizers, in addition to interaction and error components. A further example is a psychological experiment designed to permit the study of the effects of both free-versus-restricted environment and early-versus-late blindness on maze performance in the rat. Here we have two independent variables. Each variable has two categories. There are four combinations of conditions: free environment and early blindness, free environment and late blindness, restricted environment and early blindness, restricted environment and late blindness. Four groups of experimental animals may be used, and one of the four conditions applied to each group. The analysis of variance may be applied to identify parts of the variation in maze performance assignable to the different environmental and blindness conditions, and other parts as well. Experiments may be designed to permit the simultaneous study of any number of experimental variables within practical limits.

Let us proceed by considering in detail the simple case of a one-way-classification problem where the analysis of variance provides a composite test of the significance of the difference between a set of means.

15.2 NOTATION FOR ONE-WAY ANALYSIS OF VARIANCE

Consider an experiment involving k experimental treatments. The treatments may be different dosages of a drug, different methods of memorizing nonsense syllables, or different environmental variations in the rearing of experimental animals. Each treatment is applied to a different experimental group. Denote the number of members in the k groups by n_1, $n_2, \ldots, n_k$. The number of members in the jth group is n_j. The total number of members in all groups combined is $n_1 + n_2 + \cdots + n_k = N$. When the groups are of equal size we may write $n_1 = n_2 = \cdots = n_k = n = N/k$. The data may be represented as follows:

Group 1	*Group 2*	*Group k*
X_{11}	X_{12}	X_{1k}
X_{21}	X_{22}	X_{2k}
X_{31}	X_{32}	X_{3k}
.	.	.
.	.	.
.	.	.
$X_{n_1 1}$	$X_{n_2 2}$	$X_{n_k k}$
$\displaystyle\sum_{i=1}^{n_1} X_{i1}$	$\displaystyle\sum_{i=1}^{n_2} X_{i2}$	$\displaystyle\sum_{i=1}^{n} X_{ik}$

Here a system of double subscripts is used. The first subscript identifies the member of the group; the second identifies the group. Thus X_{21} represents the measurement for the second member of the first group, X_{32} repre-sents the measurement for the third member of the second group, and so on. In general, the symbol X_{ij} means the ith member of the jth group. Where the data for each group are tabulated in a separate column, the first subscript identifies the row and the second the column. The sum of measurements in the k groups is represented by

$$\sum_{i=1}^{n_1} X_{i1}, \sum_{i=1}^{n_2} X_{i2}, \quad \ldots \quad , \sum_{i=1}^{n_k} X_{ik}$$

We may denote the group means by $\overline{X}_{.1}, \overline{X}_{.2}, \ldots , \overline{X}_{.k}$. The symbol $\overline{X}_{.1}$ refers to the mean of the first column, $\overline{X}_{.2}$ the mean of the second column, and $\overline{X}_{.j}$ the mean of the jth column. The convention is to use a dot to indicate the variable subscript over which the summation extends. The mean of all the observations taken together may be represented by the sym-bol $\overline{X}_{..}$, sometimes called the grand mean. In a one-way classification the meaning associated with the various symbols is quite clear without the use of the dot notation. In discussion of one-way classification we shall there-fore simplify the notation and represent the group means by $\overline{X}_1, \overline{X}_2, \ldots ,$ $\overline{X}_k$ and the grand mean by $\overline{X}$. The dot notation is necessary in the more complex applications of the analysis of variance, and we shall return to it in Chap. 16.

The total variation in the data is represented by the sum of squares of deviations of all the observations from the grand mean. The sum of squares of deviations of the n_1 observations in the first group from the grand mean is $\sum_{i=1}^{n_1} (X_{i1} - \overline{X})^2$ and the sum of squares of the n_j observations in the jth group from the grand mean is $\sum_{i=1}^{n_j} (X_{ij} - \overline{X})^2$. For k groups each comprised of n_j observations the total sum of squares of deviations about $\overline{X}$ is

$$\sum_{j=1}^{k} \sum_{i=1}^{n_j} (X_{ij} - \overline{X})^2$$

When the meaning is clearly understood from the context, it is common practice to represent this total sum of squares by $\Sigma (X_{ij} - \overline{X})^2$, or more simply by $\Sigma (X - \overline{X})^2$.

5.3 PARTITIONING THE SUM OF SQUARES

Simple algebra may be used to demonstrate that the total sum of squares may be divided into two additive and independent parts, a *within-group* sum of squares and a *between-group* sum of squares. We proceed by writing the identity $(X_{ij} - \overline{X}) = (X_{ij} - \overline{X}_j) + (\overline{X}_j - \overline{X})$. This identity states that the deviation of a particular score from the grand mean is comprised of two parts, a deviation from the mean of the group to which the score belongs $(X_{ij} - \overline{X}_j)$ and a deviation of the group mean from the grand mean $(\overline{X}_j - \overline{X})$. We square this identity and sum over the n_j cases in the jth group to obtain

$$\sum_{i=1}^{n_j} (X_{ij} - \bar{X})^2 = \sum_{i=1}^{n_j} (X_{ij} - \bar{X}_j)^2 + \sum_{i=1}^{n_j} (\bar{X}_j - \bar{X})^2$$

$$+ 2(\bar{X}_j - \bar{X}) \sum_{i=1}^{n_j} (X_{ij} - \bar{X}_j)$$

The second term to the right requires the summation of a constant $(\bar{X}_j - \bar{X})^2$ over all n_j values of the jth group and may be written $n_j(\bar{X}_j - \bar{X})^2$. The third term to the right disappears because the sum of deviations about the mean $\bar{X}_j$ is zero. We obtain thereby

$$\sum_{i=1}^{n_j} (X_{ij} - \bar{X})^2 = \sum_{i=1}^{n_j} (X_{ij} - \bar{X}_j)^2 + n_j(\bar{X}_j - \bar{X})^2$$

This expression says that the sum of the squares of the deviations of the n_j observations in the jth group from the grand mean $\bar{X}$ is equal to the sum of squares of deviations of the observations from the group mean plus n_j times the square of the difference between the group mean and the grand mean. We now sum over the k groups to obtain

[15.1] $$\sum_{j=1}^{k} \sum_{i=1}^{n_j} (X_{ij} - \bar{X})^2 = \sum_{j=1}^{k} \sum_{i=1}^{n_j} (X_{ij} - \bar{X}_j)^2 + \sum_{j=1}^{k} n_j(\bar{X}_j - \bar{X})^2$$

The term to the left is the total sum of squares: the sum of squares of all the observations from the grand mean $\bar{X}$. The first term to the right is the sum of squares within groups: the sum of squares of deviations from the respective group means. The second and last term to the right is the sum of squares between groups: the sum of squares of deviations of the group means from the grand mean, each term $(\bar{X}_j - \bar{X})^2$ being weighted by n_j, the number of cases in the group. Thus the total sum of squares is partitioned into two additive parts, a sum of squares within groups, and a sum of squares between groups. These two parts are independent.

15.4 THE VARIANCE ESTIMATES OR MEAN SQUARES

Each sum of squares has an associated number of degrees of freedom. The total number of observations is $n_1 + n_2 + \cdots + n_k = \Sigma n_j = N$. The total sum of squares has $N - 1$ degrees of freedom. One degree of freedom is lost by taking deviations about the grand mean. $N - 1$ of these deviations are free to vary. The number of degrees of freedom associated with the within-groups sum of squares is

$$(n_1 - 1) + (n_2 - 1) + \cdots + (n_k - 1) = \sum_{j=1}^{k} n_j - k = N - k$$

The number of degrees of freedom for each group is $n_j - 1$. Hence the number of degrees of freedom for k groups is $\Sigma n_j - k$, or $N - k$. The number of degrees of freedom associated with the between-groups sum of squares is $k - 1$. We have k means, and 1 degree of freedom is lost by expressing the group means as deviations from the grand mean. The degrees of freedom are additive:

$$N - 1 = (N - k) + (k - 1)$$
$$\text{total} \qquad \text{within} \qquad \text{between}$$

The within- and between-groups sums of squares are divided by their associated degrees of freedom to obtain a within-groups variance estimate s_w^2 and a between-groups variance estimate s_b^2. Thus

[15.2] $$s_w^2 = \frac{\sum\limits_{j=1}^{k} \sum\limits_{i=1}^{n_j} (X_{ij} - \bar{X}_j)^2}{N - k}$$

[15.3] $$s_b^2 = \frac{\sum\limits_{j=1}^{k} n_j (\bar{X}_j - \bar{X})^2}{k - 1}$$

The sums of squares and degrees of freedom are additive. The variance estimates are not additive. The variance estimate is sometimes spoken of as the *mean square*.

15.5   THE MEANING OF THE VARIANCE ESTIMATES

What meaning attaches to the variance estimates s_w^2 and s_b^2? Let us assume that the k samples are drawn from populations having the same variance. The assumption is that $\sigma_1^2 = \sigma_2^2 = \cdots = \sigma_k^2 = \sigma^2$. If this assumption is tenable, the expected value of s_w^2 is σ^2; that is, $E(s_w^2) = \sigma^2$. Thus s_w^2 is an unbiased estimate of the population variance. It is an estimate obtained by combining the data for the k samples. It may be written in the form

[15.4] $$s_w^2 = \frac{\sum\limits_{i=1}^{n_1} (X_{i1} - \bar{X}_1)^2 + \sum\limits_{i=1}^{n_2} (X_{i2} - \bar{X}_2)^2 + \cdots + \sum\limits_{i=1}^{n_j} (X_{ik} - \bar{X}_k)^2}{n_1 + n_2 + \cdots + n_k - k}$$

The reader will recall that in applying the t test to determine the significance of the differences between two means for independent samples, an unbiased estimate of the population variance was obtained by combining the sums of squares about the means of the two samples and dividing this by the total number of degrees of freedom. The within-group variance s_w^2 is an estimate of precisely the same type. It is obtained by adding together the sums of squares about the k sample means and dividing this by the total number of degrees of freedom. The variance estimate used in the t test is the particular case of s_w^2 which occurs when $k = 2$.

The expected value of s_b^2 may be shown to be

[15.5] $$E(s_b^2) = \sigma^2 + \frac{\sum\limits_{j=1}^{k} (\mu_j - \mu)^2}{k - 1} \left(\frac{N - \sum\limits_{j=1}^{k} n_j^2/N}{k - 1} \right)$$

where μ_j and μ are population means. Under the null hypothesis $\mu_1 = \mu_2 = \cdots = \mu_k = \mu$, and the second term to the right of the above expression is equal to zero. Hence under the null hypothesis both s_w^2 and s_b^2 are estimates of the population variance σ^2.

That s_b^2 is an estimate of σ^2 under the null hypothesis may be illustrated by considering the particular situation where $n_1 = n_2 = \cdots = n_k = n$. The between-group variance estimate may then be written as

$$\frac{n \sum_{j=1}^{k} (\overline{X}_j - \overline{X})^2}{k-1}$$

This is n times the variance of the k means, or $ns_{\overline{x}}^2$. The error variance of the sampling distribution of the arithmetic mean for samples of size n is given by $\sigma_{\overline{x}}^2 = \sigma^2/n$. Hence $n\sigma_{\overline{x}}^2 = \sigma^2$. The quantity $ns_{\overline{x}}^2$ is an estimate of $n\sigma_{\overline{x}}^2$, hence also of σ^2. Thus s_b^2 is an estimate of σ^2.

Where the null hypothesis is false and the means of the populations from which the k samples are drawn differ one from another, the second term to the right of the expression for $E(s_b^2)$ is not equal to zero. It is a measure of the variation of the separate population means μ_j from the grand mean μ.

To test the hypothesis $H_0: \mu_1 = \mu_2 = \cdots = \mu_k$, consider the ratio s_b^2/s_w^2. This is an F ratio. Under the null hypothesis the expected value of this ratio is unity since $E(s_b^2) = E(s_w^2) = \sigma^2$. If the population means differ from each other, $E(s_b^2/s_w^2)$ will be greater than unity. If s_b^2/s_w^2 is found to be significantly greater than unity, this may be construed to be evidence for the rejection of the null hypothesis and for the acceptance of the alternative hypothesis that differences exist between the population means. The significance of the F ratio s_b^2/s_w^2 may be assessed with reference to the table of F (Table D of the Appendix) with $k-1$ degrees of freedom associated with the numerator and $N-k$ degrees of freedom associated with the denominator.

15.6 COMPUTATION FORMULAS

The calculation of the required sums of squares may be simplified by the use of computation formulas. To simplify the notation, denote the sum of all the observations in the jth group by T_j. Thus

$$\sum_{i=1}^{n_j} X_{ij} = T_j$$

Denote the sum of all observations in the k groups by T. Thus

$$\sum_{j=1}^{k} \sum_{i=1}^{n_j} X_{ij} = T$$

The computation formulas are readily obtained. The formula for the *total* sum of squares is

$$[15.6] \quad \sum_{j=1}^{k} \sum_{i=1}^{n_j} (X_{ij} - \overline{X})^2 = \sum_{j=1}^{k} \sum_{i=1}^{n_j} X_{ij}^2 - \frac{T^2}{N}$$

Thus we find the sum of squares of all observations and subtract T^2/N. The *within-groups* sum of squares is

$$[15.7] \quad \sum_{j=1}^{k} \sum_{i=1}^{n_j} (X_{ij} - \overline{X}_j)^2 = \sum_{j=1}^{k} \sum_{i=1}^{n_j} X_{ij}^2 - \sum_{j=1}^{k} \left(\frac{T_j^2}{n_j} \right)$$

The quantity T_j^2/n_j is the square of the sum of the jth group divided by the number of cases in that group. These values are calculated and summed over the k groups. The *between-groups* sum of squares is

$$[15.8] \quad \sum_{j=1}^{k} n_j (\overline{X}_j - \overline{X})^2 = \sum_{j=1}^{k} \left(\frac{T_j^2}{n_j} \right) - \frac{T^2}{N}$$

The above formulas are generally applicable to groups of unequal or equal size. In the particular case where the groups are of equal size and $n_1 = n_2 = \cdots = n_k = n$, the within-groups sum of squares may be written as

$$[15.9] \quad \sum_{j=1}^{k} \sum_{i=1}^{n} X_{ij}^2 - \frac{\sum_{j=1}^{k} T_j^2}{n}$$

and the between-groups sum of squares becomes

$$[15.10] \quad \frac{\sum_{j=1}^{k} T_j^2}{n} - \frac{T^2}{N}$$

15.7 SUMMARY

Table 15.1 presents in summary form the formulas hitherto discussed.
In summary, to test the significance of the difference between k means using the analysis of variance, the following steps are involved:

1 Partition the total sum of squares into two components, a within-groups and a between-groups sum of squares, using the appropriate computation formulas.

2 Divide these sums of squares by the associated number of degrees of freedom to obtain s_w^2 and s_b^2, the within- and between-groups variance estimates.

3 Calculate the F ratio s_b^2/s_w^2 and refer this to the table of F (Table D of the Appendix).

4 If the probability of obtaining the observed F value is small, say, less than .05 or .01, under the null hypothesis, reject that hypothesis.

Table 15.1   Analysis of variance: one-way classification summary of formulas

Source of variation

	Between	Within	Total
Sum of squares	$\sum\limits_{j=1}^{k} n_j(\bar{X}_j - \bar{X})^2$	$\sum\limits_{j=1}^{k}\sum\limits_{i=1}^{n_j} (X_{ij} - \bar{X}_j)^2$	$\sum\limits_{j=1}^{k}\sum\limits_{i=1}^{n_j} (X_{ij} - \bar{X})^2$
Degrees of freedom	$k - 1$	$N - k$	$N - 1$
Variance estimate; mean square	$s_b^2 = \dfrac{\sum\limits_{j=1}^{k} n_j(\bar{X}_j - \bar{X})^2}{k-1}$	$s_w^2 = \dfrac{\sum\limits_{j=1}^{k}\sum\limits_{i=1}^{n_j} (X_{ij} - \bar{X})^2}{N-k}$	
Expectation	$E(s_b^2) = \sigma^2 + \dfrac{\sum\limits_{j=1}^{k} (\mu_j - \mu)^2}{k-1} \left(\dfrac{N - \sum\limits_{j=1}^{k} n_j^2/N}{k-1} \right)$	$(Es_w^2) = \sigma^2$	
Computation formulas	$\sum\limits_{j=1}^{k}\left(\dfrac{T_j^2}{n_j}\right) - \dfrac{T^2}{N}$	$\sum\limits_{j=1}^{k}\sum\limits_{i=1}^{n_j} X_{ij}^2 - \sum\limits_{j=1}^{k}\left(\dfrac{T_j^2}{n_j}\right)$	$\sum\limits_{j=1}^{k}\sum\limits_{i=1}^{n_j} X_{ij}^2 - \dfrac{T^2}{N}$
Computation formulas: equal groups	$\dfrac{\sum\limits_{j=1}^{k} T_j^2}{n} - \dfrac{T^2}{N}$	$\dfrac{\sum\limits_{j=1}^{k} T_j^2}{n}$	$\sum\limits_{j=1}^{k}\sum\limits_{i=1}^{n} X_{ij}^2 - \dfrac{T^2}{N}$

15.8 ILLUSTRATIVE EXAMPLE: ONE-WAY CLASSIFICATION

Table 15.2 shows the number of nonsense syllables recalled by four groups of subjects using four different methods of presentation. Fictitious data are used here for simplicity of illustration. The sums of squares have been calculated using the computation formulas. The data are presented in summary form in Table 15.3. The number of groups is 4. The number of degrees of freedom associated with the between-groups sum of squares is $k - 1 = 4 - 1 = 3$. The number of degrees of freedom associated with the within-groups sum of squares is $N - k = 26 - 4 = 22$. The number of degrees of freedom associated with the total is $N - 1 = 26 - 1 = 25$. The between and within sums of squares are divided by the associated degrees of freedom to obtain the variance estimates s_b^2 and s_w^2.

The F ratio is $s_b^2/s_w^2 = 27.41/3.91 = 7.01$. Consulting a table of F with $df = 3$ associated with the numerator and $df = 22$ with the denominator,

Table 15.2 **Computation for the analysis of variance; one-way classification. Number of nonsense syllables correctly recalled under four methods of presentation**

		Method		
	I	*II*	*III*	*IV*
	5	9	8	1
	7	11	6	3
	6	8	9	4
	3	7	5	5
	9	7	7	1
	7		4	4
	4		4	
	2			
n_j	8	5	7	6	$N = 26$
T_j	43	42	43	18	$T = 146$
$\bar{X}_j$	5.38	8.40	6.14	3.00	$T^2/N = 819.85$
$\sum_{i=1}^{n_j} X_{ij}^2$	269	364	287	68	$\sum_{j=1}^{k} \sum_{i=1}^{n_j} X_{ij}^2 = 988$
$\dfrac{T_j^2}{n_j}$	231.13	352.80	264.14	54.00	$\sum_{j=1}^{k} \dfrac{T_j^2}{n_j} = 902.07$

Sum of squares

Between	$902.07 - 819.85 =$	82.22
Within	$988 \quad - 902.07 =$	85.93
Total	$988 \quad - 819.85 =$	168.15

Table 15.3   Analysis of variance for data of Table 15.2

Source of variation	Sum of squares	Degrees of freedom	Variance estimate
Between	82.22	3	$27.41 = s_b^2$
Within	85.93	22	$3.91 = s_w^2$
Total	168.15	25	$F = 7.01$

we find that the value of F required for significance at the .01 level is 4.82. We may safely conclude that the method of presentation affects the number of nonsense syllables recalled.

15.9   THE ANALYSIS OF VARIANCE WITH TWO GROUPS

With two groups only, the significance of the differences between means may be tested using either a t test or the analysis of variance. These procedures lead to the same result. Where $k = 2$ it may be readily shown that $\sqrt{F} = t$.

Consider a situation where $k = 2$ and $n_1 = n_2 = n$. Under these circumstances the between-groups variance estimate s_b^2 is

$$s_b^2 = \frac{n(\bar{X}_1 - \bar{X})^2 + n(\bar{X}_2 - \bar{X})^2}{2 - 1}$$

For groups of equal size the grand mean $\bar{X}$ is halfway between the two group means $\bar{X}_1$ and $\bar{X}_2$. Thus $(\bar{X}_1 - \bar{X}) = (\bar{X}_2 - \bar{X}) = \frac{1}{2}(\bar{X}_1 - \bar{X}_2)$ and $(\bar{X}_1 - \bar{X})^2 = (\bar{X}_2 - \bar{X})^2 = \frac{1}{4}(\bar{X}_1 - \bar{X}_2)^2$. We may therefore write

$$s_b^2 = \frac{n}{2}(\bar{X}_1 - \bar{X}_2)^2$$

When $k = 2$ the within-groups variance estimate s_w^2 is the unbiased variance estimate s^2, obtained by adding the two sums of squares about the means of the two samples and dividing by the total number of degrees of freedom (Sec. 15.5). Hence

[15.11] $F = \dfrac{(\bar{X}_1 - \bar{X}_2)^2}{s^2(2/n)}$

and

$$\sqrt{F} = \frac{\bar{X}_1 - \bar{X}_2}{s\sqrt{(1/n) + (1/n)}} = t$$

Thus $\sqrt{F} = t$ and $F = t^2$. To illustrate, let $n_1 = n_2 = 8$. In applying the analysis of variance with $df = 1$ associated with the numerator and $df = 14$ associated with the denominator of the F ratio, an F of 4.60 is required for significance at the .05 level. The corresponding t for $df = 14$ required for significance at the .05 level is $\sqrt{4.60} = 2.145$. The t test may be considered a particular case of the F test. It is a particular case which arises when $k = 2$.

In the above discussion we have considered two groups of equal size. The result $\sqrt{F} = t$ is, however, quite general and holds when n_1 and n_2 are unequal. For unequal groups the algebraic development is a bit more cumbersome than that given here. The grand mean does not fall midway between the two group means.

15.10 ASSUMPTIONS UNDERLYING THE ANALYSIS OF VARIANCE

In the mathematical development of the analysis of variance a number of assumptions are made. Questions may be raised about the nature of these assumptions and the extent to which the failure of the data to satisfy them leads to the drawing of invalid inferences.

One assumption is that the distribution of the dependent variable in the population from which the samples are drawn is normal. For large samples the normality of the distribution may be tested using a test of goodness of fit, although in practice this is rarely done. When the samples are fairly small, it is usually not possible to rigorously demonstrate lack of normality in the data. Unless there is reason to suspect a fairly extreme departure from normality, it is probable that the conclusions drawn from the data using an F test will not be seriously affected. In general, the effect of a departure from normality is to make the results appear somewhat more significant than they are. Consequently, where a fairly gross departure from normality occurs, a somewhat more rigorous level of confidence than usual may be employed.

A furthur assumption in the application of the analysis of variance is that the variances in the populations from which the samples are drawn are equal. This is known as homogeneity of variance. A variety of tests of homogeneity of variance may be applied. These are discussed in more advanced texts (Winer, 1962). Moderate departures from homogeneity should not seriously affect the inferences drawn from the data. Gross departures from homogeneity may lead to results which are seriously in error. Under certain circumstances a transformation of the variable, which leads to greater uniformity of variance, may be used. Under other circumstances it may be possible to use a nonparametric procedure.

A further assumption is that the effects of various factors on the total variation are additive, as distinct from, say, multiplicative. The basic model underlying the analysis of variance is that a given observation may be partitioned into independent and additive bits, each bit resulting from an identifiable source. In most situations there are no grounds to suspect the validity of this model.

With most sets of real data the assumptions underlying the analysis of variance are, at best, only roughly satisfied. The raw data of experiments frequently do not exhibit the characteristics which the mathematical models require. One advantage of the analysis of variance is that reasonable departures from the assumptions of normality and homogeneity may

occur without seriously affecting the validity of the inferences drawn from
the data.

15.11 TRANSFORMATION OF DATA

As indicated in Sec. 15.10 the appropriate use of the analysis of variance
involves assumptions of homogeneity of variance, normality, and additivity. With some data these assumptions are obviously not satisfied. Sometimes it is possible to use a simple transformation, resulting in a set of
transformed values which conform more closely to one or more of the
assumptions which the appropriate use of the analysis requires. Most
transformations are intended to bring the variances closer to equality. In
many cases the transformed values will approximate the normal form more
closely than do the original observations.

Commonly used transformations are the square root transformation,
$\sqrt{X}$, the logarithmic transformation $\log X$, the reciprocal transformation,
$1/X$, and the arc sine transformation, $\arcsin \sqrt{X}$. A decision regarding
which transformation is appropriate is made by exploring the relation
between the variances and the treatment means. The nature of this relation
determines which transformation to use.

The square root transformation replaces each of the original observations by its square root. This transformation is appropriate when the
variances are proportional to the means; that is, when $s_i^2/\overline{X}_i = c$, where c
is a constant. Of course with real data this relation will be only roughly
approximated.

Table 15.4 shows an example of a square root transformation. The original data are shown on the left. Note that the variances differ markedly one
from another. Note also that the variances are roughly proportional to the
means. The ratio $s_i^2/\overline{X}_i$ is in the neighborhood of 1.25. Transformed values
are shown at the right in Table 15.4. Note that the effect of the transformation is to reduce the disparity between the variances. The variances for the
four groups are now .34, .29, .30, .34, and the assumption of homogeneity of
variance is approximately satisfied in the transformed data.

With data containing small numbers, and with zeros, it is advisable to
add .5 to each value before taking the square root. The transformation
becomes $\sqrt{X + .5}$.

When the variances are proportional, not to the means, but to the square
of the means, a logarithmic transformation, $\log \overline{X}$, is appropriate. Here
again, of course, with real data the relation $s_i^2/\overline{X}_i^2 = c$ may be only
roughly satisfied. If some of the observations are zero and small numbers,
the suggestion is that the transformation $\log (\overline{X} + 1)$ be used. When the
standard deviations, and not the variances, are proportional to the square
of the treatment means, that is, when $s_i/\overline{X}_i^2 = c$, a reciprocal transformation, $1/X$, may prove useful.

In some experiments the observations obtained are proportions or percentages, such as the proportion of successful responses in a given number
of trials, or the percentage of correct answers in performing a series of
tasks. When the data are proportions, a relation between means and

Table 15.4 **Comparison of original and transformed data using a square root transformation**

	Original data, X				Transformed data, $\sqrt{X}$			
	I	**II**	**III**	**IV**	**I**	**II**	**III**	**IV**
	1	5	14	2	1.00	2.24	3.74	1.41
	5	9	18	7	2.24	3.00	4.24	2.65
	2	5	9	3	1.41	2.24	3.00	1.73
	6	11	21	10	2.45	3.32	4.58	3.16
	7	12	20	7	2.65	3.46	4.47	2.65
	3	5	12	5	1.73	2.24	3.46	2.24
	3	8	16	6	1.73	2.83	4.00	2.45
	27	55	110	40	13.21	19.33	27.49	16.29
$\overline{X}$	3.86	7.86	15.71	5.71	1.89	2.76	3.93	2.33
s^2	4.78	8.76	19.06	7.30	.34	.29	.30	.34

variances roughly of the kind $s_1^2 = \overline{X}_i (1 - \overline{X}_i)$ may be observed. Under this circumstance an arc sine transformation, arcsin $\sqrt{X}$, may be used. Each of the original observations is replaced by an angle whose sine is the square root of the original observation. A useful table of angles corresponding to percentages, that is, arcsin $\sqrt{\text{percentage}}$, has been prepared by C. I. Bliss and is given in Snedecor and Cochran (1967).

EXERCISES **1** How many degrees of freedom are associated with the variation in the data for (a) a comparison of two means for independent samples, each containing 20 cases, (b) a comparison of four means for independent samples, each containing 14 cases, (c) a comparison of four means for independent samples of size 10, 16, 18, and 11, respectively?

2 The following are measurements obtained for five equal groups of subjects:

Group	Measurements					$\overline{X}_j$
I	4	7	9	9	14	8.60
II	5	6	12	12	7	8.40
III	15	18	21	26	20	20.00
IV	35	27	29	30	25	29.20
V	17	26	17	20	12	18.40

Apply the analysis of variance to test the null hypothesis

$H_0: \mu_1 = \mu_2 = \mu_3 = \mu_4 = \mu_5$

3   The following are error scores on a psychomotor test for four groups of subjects tested under four experimental conditions:

Group	Error scores							$\overline{X}_j$
I	16	7	19	24	31			19.40
II	24	6	15	25	32	24	29	22.14
III	16	15	18	19	6	13	18	15.00
IV	25	19	16	17	42	45		27.33

Apply the analysis of variance to test the null hypothesis

$$H_0: \mu_1 = \mu_2 = \mu_3 = \mu_4$$

4   Apply the analysis of variance to test the significance of the difference between means for the following data:

	I	*II*	*III*
n	10	10	10
$\overline{X}_j$	7.40	8.30	10.56
$\sum\limits_{i=1}^{n} X_{ij}^2$	649	755	1,263

5   The following are test scores for two groups:

Group	Scores							$\overline{X}_j$
I	12	26	31	12	14	16	10	17.29
II	8	14	29	7	14	6		13.00

Calculate s_w^2, s_b^2, and F. Test the null hypothesis $H_0: \mu_1 = \mu_2$.

6   What assumptions underly the analysis of variance?

7   The following are experimental data for three independent groups:

Group	Data					$\overline{X}_j$
I	4	16	49	64	81	42.80
II	49	121	144	169	196	135.80
III	16	36	81	100	121	70.80

What would be an appropriate transformation on these data? Transform the data and apply an analysis of variance to the transformed values, calculating the required F ratio.

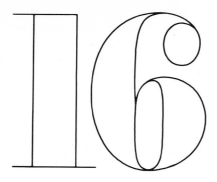

ANALYSIS OF VARIANCE:
TWO-WAY CLASSIFICATION

16.1 INTRODUCTION

Experiments may be designed to permit the simultaneous investigation of two experimental variables. Such experiments involve two bases of classification. To illustrate, assume that an investigator wishes to study the effects of two methods of presenting nonsense syllables on recall after 5 min, 1 hr, and 24 hr. One experimental variable is method of presentation, the other the interval between presentation and recall. There are six combinations of experimental conditions. One method of conducting such an experiment is to select a group of subjects and allocate these at random to the experimental conditions, an equal number being assigned to each condition. With, say, 10 subjects allocated to each experimental condition, the total number of subjects will be $2 \times 3 \times 10 = 60$. The data may be arranged in a table containing two rows and three columns. The rows correspond to methods, the columns to time intervals. The 10 observations for each group may be entered in each cell of the table. Differences in the means of the rows result from differences in recall under the two methods of presentation. Differences in the means of the columns result from differences in recall after the three time intervals.

Experiments with two-way classification may be conducted with only one sampling unit, and measurement, for each experimental condition. With one measurement for each experimental condition the total sum of squares is partitioned into three parts, a between-rows, a between-columns, and an interaction sum of squares. With more than one measurement for each experimental condition, the total sum of squares is partitioned into four parts, a between-rows, a between-columns, an interaction sum of squares, and a within-cells sum of squares. Each sum of squares has an associated number of degrees of freedom. By dividing the sums of squares by the associated degrees of freedom four variance estimates are obtained. These variance estimates are used to test the significance of the differences between row means, column means, and, with more than one measurement per cell, the interaction effect.

16.2 NOTATION FOR TWO-WAY ANALYSIS OF VARIANCE

Consider an experiment involving R experimental treatments of one variable and C experimental treatments of another variable. The number of treatment combinations is RC. Let us consider the particular case where we have one sampling unit, and one measurement, for each of the RC treatment combinations. The total number of measurements is $RC = N$. The data may be represented as follows:

	1	2	3	. . .	C	Row mean
1	X_{11}	X_{12}	X_{13}	. . .	X_{1C}	$\overline{X}_{1.}$
2	X_{21}	X_{22}	X_{23}	. . .	X_{2C}	$\overline{X}_{2.}$
3	X_{31}	X_{32}	X_{33}	. . .	X_{3C}	$\overline{X}_{3.}$
.	.	.	.	. . .	.	.
.	.	.	.	. . .	.	.
.	.	.	.	. . .	.	.
R	X_{R1}	X_{R2}	X_{R3}	. . .	X_{RC}	$\overline{X}_{R.}$
Column mean	$\overline{X}_{.1}$	$\overline{X}_{.2}$	$\overline{X}_{.3}$	. . .	$\overline{X}_{.C}$	$\overline{X}_{..}$

Double subscripts are used. The first subscript identifies the *row*; the second subscript identifies the *column*. Thus X_{32} is the measurement in the third row and the second column. Usually X_{rc} denotes the measurement in the rth row and cth column, where $r = 1, 2, \ldots, R$ and $c = 1, 2, \ldots, C$. A dot notation is used to identify means. The symbol $\overline{X}_{1.}$ refers to the mean of the first row, $\overline{X}_{2.}$ the mean of the second row, and $\overline{X}_{r.}$ the mean of the rth row. Similarly, $\overline{X}_{.1}$ refers to the mean of the first column. $\overline{X}_{.2}$ the mean of the second column, and $\overline{X}_{.c}$ the mean of the cth column. The grand mean, the mean of all N observations, is $\overline{X}_{..}$. The total sum of squares of deviations about the grand mean is given by

$$\sum_{r=1}^{R} \sum_{c=1}^{C} (X_{rc} - \overline{X}_{..})^2$$

Consider now a situation where we have n sampling units and n measurements for each of the RC treatment combinations. The total number of measurements is $nRC = N$. Where $R = 2$, $C = 3$, and $n = 3$, the data may be represented as follows:

	1	2	3	Row mean
1	X_{111} X_{112} X_{113}	X_{121} X_{122} X_{123}	X_{131} X_{132} X_{133}	$\overline{X}_{1..}$
2	X_{211} X_{212} X_{213}	X_{221} X_{222} X_{223}	X_{231} X_{232} X_{233}	$\overline{X}_{2..}$
Column mean	$\overline{X}_{.1.}$	$\overline{X}_{.2.}$	$\overline{X}_{.3.}$	

Triple subscripts are used. The first subscript identifies the *row*, the second the *column*, and the third the measurement *within* the cell. Thus X_{231} means the measurement for the first individual in the second row and the third column. In general X_{rci} denotes the measurement for the ith individual in the rth row and cth column, where $i = 1, 2, \ldots , n$. Row, column, and cell means are identified by a dot notation. The mean of all the observations in the first row is $\overline{X}_{1..}$. The mean of all the observations in the rth row is $\overline{X}_{r..}$. Similarly, the mean of the first column is $\overline{X}_{.1.}$ and of the cth column $\overline{X}_{.c.}$. The mean of all the observations in the cell corresponding to the rth row and cth column is $\overline{X}_{rc.}$. The mean of all nRC observations, the grand mean, is $\overline{X}_{...}$. The total sum of squares of all observations about the grand mean is

$$\sum_{r=1}^{R} \sum_{c=1}^{C} \sum_{i=1}^{n} (X_{rci} - \overline{X}_{...})^2$$

The sum of squares of deviations about the grand mean, both where $n = 1$ and $n > 1$, is partitioned into additive components.

16.3 PARTITIONING THE SUM OF SQUARES

With one measurement only for each of the RC treatment combinations, the total sum of squares may be partitioned into three additive components, a between-rows, a between-columns, and an interaction sum of squares. We proceed by writing the identity

$$(X_{rc} - \overline{X}_{..}) = (\overline{X}_{r.} - \overline{X}_{..}) + (\overline{X}_{.c} - \overline{X}_{..}) + (X_{rc} - \overline{X}_{r.} - \overline{X}_{.c} + \overline{X}_{..})$$

This identity states that the deviation of an observation from the grand mean may be viewed as composed of three parts, a deviation of the row mean from the grand mean, a deviation of the column mean from the

grand mean, and a remainder, or residual term, known as an interaction
term. By squaring both sides of the above identity, an expression is obtained containing six terms. This may be summed over R rows and C
columns. Three of these terms conveniently vanish, because they contain
a sum of deviations about a mean, which, of course, is zero. The resulting
total sum of squares may be written as

$$\sum_{r=1}^{R} \sum_{c=1}^{C} (X_{rc} - \bar{X}_{..})^2 = C \sum_{r=1}^{R} (\bar{X}_{r.} - \bar{X}_{..})^2 + R \sum_{c=1}^{C} (\bar{X}_{.c} - \bar{X}_{..})^2$$

$$+ \sum_{r=1}^{R} \sum_{c=1}^{C} (X_{rc} - \bar{X}_{r.} - \bar{X}_{.c} + \bar{X}_{..})^2$$

The first term to the right is C times the sum of squares of deviations of
row means from the grand mean. This is the between-rows sum of squares.
It describes the variation in row means. The second term is R times the
sum of squares of deviations of column means from the grand mean. This
is the between-columns sum of squares. It describes variation in column
means. The third term is a residual, or interaction, sum of squares. The
meaning of the interaction term is discussed in detail in Sec. 16.5.

With n measurements for each of the RC treatment combinations the
total sum of squares may be partitioned into four additive components.
These are a between-rows, a between-columns, an interaction, and a
within-cells sum of squares. In this situation we write the identity

$$(X_{rci} - \bar{X}_{...}) = (\bar{X}_{r..} - \bar{X}_{...}) + (\bar{X}_{.c.} - \bar{X}_{...})$$
$$+ (\bar{X}_{rc.} - \bar{X}_{r..} - \bar{X}_{.c.} + \bar{X}_{...}) + (X_{rci} - \bar{X}_{rc.})$$

This expression may be squared and summed over rows, columns, and
within cells. All but four terms vanish, and the resulting total sum of
squares may be written as

$$[16.1] \quad \sum_{r=1}^{R} \sum_{c=1}^{C} \sum_{i=1}^{n} (X_{rci} - \bar{X}_{...})^2 = nC \sum_{r=1}^{R} (\bar{X}_{r..} - \bar{X}_{...})^2 + nR \sum_{c=1}^{C} (\bar{X}_{.c.} - \bar{X}_{...})^2$$

$$+ n \sum_{r=1}^{R} \sum_{c=1}^{C} (\bar{X}_{rc.} - \bar{X}_{r..} - \bar{X}_{.c.} + \bar{X}_{...})^2$$

$$+ \sum_{r=1}^{R} \sum_{c=1}^{C} \sum_{i=1}^{n} (X_{rci} - \bar{X}_{rc.})^2$$

The first term to the right is descriptive of the variation of row means,
the second of column means, and the third of interaction. The fourth
term is the within-cells sum of squares. It is the sum of squares of the
deviations of observations from the means of the cells to which they belong.

16.4 VARIANCE ESTIMATES OR MEAN SQUARES

With a single entry in each cell, $n = 1$ and $RC = N$. The number of degrees
of freedom associated with the total sum of squares is $RC - 1 = N - 1$. The

numbers of degrees of freedom associated with row and column sums are $R - 1$ and $C - 1$, respectively. The number of degrees of freedom associated with the interaction sum of squares is $(R - 1)(C - 1)$. The degrees of freedom are additive, and

$$\underset{\text{total}}{N - 1} = \underset{\text{row}}{(R - 1)} + \underset{\text{column}}{(C - 1)} + \underset{\text{interaction}}{(R - 1)(C - 1)}$$

The sums of squares are divided by the associated degrees of freedom to obtain three variance estimates, or mean squares. The between-rows, between-columns, and interaction variance estimates are s_r^2, s_c^2, and s_i^2, respectively.

With n entries in each cell, where $n > 1$, the total number of observations is $nRC = N$. The number of degrees of freedom associated with the total sum of squares is $nRC - 1 = N - 1$. The numbers of degrees of freedom associated with row, column, and interaction sums of squares are $R - 1, C - 1$, and $(R - 1)(C - 1)$, respectively. The number of degrees of freedom associated with the within-cells sum of squares is $nRC - RC = RC(n - 1)$. Because the deviations are taken about the cell means, 1 degree of freedom is lost for each cell. In each cell $n - 1$ deviations are free to vary. The number of degrees of freedom for RC cells, therefore, is $RC(n - 1)$. The degrees of freedom are additive. The sums of squares are divided by the associated degrees of freedom to obtain the variance estimates, or mean squares.

Table 16.1 shows in summary form the sum of squares, degrees of freedom, and variance estimates for a two-way classification with n entries per cell.

F ratios are formed from the variance estimates and used to test the

Table 16.1 Analysis of variance for two-way classification with n entries per cell: $n > 1$

Source	Sum of squares	df	Variance estimate
Rows	$nC \sum\limits_{i=1}^{R} (\overline{X}_{r..} - \overline{X}_{...})^2$	$R - 1$	s_r^2
Columns	$nR \sum\limits_{c=1}^{C} (\overline{X}_{.c.} - \overline{X}_{...})^2$	$C - 1$	s_c^2
Interaction	$n \sum\limits_{r=1}^{R} \sum\limits_{c=1}^{C} (\overline{X}_{rc.} - \overline{X}_{r..} - \overline{X}_{.c.} + \overline{X}_{...})^2$	$(R - 1)(C - 1)$	s_i^2
Within cells	$\sum\limits_{r=1}^{R} \sum\limits_{c=1}^{C} \sum\limits_{i=1}^{n} (X_{rci} - \overline{X}_{rc.})^2$	$RC(n - 1)$	s_w^2
Total	$\sum\limits_{r=1}^{R} \sum\limits_{c=1}^{C} \sum\limits_{i=1}^{n} (X_{rci} - \overline{X}_{...})^2$	$nRC - 1$	

significance of row, column, and, where $n > 1$, interaction effects. The correct procedure here, and the interpretation of the variance estimates, depends on the statistical model appropriate for the experiment. Three models may be identified: fixed, random, and mixed. The investigator must decide which model fits his experiment. This decision determines how the variance estimates are used in the application of tests of significance to the data. Before proceeding with a discussion of these models (Sec. 16.6), the meaning of the interaction term is discussed.

16.5   THE NATURE OF INTERACTION

The algebraic partitioning of sums of squares in a two-way classification, where $n > 1$, leads to the interaction term

$$n \sum_{r=1}^{R} \sum_{c=1}^{C} (\bar{X}_{rc.} - \bar{X}_{r..} - \bar{X}_{.c.} + \bar{X}_{...})^2$$

The nature of interaction may be illustrated by example. Consider a simple agricultural experiment with two varieties of wheat and two types of fertilizer. Assume that one variety of wheat has a higher yield than the other. If the yield is uniformly higher regardless of which fertilizer is used, then there is no interaction between the two experimental varieties. If, however, one variety produces a relatively higher yield with one type of fertilizer than with the other, then the two variables may be said to interact. To illustrate further, assume that we have two methods of teaching arithmetic and two teachers. Each teacher uses the two methods on separate groups of pupils. The achievement of the pupils is measured. If one method of instruction is uniformly superior or inferior regardless of which teacher uses it, then there is no interaction between methods and teacher. If, however, one teacher obtains better results with one method than the other, and the opposite holds for the other teacher, then teachers and methods may be said to interact.

Table 16.2 shows observed cell means for a two-way classification with three categories for each of the experimental variables. The observed cell entries are means based on an equal number of cases. What are the expected cell means on the assumption of zero interaction? This situation is somewhat analogous to the calculation of expected values for contingency tables. For a contingency table we calculate expected cell frequencies. Here we are required to calculate the expected cell means on the assumption that the two experimental variables function independently.

Assuming zero interaction, certain constant differences will be maintained between cell means. In Table 16.2 the observed row mean for A is 10 points less than the row mean for B. If the interaction were zero, we should expect a constant 10-point difference to occur between means for A and B under treatments I, II, and III. A similar relationship would be expected on comparing all other rows and columns of this table. Obviously, the observed values in Table 16.2 do not exhibit this characteristic. The interaction is not zero.

Where the interaction is zero, a deviation of a cell mean from the mean

Table 16.2   Comparison of observed cell means and means expected under zero interaction

	Observed, $\bar{X}_{rc.}$					Expected, $E(\bar{X}_{rc.})$			
	I	II	III			I	II	III	
A	2	12	16	10	A	4	16	10	10
B	6	20	34	20	B	14	26	20	20
C	64	76	40	60	C	54	66	60	60
	24	36	30	30		24	36	30	30

of the row (or column) to which it belongs will be equal to the deviation of its column (or row) mean from the grand mean. If $X_{rc.}$ is a cell mean and $\bar{X}_{r..}$ and $\bar{X}_{.c.}$ are its corresponding row and column means, then under zero interaction, $\bar{X}_{rc.} - \bar{X}_{r..} = \bar{X}_{.c.} - \bar{X}_{...}$. Thus the expected value of $\bar{X}_{rc.}$ under zero interaction is given by $E(\bar{X}_{rc.}) = \bar{X}_{r..} + \bar{X}_{.c.} - \bar{X}_{...}$. These expected values have been calculated for the observed data of Table 16.2 and are shown to the right of the table. On comparing the expected values in any two rows or columns, note the constant increment or decrement. If $\bar{X}_{rc.}$ is an observed and $E(\bar{X}_{rc.})$ an expected value, the deviation of an observed from an expected value is $\bar{X}_{rc.} - \bar{X}_{r..} - \bar{X}_{.c.} + \bar{X}_{...}$. The interaction term in the analysis of variance is n times the sum of squares of deviations of the observed cell means from the expected cell means.

16.6   FINITE, RANDOM, FIXED, AND MIXED MODELS

Different authors recommend different procedures for testing row, column, and interaction effects in a two-way analysis of variance. Difficulties associated with the selection of the appropriate procedure are resolved by the recognition of a general statistical model underlying the analysis of variance. This model is referred to here as the *finite* model. Three particular cases of the finite model may be identified. These are the *random, fixed,* and *mixed* models. The models appropriate for different experiments differ. The investigator must decide which model best represents his experiment. The choice of model determines the procedure for testing row, column, and interaction effects. The choice of model depends on the nature of the variables used as the basis of classification in the experimental design.

The general finite model makes the linearity assumption that a deviation of an observation X_{rci} from the population value of the grand mean μ may be expressed in the form

[16.2] $X_{rci} - \mu = a_r + b_c + (ab)_{rc} + e_{rci}$

The four quantities to the right are in deviation form. Thus $a_r = \mu_{r..} - \mu$, a deviation of the population value of the row mean from the grand mean μ. Similarly, $b_c = \mu_{.c.} - \mu$, a deviation of a column mean from the grand mean. The interaction term $(ab)_{rc} = (\mu_{rc.} - \mu_{r..} - \mu_{.c.} + \mu)$, and the error term $e_{rci} = X_{rci} - \mu_{rc.}$. Where this model is used to represent experimental data the implicit assumption is made that treatment effects can meaningfully be partitioned into additive components for each sampling unit. Because a_r, b_c, $(ab)_{rc}$, and e_{rci} are in deviation form, they sum to zero. The population variances of the four components are σ_a^2, σ_b^2, σ_{ab}^2, and σ_e^2.

The null hypothesis under test, for example, for row effects is $H_r : \mu_{1..} = \mu_{2..} = \ldots = \mu_{R..}$. This hypothesis may be stated in the form $H_r : \sigma_a^2 = 0$. Similarly, the null hypotheses for column and interaction effects may be stated as $H_c : \sigma_b^2 = 0$ and $H_{rc} : \sigma_{ab}^2 = 0$. We wish to obtain from the experimental data information which will provide a valid test of these hypotheses.

We now consider an actual experiment involving R levels of one variable and C levels of another. The R and C levels may be regarded as samples drawn at random from two populations of levels comprised of R_p and C_p members, respectively. Thus we conceptualize two populations of levels. The levels used in a particular experiment are construed to be drawn at random from these two populations. R_p, R, C_p, and C may take any integral values, provided, of course, that $R \leqslant R_p$ and $C \leqslant C_p$. The RC treatment combinations are assigned at random to the nRC sampling units or individuals. Under these conditions, and given the basic linearity assumption, Wilk and Kempthorne (1955) have shown that the expectations of the mean squares for the general finite model are as shown in Table 16.3.

Table 16.3   **Expectation of mean squares for general finite model: two-way analysis of variance with n entries in each cell: $n > 1$**

Mean square	Expectation of mean square
Row, s_r^2	$\sigma_e^2 + \dfrac{(C_p - C)}{C_p} n\sigma_{ab}^2 + nC\sigma_a^2$
Column, s_c^2	$\sigma_e^2 + \dfrac{(R_p - R)}{R_p} n\sigma_{ab}^2 + nR\sigma_b^2$
Interaction, s_i^2	$\sigma_e^2 + n\sigma_{ab}^2$
Within cells, s_w^2	σ_e^2

Thus the mean squares provide estimates of variance components, and these are used to test the significance of row, column, and interaction effects. How these are used depends on a consideration of three particular instances of the general finite model.

Consider an experiment involving R levels of one variable and C of another, these being regarded as random samples of levels from populations comprised of R_p and C_p members. We may consider a case where R_p and C_p are very large, so that $R_p >> R$ and $C_p >> C$, where $>>$ denotes much greater than. Under these circumstances such terms as $(R_p - R)/R_p$ and $(C_p - C)/C_p$ approach unity. When this is so we have what is referred to as a *random*-model situation. The expectations for the random model are obtained by substituting $(R_p - R)/R_p = 1$ and $(C_p - C)/C_p = 1$ in the expectations of the mean squares for the general finite model given in Table 16.3. Thus the random model is a particular case of the finite model.

In psychological research, experiments where the random model is appropriate are not numerous. Satisfactory examples are not readily found. One example is an experiment where each member of a sample of R job applicants is assigned a rating by each member of a sample of C interviewers. Here both job applicants and interviewers may be viewed as samples drawn at random from populations such that $R_p >> R$ and $C_p >> C$.

In many experiments the R levels of one variable and the C levels of the other are not conceptualized as random samples. In agricultural experiments where R varieties of wheat and C varieties of fertilizer are used, the investigator is usually concerned with the yield of particular wheat varieties and with the effect of particular fertilizers on yield. He is not concerned with drawing inferences about hypothetical populations of wheat and fertilizer varieties. Both variables or factors are *fixed*. Any factor is fixed if the investigator on repeating the experiment would use the same levels of it. Under the fixed model $R = R_p$ and $C = C_p$. By substituting $(R_p - R)/R_p = 0$ and $(C_p - C)/C_p = 0$ in the expectations of the mean squares for the finite model given in Table 16.3, the expectations for the fixed model are obtained.

In psychological experiments different methods of learning, environmental conditions, methods of inducing stress, and the like, are examples of fixed factors or variables. In many experiments different levels of the experimental variable are introduced, e.g., levels of illumination, time intervals, size of brain lesion, and dosages of a drug. While the levels may be thought to constitute a representative set and interpolation between levels may be possible, such variables are usually regarded as fixed. Of course it is possible to conceptualize a study where, for example, levels of illumination or dosages of a drug are sampled at random from a population of levels or dosages. Ordinarily, however, experiments are not designed in this way.

In many experiments one basis of classification is a random factor or variable and the other is fixed. Measurements may be obtained for a sample of R individuals for each of C treatments or experimental conditions. Here one basis of classification is random and the other is fixed. This is a *mixed* model. In the mixed-model situation either $R_p = R$ and $C_p >> C$ or $R_p >> R$ and $C_p = C$. By substituting $(R_p - R)/R_p = 1$ and $(C_p - C)/C_p = 0$, or vice versa, in the expectations for the finite model of Table 16.3, we obtain the expectations for the mixed model.

Table 16.3 may be used to provide the required expectations for a two-way classification where $n = 1$. Under this circumstance no within-cells variance estimate s_w^2 is available. The expectations for row, column, and interaction effects for the random, fixed, and mixed models are obtained by writing $n = 1$ and substituting the appropriate values of $(R_p - R)/R_p$ and $(C_p - C)/C_p$.

16.7 CHOICE OF ERROR TERM

By choice of error term is meant the selection of the appropriate variance estimate for the denominator of the F ratio in testing row, column, and interaction effects. In general, in forming an F ratio, the expectation of the variance estimate in the numerator should contain one term more than the expectation of the variance estimate in the denominator, the additional term involving the effect under test. On applying this principle to the expectations of Table 16.3, the following rules may be formulated:

1 **Random model: $n > 1$** The proper error term for testing the interaction effect is s_w^2. $F_i = s_i^2/s_w^2$. The correct error term for testing row and column effects is s_i^2. $F_r = s_r^2/s_i^2$ and $F_c = s_c^2/s_i^2$.

2 **Fixed model: $n > 1$** The proper error term is s_w^2 for interaction, row, and column effects. The three F ratios are $F_i = s_i^2/s_w^2$, $F_r = s_r^2/s_w^2$, and $F_c = s_c^2/s_w^2$.

3 **Mixed model: $n > 1$** The proper error term for testing the interaction effect is s_w^2. $F_i = s_i^2/s_w^2$. When R is random and C is fixed, the proper error term for testing row effects is s_w^2. $F_r = s_r^2/s_w^2$. The proper term for testing column effects is s_i^2. $F_c = s_c^2/s_i^2$. When R is fixed and C is random, the converse procedure applies. $F_r = s_r^2/s_i^2$ and $F_c = s_c^2/s_w^2$.

4 **Random model: $n = 1$** No s_w^2 is available. The correct error term for testing both row and column effects is s_i^2. $F_r = s_r^2/s_i^2$ and $F_c = s_c^2/s_i^2$.

5 **Fixed model: $n = 1$** The point of view may be adopted that no test of either row or columns effects can be made. This point of view requires some modification. The ratio s_r^2/s_i^2 is an estimate of $(\sigma_e^2 + C\sigma_a^2)/(\sigma_e^2 + \sigma_{ab}^2)$ and will, where $\sigma_{ab}^2 > 0$, be an underestimate of $(\sigma_e^2 + C\sigma_a^2)/\sigma_e^2$. This means that if a significant result is obtained, the investigator knows a fortiori that the effect tested is significant. If the result is not significant, the probability of accepting the null hypothesis, $H_0: \sigma_a^2 = 0$, when it is false, may be high. Thus in the absence of significance no conclusions should be drawn from the data.

6 **Mixed model: $n = 1$** When R is random and C is fixed, the situation pertaining to the testing of row effects is as described above for the fixed

model, $n = 1$. The proper error term for the column effect is s_i^2. $F_c = s_c^2/s_i^2$. When C is random and R is fixed, the argument relating to the fixed model, $n = 1$, again applies. The proper error term for the row effect is s_i^2. $F_r = s_r^2/s_i^2$.

The above rules, excluding the modifications of rules 5 and 6 above, can be very simply obtained by using the following schema for the proper choice of error term:

Row $\dfrac{C_p - C}{C_p} s_i^2 + \dfrac{C}{C_p} s_w^2$

Column $\dfrac{R_p - R}{R_p} s_i^2 + \dfrac{R}{R_p} s_w^2$

Interaction s_w^2

For the random model, $(C_p - C)/C_p = 1$ and $C/C_p = 0$. The proper error term for row and column effects is s_i^2. Similarly, the proper error term for the fixed and mixed models may be obtained. When $n = 1$, all terms containing s_w^2 vanish. For the random model, s_i^2 becomes the correct error term for row and column effects. For the fixed model, no tests are possible. When rows are random and the columns are fixed, the column effect may be tested, but not the row.

16.8   POOLING SUMS OF SQUARES: $n > 1$

Under certain circumstances the within-cells and interaction sums of squares may be added together and divided by the combined degrees of freedom to obtain an estimate of variance based on a larger number of degrees of freedom. Caution should be exercised in applying this procedure.

For the fixed model, the within-cells variance estimate is the proper error term for testing interaction, row, and column effects. For the random model, the interaction variance estimate is the proper error term for testing row and column effects. These procedures are always correct. For both models, when the interaction is quite clearly *not* significant, the within-cells and interaction sums of squares may be pooled to obtain a variance estimate for the denominator of the F ratio based on a larger number of degrees of freedom. Of course, when row and column effects are clearly significant, when tested without pooling, the pooling procedure is unnecessary.

When doubt exists as to the significance of the interaction, the investigator may or may not choose to pool the sums of squares. If the interaction effect in fact exists, σ_{ab}^2 being greater than zero, and terms are pooled, the pooling may be said to be erroneous.

For the fixed model, erroneous pooling will increase the size of the error term. For the random model, erroneous pooling will decrease the size of the error term. In both instances the number of degrees of freedom is increased. Erroneous pooling will for the fixed model usually lead to too few significant results and for the random model to too many significant results.

For the mixed model, when rows are random and columns are fixed, pooling may be applied with nonsignificant interaction. In this situation erroneous pooling will tend to make the error term too large for testing row effects and too small for testing column effects, leading to too few significant effects for rows and too many for columns.

An understanding of the consequences of pooling sums of squares for fixed, mixed, and random models, when interaction does exist, that is, when $\sigma_{ab}^2 > 0$, may be obtained by examination of the expectations of the variance estimates given in Table 16.3. Quite clearly, for the fixed model, when $\sigma_{ab}^2 > 0$, combining interaction and within cells will lead to an error term whose expectation is greater than σ_e^2. Consequently, too few significant results will be obtained.

In general, it is probably advisable not to pool unless the investigator is quite confident that the interaction is not significant. For a detailed discussion of this rather troublesome problem, see Binder (1955).

16.9   COMPUTATION FORMULAS FOR SUMS OF SQUARES

Computation formulas are used to calculate the required sums of squares. A simplified notation is used. Denote the sum of all observations in the rth row by $T_{r.}$, the sum of all observations in the cth column by $T_{.c}$, the sum of all observations in the cell corresponding to the rth row and cth column by T_{rc}, and the sum of all N observations by T.

With *one* entry in each cell, the computation formulas for sums of squares are as follows:

Rows   [16.3] $\displaystyle \frac{1}{C} \sum_{r=1}^{R} T_{r.}^2 - \frac{T^2}{N}$

Columns   [16.4] $\displaystyle \frac{1}{R} \sum_{c=1}^{C} T_{.c}^2 - \frac{T^2}{N}$

Interaction   [16.5] $\displaystyle \sum_{r=1}^{R} \sum_{c=1}^{C} X_{rc}^2 - \frac{1}{C} \sum_{r=1}^{R} T_{r.}^2 - \frac{1}{R} \sum_{c=1}^{C} T_{.c}^2 + \frac{T^2}{N}$

Total   [16.6] $\displaystyle \sum_{r=1}^{R} \sum_{c=1}^{C} X_{rc}^2 - \frac{T^2}{N}$

The interaction sum of squares may be obtained by adding row and column sums and subtracting this from the total sum of squares. This provides no check on the accuracy of the calculation; consequently it is preferable to compute the interaction term directly.

Computation formulas for sums of squares with n entries in each cell are as follows:

Rows [16.7] $$\frac{1}{nC} \sum_{r=1}^{R} T_{r.}^{2} - \frac{T^2}{N}$$

Columns [16.8] $$\frac{1}{nR} \sum_{c=1}^{C} T_{.c}^{2} - \frac{T^2}{N}$$

Interaction [16.9] $$\frac{1}{n} \sum_{r=1}^{R} \sum_{c=1}^{C} T_{rc}^{2} - \frac{1}{nC} \sum_{r=1}^{R} T_{r.}^{2} - \frac{1}{nR} \sum_{c=1}^{C} T_{.c}^{2} + \frac{T^2}{N}$$

Within cells [16.10] $$\sum_{r=1}^{R} \sum_{c=1}^{C} \sum_{i=1}^{n} X_{rci}^{2} - \frac{1}{n} \sum_{r=1}^{R} \sum_{c=1}^{C} T_{rc}^{2}$$

Total [16.11] $$\sum_{r=1}^{R} \sum_{c=1}^{C} \sum_{i=1}^{n} X_{rci}^{2} - \frac{T^2}{N}$$

Here again the interaction sums of squares may be obtained by subtracting the row, column, and within-cells sums of squares from the total, although direct calculation of the interaction term is preferable.

The reader should note that the analysis of variance for two-way classification with a single entry in each cell is a particular case of the more general case with more than one entry in each cell. When $n = 1$, formulas for the latter case become the formulas for the former.

16.10 ILLUSTRATIVE EXAMPLE OF TWO-WAY CLASSIFICATION: $n > 1$

Table 16.4 shows data obtained in an animal experiment designed to study the effects of two variables on measures of performance of rats in a maze test. Three strains of rats were used, bright, mixed, and dull. A group from each strain was reared under free and restricted environmental conditions. Thus there are six groups of experimental animals with eight animals in each group. The total N is 48. The data are arranged in a 2×3 table with eight observations in each of the six cells. The row means permit a comparison of environments, and the column means a comparison of strains. Table 16.5 shows the sums, means, and sum of squares of row, column, and cell totals. The sum of squares for all the observations is also given.

Table 16.4　Data for the analysis of variance with two-way classification: $n > 1$ error scores for three strains of rats reared under two environmental conditions

Environ-		Strain				
ment	Bright		Mixed		Dull	
Free	26	14	41	82	36	87
	41	16	26	86	39	99
	28	29	19	45	59	126
	92	31	59	37	27	104
Restricted	51	35	39	114	42	133
	96	36	104	92	92	124
	97	28	130	87	156	68
	22	76	122	64	144	142

Applying the computation formulas, the calculations are as follows:

Rows
$$\frac{1}{nC} \sum_{r=1}^{R} T_{r.}^2 - \frac{T^2}{N} = \frac{5{,}944{,}837}{24} - \frac{(3{,}343)^2}{48} = 14{,}875.52$$

Columns
$$\frac{1}{nR} \sum_{c=1}^{C} T_{.c}^2 - \frac{T^2}{N} = \frac{4{,}015{,}617}{16} - \frac{(3{,}343)^2}{48} = 18{,}150.04$$

Within cells
$$\sum_{r=1}^{R} \sum_{c=1}^{C} \sum_{i=1}^{n} X_{rci}^2 - \frac{1}{n} \sum_{r=1}^{R} \sum_{c=1}^{C} T_{rc}^2 = 309{,}851 - \frac{2{,}137{,}469}{8}$$
$$= 42{,}667.38$$

Interaction
$$\frac{1}{n} \sum_{r=1}^{R} \sum_{c=1}^{C} T_{rc}^2 - \frac{1}{nC} \sum_{r=1}^{R} T_{r.}^2 - \frac{1}{nR} \sum_{c=1}^{C} T_{.c}^2 + \frac{T^2}{N}$$
$$= \frac{2{,}137{,}469}{8} - \frac{5{,}944{,}837}{24} - \frac{4{,}015{,}617}{16} + \frac{(3{,}343)^2}{48}$$
$$= 1{,}332.04$$

Total
$$\sum_{r=1}^{R} \sum_{c=1}^{C} \sum_{i=1}^{n} X_{rci}^2 - \frac{T^2}{N} = 309{,}851 - \frac{(3{,}343)^2}{48} = 77{,}024.98$$

The analysis-of-variance table for these data is given in Table 16.6. The *df* for rows is $R - 1 = 2 - 1 = 1$, for columns $C - 1 = 3 - 1 = 2$, for interaction $(R - 1)(C - 1) = (2 - 1)(3 - 1) = 2$, and for within cells $RC(n - 1) = 2 \times 3(8 - 1) = 42$. These sum to the total sum of squares $RCn - 1 = 2 \times 3 \times 8 - 1 = 47$. For these data a fixed model is appropriate

Table 16.5   Computation for data of Table 16.4

| | | Strain | | |
Environment	Bright	Mixed	Dull	Total
Free	$T_{11} = 277$ $\overline{X}_{11} = 34.63$	$T_{12} = 395$ $\overline{X}_{12} = 49.38$	$T_{13} = 577$ $\overline{X}_{13} = 72.13$	$T_{1.} = 1{,}249$ $\overline{X}_{1.} = 52.04$
Restricted	$T_{21} = 441$ $\overline{X}_{21} = 55.13$	$T_{22} = 752$ $\overline{X}_{22} = 94.00$	$T_{23} = 901$ $\overline{X}_{23} = 112.63$	$T_{2.} = 2{,}094$ $\overline{X}_{2.} = 87.25$
Total	$T_{.1} = 718$ $\overline{X}_{.1} = 44.88$	$T_{.2} = 1{,}147$ $\overline{X}_{.2} = 71.69$	$T_{.3} = 1{,}478$ $\overline{X}_{.3} = 92.38$	$T = 3{,}343$ $\overline{X}_{...} = 69.65$

$$\sum_{r=1}^{R} T_{r.}^2 = 5{,}944{,}837 \qquad \sum_{c=1}^{C} T_{.c}^2 = 4{,}015{,}617$$

$$\sum_{r=1}^{R} \sum_{c=1}^{C} T_{rc}^2 = 2{,}137{,}469 \qquad \sum_{r=1}^{R} \sum_{c=1}^{C} \sum_{i=1}^{n} X_{rci}^2 = 309{,}851$$

and s_w^2 is the proper error term for testing row, column, and interaction effects. For interaction we have

$$F_i = \frac{s_i^2}{s_w^2} = \frac{666.02}{1{,}015.89} = .656$$

This is less than unity. The expectation on the basis of the null hypothesis is unity. The interaction is somewhat less than we would ordinarily expect under the null hypothesis. We may safely conclude that there is no significant interaction between the two experimental variables. For differences

Table 16.6   Analysis of variance for data of Table 16.4

Source of variation	Sum of squares	Degrees of freedom	Variance estimate
Rows (environments)	14,875.52	1	$14{,}875.52 = s_r^2$
Columns (strains)	18,150.04	2	$9{,}075.02 = s_c^2$
Interaction	1,332.04	2	$666.02 = s_i^2$
Within cells	42,667.38	42	$1{,}015.89 = s_w^2$
Total	77,024.98	47	

$$F_i = \frac{s_i^2}{s_w^2} = .656 \qquad F_r = \frac{s_r^2}{s_w^2} = 14.64 \qquad F_c = \frac{s_c^2}{s_w^2} = 8.93$$

in environments we have

$$F_r = \frac{s_r^2}{s_w^2} = \frac{14{,}875.52}{1{,}015.89} = 14.64$$

with 1 df associated with the numerator and 42 df with the denominator. For these df the values required for significance at the 5 and 1 per cent levels are 4.07 and 7.27. We conclude that the different environments have affected the maze performance of the animals. For strains the required ratio is $F_c = s_c^2/s_w^2 = 9{,}075.02/1{,}015.89 = 8.93$ with 2 df associated with the numerator and 42 df with the denominator. Again, this difference is significant at well beyond the 1 per cent level, and the conclusion is that differences in strain affect maze performance.

16.11 UNEQUAL NUMBERS IN THE SUBCLASSES

Situations arise in educational and psychological research where the numbers of observations in the subclasses in a two-way analysis of variance are unequal. In animal experimentation in psychology, this situation may result from loss by death or accident of a number of animals during the conduct of the experiment. For the fixed model, if the cell frequencies do not depart significantly from either equality or proportionality, simple adjustments may be made to the data. Two methods will be briefly described: the method of expected equal frequencies and the method of expected proportionate frequencies. The treatment given here is based on the work of Fei Tsao (1946). These methods are also described by Bancroft (1968).

In applying the method of *expected equal frequencies* the following steps are involved:

1   Apply a χ^2 criterion to determine whether the cell frequencies depart from equality. Denote the frequency in the cell corresponding to the rth row and cth column by n_{rc}. The expected equal frequency is the average value of n_{rc}, or N/RC. Denote this by $\bar{n}$. The required χ^2 is

$$\chi^2 = \sum_{r=1}^{R} \sum_{c=1}^{C} \frac{(n_{rc} - \bar{n})^2}{\bar{n}}$$

with $RC - 1$ degrees of freedom.

2   If the cell frequencies do not depart significantly from equality at, say, the 1 per cent level, apply a simple adjustment to the sum and sum of squares for each cell by multiplying these values by $\bar{n}/n_{rc}$. Thus the adjusted cell sum is

$$\frac{\bar{n}}{n_{rc}} \sum_{r=1}^{n_{rc}} X_{rci}$$

and the adjusted cell sum of squares is

$$\frac{\bar{n}}{n_{rc}} \sum_{i=1}^{n_{rc}} X_{rci}^2$$

This adjustment estimates what the cell sum and sum of squares would be were there an equal number of cases $\bar{n}$ in each cell. Note that this adjustment does not change the cell means or the row and column means.

3 Use the adjusted cell sums and sums of squares to obtain row and column totals and the total sum of squares.

4 Proceed with the analysis of variance in the usual way, employing the computation formulas given in Sec. 16.9.

The method of expected equal frequencies is simple and may be usefully applied where the numbers of observations in the cells do not differ very much.

In situations where the numbers of observations in the cells differ, but are roughly proportionate to the marginal totals, the method of *expected proportionate frequencies* is appropriate. This method requires the following steps:

1 Apply a χ^2 criterion to determine whether the cell frequencies in the rows and columns depart significantly from proportionality. Denote the observed frequency in the cell corresponding to the rth row and cth column by n_{rc} and the marginal frequencies for rows and columns by $n_{r.}$ and $n_{.c}$, respectively. Denote the cell frequencies expected on the assumption of proportionality by $\bar{n}_{rc}$. The expected frequencies are given by

$$\bar{n}_{rc} = \frac{n_{r.}n_{.c}}{N}$$

The procedure here is identical with that used in calculating expected cell frequencies for a contingency table given the restrictions of the marginal totals. The χ^2 criterion is

$$\chi^2 = \sum_{r=1}^{R} \sum_{c=1}^{C} \frac{(n_{rc} - \bar{n}_{rc})^2}{\bar{n}_{rc}}$$

with $(R-1)(C-1)$ degrees of freedom.

2 If the cell frequencies do not depart significantly from proportionality, the sum and sum of squares for each cell are adjusted by multiplying them by $\bar{n}_{rc}/n_{rc}$. The adjusted cell sum is then

$$\frac{\bar{n}_{rc}}{n_{rc}} \sum_{i=1}^{n_{rc}} X_{rci}$$

and the adjusted cell sum of squares is

$$\frac{\bar{n}_{rc}}{n_{rc}} \sum_{i=1}^{n_{rc}} X_{rci}{}^2$$

This adjustment provides estimates of what the cell sums and sums of squares would be were the numbers in each cell proportional to the marginal totals.

3 The required sums of squares for the analysis of variance are obtained, using the *adjusted* values, by applying the following formulas:

Rows   [16.12] $\displaystyle\sum_{r=1}^{R} \left(\frac{T_{r.}{}^2}{n_{r.}}\right) - \frac{T^2}{N}$

Columns   [16.13] $\displaystyle\sum_{c=1}^{C} \left(\frac{T_{.c}{}^2}{n_{.c}}\right) - \frac{T^2}{N}$

Within cells [16.14] $\displaystyle\sum_{r=1}^{R}\sum_{c=1}^{C} \left(\frac{\bar{n}_{rc}}{n_{rc}}\sum_{i=1}^{n_{rc}} X_{rci}{}^2\right) - \sum_{r=1}^{R}\sum_{c=1}^{C}\left(\frac{T_{rc}{}^2}{\bar{n}_{rc}}\right)$

Interaction [16.15] $\displaystyle\sum_{r=1}^{R}\sum_{c=1}^{C} \left(\frac{T_{rc}{}^2}{\bar{n}_{rc}}\right) - \sum_{r=1}^{R}\left(\frac{T_{r.}{}^2}{n_{r.}}\right) - \sum_{c=1}^{C}\left(\frac{T_{.c}{}^2}{n_{.c}}\right) + \frac{T^2}{N}$

Total   [16.16] $\displaystyle\sum_{r=1}^{R}\sum_{c=1}^{C} \left(\frac{\bar{n}_{rc}}{n_{rc}}\sum_{i=1}^{n_{rc}} X_{rci}{}^2\right) - \frac{T^2}{N}$

All T's relate to adjusted values. The above formulas differ from those previously given in Sec. 16.9 only in that they make allowance for the fact that the numbers of cases in the subclasses are unequal.

4 Proceed with the analysis of variance in the usual way.

In the above procedure the within-cells sum of squares is based on the adjusted values. Arguments may be advanced for using the unadjusted values in calculating the within-cells sum of squares. For comment on this point see Gourlay (1955).

Both the methods of expected equal and expected proportionate frequencies are in some degree approximate. Departure from equal n's in the former method and from proportionality of n's in the latter method will introduce some bias in the F test, the extent of the bias being related to the magnitude of the departures. By bias here is meant that the F test produces either a larger or smaller proportion of significant F ratios than is warranted by the F distribution.

The methods of equal and proportionate frequencies are applicable to a substantial proportion of situations encountered in practice. When the frequencies differ markedly from proportionality, other methods may be applied. For a discussion of these, see Winer (1962) and Bancroft (1968).

For the random model, bias is introduced in the F test despite the proportionality of the numbers in the subclasses. From a practical viewpoint this is not an important consideration. Good examples of the random model with unequal n's are difficult to find in educational and psychological research. Of more practical importance is the fact that for the mixed model F test bias is introduced when the cell frequencies are proportional, and experiments involving this model are not infrequent. The bias is positive, the F test producing a larger proportion of significant F ratios than the F distribution warrants. For a discussion of this problem the reader is directed to Gourlay (1955).

In general, because of the complications associated with unequal frequencies, it is advisable, whenever possible, to design experiments with an equal number of cases in the subclasses, although for the fixed model proportionate numbers of cases in the subclasses will introduce no bias. The investigator will thereby avoid a number of inconvenient complexities.

16.12 ONE-FACTOR EXPERIMENTS WITH REPEATED MEASUREMENTS

Many experiments in psychology and education require the repeated measurement of the same subjects under a number of different conditions. In such experiments it is sometimes said that each subject acts as his own control. The simplest experiment of this type would be one in which the same subjects are tested under two conditions, a control condition and an experimental condition. Not infrequently the same subjects are tested under five or six different experimental conditions. Sometimes, when the same subjects are tested under a number of different treatments, the order of the presentation of treatments to subjects is randomized independently for each subject. The purpose of randomization in this case is to eliminate effects which might result from the order of the treatments. In other cases randomization is not appropriate because the different levels of the treatment variable have a natural order. This is the case where performance is measured at different time intervals, as, for example, in the study of changes in dark adaptation with time, or for different numbers of trials in a simple learning experiment. Such experiments are called *one-factor experiments with repeated measurements*.

The data resulting from such experiments may be represented as a table of numbers in which rows represent experimental subjects and columns represent treatments, that is, the representation of the data is the same as that for the two-way classification with one observation per cell. The analysis of such data involves nothing new. The data are analyzed as in the two-way classification case with one observation per cell. Three sums of squares result: sums of squares for subjects (rows), treatments (columns), and interaction.

With such data, subjects constitute a random variable and treatments are usually viewed as fixed. The model is the mixed model for $n = 1$. The expectations of the mean squares are as follows:

Mean squares	Expectation of mean squares
Subjects, s_r^2	$\sigma_e^2 + C\sigma_a^2$
Treatments, s_c^2	$\sigma_e^2 + \sigma_{ab}^2 + R\sigma_b^2$
Interactions, s_i^2	$\sigma_e^2 + \sigma_{ab}^2$

The proper error term for testing differences between treatments is s_i^2, that is, $F_c = s_c^2/s_i^2$. No unbiased test of individual differences between subjects is possible, unless it is assumed that the interaction term is zero. With nearly all sets of data this assumption is not warranted, because the performance of subjects under different pairs of treatments is correlated. Ordinarily in most experiments of this type individual differences between subjects is of limited interest anyway, because with most variables which are the object of study the investigator expects a priori substantial differences between subjects.

16.13 ILLUSTRATIVE EXAMPLE OF ONE-FACTOR EXPERIMENT WITH REPEATED MEASUREMENTS

Table 16.7 shows hypothetical data for a one-factor experiment with repeated measurements. Rows are individuals, and columns are treat-

Table 16.7 Data for the analysis of variance with two-way classification: $n = 1$, scores for a sample of subjects tested under four different conditions

Subject	Conditions A	B	C	D	T_r.	$\bar{X}_r$.
1	31	42	14	80	167	41.75
2	42	26	25	106	199	49.75
3	84	21	19	83	207	51.75
4	26	60	36	69	191	47.75
5	14	35	44	48	141	35.25
6	16	80	28	76	200	50.00
7	29	49	80	39	197	49.25
8	32	38	76	84	230	57.50
9	45	65	15	91	216	54.00
10	30	71	82	39	222	55.50
$T_{.c}$	349	487	419	715	$T = 1{,}970$	
$\bar{X}_{.c}$	34.90	48.70	41.90	71.50	$\bar{X}_{..} = 49.25$	

$$\sum_{r=1}^{R} T_r.^2 = 394{,}350 \qquad \sum_{c=1}^{C} T_{.c}^2 = 1{,}045{,}756 \qquad \sum_{r=1}^{R}\sum_{c=1}^{C} X_{rc}^2 = 122{,}984$$

ments. The data are presumed to relate to a random sample of individuals tested under different treatment conditions. This is a mixed model. One basis of classification, the columns, is fixed. The other basis of classification, the rows, is random.

Applying the appropriate computation formulas, the following sums of squares are obtained:

Rows
$$\frac{1}{C} \sum_{r=1}^{R} T_{r.}^{2} - \frac{T^2}{N} = \frac{394,350}{4} - \frac{(1,970)^2}{40} = 1,565.00$$

Columns
$$\frac{1}{R} \sum_{c=1}^{R} T_{.c}^{2} - \frac{T^2}{N} = \frac{1,045,756}{10} - \frac{(1,970)^2}{40} = 7,553.10$$

Interaction
$$\sum_{r=1}^{R} \sum_{c=1}^{C} X_{rc}^{2} - \frac{1}{C} \sum_{r=1}^{R} T_{r.}^{2} - \frac{1}{R} \sum_{c=1}^{C} T_{.c}^{2} + \frac{T^2}{N}$$

$$= 122,984 - \frac{394,350}{4} - \frac{1,045,756}{10}$$

$$+ \frac{(1,970)^2}{40} = 16,843.40$$

Total
$$\sum_{r=1}^{R} \sum_{c=1}^{C} X_{rc}^{2} - \frac{T^2}{N} = 122,984 - \frac{(1,970)^2}{40} = 25,961.50$$

Table 16.8 summarizes the analysis-of-variance data for this example. Because this is a mixed model with $n = 1$ and $F_r = s_r^2/s_i^2 = .279$, no meaningful test of row effects is possible. The proper error term for column effects is s_i^2. The F ratio for column effects is found to be 4.04. The F ratios required for significance with 3 and 27 degrees of freedom associated with the numerator and denominator, respectively, are 2.96 at the 5 per cent and 4.60 at the 1 per cent levels. Thus the column differences are significant at the 5 per cent level but fall short of significance at the 1 per cent level.

Table 16.8   **Analysis of variance for data of Table 16.7**

Source of variation	Sum of squares	Degrees of freedom	Variance estimate
Rows	1,565.00	9	$173.89 = s_r^2$
Columns	7,553.10	3	$2,517.70 = s_c^2$
Interaction	16,843.40	27	$623.83 = s_i^2$
Total	25,961.50		

$$F_c = \frac{s_c^2}{s_i^2} = 4.04 \qquad F_r = \frac{s_r^2}{s_i^2} = .279$$

EXERCISES **1** In an experiment involving double classification with 10 observations in each cell, the following cell and marginal means were obtained:

	C_1	C_2	C_3	
R_1	8.3	3.2	17.4	9.6
R_2	12.5	4.6	12.6	9.9
	10.4	3.9	15.0	9.8

Compute (*a*) the cell means expected under zero interaction and (*b*) the interaction sum of squares.

2 The following are measurements made on a sample of 12 subjects under three experimental conditions:

		Condition	
Subject	C_1	C_2	C_3
1	8	7	15
2	19	14	20
3	7	9	6
4	23	20	18
5	14	26	12
6	6	14	15
7	5	9	20
8	22	25	20
9	11	15	16
10	4	12	8
11	13	18	20
12	8	6	28
$T_{.c}$	140	175	198
$\bar{X}_{.c}$	11.67	14.58	16.50

Obtain the sums of squares and the variance estimates. Test the column means on the assumption that experimental condition is a fixed variable.

3 The following are data for a double-classification experiment involving two fixed variables:

	C_1		C_2		C_3	
	29	31	23	62	17	32
R_1	26	50	31	60	18	49
	42	25	18	20	50	58
	17	62	35	83	17	28
R_2	27	62	50	42	14	58
	50	29	62	19	49	62

Apply the analysis of variance to test the significance of row, column, and interaction effects.

4 The following are data with unequal numbers in the subclasses:

	C_1			C_2		
	8	9	20	6	11	6
R_1	5	16	11	2	4	
	23	4	2	1	3	
	8	20		11	15	6
R_2	14	16		12	18	3
	12	15		6	2	

Apply the analysis of variance to test row, column, and interaction effect, on the assumption that the two experimental variables are fixed.

ANALYSIS OF VARIANCE: THREE-WAY CLASSIFICATION

17.1 INTRODUCTION

Many experiments involve the simultaneous study of more than two independent variables or factors. For example, an experiment may involve two levels or categories of one factor, three of a second factor, and five of a third factor, with n subjects assigned at random to each of the $2 \times 3 \times 5$ groups. Such an experiment may be spoken of as a "$2 \times 3 \times 5$ factorial experiment." The data from such an experiment may be conceptualized as a three-dimensional cube of numbers containing two rows, three columns, and five layers, with n observations in each of the 30 different cells of which the cube may be thought to be comprised. Another experiment may involve two levels or categories of one factor and three levels or categories of another factor, with each of the 2×3 conditions being administered to each of N subjects. This is a repeated-measurement design, with each subject receiving the six different combinations of experimental treatments. Here again the data may be represented by a cube of numbers with one observation or measurement in each of the $2 \times 3 \times N$ cells.

The analysis and interpretation of data resulting from such experiments are a direct extension of the analysis and interpretation of data for two-way classification. In a two-factor experiment with n observations in each cell the total sum of squares is divided into four parts, a between-rows, a between-columns, an interaction, and a within-cells sum of squares. In a three-way classification, or three-factor, experiment with n observations per cell, the total sum of squares is partitioned into eight parts, three sums of squares for main effects, four interaction sums of squares, and a within-cells sum of squares. Each sum of squares has an associated number of degrees of freedom. Sums of squares are, as previously, divided by their associated degrees of freedom to obtain variance estimates, or mean squares, which are used to test the significance of main effects and interactions. Although such analysis and interpretation are clearly more complex than for two-way classification, the essential ideas are the same.

17.2 NOTATION FOR THREE-WAY ANALYSIS OF VARIANCE

Consider an experiment involving R levels of one factor, C levels of a second factor, and L levels of a third factor. The number of treatment combinations is RCL. Consider the particular case where we have one measurement, or observation, for each of the RCL combinations, the total number of measurements being N. The data for the first layer of numbers may be represented as follows:

		1	*2*	*3*	$\cdots$	*C*	*Row mean*
	1	X_{111}	X_{121}	X_{131}	$\cdots$	X_{1C1}	$\overline{X}_{1.1}$
	2	X_{211}	X_{221}	X_{231}	$\cdots$	X_{2C1}	$\overline{X}_{2.1}$
Layer 1	*3*	X_{311}	X_{321}	X_{331}	$\cdots$	X_{3C1}	$\overline{X}_{3.1}$
	$\cdot$	$\cdot$	$\cdot$	$\cdot$	$\cdots$	$\cdot$	$\cdot$
	$\cdot$	$\cdot$	$\cdot$	$\cdot$	$\cdots$	$\cdot$	$\cdot$
	R	X_{R11}	X_{R21}	X_{R31}	$\cdots$	X_{RC1}	$\overline{X}_{R.1}$
Column mean		$\overline{X}_{.11}$	$\overline{X}_{.21}$	$\overline{X}_{.31}$	$\cdots$	$\overline{X}_{.C1}$	$\overline{X}_{..1}$

Here the first subscript identifies the row, the second the column, and the third the layer. Thus, for example, X_{321} denotes the observation in the third row and second column of the first layer. The mean $\overline{X}_{1.1}$ is the mean of the first row of the first layer, $\overline{X}_{.11}$ is the mean of the first column of the first layer, and $\overline{X}_{..1}$ is the mean of all the observations in the first layer.

Notation for the second layer is as follows:

		1	*2*	*3*	$\cdots$	*C*	*Row mean*
	1	X_{112}	X_{122}	X_{132}	$\cdots$	X_{1C2}	$\overline{X}_{1.2}$
	2	X_{212}	X_{222}	X_{232}	$\cdots$	X_{2C2}	$\overline{X}_{2.2}$
Layer 2	*3*	X_{312}	X_{322}	X_{332}	$\cdots$	X_{3C2}	$\overline{X}_{3.2}$
	$\cdot$	$\cdot$	$\cdot$	$\cdot$	$\cdots$	$\cdot$	$\cdot$
	$\cdot$	$\cdot$	$\cdot$	$\cdot$	$\cdots$	$\cdot$	$\cdot$
	R	X_{R12}	X_{R22}	X_{R32}	$\cdots$	X_{RC2}	$\overline{X}_{R.2}$
Column mean		$\overline{X}_{.12}$	$\overline{X}_{.22}$	$\overline{X}_{.32}$	$\cdots$	$\overline{X}_{.C2}$	$\overline{X}_{..2}$

Similarly the third, fourth, and Lth layers may be considered. In general X_{rcl} denotes a measurement for the rth row, the cth column, and the lth layer. The reader should note that R denotes the number of rows, C the number of columns, and L the number of layers. The symbol r denotes the rth row, where r may take the values 1, 2, . . . , R. Similarly c and l denote the cth column and the lth layer, respectively.

The grand mean of all the $RCL = N$ observations is $\bar{X}_{...}$. The total sum of squares of deviations about the grand mean is given by

$$\sum_{r=1}^{R} \sum_{c=1}^{C} \sum_{l=1}^{L} (X_{rcl} - \bar{X}_{...})^2$$

In many experiments there are n sampling units and measurements in each of the RCL treatment combinations. The total number of measurements is then $nRCL = N$. The notation for such data in the particular case where $R = 2$, $C = 3$, $L = 2$, and $n = 3$ may be represented as follows:

		1	*2*	*3*	
Layer 1	*1*	X_{1111} X_{1112} X_{1113}	X_{1211} X_{1212} X_{1213}	X_{1311} X_{1312} X_{1313}	$\bar{X}_{1.1.}$
	2	X_{2111} X_{2112} X_{2113}	X_{2211} X_{2212} X_{2213}	X_{2311} X_{2312} X_{2313}	$\bar{X}_{2.1.}$
		$\bar{X}_{.11.}$	$\bar{X}_{.21.}$	$\bar{X}_{.31.}$	$\bar{X}_{..1.}$

		1	*2*	*3*	
Layer 2	*1*	X_{1121} X_{1122} X_{1123}	X_{1221} X_{1222} X_{1223}	X_{1321} X_{1322} X_{1323}	$\bar{X}_{1.2.}$
	2	X_{2121} X_{2122} X_{2123}	X_{2221} X_{2222} X_{2223}	X_{2321} X_{2322} X_{2323}	$\bar{X}_{2.2.}$
		$\bar{X}_{.12.}$	$\bar{X}_{.22.}$	$\bar{X}_{.32.}$	$\bar{X}_{..2.}$

Quadruple subscripts are used. The first identifies the row, the second the column, the third the layer, and the fourth the measurement within the cell. Thus, for example, the symbol X_{1323} indicates the third measurement in the cell, corresponding to the first row, the third column, and the second layer. In general X_{rcli} denotes the ith measurement in the rth row and cth column of the lth layer, where $i = 1, 2, \ldots, n$. Row and column means for each layer are shown. Thus $\bar{X}_{1.1.}$ is the mean for the first row of the first layer, $\bar{X}_{.22.}$ is the mean for the second column of the second layer, and so on. The grand mean, the mean of all $nRCL$ observations, is $\bar{X}_{....}$. The total sum of squares for a triple-classification experiment with n observations per cell may be denoted by

$$\sum_{r=1}^{R} \sum_{c=1}^{C} \sum_{l=1}^{L} \sum_{i=1}^{n} (X_{rcli} - \bar{X}_{....})^2$$

The sum of squares, both in the case where $n = 1$ and where $n > 1$, may be partitioned into additive sums of squares.

17.3   PARTITIONING THE SUM OF SQUARES

With a single measurement in each of the RCL-treatment combinations, the total sum of squares may be partitioned into seven additive parts, three main effects for rows, columns, and layers, three first-order interaction terms, and one second-order interaction term. The first-order interaction terms are row by column, row by layer, and column by layer. The second-order interaction term is row by column by layer. The precise meaning which attaches to each of these terms is described in detail in Sec. 17.5. With more than one observation per cell, $n > 1$, the total sum of squares is partitioned into eight additive parts: three main effects, three first-order interaction terms, one second-order interaction term, and one within-cells sum of squares.

   The procedure used in partitioning the sum of squares is directly analogous to that used in the two-way classification case, although somewhat more complex. For $n > 1$ we begin with the rather involved identity,

$$\begin{aligned}
(X_{rcli} - \bar{X}_{....}) = & (\bar{X}_{r...} - \bar{X}_{....}) + (\bar{X}_{.c..} - \bar{X}_{....}) \\
& + (\bar{X}_{..l.} - \bar{X}_{....}) + (\bar{X}_{rc..} - \bar{X}_{r...} - \bar{X}_{.c..} + \bar{X}_{....}) \\
& + (\bar{X}_{r.l.} - \bar{X}_{r...} - \bar{X}_{..l.} + \bar{X}_{....}) + (\bar{X}_{.cl.} - \bar{X}_{.c..} - \bar{X}_{..l.} \\
& + \bar{X}_{....}) + (\bar{X}_{rcl.} - \bar{X}_{rc..} - \bar{X}_{r.l.} - \bar{X}_{.cl.} + \bar{X}_{r...} + \bar{X}_{.c..} \\
& \qquad\qquad + \bar{X}_{..l.} - \bar{X}_{....}) + (X_{rcli} - \bar{X}_{rcl.})
\end{aligned}$$

   As in the two-way classification case both sides are squared, summed over R rows, C columns, and L layers, and over the n observations in each cell. Certain terms contain a sum of deviations about a mean. These vanish, and the eight sums of squares as shown in Table 17.1 result. Note that three sums of squares for main effects are obtained, three first-order interaction sums of squares, one second-order interaction sum of squares, and one within-cells sum of squares. With one observation in each cell the required sums of squares are obtained from Table 17.1 simply by writing $n = 1$. No within-cells sum of squares exists, and in this case the total sum of squares is partitioned into seven, and not eight, additive parts.

Table 17.1 Analysis of variance for three-way classification with n entries per cell: $n > 1$

Source	Sum of squares	df	Variance estimate
Rows	$nCL\sum^{R}(\bar{X}_{r..} - \bar{X}_{....})^2$	$R-1$	s_r^2
Columns	$nRL\sum^{C}(\bar{X}_{.c.} - \bar{X}_{....})^2$	$C-1$	s_c^2
Layers	$nCL\sum^{L}(\bar{X}_{..l} - \bar{X}_{....})^2$	$L-1$	s_l^2
$R \times C$	$nL\sum^{R}\sum^{C}(\bar{X}_{rc.} - \bar{X}_{r..} - \bar{X}_{.c.} + \bar{X}_{....})^2$	$(R-1)(C-1)$	s_{rc}^2
$R \times L$	$nC\sum^{R}\sum^{L}(\bar{X}_{r.l} - \bar{X}_{r..} - \bar{X}_{..l} + \bar{X}_{....})^2$	$(R-1)(L-1)$	s_{rl}^2
$C \times L$	$nR\sum^{C}\sum^{L}(\bar{X}_{.cl} - \bar{X}_{.c.} - \bar{X}_{..l} + \bar{X}_{....})^2$	$(C-1)(L-1)$	s_{cl}^2
$R \times C \times L$	$n\sum^{R}\sum^{C}\sum^{L}(\bar{X}_{rcl} - \bar{X}_{rc.} - \bar{X}_{r.l} - \bar{X}_{.cl} + \bar{X}_{r..} + \bar{X}_{.c.} + \bar{X}_{..l} - \bar{X}_{....})^2$	$(R-1)(C-1)(L-1)$	s_{rcl}^2
Within cells	$\sum^{n}\sum^{R}\sum^{C}\sum^{L}(X_{rcli} - \bar{X}_{rcl})^2$	$RCL(n-1)$	s_w^2
Total	$\sum^{n}\sum^{R}\sum^{C}\sum^{L}(X_{rcli} - \bar{X}_{....})^2$	$nRCL-1$	

17.4 DEGREES OF FREEDOM AND MEAN SQUARES

As shown in Table 17.1 the number of degrees of freedom for rows is $R - 1$, for columns $C - 1$, and for layers $L - 1$. The first-order interaction for row-by-column has $(R - 1)(C - 1)$ degrees of freedom. Similarly the row-by-layer and column-by-layer interactions have $(R - 1)(L - 1)$ and $(C - 1)(L - 1)$ degrees of freedom, respectively. The second-order interaction term has $(R - 1)(C - 1)(L - 1)$ degrees of freedom. Each cell has associated with it $n - 1$ degrees of freedom. Since there are RCL cells, the total number of degrees of freedom associated with the within-cells sum of squares is $RCL(n - 1)$. The degrees of freedom are directly additive, that is, it may be shown that

$$N - 1 = nRCL - 1 = (R - 1) + (C - 1) + (L - 1)$$
$$+ (R - 1)(C - 1) + (R - 1)(L - 1) + (C - 1)(L - 1)$$
$$+ (R - 1)(C - 1)(L - 1) + RCL(n - 1)$$

For $n = 1$ the degrees of freedom are the same as for $n > 1$ except that no within-cells sum of squares exists.

Sums of squares are divided by their associated number of degrees of freedom to obtain variance estimates, or mean squares. As previously, F ratios are formed using these mean squares to obtain tests of significance for main effects and interaction effects. As in the two-way classification case, the correct choice of error term, the appropriate mean square to insert in the denominator of the F ratio, depends on whether the model appropriate to the experiment is fixed, random, or mixed.

17.5 INTERACTION IN THREE-WAY ANALYSIS OF VARIANCE

As stated above, partitioning a sum of squares in a three-way analysis of variance results in four interaction sums of squares, $R \times C$, $R \times L$, $C \times L$, and $R \times C \times L$. What meaning may be attached to these sums of squares? How may they be interpreted? The answers to these questions may be clarified by a hypothetical example. The following are cell means for a $3 \times 3 \times 2$ factorial design with n observations per cell.

	Layer 1 Columns					**Layer 2** Columns		
	10	20	6			16	26	12
Rows	5	15	1		**Rows**	11	21	7
	15	25	11			21	31	17

The data to the left are means for the first layer, and those to the right are means for the second layer. These data may be visualized as a $3 \times 3 \times 2$ cube of numbers with the second layer superimposed on the first as shown in Fig. 17.1. The means shown in Fig. 17.1 are means on the surface of the cube obtained by averaging cell means over rows, columns, and layers, a procedure which is, of course, correct only when the cell means are based on equal n's. Thus in Fig. 17.1 the mean of 13 for the first row and column is obtained by averaging the means for the first row and column over the two layers. Thus the mean of 13 is the average of the two-cell means 10 and 16. Likewise the mean of 10 for the first column and the first layer is obtained by averaging over rows. Thus 10 is the average of the three-cell means for the first layer, 10, 5, 15. All other means in Fig. 17.1 are similarly obtained. Fig. 17.1 also shows certain marginal means, or means along the edge of the cube. These means are obtained by averaging over rows, columns, and layers. The grand mean, the mean of all the observations taken together, is also shown, which in this case is 15.

The first-order interactions in a three-way analysis of variance, that is, the $R \times C$, $R \times L$, and $C \times L$ interactions, are concerned with the parallelism, or its absence, of the means on the surface of the cube. The rationale is essentially the same as that used in the double-classification case. Consider for example the means obtained by averaging over layers in Fig. 17.1 These means are

		Columns			
		I	II	III	
	I	13	23	9	15
Rows	II	8	18	4	10
	III	18	28	14	20
		13	23	9	15

Here we note that all sets of means for both rows and columns are parallel. For example, means for the first row, 13, 23, and 9, are uniformly five points higher than the means for the second row, 8, 18, 4. Likewise, the means for the first column, 13, 8, 18, are uniformly 10 points less than the means for the second column, 23, 18, and 28, and so on. In Fig. 17.1 the $R \times C$ interaction is zero.

As in the two-way classification case the values of the cell means for rows and columns expected in the case of zero interaction are given by

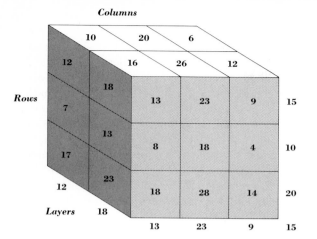

Fig. 17.1 Representation of data for three-way classification as a cube of numbers.

$E(\overline{X}_{rc..}) = \overline{X}_{r...} + \overline{X}_{.c..} - \overline{X}_{....}$. The observed means are $\overline{X}_{rc..}$. The differences between the observed and expected are $\overline{X}_{rc..} - \overline{X}_{r...} - \overline{X}_{.c..} + \overline{X}_{....}$. The $R \times C$ interaction sum of squares is given by

$$nL \sum_{}^{R} \sum_{}^{C} (\overline{X}_{rc..} - \overline{X}_{r...} - \overline{X}_{.c..} + \overline{X}_{....})^2$$

This is seen to be nL times the sum of squares of these differences summed over rows and columns, and is simply an over-all measure of the departures of the means from parallelism. Note that the $R \times C$ interaction is concerned with parallelism among means obtained by averaging over layers. The $R \times L$ and $L \times C$ interactions involve means obtained by averaging over columns and rows, respectively.

The $R \times C \times L$ interaction in a three-way analysis of variance is concerned with the similarity, or otherwise, of the interaction, whether it be zero or not, between each pair of variables at different levels of the third variable. Consider the following means for a $2 \times 2 \times 2$ factorial experiment:

	L_1			L_2	
	C_1	C_2		C_1	C_2
R_1	10	30	R_1	20	40
R_2	10	20	R_2	20	30

Inspection of the means for L_1 indicate that the means are not parallel and an interaction is present. Likewise inspection of the means for L_2 indicates again that the means are not parallel and an interaction is present. We note, however, that for L_1 the difference between R_1 and R_2 for C_1 is zero. The corresponding difference for L_2 is also zero. Note also that for L_1 the difference between R_1 and R_2 for C_2 is 10. The corresponding difference for L_2 is also 10. It is appropriate to say here that the interaction between rows and columns is the same, or geometrically similar, for the two layers. In this case the triple interaction will be zero. In this illustrative example it may be ascertained that the $R \times C$ interaction is nonzero whereas the $R \times L$, the $C \times L$, and the $R \times C \times L$ interactions are zero.

The $R \times C \times L$ interaction is a measure of the differences between a set of observed values and a set of expected values. The expected values are those we would obtain if the interactions were the same between all pairs of variables at different levels of the third variable.

17.6 EXPECTATIONS OF MEAN SQUARES IN THREE-WAY ANALYSIS OF VARIANCE

As in the two-way analysis of variance the correct choice of error term in a three-way analysis of variance depends on the nature of the variables which serve as the basis of classification in the experimental design. All three variables may be *fixed*, all three may be *random*, any two may be fixed and the third random, or any one may be fixed and the other two random. Thus we may have a fixed, random, or mixed model.

The expectations of mean squares for the general finite model are shown in Table 17.2. The reader should note that in Table 17.2 the quantity σ_a^2 is the variance in the population for row means. The null hypothesis under test for rows is $H_r: \mu_{1...} = \mu_{2...} = \ldots = \mu_{R...}$. The variance of these means is σ_a^2, and the null hypothesis for rows may be stated as $H_r: \sigma_a^2 = 0$. Likewise σ_b^2 and σ_c^2 are the variances in the population for column and layer means, respectively.

From the general model of Table 17.2 we can deduce the expectations of the mean squares for the fixed model, the random model, or any of the mixed models. For the fixed model $R = R_p$, $C = C_p$, and $L = L_p$. By substituting $(R_p - R)/R_p = 0$, $(C_p - C)/C_p = 0$, and $(L - L_p)/L_p = 0$ in the general finite model, the expectations for the fixed model are obtained. These expectations are shown in Table 17.3.

For the random model the expectations of the mean squares are obtained by substituting $(R_p - R)/R_p = 1$, $(C_p - C)/C_p = 1$, and $(L_p - L)/L_p = 1$ in the expectation for the general finite model. These expectations are shown in Table 17.4.

For the mixed model where rows are random and columns and layers are fixed, the expectations of the mean squares are obtained by substituting $(R_p - R)/R_p = 1$, $(C_p - C)/C_p = 0$, and $(L_p - L)/L_p = 0$ in the expectation for the general finite model given in Table 17.2. The expectations for this

Table 17.2 Expectations of mean squares for general finite model: three-way analysis with n entries in each cell

Mean square	Expectation of mean square
Rows, s_r^2	$\sigma_e^2 + \dfrac{L_p-L}{L_p}\dfrac{C_p-C}{C_p}n\sigma_{abc}^2 + \dfrac{C_p-C}{C_p}Ln\sigma_{ab}^2 + \dfrac{L_p-L}{L_p}Cn\sigma_{ac}^2 + LCn\sigma_a^2$
Columns, s_c^2	$\sigma_e^2 + \dfrac{R_p-R}{R_p}\dfrac{L_p-L}{L_p}n\sigma_{abc}^2 + \dfrac{R_p-R}{R_p}Ln\sigma_{ab}^2 + \dfrac{L_p-L}{L_p}Rn\sigma_{bc}^2 + RLn\sigma_b^2$
Layers, s_l^2	$\sigma_e^2 + \dfrac{R_p-R}{R_p}\dfrac{C_p-C}{C_p}n\sigma_{abc}^2 + \dfrac{R_p-R}{R_p}Cn\sigma_{ac}^2 + \dfrac{C_p-C}{C_p}Rn\sigma_{bc}^2 + RCn\sigma_c^2$
$R \times C$, s_{rc}^2	$\sigma_e^2 + \dfrac{L_p-L}{L_p}n\sigma_{abc}^2 + Ln\sigma_{ab}^2$
$R \times L$, s_{rl}^2	$\sigma_e^2 + \dfrac{C_p-C}{C_p}n\sigma_{abc}^2 + Cn\sigma_{ac}^2$
$C \times L$, s_{cl}^2	$\sigma_e^2 + \dfrac{R_p-R}{R_p}n\sigma_{abc}^2 + Rn\sigma_{bc}^2$
$R \times C \times L$, s_{rcl}^2	$\sigma_e^2 + n\sigma_{abc}^2$
Within cells, s_w^2	σ_e^2

Table 17.3   Expectation of mean squares for fixed model: three-way analysis of variance with n entries in each cell

Mean square	Expectation of mean squares
Rows, s_r^2	$\sigma_e^2 + LCn\sigma_a^2$
Columns, s_c^2	$\sigma_e^2 + RLn\sigma_b^2$
Layers, s_l^2	$\sigma_e^2 + RCn\sigma_c^2$
$R \times C$, s_{rc}^2	$\sigma_e^2 + Ln\sigma_{ab}^2$
$R \times L$, s_{rl}^2	$\sigma_e^2 + Cn\sigma_{ac}^2$
$C \times L$, s_{cl}^2	$\sigma_e^2 + Rn\sigma_{bc}^2$
$R \times C \times L$, s_{rcl}^2	$\sigma_e^2 + n\sigma_{abc}^2$
Within cells, s_w^2	σ_e^2

Table 17.4   Expectation of mean squares for random model: three-way analysis of variance with n entries in each cell

Mean square	Expectation of mean squares
Rows, s_r^2	$\sigma_e^2 + n\sigma_{abc}^2 + Ln\sigma_{ab}^2 + Cn\sigma_{ac}^2 + LCn\sigma_a^2$
Columns, s_c^2	$\sigma_e^2 + n\sigma_{abc}^2 + Ln\sigma_{ab}^2 + Rn\sigma_{bc}^2 + RLn\sigma_b^2$
Layers, s_l^2	$\sigma_e^2 + n\sigma_{abc}^2 + Cn\sigma_{ac}^2 + Rn\sigma_{bc}^2 + RCn\sigma_c^2$
$R \times C$, s_{rc}^2	$\sigma_e^2 + n\sigma_{abc}^2 + Ln\sigma_{ab}^2$
$R \times L$, s_{rl}^2	$\sigma_e^2 + n\sigma_{abc}^2 + Cn\sigma_{ac}^2$
$C \times L$, s_{cl}^2	$\sigma_e^2 + n\sigma_{abc}^2 + Rn\sigma_{bc}^2$
$R \times C \times L$, s_{rcl}^2	$\sigma_e^2 + n\sigma_{abc}^2$
Within cells, s_w^2	σ_e^2

particular mixed model are shown in Table 17.5. The expectations for other varieties of the mixed-model case can be similarly obtained.

17.7 CHOICE OF ERROR TERM

In forming F ratios the expectation of the mean square in the numerator should contain one more term than the expectation of the mean square in the denominator, all other terms in both the numerator and the denominator being the same. Applying this rule to the expectations shown in Tables 17.3, 17.4, and 17.5 indicates how we should proceed in the forming of F ratios.

Table 17.5 Expectation of mean squares for mixed model: rows are random, columns and layers are fixed

Mean square	Expectation of mean squares
Rows, s_r^2	$\sigma_e^2 + LCn\sigma_a^2$
Columns, s_c^2	$\sigma_e^2 + Ln\sigma_{ab}^2 + RLn\sigma_b^2$
Layers, s_l^2	$\sigma_e^2 + Cn\sigma_{ac}^2 + RCn\sigma_c^2$
$R \times C$, s_{rc}^2	$\sigma_e^2 + Ln\sigma_{ab}^2$
$R \times L$, s_{rl}^2	$\sigma_e^2 + Cn\sigma_{ac}^2$
$C \times L$, s_{cl}^2	$\sigma_e^2 + n\sigma_{abc}^2 + Rn\sigma_{bc}^2$
$R \times C \times L$, s_{rcl}^2	$\sigma_e^2 + n\sigma_{abc}^2$
Within cells, s_w^2	σ_e^2

1 Fixed model: $n > 1$. The correct error term for testing all main effects and all interaction effects is s_w^2. All F ratios are formed with s_w^2 in the denominator.

2 Random model: $n > 1$. The correct error term for testing the $R \times C \times L$ interaction is s_w^2. The correct error term for testing the $R \times C$, $R \times L$, and $C \times L$ interactions is s_{rcl}^2. No exact tests of the main effects can be made, although approximate F tests can be used. For a discussion of such approximate tests, see Scheffé (1959).

3 Mixed model: $n > 1$. Inspection of the expectations in the particular mixed-model case where rows are random and columns and layers are fixed indicates that the correct error term for testing rows, $R \times C$, $R \times L$, and $R \times C \times L$ interactions is the within-cells variance estimate, s_w^2. The correct error term for testing the column effect is the $R \times C$ interaction mean square, s_{rc}^2, and for the layer effect the $R \times L$ interaction mean square, s_{rl}^2. The correct error term for testing the $C \times L$ interaction is the $R \times C \times L$ interaction mean square, s_{rcl}^2. The correct error term for other examples of the mixed-model case will, of course, differ from the above, but may be deduced from the general finite model.

17.8 COMPUTATION FORMULAS FOR SUMS OF SQUARES

As previously, computation formulas are used to calculate the required sums of squares. The sum of all N observations is denoted by T. The sum of all the observations for rows, summed within cells and then over columns and blocks, is $T_{r..}$. If we conceptualize the data as comprising a cube of observations, $T_{r..}$ are totals along the edge of the cube. Likewise $T_{.c.}$ are totals summed over rows and layers, and $T_{..l}$ are totals summed over rows and columns.

The quantities $T_{rc..}$, $T_{r.l.}$, and $T_{.cl.}$ are totals summed within cells, and then over layers, columns, and rows, respectively. These are totals on the surface of the cube. $T_{rcl.}$ is an individual cell total corresponding to the rth row, the cth column, and the lth layer.

With n entries in each cell the computation formulas for the sums of squares are as follows:

Rows   [17.1] $\dfrac{1}{nCL} \sum\limits^{R} T_{r...}^2 - \dfrac{T^2}{N}$

Columns   [17.2] $\dfrac{1}{nRL} \sum\limits^{C} T_{.c..}^2 - \dfrac{T^2}{N}$

Layers   [17.3] $\dfrac{1}{nRC} \sum\limits^{L} T_{..l.}^2 - \dfrac{T^2}{N}$

R × C Interaction

[17.4] $\dfrac{1}{nL} \sum\limits^{R} \sum\limits^{C} T_{rc..}^2 - \dfrac{1}{nLC} \sum\limits^{R} T_{r...}^2 - \dfrac{1}{nRL} \sum\limits^{C} T_{.c..}^2 + \dfrac{T^2}{N}$

R × L Interaction

[17.5] $\dfrac{1}{nC} \sum\limits^{R} \sum\limits^{L} T_{r.l.}^2 - \dfrac{1}{nLC} \sum\limits^{R} T_{r...}^2 - \dfrac{1}{nRC} \sum\limits^{L} T_{..l.}^2 + \dfrac{T^2}{N}$

C × L Interaction

[17.6] $\dfrac{1}{nR} \sum\limits^{C} \sum\limits^{L} T_{.cl.}^2 - \dfrac{1}{nLR} \sum\limits^{C} T_{.c..}^2 - \dfrac{1}{nRC} \sum\limits^{L} T_{..l.}^2 + \dfrac{T^2}{N}$

R × C × L Interaction

[17.7] $\dfrac{1}{n} \sum\limits^{R} \sum\limits^{C} \sum\limits^{L} T_{rcl.}^2 - \dfrac{1}{nL} \sum\limits^{R} \sum\limits^{C} T_{rc..}^2 - \dfrac{1}{nC} \sum\limits^{R} \sum\limits^{L} T_{r.l.}^2 - \dfrac{1}{nR} \sum\limits^{C} \sum\limits^{L} T_{.cl.}^2$

$+ \dfrac{1}{nCL} \sum\limits^{R} T_{r...}^2 + \dfrac{1}{nRL} \sum\limits^{C} T_{.c..}^2 + \dfrac{1}{nRC} \sum\limits^{L} T_{..l.}^2 - \dfrac{T^2}{N}$

Within cells   [17.8] $\sum\limits^{R} \sum\limits^{C} \sum\limits^{L} \sum\limits^{n} X_{rcli}^2 - \dfrac{1}{n} \sum\limits^{R} \sum\limits^{C} \sum\limits^{L} T_{rcl.}^2$

Total   [17.9] $\sum\limits^{R} \sum\limits^{C} \sum\limits^{L} \sum\limits^{n} X_{rcli}^2 - \dfrac{T^2}{N}$

Although some of the above formulas are cumbersome in appearance, their application is really quite simple and involves the calculation of only nine different terms from the data, as can be seen in the illustrative example in the following section. The most cumbersome term of all is the triple interaction term, which is sometimes obtained very simply by subtracting from the total sum of squares the sum of all the other sums of squares. The reader should note also that if computational formulas for triple classification for $n = 1$ are required, these are directly obtained by writing $n = 1$ in the computation formulas given above.

17.9 ILLUSTRATIVE EXAMPLE OF THREE-WAY CLASSIFICATION

Table 17.6 shows data for a $2 \times 3 \times 2$ factorial experiment. Early- and late-blinded animals belonging to three strains, bright, mixed, and dull, were reared under two environmental conditions, free and restricted.

Table 17.6 Data for the analysis of variance with three-way classification: $n > 1$, error scores for early- and late-blinded rats for three strains reared under two environmental conditions: cell totals shown in parentheses

Early blinded

| Environment | Strain | | |
	Bright	Mixed	Dull
Free	27 22 45 18 76 33 (221)	31 37 52 45 86 66 (317)	55 62 76 85 104 126 (508)
Restricted	55 40 81 50 36 70 (332)	77 76 98 68 42 104 (465)	132 104 96 70 89 142 (633)

Late blinded

| Environment | Strain | | |
	Bright	Mixed	Dull
Free	61 39 76 60 46 59 (341)	61 71 82 92 103 105 (514)	140 122 99 92 68 101 (622)
Restricted	88 92 95 103 51 73 (502)	100 120 120 131 89 76 (636)	142 150 96 105 80 125 (698)

Twelve groups of animals were used, each group comprised of six animals. In this example rows are environments, columns are strains, and layers are early- versus late-blinded. In Table 17.6 the cell totals are shown in parentheses.

For computational purposes it is necessary to write down the totals for rows by columns summed over layers, rows by layers summed over columns, and layers by columns summed over rows. If these data are visualized as a cube of numbers, these are the totals on the surface of the cube. The totals for rows by columns summed over layers, $T_{rc..}$, are as follows:

$$T_{rc..}$$
Columns

			$T_{r...}$
562	831	1,130	2,523
834	1,101	1,331	3,266

Rows (label at left)

$T_{.c..}$ 1,396 1,932 2,461 $5{,}789 = T_{....}$

The totals above are obtained very simply by adding the cell totals over layers, that is, for the early- and late-blinded animals. Thus $221 + 341 = 562$, $317 + 514 = 831$, and so on. Totals for rows, $T_{r...}$, and for columns, $T_{.c..}$, are also shown, together with the grand total, $T_{....}$. Totals for rows by layers summed over columns, $T_{r.l.}$, are as follows:

$$T_{r.l.}$$
Layers

		$T_{r...}$
1,046	1,477	2,523
1,430	1,836	3,266

Rows (label at left)

$T_{..l.}$ 2,476 3,313 $5{,}789 = T_{....}$

Here 1,046 is obtained by summing the cell totals for the first layer over columns; thus, $221 + 317 + 508 = 1{,}046$. Likewise $332 + 465 + 633 = 1{,}430$, and so on. Totals for layers by columns summed over rows, $T_{.cl.}$, are as follows:

$T_{.ct.}$
Columns

$T_{..l.}$

553	782	1,141	2,476
843	1,150	1,320	3,313

Layers

$T_{.c..}$ 1,396   1,932   2,461 $5,789 = T_{....}$

Here 553 is obtained by summing the cell totals for the first column over rows; thus, $221 + 332 = 553$, $317 + 465 = 782$, and so on.

From the data as arranged above *nine* quantities are calculated which are the different terms in the computation formulas. These quantities in this illustrative example are as follows:

$$\frac{1}{nCL} \sum^{R} T_{r...}^2 = \frac{1}{6 \times 3 \times 2} \times 17{,}032{,}285 = 473{,}119$$

$$\frac{1}{nRL} \sum^{C} T_{.c..}^2 = \frac{1}{6 \times 2 \times 2} \times 11{,}737{,}961 = 489{,}082$$

$$\frac{1}{nRC} \sum^{L} T_{..l.}^2 = \frac{1}{6 \times 2 \times 3} \times 17{,}106{,}545 = 475{,}182$$

$$\frac{1}{nL} \sum^{R} \sum^{C} T_{rc..}^2 = \frac{1}{6 \times 2} \times 5{,}962{,}623 = 496{,}885$$

$$\frac{1}{nC} \sum^{R} \sum^{L} T_{r.l.}^2 = \frac{1}{6 \times 3} \times 8{,}691{,}441 = 482{,}858$$

$$\frac{1}{nR} \sum^{C} \sum^{L} T_{.cl.}^2 = \frac{1}{6 \times 2} \times 5{,}994{,}763 = 499{,}564$$

$$\frac{1}{n} \sum^{R} \sum^{C} \sum^{L} T_{rcl.}^2 = \frac{1}{6} \times 3{,}045{,}597 = 507{,}600$$

$$\sum^{R} \sum^{C} \sum^{L} \sum^{n} X_{rcli}^2 = 536{,}369$$

$$\frac{T^2}{N} = \frac{1}{72} \times 33{,}512{,}521 = 465{,}452$$

In the above computation the quantity $\sum^{R} T_{r...}^2 = 2{,}523^2 + 3{,}266^2 = 17{,}032{,}285$, $\sum^{C} T_{.c..}^2 = 1{,}396^2 + 1{,}932^2 + 2{,}461^2 = 11{,}737{,}961$, and so on. Applying the computation formulas the required sums of squares are then as follows:

Rows
$$\frac{1}{nLC} \sum^{R} T_{r...}^2 - \frac{T^2}{N} = 473{,}119 - 465{,}452 = 7{,}667$$

Columns
$$\frac{1}{nRL} \sum^{C} T_{.c..}^2 - \frac{T^2}{N} = 489{,}082 - 465{,}452 = 23{,}630$$

Layers
$$\frac{1}{nRC} \sum^{L} T_{..l.}^2 - \frac{T^2}{n} = 475{,}182 - 465{,}452 = 9{,}730$$

R × C Interaction
$$\frac{1}{nL} \sum^{R} \sum^{C} T_{rc..}^2 - \frac{1}{nLC} \sum^{R} T_{r...}^2 - \frac{1}{nRL} \sum^{C} T_{.c..}^2 + \frac{T^2}{N}$$
$$= 496{,}885 - 473{,}119 - 489{,}082 + 465{,}452 = 136$$

R × L Interaction
$$\frac{1}{nC} \sum^{R} \sum^{L} T_{r.l.}^2 - \frac{1}{nLC} \sum^{R} T_{r...}^2 - \frac{1}{nRC} \sum^{L} T_{..l.}^2 + \frac{T^2}{N}$$
$$= 482{,}858 - 473{,}119 - 475{,}182 + 465{,}452 = 9$$

C × L Interaction
$$\frac{1}{nR} \sum^{C} \sum^{L} T_{.cl.}^2 - \frac{1}{nLR} \sum^{C} T_{.c..}^2 - \frac{1}{nRC} \sum^{L} T_{..l.}^2 + \frac{T^2}{N}$$
$$= 499{,}564 - 489{,}082 - 475{,}182 + 465{,}452 = 752$$

R × C × L Interaction
$$\frac{1}{n} \sum^{R} \sum^{C} \sum^{L} T_{rcl.}^2 - \frac{1}{nL} \sum^{R} \sum^{C} T_{rc..}^2 - \frac{1}{nC} \sum^{R} \sum^{L} T_{r.l.}^2 - \frac{1}{nR} \sum^{C} \sum^{L} T_{.cl.}^2$$
$$+ \frac{1}{nCL} \sum^{R} T_{r...}^2 + \frac{1}{nRL} \sum^{C} T_{.c..}^2 + \frac{1}{nRC} \sum^{L} T_{..l.}^2 - \frac{T^2}{N}$$
$$= 507{,}600 - 496{,}885 - 482{,}858 - 499{,}564 + 473{,}119$$
$$+ 489{,}082 + 475{,}182 - 465{,}452 = 224$$

Within cells
$$\sum^{R} \sum^{C} \sum^{L} \sum^{n} X_{rcli}^2 - \frac{1}{n} \sum^{R} \sum^{C} \sum^{L} T_{rcl.}^2 = 536{,}369 - 507{,}600 = 28{,}769$$

Total
$$\sum^{R} \sum^{C} \sum^{L} \sum^{n} X_{rcli}^2 - \frac{T^2}{N} = 536{,}369 - 465{,}452 = 70{,}917$$

The analysis of variance table for these data is given in Table 17.7. The df for rows is $R - 1 = 2 - 1 = 1$; for columns, $C - 1 = 3 - 1 = 2$; for layers, $L - 1 = 2 - 1 = 1$; for $R \times C$ interactions, $(R - 1)(C - 1) = (2 - 1)(3 - 1) = 2$; for $R \times L$ interactions, $(R - 1)(L - 1) = (2 - 1)(2 - 1) = 1$; for $C \times L$ interactions, $(C - 1)(L - 1) = (3 - 1)(2 - 1) = 2$; for $R \times C \times L$ interactions, $(R - 1)(C - 1)(L - 1) = (2 - 1)(3 - 1)(2 - 1) = 2$; and for

Table 17.7   Analysis of variance for the data of Table 17.6

Source of variation	Sum of squares	Degrees of freedom	Variance estimate
Rows (environments)	7,667	1	$7,667 = s_r^2$
Columns (strain)	23,630	2	$11,815 = s_c^2$
Layers (early vs. late)	9,730	1	$9,730 = s_l^2$
$R \times C$	136	2	$68 = s_{rc}^2$
$R \times L$	9	1	$9 = s_{rl}^2$
$C \times L$	752	2	$376 = s_{cl}^2$
$R \times C \times L$	224	2	$112 = s_{rcl}^2$
Within cells	28,769	60	$479 = s_w^2$
Total	70,917	71	

within cells, $RCL(n-1) = 12 \times 5 = 60$. In this illustrative sample all three variables may be viewed as fixed, and s_w^2 becomes the appropriate error term for testing all effects. The following F ratio may be calculated:

$$F_r = \frac{s_r^2}{s_w^2} = \frac{7,667}{479} = 16.01 \qquad p < .01$$

$$F_c = \frac{s_c^2}{s_w^2} = \frac{11,815}{479} = 24.67 \qquad p < .01$$

$$F_l = \frac{s_l^2}{s_w^2} = \frac{9,730}{479} = 20.31 \qquad p < .01$$

$$F_{rc} = \frac{s_{rc}^2}{s_w^2} = \frac{68}{479} = .14 \qquad p > .05$$

$$F_{rl} = \frac{s_{rl}^2}{s_w^2} = \frac{9}{479} = .02 \qquad p > .05$$

$$F_{cl} = \frac{s_{cl}^2}{s_w^2} = \frac{376}{479} = .78 \qquad p > .05$$

$$F_{rcl} = \frac{s_{rcl}^2}{s_w^2} = \frac{112}{479} = .23 \qquad p > .05$$

In this illustrative example the difference in environments, strains, and early versus late blinded are highly significant. All interaction effects are not significant. In fact all interaction terms are somewhat less than would ordinarily be expected on a chance basis.

17.10   UNEQUAL NUMBERS IN THE SUBCLASSES

Difficulties associated with unequal numbers in the subclasses discussed in Sec. 16.11 apply also in the three-way classification cases. For small departures from equality the method of expected equal frequencies as discussed in Sec. 16.11 may be used. The least-squares solution to the problem of unequal n's in the subclasses is described by Bancroft (1968). In general it is advisable to avoid unequal n's wherever possible.

17.11   TWO-FACTOR EXPERIMENTS WITH REPEATED MEASUREMENTS

In Secs. 16.12 and 16.13 one-factor experiments with repeated measurements were considered. On occasion experiments are encountered that involve two factors with repeated measurements. Given R levels of one treatment and C levels of another, each subject may be tested under each of the RC treatments. If $R = 2$ and $C = 2$, the levels of R being R_1 and R_2 and of C being C_1 and C_2, there are four treatment combinations, R_1C_1, R_1C_2, R_2C_1, and R_2C_2. Each of N subjects might receive all of the four treatments, the presentations being possibly, although not necessarily, arranged in random order for each subject.

Such data constitute an RCN block of numbers. Rows and columns are treatments, and layers are experimental subjects. These data are analyzed as in the triple-classification case with one observation in each cell. Use the computation formulas given in Sec. 17.8, writing $n = 1$. Seven sums of squares result: row, column, subjects, $R \times C$, $R \times S$, $C \times S$, and $R \times C \times S$. There is, of course, no within-cells sum of squares.

The model here is a mixed model with $n = 1$. Rows and columns will ordinarily be fixed variables. Layers, or subjects, is a random variable. For this model the expectation of the sums of squares and the degrees of freedom are as follows:

Mean square	Expectation of mean square	df
Rows, s_r^2	$\sigma_e^2 + C\sigma_{ac}^2 + NC\sigma_a^2$	$R - 1$
Columns, s_c^2	$\sigma_e^2 + R\sigma_{bc}^2 + NR\sigma_b^2$	$C - 1$
Subjects, s_s^2	$\sigma_e^2 + RC\sigma_c^2$	$N - 1$
$R \times C$, s_{rc}^2	$\sigma_e^2 + \sigma_{abc}^2 + N\sigma_{ab}^2$	$(R - 1)(C - 1)$
$R \times S$, s_{rs}^2	$\sigma_e^2 + C\sigma_{ac}^2$	$(R - 1)(N - 1)$
$C \times S$, s_{cs}^2	$\sigma_e^2 + R\sigma_{bc}^2$	$(C - 1)(N - 1)$
$R \times C \times S$, s_{rcs}^2	$\sigma_e^2 + \sigma_{abc}^2$	$(R - 1)(C - 1)(N - 1)$

Inspection of these expectations indicates that the appropriate error term for testing row effects is the $R \times S$ mean square, $F = s_r^2/s_{rs}^2$. The appropriate error term for testing column effects is the $C \times S$ mean square, $F_c = s_c^2/s_{cs}^2$. The appropriate error term for testing $R \times C$ interaction is the $R \times C \times S$ mean square, $F_{rc} = s_{rc}^2/s_{rcs}^2$. Unless the $R \times S$, $C \times S$, and $F \times C \times S$ interactions are assumed to be zero, which with most sets of data will not be the case, no unbiased tests of difference between subjects, or $R \times S$ or $C \times S$ interactions, can be made. These are ordinarily not of interest.

EXERCISES **1** In a $5 \times 4 \times 3$ factorial experiment with 10 observations per cell, calculate the number of degrees of freedom associated with (a) the main effects, (b) interaction sums of squares, and (c) within-cells sums of squares.

2 Given the following cell means with five observations per cell:

	L_1				L_2	
	C_1	C_2			C_1	C_2
R_1	5	15		R_1	10	20
R_2	10	5		R_2	15	10

Calculate the $R \times C$, $R \times L$, $C \times L$, and $R \times C \times L$ interactions.

3 In a three-way analysis of variance what is the correct error term for testing (a) the first-order interaction mean squares when all three variables are random, (b) the column sum of squares when rows are random and columns and layers are fixed, (c) the second-order interaction when all three variables are fixed.

4 The following are data for a $2 \times 2 \times 2$ factorial experiment with six observations in each of the eight experimental treatments.

	L_1					L_2			
	C_1		C_2			C_1		C_2	
R_1	5	8	8	19	R_1	7	18	25	35
	2	10	10	17		10	26	31	40
	4	3	23	16		25	31	16	18
R_2	22	16	4	8	R_2	31	22	35	18
	15	10	10	5		8	19	41	16
	8	26	11	12		15	27	32	20

Calculate (*a*) the sums of squares and (*b*) the mean squares for these data. On the assumption that the three variables are fixed, (*c*) test the significance of the main effects and interaction effects.

5 The following are data for a $3 \times 3 \times 2$ factorial experiment with nine observations in each of the 18 combinations of experimental treatments.

Layer 1

	C_1			C_2			C_3		
R_1	5	4	11	6	7	15	12	17	18
	1	2	1	2	1	5	13	16	4
	6	9	5	8	12	10	18	21	17
R_2	8	8	10	10	18	19	25	16	12
	4	2	10	15	17	5	17	18	13
	10	10	8	19	10	18	14	9	31
R_3	10	19	6	12	21	31	15	31	22
	15	16	18	11	19	32	10	19	11
	17	8	20	13	19	27	22	35	16

Layer 2

	C_1			C_2			C_3		
R_1	15	12	5	22	18	5	6	5	12
	18	16	18	17	17	4	2	4	2
	10	19	16	9	6	3	9	10	6
R_2	17	18	25	12	17	11	6	9	7
	20	22	16	20	16	12	3	5	2
	15	19	29	18	10	10	4	8	12
R_3	25	22	19	16	10	31	9	18	5
	16	31	12	12	19	30	10	15	12
	19	34	17	12	19	26	17	7	9

Calculate (*a*) the sums of squares and (*b*) the mean squares for these data. On the assumption that the three variables are fixed, (*c*) test the significance of the main effects and interaction effects.

6 The following are data for a repeated-measurements experiment in which four experimental subjects are tested under four treatment combinations.

Subject 1

	C_1	C_2
R_1	5	10
R_2	17	19

Subject 2

	C_1	C_2
R_1	9	19
R_2	20	22

Subject 3

	C_1	C_2
R_1	8	22
R_2	23	28

Subject 4

	C_1	C_2
R_1	20	35
R_2	19	21

Apply an analysis of variance to these data, and test the significance of row and column effects.

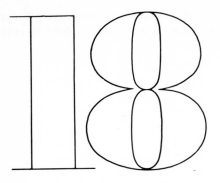

MULTIPLE COMPARISONS

18.1 INTRODUCTION

On comparing k means the analysis of variance may lead to a significant F test. A meaningful interpretation of the data may require a comparison of pairs of means. The differences between some pairs may be significant, while those between others may not be. The investigator may wish to compare selected pairs of means, or to compare every mean with every other mean, there being $k(k-1)/2$ such comparisons.

A variety of methods exist for making such comparisons. One method involves the use of a simple t test to compare pairs of means. This method has limited application. Other methods in common use, using either the F test or a statistic known as the "studentized range," have been developed by Scheffé (1953), Tukey (1949, 1953), Newman (1939), Keuls (1952), and Duncan (1955, 1957). Useful summaries and comparisons of these methods are given by Winer (1962) and Bancroft (1968). Different methods for making multiple comparisons adopt different criteria for the rejection of the null hypothesis. The selection of an appropriate criterion in this context is a thorny problem, and no widely accepted method exists for deciding between different multiple-comparison procedures. The problem has been discussed by Tukey (1953), who drew a distinction between *per-comparison* error rate and *experimentwise* error rate. The *per-comparison* error rate is the number of comparisons falsely declared significant divided by the total number of comparisons. The *experimentwise* error rate is the number of experiments with at least one difference falsely declared significant divided by the total number of experiments. For further elaboration of different error rates in the multiple-comparison problem the reader is referred to Ryan (1959, 1960).

A distinction is commonly made between *a priori* comparisons and *a posteriori* or *postmortem* comparisons. A priori comparisons are formulated prior to, and quite apart from, an inspection of the data. For example, given an experiment involving the comparison of four means, the investigator may decide in advance to test $\bar{X}_1 - \bar{X}_4$. A posteriori comparisons are formulated after, and may be suggested by, inspection of the data. From a logical viewpoint a priori comparisons may be applied whether or not the F test has led to the rejection of the null hypothesis. A posteriori tests are applied only following a significant F test. From a practical point of view the utility of the distinction between a priori and a posteriori comparisons may be questioned. Ordinarily in the analysis of most sets of experimental data no way exists for ascertaining whether certain comparisons were formulated on an a priori basis or not.

When more than one comparison is made, a distinction is drawn between orthogonal and nonorthogonal comparisons. Orthogonal comparisons are independent of each other. It is convenient to conceptualize a comparison as a weighted sum of means. Thus for a set of four means the difference $\bar{X}_1 - \bar{X}_2$ is a weighted sum of $\bar{X}_1$, $\bar{X}_2$, $\bar{X}_3$, and $\bar{X}_4$, the weights being 1, -1, 0, and 0. Also the difference $\bar{X}_3 - \bar{X}_4$ is a weighted sum, the weights being 0, 0, 1, and -1. In general, for equal n's two comparisons are said to be orthogonal when the sum of products of the paired weights is zero. The comparisons $\bar{X}_1 - \bar{X}_2$ and $\bar{X}_3 - \bar{X}_4$ are orthogonal because $(1)(0) + (-1)(0) + (0)(1) + (0)(-1) = 0$. If the sum of products of weights is not zero, the comparisons are not orthogonal. They are not independent of each other. For further discussion of orthogonality see Chap. 19.

18.2 MULTIPLE COMPARISONS USING THE t TEST

For orthogonal a priori comparisons between pairs of means we apply a simple t test to the differences between pairs of means, using the within-group variance estimate, s_w^2. This estimate is based on a larger number of degrees of freedom than a variance estimate obtained from two groups only. The required value of t is given by:

[18.1] $$t = \frac{\bar{X}_1 - \bar{X}_2}{\sqrt{s_w^2/n_1 + s_w^2/n_2}}$$

To illustrate, from the data of Table 15.2 the means for samples I and II are 5.38 and 8.40, respectively. The within-group variance, based on 22 degrees of freedom, is 3.91. The numbers in the two groups are 8 and 5. The t test is then

$$t = \frac{5.38 - 8.40}{\sqrt{3.91/8 + 3.91/5}} = -2.68$$

We consult a table of t for $df = 22$. The values required for significance at the .05 and .01 levels are 2.07 and 2.82, respectively. The observed t is between the .05 and .01 levels. Because of its restriction to a priori orthogonal comparisons, the t test method has limited application.

18.3   MULTIPLE COMPARISONS USING THE F TEST

The method described here is due to Scheffé. This method uses the criterion that the probability of rejecting the null hypothesis when it is true, a Type I error, should not exceed .01 or .05, for example, for any of the comparisons made. This is a very rigorous criterion. To apply the method, follow these steps. *First*, calculate F ratios using the square of the values of t given by formula [18.1]. Here we recall from Sec. 15.9 that for $k = 2$, $F = t^2$. *Second*, consult a table of F and obtain the value of F required for significance at the .05 or .01, or any desired level, for $df_1 = k - 1$ and $df_2 = N - k$. *Third*, calculate a quantity F', which is $k - 1$ times the F required for significance at the desired significance level; that is, $F' = (k - 1)F$. *Fourth*, compare the values of F and F'. For any difference to be significant at the required level, F must be greater than or equal to F'.

The above method may be illustrated with reference to the data of Table 15.2. The means for the four groups are $\bar{X}_1 = 5.38$, $\bar{X}_2 = 8.40$, $\bar{X}_3 = 6.14$, and $\bar{X}_4 = 3.00$; also, $n_1 = 8$, $n_2 = 5$, $n_3 = 7$, and $n_4 = 6$. The values of F obtained using the square of formula [18.1] are as follows:

Comparison	F
I, II	7.18
I, III	.55
I, IV	4.97
II, III	3.81
II, IV	20.34
III, IV	8.18

The values of F required for significance at the .05 and .01 levels, respectively, for $df_1 = 3$ and $df_2 = 22$ are 3.05 and 4.82. The values of F' required for significance at these levels are 9.15 and 14.46. The only comparison that achieves significance is that between groups II and IV. For this comparison the significance level is less than .01.

The Scheffé method may be used not only to compare means two at a time, but to make any comparison at all. For example, given four groups, we may wish to compare the mean for the first two groups combined, $\bar{X}_{1+2} = (n_1\bar{X}_1 + n_2\bar{X}_2)/(n_1 + n_2)$, with the mean for the second two groups, $\bar{X}_{3+4} = (n_3\bar{X}_3 + n_4\bar{X}_4)/(n_3 + n_4)$. The F ratio in this case is

[18.2] $\quad F = \dfrac{(\overline{X}_{1+2} - \overline{X}_{3+4})^2}{s_w^2/(n_1 + n_2) + s_w^2/(n_3 + n_4)}$

For a discussion of the application of the Scheffé method to any or all possible comparisons see Edwards (1968).

The Scheffé method is more rigorous than other multiple comparison methods with regard to Type I error. It will lead to fewer significant differences. It is easy to apply. No special problems arise because of unequal n's. It uses the readily available F test. The criterion it employs in the evaluation of the null hypothesis is simple and readily understood. It is not seriously affected by violations of the assumptions of normality and homogeneity of variance, unless these are gross. It can be used for making any comparison the investigator wishes to make.

Concern may attach to the fact that the Scheffé procedure is more rigorous than other procedures, and will lead to fewer significant results. Because this is so, the investigator may choose to employ a less rigorous significance level in using the Scheffé procedure; that is, the .10 level may be used instead of the .05 level. This is Scheffé's recommendation (1959). Readily available tables of F do not ordinarily contain critical values at the .10 level. Tables showing the .10 critical values are given in Fisher and Yates (1963), and also in Winer (1962).

18.4 MULTIPLE COMPARISONS USING THE STUDENTIZED RANGE

A number of methods for making multiple comparisons use a statistic known as the *studentized range*. Given k treatment means based on equal n's, the studentized range is simply the difference between the largest and the smallest treatment means divided by an estimate of the standard error associated with a single treatment mean; that is,

[18.3] $\quad Q = \dfrac{\overline{X}_{\max} - \overline{X}_{\min}}{s_{\overline{x}}} = \dfrac{\overline{X}_{\max} - \overline{X}_{\min}}{\sqrt{s_w^2/n}}$

For different values of k, and different df associated with s_w^2, the sampling distribution of Q is known. Table L shows the 95 and 99 percentile points for the distribution of Q for different k and df. These are the values required for the rejection of the null hypothesis at the .05 and .01 levels. The null hypothesis in this case is $H_0{:}\mu_1 = \mu_2 = \cdots = \mu_k$ or that the samples came from a single population with mean μ. The Q test could quite clearly be used as an alternate to the usual F test.

Multiple-comparison methods based on the studentized ranges compare the observed studentized ranges for pairs of means with determined criterion values of the studentized range. If the observed value of Q exceeds the criterion value of Q, the difference is said to be significant. Some of these methods are sequential. They involve ordering the means from low to high. The size of the difference required for significance changes with the separation of the means in the rank order. Other methods use a fixed interval:

that is, a fixed criterion value of Q for all comparisons is used, regardless of the separation of the means in the rank order. All multiple-comparison methods using the studentized range apply only to equal n's. This detracts from their usefulness.

One commonly used multiple-comparison method using the studentized range is the Newman-Keuls method. This method uses the criterion that the probability of rejecting the null hypothesis when it is true should not exceed .01 or .05 for all ordered pairs, regardless of the number of steps they are apart. To apply this method the means are ranked from low to high. The studentized ranges are obtained for all $k(k-1)/2$ pairs of means, using $s_{\bar{x}} = \sqrt{s_w^2/n}$. Criterion values of Q at the .01 or .05 levels for $k = 2, 3, \ldots, k$, are obtained from Table L for the df associated with s_w^2. Denote these criterion values as $Q_2, Q_3, \ldots, Q_k$. Q_2 is the criterion value for comparing two adjacent means, Q_3 for means separated by one intervening rank, Q_4 for means separated by two intervening ranks, and Q_k for means separated by $k-2$ intervening ranks. Comparisons of observed Q's with criterion values are made in a particular sequential manner. Values in the first row of the table of Q are compared with the criterion values, proceeding from right to left until a nonsignificant difference is found. No further comparisons are made in that row. This procedure is then applied to the second row, third row, and so on.

To illustrate, consider the following means for five groups of equal size, $n = 6$.

	I	II	III	IV	V
n	6	6	6	6	6
$\bar{X}_j$	12.56	9.54	3.00	6.45	8.32

The analysis of variance table for these data is:

Source	Sum of squares	df	Variance estimate *Mean Square*
Between	304.02	4	$76.01 = s_b^2$
Within	262.50	25	$10.50 = s_w^2$
Total	566.52	29	

The overall F here is $F = 76.01/10.50 = 7.24$, which is significant at better than the .01 level. The means are arranged in rank order from $\bar{X}_3 = 3.00$ to $\bar{X}_1 = 12.56$. The order of the means is $\bar{X}_3, \bar{X}_4, \bar{X}_5, \bar{X}_2, \bar{X}_1$. The difference between every mean and every other mean is calculated. These differences are divided by $s_{\bar{x}} = \sqrt{s_w{}^2/n} = \sqrt{10.50/6} = 1.324$ to obtain the studentized ranges Q, as follows:

Table of Q

	III	IV	V	II	I
III		2.61	4.02	4.92	7.22
IV			1.41	2.33	4.61
V				.92	3.20
II					2.28
I					

If the .05 significance level is adopted, criterion values of Q for a k of 2, 3, 4, and 5 and $df = 25$ are obtained from Table L. These are $Q_2 = 2.91$, $Q_3 = 3.52$, $Q_4 = 3.89$, and $Q_5 = 4.16$. First, we compare the top row of the above table of Q with the criterion values. The larger value 7.22 involves the comparison of III with I. The criterion here is $Q_5 = 4.16$. The observed value exceeds the criterion value and is declared significant. The next largest value in this row is 4.92, which is compared with $Q_4 = 3.89$ and is found to be significant. The next largest is 4.02, which is compared with $Q_3 = 3.52$ and is significant. The next value is 2.61, which when compared with 2.91 is not significant. We now proceed to the second row of the Q table, and compare 4.61 with $Q_4 = 3.89$, which is significant. The next value, 2.33, is not significant and comparisons for that row are discontinued. Proceeding similarly, no other values in the table attain significance. Thus, in this example, the differences between III and I, between III and II, between III and IV, and between IV and I are significant at the .05 level. The reader should note that the above procedure requires that comparisons be discontinued for any row at the first nonsignificant Q value. This rule prevents the making of an inconsistent decision, that is, declaring that a larger difference is not significant whereas a smaller difference adjacent to it is significant. This circumstance could arise on occasion with the Newman-Keuls method if all Q values were compared directly with the criterion values.

The Duncan method, sometimes called the *Duncan new multiple-range test*, is a sequential test which is similar to the Newman-Keuls test. It differs, however, in the level of significance used. The Newman-Keuls test uses a significance level of .05 or .01 for $k = 2, 3, \ldots, k$. The Duncan method uses a significance level $1 - (1 - \alpha)^{k-1}$, where α is ordinarily .05 or .01. Thus for $\alpha = .01$ for two adjacent means, $k = 2$ and the significance level is $1 - .99 = .01$. For means separated by an intervening mean, $k = 3$ and the significance level is $1 - (.99)^2 = .02$. For means separated by two intervening means, $k = 4$ and the significance level is $(1 - .99)^3 = .03$.

Thus the farther the means are apart in the rank ordering, the more lenient the standard of significance. The Duncan method requires a smaller studentized range for significance than does the Newman-Keuls method, except when $k = 2$. The method will yield more significant results than the Newman-Keuls method. The Duncan method requires the use of special tables. These tables are given by Edwards (1968), who provides a detailed description of this test.

Another multiple-comparison method is due to Tukey and has been called the *honestly significant difference method*. It uses a single criterion value of the studentized range, regardless of the separation of means in the rank order. It is a nonsequential method. The criterion value is Q_k, the value required for significance at the .05 or .01 levels for k and df, where k is the total number of means in the set and, as before, df is the number of degrees of freedom associated with s_w^2. This method will lead to fewer significant differences than either the Newman-Keuls or the Duncan method.

All multiple-comparison procedures that use the studentized range are limited in their application because they apply only to groups composed of equal n's. Bancroft (1968) has suggested an intuitive modification for dealing with unequal sample size. His proposal is to replace the n used in $s_{\bar{x}} = \sqrt{s_w^2/n}$ by $\bar{n}_h$, that is, $s_{\bar{x}} = \sqrt{s_w^2/\bar{n}_h}$, where $\bar{n}_h$ is the harmonic mean of the number of observations in the groups, that is:

$$[18.4] \quad \bar{n}_h = \frac{k}{(1/n_1) + (1/n_2) + \cdots + (1/n_k)}$$

Clearly this method should not be used where the n's differ appreciably from each other.

18.5 CONCLUDING OBSERVATIONS

The problem of choosing a particular multiple-comparison procedure to apply to a particular set of experimental data has never been properly resolved. This choice is made by each investigator in the light of considerations which appear relevant to him. One consideration here is clearly the relative importance which attaches to Type I and Type II errors. The reader will recall that Type I error is the probability of asserting that a difference exists when no such difference exists. A Type II error is the probability of asserting that there is no difference when a difference does in fact exist. In terms of per-comparison Type I error, multiple-comparison procedures may be ordered from low to high as follows: Scheffé, Tukey, Newman-Keuls, and Duncan. The Scheffé method in any experiment will lead to the smallest number of significant differences whereas the Duncan method will lead to the largest number. In terms of Type II error, the order of the procedure is the reverse: Duncan, Newman-Keuls, Tukey, and Scheffé. Many investigators may wish to compromise between Type I and

Type II errors and may choose to use the Scheffé procedure with a significance level greater than .05 say .10, or the Newman-Keuls procedure.

EXERCISES **1** The following are means for five groups of experimental subjects:

	I	*II*	*III*	*IV*	*V*
n	5	5	5	5	5
$\overline{X}_j$	8.60	8.40	20.00	29.20	18.40

The within-group mean square $s_w^2 = 16.32$, and the F ratio is 23.81. Test the significance of the difference between means for groups IV and V, using a t test procedure.

2 Compare means two at a time for the data of Exercise 1 above, using Scheffé's procedure. Indicate which comparisons are significant at the .01 level.

3 Apply the Newman-Keuls procedure to the data of Exercise 1 and indicate which comparisons are significant at the .01 level.

4 Apply Scheffé's procedure to the data of Exercise 3, Chap. 15. Indicate which comparisons are significant at the .01 level.

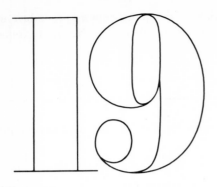

TREND ANALYSIS

19.1 INTRODUCTION

In experiments where the treatment, or independent, variable is nominal, the analysis of the data cannot be extended beyond an F test applied to the group means and the comparison of means either two at a time or in subgroups. In some experiments the treatment variable is of the interval or ratio type. Examples are experiments on the behavioral effects of different dosages of a drug, different periods of practice in learning a task, or different numbers of reinforcements in conditioning or extinction. With such experiments the investigator may extend his analysis to an examination of certain characteristics of the shape of the relation between the treatment variable and the experimental variable. Questions of the following type may be raised. Do the group means increase significantly in a linear fashion with increase in the treatment variable? Is a straight line a good fit to the group means, or do significant deviations from linearity exist? Do the group means increase and then decrease with increase in the treatment variable in an inverted U fashion? Trend analysis as described in this chapter provides answers to questions of this kind. It is an application of the analysis of variance.

In most applications of trend analysis the investigator is usually not interested in the development of a precise equation to describe the functional relation between the treatment variable and the experimental variable, although such an equation could readily be obtained. Usually concern is with questions of significance.

The procedures described in this chapter assume that the levels of the treatment variable are *equally spaced*. Equal spacing is not a necessary condition for the application of methods of trend analysis. It leads, however, to simplification in computation.

19.2 LINEAR TREND: PARTITIONING THE SUM OF SQUARES

Consider an experiment in which k equally spaced treatments are administered to k groups, composed of $n_1, n_2, \ldots, n_k$ members. In applying an analysis of variance for one-way classification to such data, the total sum of squares is partitioned into two parts, a within-groups and a between-groups sum of squares. These are used in an F test of the hypothesis, $H_0: \mu_1 = \mu_2 = \cdots = \mu_k$. The group means $\bar{X}_1, \bar{X}_2, \ldots, \bar{X}_k$ may, however, show some systematic tendency to either increase or decrease in a linear fashion with increase or decrease in the treatment variable. Consequently, the investigator may choose to concern himself with questions of linear trend.

A linear regression line may be fitted to the group means. This line may be used to predict the group means from values of the treatment variable. The mean for group j is $\bar{X}_j$. Denote the corresponding value predicted from the regression line by X'_j. The total sum of squares may now be partitioned into three independent parts. We begin by writing an identity

[19.1] $(X_{ij} - \bar{X}) = (X_{ij} - \bar{X}_j) + (\bar{X}_j - X'_j) + (X'_j - \bar{X})$

This identity states that the deviation of a particular score X_{ij} from the grand mean $\bar{X}$ may be viewed as composed of three parts: (1) a deviation of the score X_{ij} from the mean of the group $\bar{X}_j$ to which it belongs; (2) a deviation of $\bar{X}_j$ from the value predicted from the linear regression line; and (3) a deviation of the predicted value X'_j from the grand mean. This identity is squared, summed over the n_j cases in the jth group, and summed over the k groups. The cross-product terms vanish, and we obtain

[19.2] $$\sum_{j=1}^{k} \sum_{i=1}^{n_j} (X_{ij} - \bar{X})^2 = \sum_{j=1}^{k} \sum_{i=1}^{n_j} (X_{ij} - \bar{X}_j)^2$$
$$+ \sum_{j=1}^{k} n_j(\bar{X}_j - X'_j)^2 + \sum_{j=1}^{k} n_j(X'_j - \bar{X})^2$$

Thus the total sum of squares is partitioned into three parts. The first part is a within-groups sum of squares, and has associated with it $N - k$ degrees of freedom. The second part is a sum of squares of deviations of array means from linear regression. If the group means fall exactly on a straight line, this sum of squares is zero. It has associated with it $k - 2$ degrees of freedom. The third part is a sum of squares of deviations of the predicted values, X'_j, from the grand mean. It has associated with it one degree of freedom. This sum of squares is closely related to the slope b_{xy} of the regression line. It may be shown that

[19.3] $$\sum_{j=1}^{k} n_j(X'_j - \bar{X})^2 = Nb_{xy}{}^2 s_y{}^2$$

where s_y^2 is the variance of the treatment variable. For a particular experiment, the quantity Ns_y^2 exhibits no freedom of variation. It may be viewed as fixed. Variation in the sum of squares depends, therefore, on variation in one quantity only, b_{xy}, the slope of the line. Consequently, the sum of squares has associated with it one degree of freedom.

The three sums of squares are divided by their associated degrees of freedom. Three variance estimates, or mean squares, are obtained. These are s_w^2, the within-groups mean square; s_d^2, the mean square for deviations from the linear regression line; and s_b^2, the linear regression mean square. Two F ratios may be obtained. The first is $F_1 = s_b^2/s_w^2$, with one degree of freedom associated with the numerator and $N - k$ degrees of freedom associated with the denominator. This provides a test of linear trend. It tests whether the slope of the regression line is significantly different from zero. The second F ratio is $F_2 = s_d^2/s_w^2$ with $k - 2$ degrees of freedom associated with the numerator and $N - k$ with the denominator. This provides a test of departures from linearity. It tests whether the variance of deviations from linearity is significantly different from zero. If F_2 is significant, we may conclude that a straight regression line is not a good fit to the set of group means.

19.3 COMPUTATION FORMULAS FOR LINEAR TREND

The calculation involved in the analysis of data for trend with equally spaced treatment levels may be simplified by using a computation variable. In the process of calculation the values of the computation variable are assigned to different levels of the treatment variable. For $k = 3$ to $k = 7$ the computation variables, denoted by c_{1j}, are as follows:

k						
3	−1	0	1			
4	−3	−1	1	3		
5	−2	−1	0	1	2	
6	−5	−3	−1	1	3	5
7	−3	−2	−1	0	1	2	3

Thus for $k = 3$, c_{1j} is −1, 0, 1. These are, in fact, coefficients for orthogonal polynomials in the linear case, and are discussed later in this chapter. Values for $k > 7$ may be obtained from Table J of the Appendix. For the present these coefficients may be viewed as a convenient computational device, analogous perhaps to the computation variable used in calculating a mean or standard deviation from grouped data.

The computation formulas for the total and within-groups sums of squares are the same as for one-way classification, and are given in Chap. 15.

For unequal n's the sum of squares for *linear regression* is given by

$$[19.4] \quad \sum_{j=1}^{k} n_j (X'_j - \overline{X})^2 = \frac{\left(\sum_{j=1}^{k} c_{1j}T_j - T \sum_{j=1}^{k} n_j c_{1j}/N\right)^2}{\sum_{j=1}^{k} n_j c_{1j}^2 - \left(\sum_{j=1}^{k} n_j c_{1j}\right)^2 /N}$$

The *deviation* sum of squares may be obtained by substracting the within and linear regression sums of squares from the total sum of squares. For equal n's the sum of squares for linear regression reduces to

$$[19.5] \quad n \sum_{j=1}^{k} (X'_j - \overline{X})^2 = \frac{\left(\sum_{j=1}^{k} c_{1j}T_j\right)^2}{n\Sigma c_{1j}^2}$$

19.4 ILLUSTRATIVE EXAMPLE OF ANALYSIS FOR LINEAR TREND

The computation required for an analysis of variance for linear trend is illustrated in Table 19.1. The data are those used previously in Table 15.2. The total sum of squares is 168.15, as before, and the within-groups sum is 85.93. The linear regression sum of squares is

$$\frac{\left(\sum_{j=1}^{k} c_{1j}T_j - T \sum_{j=1}^{k} n_j c_{1j}/N\right)^2}{\sum_{j=1}^{k} n_j c_{1j}^2 - \left(\sum_{j=1}^{k} n_j c_{1j}\right)^2 /N} = \frac{[-74 - (146 \times -4)/26]^2}{138 - 16/26} = 19.34$$

Table 19.1 Computation for linear trend analysis with unequal n's using the data of Table 15.2

	Method				
	1	**2**	**3**	**4**	$N = 26$ $T = 146$ $T^2/N = 819.85$
$\overline{X}_j$	5.38	8.40	6.14	3.00	$\sum_{j=1}^{k} n_j c_{1j} = -4$
n_j	8	5	7	6	
T_j	43	42	43	18	$\sum_{j=1}^{k} c_{1j}T_j = -74$
c_{1j}	-3	-1	1	3	$\sum_{j=1}^{k} n_j c_{1j}^2 = 138$
$\sum_{i=1}^{n_j} X_{ij}^2$	269	364	287	68	$\sum_{j=1}^{k} \sum_{i=1}^{n_j} X_{ij}^2 = 988$
$\dfrac{T_j^2}{n_j}$	231.13	352.80	264.14	54.00	$\sum_{j=1}^{k} \dfrac{T_j^2}{n_j} = 902.07$

Table 19.2   Linear trend analysis for the data of Table 19.1

Source of variation	Sum of squares	Degrees of freedom	Variance estimate
Linear regression	19.34	1	$s_b^2 = 19.34$
Deviation	62.88	2	$s_d^2 = 31.44$
Within	85.93	22	$s_w^2 = 3.91$
Total	168.15	25	

$$F_1 = \frac{19.34}{3.91} = 4.95 \qquad F_2 = \frac{31.44}{3.91} = 8.04$$

The deviation sum of squares, obtained by subtraction, is $168.15 - 85.93 - 19.34 = 62.88$. The F ratio for linear regression is significant at a little better than the .05 level. The F ratio for deviations from linear regression is significant at better than the .01 level. If the .05 level is accepted as a suitable level of significance, the conclusion is that the group means show a significant linear trend. Because of the large deviation component, however, the conclusion is that a straight regression line is not a good fit to the data. The relation between the treatment and the experimental variables may be represented more appropriately by a nonlinear regression line.

19.5 POLYNOMIAL REGRESSION

The procedures described thus far deal with linear regression. A nonlinear regression line may describe more appropriately the relation between the experimental and the treatment variables. Although equations of many kinds may be used to describe nonlinear trend, one general equation is a polynomial of degree m as follows:

[19.6] $X' = a + b_1 Y + b_2 Y^2 + \cdots + b_m Y^m$

In this equation a is the point where the curve intercepts the X axis, b_1, b_2, . . . , b_m are regression coefficients or weights, and Y, Y^2, . . . , Y^m are the powers of Y. Note that the equation for a straight line is a particular instance of this equation which occurs when all terms to the right of $b_1 Y$ vanish. A straight line is a polynomial of the first degree. If all terms to the right of $b_2 Y^2$ vanish, a second-degree polynomial, or *quadratic* equation, results. Similarly, higher-order polynomials, cubic equations, quartic equations, and the like may be considered. The method of calculating the coefficients b_1, b_2, . . . , b_m is quite simple, although the details of it need not concern us here. The coefficients may be obtained by the method of least squares using the multiple regression method described in Chap. 26 of this book. The powers of Y, that is, Y, Y^2, . . . , Y^m, are used as the independent variables, and b_1, b_2, . . . , b_m are multiple regression weights.

19.6 ORTHOGONAL POLYNOMIALS

Of particular interest in the analysis of data for trend is a particular type of polynomial known as an orthogonal polynomial. In the polynomial equation of formula [19.6] Y, Y^2, . . . , Y^m are not independent of each other. Consequently it is not possible to partition a sum of squares on X' into a number of independent components each associated in turn with Y, Y^2, . . . , Y^m. A polynomial equation of the type shown in formula [19.6] can be expressed, or represented, in a form such that the successive terms are, in fact, independent or uncorrelated. Such an equation may be written as

[19.7] $$X'_j = a + b_1 c_{1j} + b_2 c_{2j} + \cdots + b_m c_{mj}$$

In this equation the b's are regression coefficients, and the c's are orthogonal sets of coefficients.

To illustrate the nature of the c's in the above equation, consider the two sets of numbers

c_{1j}	-1	0	1
c_{2j}	1	-2	1

Note that $\Sigma c_{1j} = \Sigma c_{2j} = 0$; also $\Sigma c_{1j} c_{2j} = 0$. Because the sum of products is zero, c_{1j} and c_{2j} are said to be orthogonal. They are independent of each other, or uncorrelated. Again consider the following:

c_{1j}	-3	-1	1	3
c_{2j}	1	-1	-1	1
c_{3j}	-1	3	-3	1

Here $\Sigma c_{1j} = \Sigma c_{2j} = \Sigma c_{3j} = 0$.   Also, $\Sigma c_{1j} c_{2j} = \Sigma c_{1j} c_{3j} = \Sigma c_{2j} c_{3j} = 0$.   The three sets of numbers are spoken of as coefficients for orthogonal polynomials. The set c_{1j} is the set of linear coefficients; c_{2j}, the quadratic; and c_{3j}, the cubic. The number of times the signs change determines the degree of the polynomial. For the coefficients above, one sign change occurs in the linear set, two in the quadratic, and three in the cubic.

A description of the procedure for obtaining sets of orthogonal coefficients is beyond the scope of the present discussion. The procedure is described in Kendall (1943) and elsewhere. Table J of the Appendix shows a short table of coefficients for orthogonal polynomials. A more extended table will be found in Fisher and Yates (1963).

19.7 PARTITIONING A SUM OF SQUARES USING ORTHOGONAL POLYNOMIALS

In the discussion on linear regression in Sec. 19.2 the values of X'_j of equation [19.1] were obtained by using a linear regression equation. In polynomial regression the values of X'_j may be obtained by using a polynomial regression equation as shown in formula [19.6]. With orthogonal polynomials the values of X'_j are obtained by using an orthogonal polynomial as shown in formula [19.7]. Because for equal n's $a = \overline{X}$, we may write

[19.8] $(X'_j - \overline{X}) = b_1 c_{1j} + b_2 c_{2j} + \cdots + b_m c_{mj}$

By substitution in equation [19.1]

[19.9] $(X_{ij} - \overline{X}) = (X_{ij} - \overline{X}_j) + (\overline{X}_j - X'_j) + b_1 c_{1j} + b_2 c_{2j} + \cdots + b_m c_{mj}$

This equation is squared, summed over the n cases in the jth group, the n's for the k groups being considered equal, and summed over the k groups. The cross-product terms vanish, and we obtain

[19.10] $$\sum_{j=1}^{k} \sum_{i=1}^{n} (X_{ij} - \overline{X})^2 = \sum_{j=1}^{k} \sum_{i=1}^{n} (X_{ij} - \overline{X}_j)^2 + n \sum_{j=1}^{k} (\overline{X}_j - X'_j)^2 + nb_1^2 \sum_{j=1}^{k} c_{1j}^2$$
$$+ nb_2^2 \sum_{j=1}^{k} c_{2j}^2 + \cdots + nb_m^2 \sum_{j=1}^{k} c_{mj}^2$$

Thus the total sum of squares is partitioned into a within-group sum of squares, a sum of squares of deviations of group means from the values X'_j obtained from the orthogonal polynomial regression equation, and a series of sums of squares each associated with a regression component, linear, quadratic, cubic, quartic, and so on.

The within-group sum of squares has $k(n-1)$ degrees of freedom. The deviation sum of squares has $k-2$ degrees of freedom for a linear equation, $k-3$ for a quadratic equation, $k-4$ for a cubic equation, and $k-m-1$ degrees of freedom in the general case where m is the degree of the polynomial. Each regression sum of squares has one degree of freedom. F ratios may be formed to test the deviation and the regression components.

To illustrate the meaning of the various regression components, consider five experiments, A, B, C, D, E. Each experiment has five equally spaced treatments, I, II, III, IV, V. Let the arithmetic means for the five experiments be as follows:

	I	II	III	IV	V
A	20	25	30	35	40
B	20	25	30	25	20
C	20	28	34	38	40
D	20	25	30	25	30
E	20	25	20	25	20

For experiment A all means fall exactly on a straight line. A linear regression component will be obtained, but no higher-order or deviation components. For experiment B the means increase and then decrease. A quadratic regression component will be obtained, but no linear component because the means show no over-all tendency to either increase or decrease in a linear fashion. For experiment C the means show an overall tendency to increase, but not in a linear fashion. A decreasing increment is observed from one mean to the next. Here a linear regression component corresponding to the over-all increase in means will result; a quadratic component will also result, which will reflect, as it were, the bending nature of the relation. Experiment D illustrates a cubic relation. A linear component will also be obtained which reflects the over-all tendency of the means to increase. A quadratic component will reflect the asymmetrical nature of the cubic relation. Experiment E illustrates a quartic relation. For this experiment the linear, quadratic, and cubic components will be zero.

19.8 COMPUTATION FORMULAS FOR TREND ANALYSIS USING ORTHOGONAL POLYNOMIALS

The computation formulas for the total and within-groups sum of squares are the same as for one-way classification, and are given in Chap. 15. For equal n's the various regression components are given by

$$[19.11] \quad nb_1^2 \sum_{j=1}^{k} c_{1j}^2 = \frac{\left(\sum_{j=1}^{k} c_{1j}T_j \right)^2}{n \sum_{j=1}^{k} c_{1j}^2}$$

$$[19.12] \quad nb_2^2 \sum_{j=1}^{k} c_{2j}^2 = \frac{\left(\sum_{j=1}^{k} c_{2j}T_j \right)^2}{n \sum_{j=1}^{k} c_{2j}^2}$$

and so on.

19.9 ILLUSTRATIVE EXAMPLE OF TREND ANALYSIS USING ORTHOGONAL POLYNOMIALS

Table 19.3 shows an illustrative example of the computation required for a trend analysis using orthogonal polynomials. In this example $n = 8$, the groups being of equal size, and $k = 4$. Table 19.4 shows the corresponding analysis of variance table. In this example the linear component is significant with $p < .01$. Neither the quadratic nor the cubic regression components are significant. The reader will note that the deviation component is zero. This occurs because a cubic equation will always fit four points

Table 19.3 Illustrative example of computation for trend analysis using orthogonal polynomials

	Treatment				
	I	*II*	*III*	*IV*	
	5	4	5	14	
	7	12	8	7	$T = 294$
	6	8	17	6	$T^2/N = 2{,}701.13$
	3	7	26	12	$\sum T_j^2/n = 2{,}995.75$
	9	7	14	15	$\sum_{j=1}^{k}\sum_{i=1}^{n} X_{ij}{}^2 = 3{,}524$
	7	6	8	19	$\sum_{j=1}^{k} c_{1j}T_j = 192$
	4	4	13	8	$\sum_{j=1}^{k} c_{2j}T_j = -24$
	2	9	11	11	
n	8	8	8	8	$\sum_{j=1}^{k} c_{3j}T_j = -86$
T_j	43	57	102	92	
$\bar{X}_j$	5.38	7.13	12.75	11.50	$\sum_{j=1}^{k} c_{1j}{}^2 = 20$
c_{1j}	−3	−1	1	3	
c_{2j}	1	−1	−1	1	$\sum_{j=1}^{k} c_{2j}{}^2 = 4$
c_{3j}	−1	+3	−3	1	
$\sum_{i=1}^{n} X_{ij}{}^2$	269	455	1,604	1,196	$\sum_{j=1}^{k} c_{3j}{}^2 = 20$

Sum of squares

Linear	$192^2/(8 \times 20)$	$= 230.40$
Quadratic	$-24^2/(8 \times 4)$	$= \ \ 18.00$
Cubic	$-86^2/(8 \times 20)$	$= \ \ 46.23$
Deviation	$822.87 - 528.24 - 230.40$	
	$\qquad\qquad\qquad\quad - 18.00 - 46.23 =$	$.00$
Within	$3{,}524.00 - 2{,}995.76$	$= 528.24$
Total	$3{,}524 - 2{,}701.13$	$= 822.87$

Table 19.4   Analysis of variance using trend analysis for data of Table 19.3

Source of variation	Sum of squares	Degrees of freedom	Variance estimate
Linear regression	230.40	1	$s_b{}^2 = 230.40$
Quadratic regression	18.00	1	$s_q{}^2 = 18.00$
Cubic regression	46.23	1	$s_c{}^2 = 46.23$
Deviation	.00	0	
Within	528.24	28	$s_w{}^2 = 18.87$
Total	822.87	31	

$$F_1 = \frac{230.40}{18.87} = 12.21 \qquad F_2 = \frac{18.00}{18.87} = .95 \qquad F_3 = \frac{46.23}{18.87} = 2.45$$

exactly, just as a linear equation will always fit two points exactly, and a quadratic equation three points. In this example with $k = 4$ the analysis has been carried as far as possible.

19.10   TREND ANALYSIS WITH UNEQUAL n's

If a single trend component only is under study, whether linear, quadratic, cubic, or of a higher order, the n's may be equal or unequal. If the intent of the investigator is to examine only the linear and quadratic components and no higher-order components, the experiment should be designed such that the n's are either equal or constitute a symmetrical set, such as, for example, 4, 10, 4, for $k = 3$, or 10, 20, 20, 10, for $k = 4$. The linear and quadratic components are orthogonal when $\Sigma n_j c_{1j} c_{2j} = 0$. For orthogonal components with unequal n's the linear regression sum of squares is given by $\Sigma (c_{1j} T_j)^2 / \Sigma n_j c_{1j}{}^2$, and the quadratic by $\Sigma (c_{2j} T_j)^2 / \Sigma n_j c_{2j}{}^2$.

For higher-order trend analysis it is advisable to design the experiment such that the n's are equal. The reason for this is that the various score components, $b_1 c_{1j}$, $b_2 c_{2j}$, and the like, in the equation for an orthogonal polynomial, are orthogonal for equal n's. For unequal n's the components will not be orthogonal, except under rather special circumstances. The first three components form an orthogonal set when $\Sigma n_j c_{1j} c_{2j} = \Sigma n_j c_{1j} c_{1j} c_{3j} = \Sigma n_j c_{2j} c_{3j} = 0$. Clearly for unequal n's this set of components will not ordinarily be orthogonal.

In practical experimental work, situations arise where the n's are unequal. If the departures from equality are not gross, methods may be used identical to those described in Chap. 16 for making adjustments for unequal n's in the analysis of variance for two-way classification.

19.11 TREND ANALYSIS: CORRELATED DATA

The methods of trend analysis described above have application to experimental data obtained from k independent groups. These methods may be extended to correlated data with measurements obtained on a group of n subjects under k conditions. Such data are illustrated in Table 16.7, Sec. 16.13. The columns represent four conditions and the rows 10 subjects. Variance estimates are obtained for rows, columns, and interaction. The difference between column means is tested using $F = s_c^2/s_i^2$, the denominator being the interaction variance estimate.

If the differences between the experimental conditions are equally spaced, it is appropriate to analyze the column means for trend using the method of orthogonal polynomials. The method is essentially the same as that used for k independent groups. The one difference is that for k independent groups the within-groups variance estimate, s_w^2, is used as the error term to test the regression and deviation components, whereas for correlated data the interaction sum of squares s_i^2 is used in the denominator of the F ratio.

Thus the F test for linear regression is $F = s_b^2/s_i^2$ with $df_1 = 1$ and $df_2 = (R - 1)(C - 1) = (n - 1)(k - 1)$. The test for quadratic regression is $F = s_q^2/s_i^2$, and so on. The test for the deviation component is $F = s_d^2/s_i^2$ with $df_1 = C - m - 1$, m being the number of regression components removed, and $df_2 = (N - 1)(k - 1)$.

19.12 EXTENDED APPLICATIONS OF TREND ANALYSIS

The methods of trend analysis described in this chapter have application to experiments involving one-way classification. The methods may be applied to two-way, and higher-order, factorial experiments. With a two-way factorial experiment of the type described in Chap. 16, both row and column means may be analyzed for trend, using orthogonal polynomials as in the one-way classification case. The various regression and deviation components for rows and columns may be tested using s_w^2 in the denominator of the F ratio. Also in factorial experiments the interaction between treatments may be analyzed for trend. In a two-way factorial experiment with R rows and C columns the analysis of interaction for trend is a method for studying the response surface which the RC cell means define. Do all cell means fall in a plane? Do they form a response surface with quadratic characteristics? Questions of this general kind, and others, can be answered by the analysis of interaction for trend. Such methods are described in detail by Winer (1962).

EXERCISES 1 Consider the following data:

	I	II	III
n	20	20	20
$\bar{X}_j$	7.40	10.50	12.65
$\sum_{i=1}^{n} X_{ij}^2$	2,156	2,150	2,760

Calculate (*a*) the within-group, linear regression, and deviation sums of squares, (*b*) the three variance estimates, (*c*) F ratios for testing the significance of linear regression and the deviations from linear regression.

2 Obtain the slope of the linear regression line for the data of Exercise 1 above, assuming a unit difference between the levels of the treatment variable.

3 What are the coefficients for orthogonal polynomials c_{1j}, c_{2j}, c_{3j} for $k = 6$?

4 Consider the following data:

	I	II	III	IV	V
n	10	10	10	10	10
$\bar{X}_j$	5.50	8.65	10.43	4.86	9.50
$\sum X_{ij}^2$	305	875	1050	305	905

Calculate the linear regression, quadratic regression, cubic regression, deviation, and within-groups variance estimates and apply the appropriate tests of significance.

5 What difficulties will attach to the analysis of experimental data for quadratic, cubic, and higher-order trend, when the n's in the k groups are unequal?

ANALYSIS OF COVARIANCE

20.1 INTRODUCTION

One object of experimental design is to ensure that the results observed may be attributed within limits of error to the treatment variable and to no other causal circumstance. For example, the assignment of subjects to groups at random and the matching of subjects are experimental procedures the purpose of which is to ensure freedom from bias. Situations arise, however, where one or more variables are uncontrolled because of practical limitations associated with the conduct of the experiment. A statistical, rather than an experimental, method may be used to "control" or "adjust for" the effects of one or more uncontrolled variables, and permit, thereby, a valid evaluation of the outcome of the experiment. The analysis of covariance is such a method.

To illustrate, an investigator may wish to compare three different methods of learning French. Each method is applied to a different group of subjects. Mean scores on a test of French achievement, following a period of instruction, are obtained for the three groups. If subjects had been assigned to the three groups at random, the means for the three groups could be directly compared. Practical considerations may, however, prevent the assignment of subjects to groups at random. The investigator may be required to use three existing classes of pupils. These classes may differ in intelligence, and intelligence may be correlated with French achievement. Thus the investigator does not know the extent to which the differences in French achievement result from the different methods of instruction or from differences in the intelligence of the three classes. Intelligence in this situation is an uncontrolled variable. If measures of intelligence are available, the analysis of covariance may be used to compare the differences in French achievement between classes, with the influence of intelligence, as it were, statistically controlled.

Consider another example. The effect of two drugs on motor performance is under study. Two groups of subjects are tested in an initial predrug condition and under the influence of one of the drugs. The initial level of motor performance for the two groups may be different. Initial level is an uncontrolled variable. Part of the differences in motor performance under the drug condition may be due to differences in initial level. The analysis of covariance may be used to remove the bias introduced by differences in initial level and permit the making of unbiased comparisons between drug effects. The analysis of covariance is quite commonly used in drug studies in just this way.

In psychology and education primary interest in the analysis of covariance rests in its use as a procedure for the statistical control of an uncontrolled variable. It may, however, serve other purposes, such as testing the homogeneity of a set of regression coefficients and related hypotheses.

In applications of the analysis of covariance the influence of the uncontrolled variable, sometimes called the *covariate* or the *concomitant variable*, is usually removed by a simple linear regression method, and the residual sums of squares are used to provide variance estimates which in turn are used to make tests of significance. The reader is advised at this time to review his knowledge of simple linear regression.

20.2 NOTATION

An application of a simple analysis of covariance requires paired observations on k groups of experimental subjects. The number of pairs of observations in the k groups is denoted by $n_1, n_2, \ldots, n_k$. The paired observations are assumed to be paired samples drawn from k populations. The data may be represented as follows:

	Group 1		*Group 2*		*Group k*	
	Y_{11}	X_{11}	Y_{12}	X_{12}	Y_{1k}	X_{1k}
	Y_{21}	X_{21}	Y_{22}	X_{22}	Y_{2k}	X_{2k}
	Y_{31}	X_{31}	Y_{32}	X_{32}	Y_{3k}	X_{3k}
	.	.	.	.	.	.
	.	.	.	.	.	.
	.	.	.	.	.	.
	$Y_{n_1 1}$	$X_{n_1 1}$	$Y_{n_2 2}$	$X_{n_2 2}$	$Y_{n_k k}$	$X_{n_k k}$
Means	$\bar{Y}_1$	$\bar{X}_1$	$\bar{Y}_2$	$\bar{X}_2$	$\bar{Y}_k$	$\bar{X}_k$

In this notation X is the variable under study, the dependent variable, whereas Y is the uncontrolled variable, or covariate. In the examples of the previous section, either X is a measure of French achievement and Y is a measure of intelligence, or X is a measure of motor performance under the influence of a drug and Y is a corresponding measure obtained under the initial predrug condition. The analysis of covariance enables a comparison of the group means of X adjusted for differences in the means of Y. The group means on Y and X are represented by $\bar{Y}_1, \bar{Y}_2, \ldots, \bar{Y}_k$ and $\bar{X}_1, \bar{X}_2, \ldots, \bar{X}_k$. A dot notation $\bar{Y}_{.1}, \bar{Y}_{.2}$, and so on, where the dot represents the variable subscript, would perhaps be more appropriate. For convenience, however, we have chosen to use a notation without the dot. The grand means of Y and X are $\bar{Y}$ and $\bar{X}$, respectively.

In the analysis of covariance, sums of products are considered. The sum of products for the observations in the jth group is denoted by

$$\sum_{i=1}^{n} (X_{ij} - \bar{X}_j)(Y_{ij} - \bar{Y}_j)$$

The sum of products for all observations in the k groups, that is, the total sum of products, is

$$\sum_{j=1}^{k} \sum_{i=1}^{n_j} (X_{ij} - \bar{X})(Y_{ij} - \bar{Y})$$

20.3 PARTITIONING A SUM OF PRODUCTS

Before proceeding further with the discussion, it is useful to observe that with paired observations on k groups the total sum of products may be partitioned into within-groups and between-groups sums of products in a manner analogous to that in which a total sum of squares is partitioned into within-groups and between-groups sums of squares. As in the analysis of variance for one-way classification we may write

$$(X_{ij} - \bar{X}) = (X_{ij} - \bar{X}_j) + (\bar{X}_j - \bar{X})$$
$$(Y_{ij} - \bar{Y}) = (Y_{ij} - \bar{Y}_j) + (\bar{Y}_j - \bar{Y})$$

These are multiplied, summed over the n_j cases in the jth group, and summed over k groups. When this is done, two cross-product terms to the right vanish in the summation process, and we obtain

$$[20.1] \quad \sum_{j=1}^{k} \sum_{i=1}^{n_j} (X_{ij} - \bar{X})(Y_{ij} - \bar{Y}) = \sum_{j=1}^{k} \sum_{i=1}^{n_j} (X_{ij} - \bar{X}_j)(Y_{ij} - \bar{Y}_j)$$
$$+ \sum_{j=1}^{k} n_j (\bar{X}_j - \bar{X})(\bar{Y}_j - \bar{Y})$$

The term to the left is a total sum of products of deviations about the grand means $\bar{X}$ and $\bar{Y}$. The first term to the right is the sum of products of deviations about the group means. It is the within-groups sum of products. The second term to the right is the sum of products of deviations between groups. It is the between-groups sum of products.

20.4 REGRESSION LINES

With data consisting of paired observations for k groups, a number of different regression lines may be identified. The slope of the regression line used in predicting X from a knowledge of Y, as given in Chap. 8, is

$$[20.2] \quad b_{xy} = r\frac{s_x}{s_y} = \frac{\displaystyle\sum_{i=1}^{N} (X_{ij} - \bar{X})(Y_{ij} - \bar{Y})}{\displaystyle\sum_{i=1}^{N} (Y_{ij} - \bar{Y})^2}$$

This slope is obtained by dividing a sum of products by a sum of squares. If we divide the total within-groups and between-groups sums of products by the corresponding sums of squares, three regression coefficients are obtained. These are the slopes of three different regression lines.

The *first* is the total over-all regression line for predicting X from a knowledge of Y based on all the observations put together. The slope of this line is

$$[20.3] \quad b_t = \frac{\displaystyle\sum_{j=1}^{k}\sum_{i=1}^{n_j} (X_{ij} - \bar{X})(Y_{ij} - \bar{Y})}{\displaystyle\sum_{j=1}^{k}\sum_{i=1}^{n_j} (Y_{ij} - \bar{Y})^2}$$

A *second* regression line is the over-all within-groups regression line. In predicting X from a knowledge of Y we may consider each of the k groups separately. Each group has its own within-group regression line with slope b_j. Information from these k separate regression lines may be pooled to obtain an over-all within-groups regression line whose slope is given by

$$[20.4] \quad b_w = \frac{\displaystyle\sum_{j=1}^{k}\sum_{i=1}^{n_j} (X_{ij} - \bar{X}_j)(Y_{ij} - \bar{Y}_j)}{\displaystyle\sum_{j=1}^{k}\sum_{i=1}^{n_j} (Y_{ij} - \bar{Y}_j)^2}$$

The numerator of this equation is the within-groups sum of products, and the denominator is the within-groups sum of squares for Y. It should be noted here that the slopes of the individual group regression lines, b_1, b_2, $\ldots$, b_k, are estimates of population parameters β_1, β_2, $\ldots$, β_k. The pooling process used to obtain b_w involves an assumption of homogeneity of slope, that is, that $\beta_1 = \beta_2 = \cdots = \beta_k$. This is a basic assumption in the analysis of covariance.

A *third* regression line may be considered with slope b_b obtained by dividing the between-groups sum of products by the between-groups sum of squares for Y. The slope of this line is

$$[20.5] \quad b_b = \frac{\displaystyle\sum_{j=1}^{k} n_j(\bar{X}_j - \bar{X})(\bar{Y}_j - \bar{Y})}{\displaystyle\sum_{j=1}^{k} n_j(\bar{Y}_j - \bar{Y})^2}$$

Of particular interest in the analysis of covariance is the within-groups regression equation. This equation has the form

[20.6] $X_{ij}'' = b_w(Y_{ij} - \bar{Y}_j) + \bar{X}_j$

where b_w is the within-groups regression slope. As will be shown, the analysis of covariance makes use of the sum of squares of residuals of X about this regression line.

20.5 ADJUSTING THE SUM OF SQUARES OF X

Given measurements on Y and X for k groups the total sum of squares for both Y and X may be partitioned into within-groups and between-groups sums of squares, using the analysis of variance methods described in Chap. 15. Also the total sum of products may be partitioned into within-groups and between-groups sums of products, as described in Sec. 20.3. How may the sum of squares on X be adjusted to allow for, or to remove, the influence of the variation of the uncontrolled variable Y?

Let us first consider, in general, the problem of calculating a sum of squares of residuals of X about the linear regression line used in predicting X from a knowledge of Y. The equation for this linear regression line may be written $X_i' = b_{xy}(Y_i - \bar{Y}) + \bar{X}$, where X_i' is a predicted value, and b_{xy} is the slope of the line. The sum of squares of residuals about this line is $\sum_{i=1}^{N} (X_i - X_i')^2$. By substituting $b_{xy}(Y_i - \bar{Y}) + \bar{X}$ for X_i' in this sum of squares and using simple algebra, it is readily shown that the sum of squares of residuals is

[20.7] $$\sum_{i=1}^{N} (X_i - X_i')^2 = \sum_{i=1}^{N} [(X_i - \bar{X}) - b_{xy}(Y_i - \bar{Y})]^2$$

$$= \sum_{i=1}^{N} (X_i - \bar{X})^2 - \frac{\left[\sum_{i=1}^{N} (X_i - \bar{X})(Y_i - \bar{Y})\right]^2}{\sum_{i=1}^{N} (Y - \bar{Y})^2}$$

This latter expression says that in order to obtain a sum of squares of residuals on X, we subtract from the sum of squares of X a quantity which is equal to the square of the sum of products of X and Y divided by the sum of squares of Y. This quantity will be either zero or positive. Thus in the nonzero case this procedure will always reduce the size of the sum of squares. A residual sum of squares obtained in this way is sometimes called a reduced sum of squares.

In the analysis of covariance we proceed by calculating an adjusted *total* sum of squares on X, using a regression line with slope b_t based on all the observations for the k groups put together. This adjusted total sum of squares is given by

[20.8] $\displaystyle\sum_{j=1}^{k}\sum_{i=1}^{n_j}(X_{ij}-X'_{ij})^2=\sum_{j=1}^{k}\sum_{i=1}^{n_j}[(X_{ij}-\bar{X})-b_t(Y_{ij}-\bar{Y})]^2$

$$=\sum_{j=1}^{k}\sum_{i=1}^{n_j}(X_{ij}-\bar{X})^2-\frac{\left[\displaystyle\sum_{j=1}^{k}\sum_{i=1}^{n_j}(X_{ij}-\bar{X})(Y_{ij}-\bar{Y})\right]^2}{\displaystyle\sum_{j=1}^{k}\sum_{i=1}^{n_j}(Y_{ij}-\bar{Y})^2}$$

The next step is to calculate an adjusted *within-groups* sum of squares using the within-group regression line with slope b_w. This sum of squares is given by

[20.9] $\displaystyle\sum_{j=1}^{k}\sum_{i=1}^{n_j}(X_{ij}-X''_{ij})^2=\sum_{j=1}^{k}\sum_{i=1}^{n_j}[(X_{ij}-\bar{X}_j)-b_w(Y_{ij}-\bar{Y}_j)]^2$

$$=\sum_{j=1}^{k}\sum_{i=1}^{n_j}(X_{ij}-\bar{X}_j)^2-\frac{\left[\displaystyle\sum_{j=1}^{k}\sum_{i=1}^{n_j}(X_{ij}-\bar{X}_j)(Y_{ij}-\bar{Y}_j)\right]^2}{\displaystyle\sum_{j=1}^{k}\sum_{i=1}^{n_j}(Y_{ij}-\bar{Y}_j)^2}$$

Thus to calculate an adjusted within-groups sum of squares on X, we subtract from the within-groups sum of squares on X a quantity which is equal to the within-groups sum of products, squared, divided by the within-groups sum of squares on Y. The adjusted sum of squares for *between groups* is now obtained by subtracting the adjusted within-groups sum of squares on X from the adjusted total sum of squares.

20.6   DEGREES OF FREEDOM AND VARIANCE ESTIMATES

The numbers of degrees of freedom associated with the unadjusted and adjusted sums of squares on X are as follows:

	Unadjusted X	*Adjusted X*
Between	$k-1$	$k-1$
Within	$N-k$	$N-k-1$
Total	$N-1$	$N-2$

The number of degrees of freedom associated with the adjusted total sum of squares on X is $N-2$. This sum of squares consists of squared residuals about a linear regression line, and as such has $N-2$, and not $N-1$, degrees of freedom. Thus 1 degree of freedom is lost in the adjustment process. The number of degrees of freedom associated with the adjusted within-groups sum of squares is $N - k - 1$. The number of degrees of freedom associated with the adjusted between-groups sum of squares is $k - 1$, and is unchanged because the between-groups regression line did not enter into the calculation of the between-groups sum of squares, this sum of squares being obtained by subtraction.

The adjusted sums of squares on X are now divided by their associated degrees of freedom to obtain within-groups and between-groups variance estimates s_w^2 and s_b^2. The interpretation of these variance estimates is the same as in the analysis of variance, except that the null hypothesis under test relates to adjusted treatment means, that is, means that are free of the linear effect of the covariate.

To test the significance of the difference between the adjusted means of X, an F ratio s_b^2/s_w^2 is obtained. This ratio is interpreted with $df = k - 1$ associated with the numerator and $df = N - k - 1$ associated with the denominator.

20.7 COMPUTATION FORMULAS

Computation formulas for sums of squares for X and Y are given in Sec. 15.6. To simplify the notation for the computation formulas for sums of products, denote the sums of all the observations in the jth group for X and Y by T_{x_j} and T_{y_j}, respectively. Thus

$$\sum_{i=1}^{n_j} X_{ij} = T_{x_j} \qquad \sum_{i=1}^{n_j} Y_{ij} = T_{y_j}$$

Denote the sums of X and Y for all the observations together in the k groups by T_x and T_y. Thus

$$\sum_{j=1}^{k} \sum_{i=1}^{n_j} X_{ij} = T_x \qquad \sum_{j=1}^{k} \sum_{i=1}^{n_j} Y_{ij} = T_y$$

The sum of products for the jth group may be represented by

$$\sum_{i=1}^{n_j} X_{ij} Y_{ij} = T_{x_j y_j}$$

and the sum of products for all observations in the k groups by

$$\sum_{j=1}^{k} \sum_{i=1}^{n_j} X_{ij} Y_{ij} = T_{xy}$$

The computation formula for the *total* sum of products is

[20.10] $$\sum_{i=1}^{k} \sum_{i=1}^{n_j} (X_{ij} - \bar{X})(Y_{ij} - \bar{Y}) = T_{xy} - \frac{T_x T_y}{N}$$

The *within-groups* sums of products may be obtained by

[20.11] $$\sum_{j=1}^{k} \sum_{i=1}^{n_j} (X_{ij} - \bar{X}_j)(Y_{ij} - \bar{Y}_j) = T_{xy} - \sum_{j=1}^{k} \frac{T_{x_j} T_{y_j}}{n_j}$$

The *between-groups* sums of products is

[20.12] $$\sum_{j=1}^{k} n_j(\overline{X}_j - \overline{X})(\overline{Y}_j - \overline{Y}) = \sum_{j=1}^{k} \frac{T_{x_j}T_{y_j}}{n_j} - \frac{T_x T_y}{N}$$

The above formulas are applicable to groups of unequal or equal size. In the particular case where $n_1 = n_2 = \cdots = n$ we may, of course, write the term

[20.13] $$\sum_{j=1}^{k} \frac{T_{x_j}T_{y_j}}{n_j} = \frac{\sum_{j=1}^{k} T_{x_j}T_{y_j}}{n}$$

20.8 SUMMARY

In summary, to test the significance of the difference between k adjusted means on X using the analysis of covariance, the following steps are involved:

1 Partition the total sum of squares on both Y and X into two components, a within-groups and a between-groups sum of squares, using the usual analysis of variance formulas.

2 Partition the total sum of products into two components, a within-groups and a between-groups sums of products.

3 Calculate an adjusted total sum of squares on X to remove the linear effects of the covariate Y.

4 Calculate an adjusted within-groups sum of squares on X using the within-groups regression of X on Y.

5 Calculate an adjusted between-groups sum of squares by subtraction; that is, subtract the adjusted within-groups sum of squares from the adjusted total sum of squares.

6 Obtain the variance estimates s_w^2 and s_b^2 by dividing the adjusted within-groups sum of squares on X by $df = N - k - 1$, and the between-groups by $df = k - 1$.

7 Test the significance of the adjusted means on X by referring $F = s_b^2/s_w^2$ to a table of F.

20.9 ILLUSTRATIVE EXAMPLE: ANALYSIS OF COVARIANCE

Table 20.1 shows an artificial example illustrating the analysis of covariance. Information is available on two variables for $k = 3$ with unequal n's. The treatment means are $\overline{X}_1 = 11.70$, $\overline{X}_2 = 8.75$, and $\overline{X}_3 = 6.17$. Although these means differ appreciably, the means for the covariate Y also differ, these being $\overline{Y}_1 = 17.60$, $\overline{Y}_2 = 9.25$, and $\overline{Y}_3 = 6.83$. If a correlation

Table 20.1   Computation for the analysis of covariance: paired observations on Y and X for three groups

	Group I		Group II		Group III		
	Y	X	Y	X	Y	X	$N = 24$
	5	5	6	5	4	7	$T_y = 291$
	11	6	6	7	5	8	$\overline{Y} = 12.13$
	12	9	7	12	7	3	$T_y^2/N = 3,528.38$
	26	12	12	10	9	4	$\sum_{j=1}^{k}\sum_{i=1}^{n_j} Y_{ij}^2 = 4,777$
	28	15	14	8	10	10	
	24	16	9	4	6	5	$\sum_{j=1}^{k}\dfrac{T_{y_j}^2}{n_j} = 4,062.27$
	11	18	8	15			
	27	20	12	9			$T_x = 224$
	12	4					$T_x^2/N = 2,090.67$
	20	12					$\sum_{j=1}^{k}\sum_{i=1}^{n_j} X_{ij}^2 = 2,618$
n_j	10		8		6		$\sum_{j=1}^{k}\dfrac{T_{x_j}^2}{n_j} = 2,209.57$
$T_{y_j} T_{x_j}$	176	117	74	70	41	37	
$\overline{Y}\ \overline{X}$	17.60	11.70	9.25	8.75	6.83	6.17	$T_{xy} = 3,248$
$\sum_{i=1}^{n_j} Y_{ij}^2 \sum_{i=1}^{n_j} X_{ij}^2$	3,720	1,651	750	704	307	263	$\sum_{j=1}^{k}\dfrac{T_{y_j}T_{x_j}}{n_j} = 2,959.53$
$T_{x_j y_j}$	2,341		652		255		$\dfrac{T_x T_y}{N} = 2,716.00$
$\dfrac{T_{y_j}^2}{n_j}\ \dfrac{T_{x_j}^2}{n_j}$	3,097.60		684.50		280.17		
	1,368.90		612.50		228.17		

Sum of squares

	Y	X
Between	$4,062.27 - 3,528.38 = 533.89$	$2,209.57 - 2,090.67 = 118.90$
Within	$4,777.00 - 4,062.27 = 714.73$	$2,618.00 - 2,209.57 = 408.43$
Total	$4,777.00 - 3,528.38 = 1,248.62$	$2,618.00 - 2,090.67 = 527.33$

Sum of products

Between	$2,959.53 - 2,716.00 = 243.53$
Within	$3,248.00 - 2,959.53 = 288.47$
Total	$3,248.00 - 2,716.00 = 532.00$

of some magnitude exists between Y and X, we might anticipate that a substantial part of the variation in the X means will result from the differences in the Y means. All terms necessary for the direct calculation of sums of squares on X and Y and sums of products are given in Table 20.1.

Table 20.2 summarizes the analysis of covariance. The adjusted total sum of squares for X is $527.33 - 532.00^2/1,248.62 = 300.66$. The adjusted within-group sum of squares is $408.43 - 288.47^2/714.73 = 292.00$. The adjusted between-groups sum of squares is $300.66 - 292.00 = 8.66$. The variance estimates are $s_b^2 = 4.33$ and $s_w^2 = 14.60$, and $F = 4.33/14.60 = .30$. This ratio is not significant, and, indeed, it falls substantially short of the value of F of unity expected under the null hypothesis. Quite clearly almost all the variation in the X means can be attributed to the influence of the uncontrolled variable Y.

It is of interest here to calculate directly the adjusted means on X. These adjusted values are given by

$$\bar{X}_j'' = b_w(\bar{Y} - \bar{Y}_j) + \bar{X}_j$$

In this sample $b_w = 288.47/714.73 = .404$. The three adjusted means are

$$\bar{X}_1' = .404(12.13 - 17.60) + 11.70 = 9.49$$
$$\bar{X}_2'' = .404(12.13 - 9.25) + 8.75 = 9.91$$
$$\bar{X}_3'' = .404(12.13 - 6.83) + 6.17 = 8.31$$

These adjusted means vary very little one from another, a fact which is clearly reflected in the small F ratio. We may safely conclude that the differences between the unadjusted means for X are due largely to the effects of Y.

Table 20.2 Analysis of covariance for data of Table 20.1

	Source of variation		
	Between	Within	Total
Sum of squares: Y	533.89	714.73	1,248.62
Sum of squares: X	118.90	408.43	527.33
Sum of products	243.52	288.47	532.00
Degrees of freedom	2	21	23
Adjusted sum of squares: X	8.66	292.00	300.66
Degrees of freedom for adjusted sum of squares	2	20	22
Variance estimates	$s_b^2 = 4.33$	$s_w^2 = 14.60$	

$F = 4.33/14.60 = .30$ $p > .05$

20.10 HOMOGENEITY OF REGRESSION COEFFICIENTS

For certain purposes it may be a matter of interest to test the hypothesis that the slopes of the regression lines within the k groups are the same; that is, $H_0 : \beta_1 = \beta_2 = \cdot \cdot \cdot = \beta_k = \beta_w$. This hypothesis is assumed to be true in any application of the analysis of covariance.

To test this hypothesis, we require that the sum of squares for Y and X and the sum of products for each of the k treatments be given separately. These values are given by

$$\sum_{i=1}^{n_j} (X_{ij} - \bar{X}_j)^2 = \sum_{i=1}^{n_j} X_{ij}^2 - \frac{T_{x_j}^2}{n_j}$$

$$\sum_{i=1}^{n_j} (Y_{ij} - \bar{Y}_j)^2 = \sum_{i=1}^{n_j} Y_{ij}^2 - \frac{T_{y_j}^2}{n_j}$$

$$\sum_{i=1}^{n_j} (Y_{ij} - \bar{Y}_j)(X_{ij} - \bar{X}_j) = \sum_{i=1}^{n_j} X_{ij}Y_{ij} - \frac{T_{x_j}T_{y_j}}{n_j}$$

The next step is to calculate separately for each group an adjusted sum of squares on X in the manner previously described. This is done by subtracting from the sum of squares for X a quantity equal to the square of the sum of products divided by the sum of squares for Y. These adjusted sums of squares are summed over the k groups to obtain

[20.14]
$$\sum_{j=1}^{k} \left\{ \sum_{i=1}^{n_j} (X_{ij} - X_j)^2 - \frac{\left[\sum_{i=1}^{n_j} (Y_{ij} - \bar{Y}_j)(X_{ij} - \bar{X}_j) \right]^2}{\sum_{i=1}^{n_j} (Y_{ij} - \bar{Y}_j)^2} \right\} = A$$

For convenience we note this term by A. This term is a sum of squares of residuals about the k individual regression lines. It has associated with it $\sum_{}^{k} (n_j - 2) = \sum_{}^{k} n_j - 2k = N - 2k$ degrees of freedom. The number of degrees of freedom for each regression line is $n_j - 2$, and there are k such lines. We now obtain an adjusted within-groups sum of squares on X for all observations in the k groups in the manner described previously in this chapter. Denote this sum of squares by B. This sum of squares has $N - k - 1$ degrees of freedom.

To test for homogeneity of regression the following F ratio is calculated:

[20.15] $$F = \frac{(B - A)/(k - 1)}{A/(N - 2k)}$$

with $k - 1$ and $N - 2k$ degrees of freedom associated with the numerator and denominator, respectively.

The general rationale underlying the above F ratio is that when $H_0 : \beta_1 = \beta_2 = \cdot \cdot \cdot = \beta_k$ is true, the sum of squares of residuals about the individual group regression lines with slope b_j will be the same, within the limits of sampling error, as the sum of squares of residuals about a single

Table 20.3 Computation for test of homogeneity of regression coefficients using data of Table 20.1

| Group | Sum of squares | | Sum of products | Adjusted sum of squares for X |
	Y	X		
1	622.40	282.10	281.80	154.51
2	63.50	91.50	4.50	91.19
3	26.83	34.83	2.17	34.65
Total	714.73	408.43	288.47	280.35 = A

$A = 280.35$ $B = 408.43 - 288.47^2/714.73 = 292.00$

$$F = \frac{(292.00 - 280.35)/(3-1)}{280.35/(24-6)} = \frac{5.83}{15.58} = .37$$

within-groups regression line with slope b_w. Under this circumstance $B - A$ will depart from zero only because of the presence of sampling error, and the expected value of the F ratio is unity. When $H_0: \beta_1 = \beta_2 = \cdots = \beta_k$ is not true, B will tend to be greater than A and the expected value of F will be greater than unity.

Table 20.3 uses the data of Table 20.1 to illustrate a test of homogeneity of regression. Sums of squares on Y and X, sums of products, and adjusted sums of squares for X have been calculated for each of the three groups separately. The adjusted sums of squares are summed to obtain A. The Y, X, and sum-of-products columns are summed to obtain within-groups sums of squares and products. An adjusted within-groups sum of squares on X is calculated, and denoted by B. An F ratio is calculated and is found to be .37, which is, of course, not significant.

In this example the regression slopes for the three groups are respectively $b_1 = .45$, $b_2 = .07$, and $b_3 = .08$. Also, $b_w = .404$. Given the small samples used in this illustrative example, the differences between such coefficients would not prove to be significant. In fact, the differences are less than chance expectation.

20.11 THE EXTENDED USE OF THE ANALYSIS OF COVARIANCE

The analysis of covariance, as described in this chapter, assumes that the regression of X on Y is linear. The method may be extended to deal with situations where the regression is nonlinear. Also, we have considered situations involving only one uncontrolled variable or covariate. The analysis of covariance may be used with more than one uncontrolled variable. This involves the use of a multiple regression method of the type described in Chap. 26. Our description of the analysis of covariance applies to single-factor experiments. The method may be adapted for use with two-way, and higher-order, factorial experiments. For a more comprehensive discussion of the analysis of covariance the reader is referred to Winer (1962).

1 The following are paired observations for three experimental groups:

	I		II		III	
	Y	X	Y	X	Y	X
	2	7	8	15	15	30
	5	6	12	24	16	35
	7	9	15	25	20	32
	9	15	18	19	24	38
	10	12	19	31	30	40
$\bar{Y}, \bar{X}$	6.60	9.80	14.40	22.80	21.00	35.00

In this example Y is the covariate or concomitant variable. Calculate the adjusted total, within-groups, and between-groups sums of squares on X, and test the significance of the differences between the adjusted means on X using the appropriate F ratio.

2 For the data of Exercise 1 above calculate: (*a*) the slope of the over-all regression line b_t for predicting X from Y; (*b*) the slope of the over-all within-groups regression line b_w; (*c*) the slope of the between-groups regression line b_b.

3 Calculate for Exercise 1 above the adjusted means on X.

4 Test the homogeneity of the slopes of the three within-groups regression lines in Exercise 1 above.

NONPARAMETRIC STATISTICS

THE STATISTICS OF RANKS

21.1 INTRODUCTION

Ordinal, or rank-order, data may arise in a number of different ways. Quantitative measurements may be available, but ranks may be substituted to reduce arithmetical labor or to make some desired form of calculation possible. For example, measurements of height and weight may be obtained for a group of school children. A correlation between the paired measurements could readily be calculated. The investigator may, however, choose to substitute ranks for the measurements and to calculate a correlation between the paired ranks. In many situations where ranking methods are used, quantitative measurements are not available. The measuring operations used may be such that no comparative statements about the intervals between members are possible. For example, employees may be rank-ordered by supervisors on job performance. School children may be ranked by teachers on social adjustment. Different teas may be rank-ordered by experienced judges on taste, or participants in a beauty contest may be rank-ordered by judges on pulchritude. In such cases the data are comprised of sets of ordinal numbers, 1st, 2d, 3d, . . . , Nth. These are replaced by the cardinal numbers 1, 2, 3, . . . , N for purposes of calculation. The substitution of cardinal numbers for ordinal numbers always assumes equality of intervals. The difference between the first and second member is assumed equal to the difference between the second and third, and so on. This assumption underlies all coefficients of rank correlation. Because of difficulties associated with the measurement of psychological variables, statistical methods for handling rank-order data are of particular interest to psychologists.

Although rank correlation methods have been in use for many years,

303

more recently extensive use has been made of ranks in dealing with many other statistical problems. Ranks are used, for example, in tests for comparing two correlated or independent samples, these being the analogues of t tests, and for a variety of other purposes. Such tests are described as *nonparametric* or *distribution free*, and are discussed in some detail in Chapter 22 to follow. Rank correlation methods are usually classed as nonparametric procedures.

21.2 INTEGERS

Ranks are represented by the integers 1, 2, 3, . . . , N. These are denoted by $X_1, X_2, X_3, \ldots , X_N$. The sum, and sum of squares, of the first N integers may be shown to be as follows:

[21.1] $$\sum_{i=1}^{N} X_i = \frac{N(N+1)}{2}$$

[21.2] $$\sum_{i=1}^{N} X_i^2 = \frac{N(N+1)(2N+1)}{6}$$

The mean of the first N integers is $\overline{X} = (N+1)/2$ and the variance, obtained by dividing the sum of squares about the mean by N, is $s^2 = (N^2 - 1)/12$. The mean is a simple function of the variance, $\overline{X} = 6s^2/(N-1)$. A knowledge of the above relations is sometimes useful.

21.3 MEASURES OF DISARRAY

Consider N individuals $A_1, A_2, A_3, \ldots , A_N$ ranked on two variables X and Y. The rankings on X may be denoted as $X_1, X_2, X_3, \ldots , X_N$ and those on Y as $Y_1, Y_2, Y_3, \ldots , Y_N$. Let the ranks for five individuals on X and Y be as follows:

X	1	2	3	4	5
Y	1	4	3	5	2

The X ranks are in their natural order. The Y ranks are not in their natural order. They exhibit a degree of disarray with respect to X. How may a measure of disarray in this situation be defined?

One commonly used measure of disarray is the sum of squares of differences between the paired ranks. This quantity may be designated by Σd^2. In the above example $\Sigma(X - Y)^2 = \Sigma d^2 = (0)^2 + (-2)^2 + (0)^2 + (-1)^2 + (3)^2 = 14$. It is of interest to consider the minimum and maximum values of Σd^2. When members are ranked in the same order on both X and Y, $\Sigma d^2 = 0$ and is a minimum. Thus if the ranks on X are 1, 2, 3, 4, 5 and on Y 1, 2, 3, 4, 5, the differences are all zero. If the paired ranks are in an inverse order, the maximum degree of disarray, Σd^2 is a maximum. Thus if the ranks on X are 1, 2, 3, 4, 5 and on Y 5, 4, 3, 2, 1, the differences d are $-4, -2, 0, 2$, and 4 and $\Sigma d^2 = 40$. No arrangement of Y with respect to X will produce a larger value of Σd^2. It may be shown that the maximum value of Σd^2 is given by

[21.3] $$\Sigma d^2_{\max} = \frac{N(N^2 - 1)}{3}$$

If ranks on Y are arranged at random with respect to X, the expected value of Σd^2 is simply one-half $\Sigma d^2_{\max}$ or

[21.4] $$E(\Sigma d^2) = \frac{N(N^2 - 1)}{6}$$

The statistic Σd^2 is only one of a number of measures of disarray which might be defined.

Another measure of disarray is the statistic S. Consider again the paired ranks, on X 1, 2, 3, 4, 5 and on Y 1, 4, 3, 5, 2. The X ranks are in their natural order and the Y ranks exhibit a degree of disarray with respect to X. To calculate S we compare each rank on Y with every other rank, there being $N(N-1)/2$ such comparisons for N ranks. If a pair is ranked in its natural order, say 1 and 4, a weight $+1$ is assigned. If a pair is ranked in an inverse order, say 4 and 3, a weight -1 is assigned. The statistic S is the sum of such weights over $N(N-1)/2$ such comparisons. In the example above the weights are $+1, +1, +1, +1, -1, +1, -1, +1, -1, -1$, and $S = 2$. The maximum and minimum values of S may be considered. The maximum value occurs when both sets of ranks are in a natural order, all weights being $+1$, and is $N(N-1)/2$. The minimum value of S occurs when both sets of ranks are in an inverse order, all weights being -1, and is $-N(N-1)/2$. When ranks are arranged at random with respect to each other, the expected value of S is zero.

Both Σd^2 and S are used in the definition of measures of rank correlation and for other purposes as well.

21.4 SPEARMAN'S COEFFICIENT OF RANK CORRELATION ρ

The measure of disarray, Σd^2, is used in the definition of Spearman's coefficient of rank correlation. A coefficient of rank correlation is a statistic defined in such a way as to take a value of $+1$ when the paired ranks are in the same order, a value of -1 when the ranks are in an inverse order, and an expected value of zero when the ranks are arranged at random with respect to each other. A definition of rank correlation which meets these requirements is:

[21.5] $$\rho = 1 - \frac{2\Sigma d^2}{\Sigma d^2_{\max}}$$

where ρ is the Greek letter rho. When the paired ranks are in the same

Table 21.1 Calculation of Spearman's coefficient of rank correlation

	Rank		Difference	
Individual	X	Y	d	d²
A_1	1	6	−5	25
A_2	2	3	−1	1
A_3	3	7	−4	16
A_4	4	2	2	4
A_5	5	1	4	16
A_6	6	8	−2	4
A_7	7	4	3	9
A_8	8	9	−1	1
A_9	9	5	4	16
A_{10}	10	10	0	0
Total			0	$\Sigma d^2 = 92$

$$\rho = 1 - \frac{6 \times 92}{10(100 - 1)} = .442$$

order, $\Sigma d^2 = 0$ and $\rho = 1$. When the paired ranks are in an inverse order, $\Sigma d^2 = \Sigma d^2_{\max}$ and $\rho = -1$. In the case of independence, $2\Sigma d^2 = \Sigma d^2_{\max}$ and $\rho = 0$. By substituting $\Sigma d^2_{\max} = N(N^2 - 1)/3$ in the above formula we obtain

[21.6] $$\rho = 1 - \frac{6\Sigma d^2}{N(N^2 - 1)}$$

This is the usual way of writing Spearman's coefficient of rank correlation.
Pearson's product-moment correlation coefficient may be written in the form:

[21.7] $$r = \frac{\Sigma(X - \bar{X})(Y - \bar{Y})}{\sqrt{\Sigma(X - \bar{X})^2 \, \Sigma(Y - \bar{Y})^2}}$$

Spearman's ρ is a particular case of the above formula. It is the particular case that arises when the variables are the first consecutive untied integers. If the above formula is applied directly to paired ranks, the result is identical with that obtained by applying the formula for ρ.

The calculation of ρ is illustrated in Table 21.1. The calculation is simple. We find the differences between the paired ranks, square them, sum to obtain Σd^2, and then apply the formula for ρ.

21.5 SPEARMAN'S ρ WITH TIED RANKS

In arranging the members of a group in order, a judge may be unable to discriminate between certain members. Where measurements are replaced

Table 21.2   Calculation of Spearman's coefficient of rank correlation with tied ranks

	Rank		Difference	
Individual	X	Y	d	d^2
A_1	1	8	−7	49.00
A_2	2.5	6.5	−4	16.00
A_3	2.5	4.5	−2	4.00
A_4	4.5	2	2.5	6.25
A_5	4.5	1	3.5	12.25
A_6	6	3	3	9.00
A_7	8	4.5	3.5	12.25
A_8	8	6.5	1.5	2.25
A_9	8	9	−1	1.00
A_{10}	10	10	0	.00
Total				$\Sigma d^2 = 112.00$

$$\rho = 1 - \frac{6 \times 112}{10(100 - 1)} = .321$$

by ranks, certain measurements may be equal. These circumstances give rise to tied ranks. If we attempt to replace the numbers 14, 19, 19, 22, 23, 23, 23, 25 by ranks, we observe immediately that 19 occurs twice and 23 three times. Under these circumstances we assign to each member the average rank which the tied observations occupy. Thus 14 is ranked 1, the two 19's are ranked 2.5 and 2.5, the 22 is ranked 4, the three 23's are ranked 6, 6, and 6, and 25 is ranked 8. Having replaced the tied ranks by their average rank, we proceed as before in the calculation of ρ. A calculation with tied ranks is illustrated in Table 21.2. If the ties are numerous, this type of adjustment for tied ranks may not prove altogether satisfactory.

The development of ρ from the ordinary product-moment r assumes that the ranks are the first N integers. Where tied ranks occur this is not so. Where a substantial number of tied ranks is found, the departure of the sum of squares of ranks from the sum of squares of the first N integers will be appreciable and the value of ρ will be thereby affected. While other procedures for correcting for ties may be used, one convenient approach is to calculate an ordinary product-moment correlation for the paired observations where average ranks have been substituted for ties.

21.6   TESTING THE SIGNIFICANCE OF SPEARMAN'S ρ

The problem of developing a test of significance for ρ is approached by considering the N factorial possible arrangements of Y with respect to a fixed ranking on X. Each arrangement is considered equally probable. For each arrangement either the quantity Σd^2 or ρ may be calculated. A frequency distribution can be made either of the Σd^2 or of the values of ρ. This is a null distribution and serves as a model for evaluating a particular observed value of Σd^2, or ρ, against the null hypothesis, H_0, that no association between X and Y exists. If the particular observed arrangement of Y with respect to X, as reflected in a low Σd^2, or a high ρ, is improbable, say $p < .05$ or $.01$, then the null hypothesis is rejected. To illustrate, for $N = 2$, if X has the ranks 1, 2, only two arrangements of Y with respect to X are possible: 1, 2 and 2, 1. Only two values of ρ are possible, $+1$ and -1. For $N = 3$, if X has the ranks 1, 2, 3, there are six possible arrangements of Y and, as it turns out, four different values of ρ, -1, $-1/2$, $+1/2$, and $+1$. The sampling distribution of Σd^2 has been studied by Kendall (1943) and others. For $N = 7$ or 8 the distribution has a somewhat jagged or serrated appearance. The distribution is always symmetrical. As N increases in size, the distribution approaches the normal form.

Table G of the Appendix shows critical values of ρ for different values of N required for significance at various levels. Observe that for a small N, values of ρ of very substantial size must be obtained before we have adequate grounds for rejecting the hypothesis that no association exists between the rankings. For $N = 10$ we require a ρ equal to or greater than $.564$ before we can argue that a significant association exists in a positive direction at the 5 per cent level.

With $N = 10$ or greater, we may test the significance of ρ by using a t given by

[21.8] $t = \rho \sqrt{\dfrac{N-2}{1-\rho^2}}$

This quantity has a t distribution with $N - 2$ degrees of freedom. For example, where $N = 10$ and $\rho = .564$, $t = 1.93$. For eight degrees of freedom, the value of t at the $.05$ level is 2.31. For a two-tailed test we have insufficient grounds for arguing that the observed ρ is significantly different from zero. For a one-tailed test the observed ρ is significant at about the 5 per cent level.

21.7   KENDALL'S COEFFICIENT OF RANK CORRELATION

An alternative form of rank correlation, τ, or tau, has been developed by Kendall (1952, 1955). In the definition of Kendall's tau use is made of the measure of disarray, S. The maximum possible value of S is $N(N-1)/2$. Kendall's coefficient of rank correlation, τ, is defined as the obtained value of S divided by its maximum possible value, that is,

[21.9] $\tau = \dfrac{S}{\frac{1}{2}N(N-1)}$

The statistic τ has a value -1 when the paired ranks are in an inverse order, and a value of $+1$ when the paired ranks are in the same order. For X of 1, 2, 3, 4, and 5 and Y of 1, 4, 3, 5, 2, $S = 2$, $N = 5$, and $\tau = 2/10 = .20$.

21.8 KENDALL's τ WITH TIED RANKS

If ties occur, the convention is adopted, as in the calculation of Spearman's ρ, of replacing the tied values by the average rank. A comparison of two tied values on Y receives a weight of zero. If ties occur on X, a comparison of the corresponding paired Y values will also receive a weight of zero, regardless of whether the paired Y values are tied. Consider the following example with no ties on X and one tied pair on Y.

X	1	2	3	4	5	6
Y	2	3	4.5	4.5	1	6

On comparing each rank on Y with every other rank, and assigning a $+1$ for a pair in their natural order, a -1 for a pair in an inverse order, and a 0 for a tie, we obtain $+1, +1, +1, -1, +1, +1, +1, -1, +1, 0, -1, +1, -1, +1, +1$, and $S = 6$. Consider another example with tied values on both X and Y.

X	1.5	1.5	3	5	5	5
Y	2	3	4.5	4.5	1	6

Here the comparison on Y of 2 with 3 receives a weight of zero, because the order of the first two paired values on X is arbitrary. Similarly, comparisons on Y involving the last three values will receive weights of zero, because of the triplet of ties on X. In the above example the weights are $0, +1, +1, -1, +1, +1, +1, -1, +1, 0, -1, +1, 0, 0, 0$, and $S = 4$.

To calculate tau with tied ranks, S is calculated in the manner described above, and the following formula applied:

$$[21.10] \quad \tau = \frac{S}{\sqrt{\left[\frac{1}{2}N(N-1) - T_x\right]\left[\frac{1}{2}N(N-1) - U_y\right]}}$$

In this formula $T_x = \frac{1}{2}\sum^{m} t(t-1)$ and $U_y = \frac{1}{2}\sum^{r} u(u-1)$. One ranking contains m sets of t ties; the other ranking, r sets of u ties. To illustrate, consider the example immediately above with ties on both X and Y. Here $T_x = \frac{1}{2}[2(2-1) + 3(3-1)] = 4$; also $U_y = \frac{1}{2}[2(2-1)] = 1$. In this example $N = 6$, and tau is as follows:

$$\tau = \frac{4}{\sqrt{\left[\frac{1}{2} \times 6(6-1) - 4\right]\left[\frac{1}{2} \times 6(6-1) - 1\right]}} = .31$$

21.9 THE SIGNIFICANCE OF S AND τ

The sampling distribution of S is obtained by considering the N factorial arrangements of Y in relation to X. A value of S may be determined for each of the N arrangements. The distribution of these N factorial values is the sampling distribution of S. This distribution is symmetrical. Frequencies taper off systematically from the maximum value toward the tails. The distribution of S rapidly approaches the normal form. For $N \geqslant 10$ the normal approximation to the exact distribution is very close. The exact sampling distributions of S for $N = 4$ to $N = 10$ are given by Kendall (1955).

In testing the significance of the association between paired ranks, it is more convenient to apply a test directly to S rather than to τ. The variance of the sampling distribution of S without ties is given by

$$[21.11] \quad \sigma_s^2 = \frac{N(N-1)(2N+5)}{18}$$

If the normal approximation to the exact sampling distribution of S is used, a correction for continuity should be applied. This is done by subtracting unity for the absolute value of S. To apply a significance test, we divide S, corrected for continuity, by the standard deviation of the sampling distribution to obtain the normal deviate z, as follows:

$$[21.12] \quad z = \frac{|S| - 1}{\sqrt{N(N-1)(2N+5)/18}}$$

As usual, 1.96 and 2.58 are required for significance at the .05 and .01 levels, respectively, for a nondirectional test. To illustrate, consider the paired ranks:

X	1	2	3	4	5	6
Y	2	4	3	5	1	6

Here the weights are $+1, +1, +1, -1, +1, -1, +1, -1, +1, +1, -1, +1,$ $-1, +1, +1,$ and $S = 5$. Hence

$$z = \frac{5 - 1}{\sqrt{6(6-1)(2 \times 6 + 5)/18}} = \frac{4}{5.323} = .751$$

Here the association between the paired ranks is clearly not significant.

Because with problems involving ranks ties are very common, it is useful to know the variance of the sampling distribution of S when ties occur. If ties occur in one set of ranks, and not in the other, the variance of S becomes

$$[21.13] \quad \sigma_s^2 = \tfrac{1}{18} \left[N(N-1)(2N+5) - \sum^{m} t(t-1)(2t+5) \right]$$

One set of ranks contains m sets of t ties. Note that the effect of ties is to reduce the variance σ_s^2. Examination of the above formula shows that the effect of ties is to reduce the variance by unity for each tied pair, by 3.67 for each triplet of ties, and by 8.67 for each quadruplet of ties. Thus we have available a very convenient correction procedure.

If ties occur in both sets of ranks, the variance of S is given by

[21.14] $\sigma_s{}^2 = \frac{1}{18}\left[N(N-1)(2N+5) - \sum\limits_{}^{m} t(t-1)(2t+5)\right.$

$$\left. - \sum\limits_{}^{r} u(u-1)(2u+5)\right] + \frac{1}{9N(N-1)(N-2)}$$

$$\left[\sum\limits_{}^{m} t(t-1)(t-2)\right]\left[\sum\limits_{}^{r} u(u-1)(u-2)\right]$$

$$+ \frac{1}{2N(N-1)}\left[\sum\limits_{}^{m} t(t-1)\right]\left[\sum\limits_{}^{r} u(u-1)\right]$$

In this formula one set of ranks contains m sets of t ties: the other set of ranks, r sets of u ties. This formula is the more general form for the variance of the distribution of S. As in the untied case, to apply a test of significance, we divide S, corrected for continuity, that is, $|S| - 1$, by the standard deviation of the sampling distribution σ_s to obtain the normal deviate z.

21.10   RANK CORRELATION WHEN ONE VARIABLE IS A DICHOTOMY

The statistic S, and the rank correlation coefficient tau, may be calculated when one set of ranks is a dichotomy. Consider an example given by Kendall (1955) for 15 boys and girls ranked on an examination. Here sex is a dichotomous variable. Let X be ranks on the examination, and Y sex.

X	1	2	3	4	5	6	7	8	9	10	11	12	13	14	15
Y	B	B	G	B	G	G	B	B	B	G	B	G	B	G	G

Here the eight boys may be considered tied on Y; also the seven girls. Adopting the average-rank procedure, a rank of 4.5 may be assigned to each of the eight boys and a rank of 12 to each of the seven girls. The rank 4.5 is the average of the ranks 1 to 8, and the rank 12 is the average of the ranks 9 to 15. Thus we may write

X	1	2	3	4	5	6	7	8	9	10	11	12	13	14	15
Y	$4\frac{1}{2}$	$4\frac{1}{2}$	12	$4\frac{1}{2}$	12	12	$4\frac{1}{2}$	$4\frac{1}{2}$	$4\frac{1}{2}$	12	$4\frac{1}{2}$	12	$4\frac{1}{2}$	12	12

The statistic S calculated from these ranks is

$$S = 7 + 7 - 6 + 6 - 5 - 5 + 4 + 4 + 4 - 2 + 3 - 1 + 2 + 0 = 18$$

The coefficient tau is obtained using formula [21.10], and is

$$\tau = \frac{18}{\sqrt{\left[\frac{1}{2} \times 15(15-1)\right]\left[\frac{1}{2} \times 15(15-1) - 49\right]}} = .235$$

where $T_x = 0$ and $U_y = \frac{1}{2}(8+7) + \frac{1}{2}(7 \times 6)$. In this example the rank 4.5 has been assigned to boys and the rank 12 to girls, resulting in a positive correlation. This, of course, is arbitrary. The ranks could have been reversed and a negative correlation obtained.

To test the significance of the association, the standard error σ_s is calculated and a normal deviate z obtained. In the present illustrative example X contains no ties, Y contains two groups of ties, and σ_s may be obtained using the square root of formula [21.13] as follows:

$$\sigma_s = \sqrt{\tfrac{1}{18}(15 \times 14 \times 35 - 8 \times 7 \times 21 - 7 \times 6 \times 19)} = 17.28$$

Subtracting unity from the absolute value of S as a continuity correction and dividing by σ_s, we obtain the normal deviate

$$z = (|S| - 1)/\sigma_s = 17/17.28 = .984$$

Clearly the association between ranks on the examination and sex is not significant.

When one variable is a dichotomy and no ties occur in the X ranks, S is reduced by unity as a continuity correction. When one variable is a dichotomy and the other contains m groupings of values of extent t_1, the absolute value of S is reduced by $(2N - t_1 - t_m)/2(m - 1)$, where $t_1 \geqslant 1$. Here t_1 and t_m are the number of tied values in the first and last groups. The reader should note also that when ties occur in the X variable, formula [21.14] and not formula [21.13] should be used in calculating σ_s.

21.11 COMPARISON OF ρ AND τ

The coefficients ρ and τ, although used for the same purpose, are on a somewhat different scale. If ρ and τ are calculated on the same data, the absolute value of ρ will be larger than the corresponding value of τ. Values of ρ and τ are highly correlated in samples from a bivariate normal population. When the correlation in the population is zero, the product-moment correlation between ρ and τ is .980 for $N = 5$ and approaches 1 as N approaches infinity. For practical purposes the correlation between the two statistics can be regarded as close to 1.

On considering the N factorial arrangements of one set of ranks in relation to another, the distributions of both ρ and τ are symmetrical and tend to the normal form for large N. The distribution of τ tends to approach the normal form more rapidly than that of ρ. The exact distributions are known for higher values of τ than of ρ. In general τ as a statistic is more amenable to mathematical manipulation than ρ. Problems resulting from tied values are more readily solved. Also, the measure of disarray, S, used in the definition of τ, seems to have a degree of generality about it that does not characterize Σd^2. S has a number of applications apart from its use in correlation.

21.12 THE COEFFICIENT OF CONCORDANCE W

For data comprised of m sets of ranks, where $m > 2$, a descriptive measure of the agreement or concordance between the m sets is provided by Kendall's coefficient of concordance W. The data of Table 21.3 consist of six

Table 21.3   Ranks assigned to six job applicants by four interviewers

Interviewer	Applicant					
	a	*b*	*c*	*d*	*e*	*f*
A	6	4	1	2	3	5
B	5	3	1	2	4	6
C	6	4	2	1	3	5
D	3	1	4	5	2	6
R_j	20	12	8	10	12	22

ranks assigned by four judges. These data were obtained in an investigation on interviewing technique. Four interviewers were required to interview six job applicants and rank order them on suitability for employment. If perfect agreement were observed between the four interviewers, one applicant would be assigned a 1 by all four. The sum of his ranks would be 4. Another applicant would be assigned a 2 by all four interviewers. The sum of his ranks would be 8. The sum of ranks for the six applicants would be 4, 8, 12, 16, 20, and 24, not necessarily in that order. In general, when perfect agreement exists among ranks assigned by m judges to N members, the rank sums are $m, 2m, 3m, 4m, \ldots, Nm$. The total sum of N ranks for m judges is $mN(N+1)/2$, and the mean rank sum is $m(N+1)/2$.

The degree of agreement between judges reflects itself in the variation in the rank sums. When all judges agree, this variation is a maximum. Disagreement between judges reflects itself in a reduction in the variation of rank sums. For maximum disagreement the rank sums will tend to be more or less equal. This circumstance provides the basis for the definition of a coefficient of concordance.

Let R_j represent the rank sum of the jth individual. The sum of squares of rank sums for N individuals is

[21.15] $$S = \Sigma \left(R_j - \frac{\Sigma R_j}{N} \right)^2$$

The maximum value of this sum of squares occurs when perfect agreement exists between judges and is equal to $m^2(N^3 - N)/12$. The coefficient of concordance W is defined as the ratio of S to the maximum possible value of S and is

[21.16] $$W = \frac{12S}{m^2(N^3 - N)}$$

When perfect agreement exists between judges, $W = 1$. When maximum disagreement exists, $W = 0$. W does not take negative values. With more than two judges complete disagreement cannot occur. For example, if A and B are in complete disagreement and A and C are also in complete disagreement, then B and C must be in complete agreement.

In the example of Table 21.3 the rank totals are 20, 12, 8, 10, 12, and 22. The sum of ranks is 84. The mean rank total, the rank sum expected in the case of independence, is $\frac{84}{6} = 14$. The sum of squares of deviations about this mean is

$$S = (20 - 14)^2 + (12 - 14)^2 + (8 - 14)^2$$
$$+ (10 - 14)^2 + (12 - 14)^2 + (22 - 14)^2 = 160$$

In our example $m = 4$ and $N = 6$ and the coefficient of concordance is

$$W = \frac{12 \times 160}{4^2(6^3 - 6)} = .571$$

The concordance among m sets of ranks may be described by calculating Spearman rank-order correlation coefficients between all possible pairs of ranks and finding the average value, denoted by $\bar{\rho}$. This average is related to W. The relation is given by

[21.17] $\bar{\rho} = \dfrac{mW - 1}{m - 1}$

For the particular case where $m = 2$ the relation is $\rho = 2W - 1$. For $W = 0$, $\rho = -1$, for $W = .5$, $\rho = 0$, and for $W = 1$, $\rho = 1$.

21.13   THE COEFFICIENT OF CONCORDANCE WITH TIED RANKS

Where tied ranks occur, proceed as before and assign to each member the average rank which the tied observations occupy. If the ties are not numerous, we may compute W directly from the data without further adjustment. If the ties are numerous, a correction factor is calculated for each set of ranks. This correction factor is

[21.18] $T = \dfrac{\Sigma(t^3 - t)}{12}$

For example, if the ranks on X are 1, 2.5, 2.5, 4, 5, 6, 8, 8, 8, 10, we have two groups of ties, one of two ranks and one of three ranks. The correction factor for this set of ranks for X is

$$T = \frac{(2^3 - 2) + (3^3 - 3)}{12} = 2.5$$

A correction factor T is calculated for each of the m sets of ranks, and these are added together over the m sets to obtain ΣT. We then apply a formula for W in which this correction factor is incorporated. The formula is

[21.19] $W = \dfrac{S}{\frac{1}{12} m^2(N^3 - N) - m\Sigma T}$

The application of this correction tends to increase the size of W. The correction has a small effect unless ties are quite numerous.

21.14   THE SIGNIFICANCE OF THE COEFFICIENT OF CONCORDANCE W

For N of 7 or less, values of W required for significance at the 5 and 1 per cent levels have been tabulated by Friedman (1940) and are reproduced in Kendall (1955) and Siegel (1956). A useful adaptation of these tables is given by Edwards (1967). Critical values of W depend both on m, the number of sets of ranks, and on N, the number of ranks in each set. For N greater than 7, a χ^2 test may be applied. Calculate the quantity

[21.20] $$\chi^2 = m(N-1)W$$

This has a chi-square distribution with $N-1$ degrees of freedom. For the data of Table 21.3, $S = 160$, $W = .571$, $m = 4$, and $N = 6$. Reference to Edwards' table provides critical values of .505 and .621 for significance at the 5 and 1 per cent levels. If we apply the chi-square test to the same data, we obtain

$$\chi^2 = 4(6-1).571 = 11.42$$

For $df = 6 - 1 = 5$ the values of χ^2 required for significance are 11.07 and 15.09 at the 5 and 1 per cent levels, and as before we are led to the conclusion of significant association at the 5 per cent level. Of course, in this case the tabled values are to be preferred because N is less than 7. For N less than 7 the chi-square test will provide a very rough estimate of the required probabilities. Other procedures for testing the significance of W exist. For a more thorough discussion of this problem see Edwards (1967).

21.15   THE COEFFICIENT OF CONSISTENCE K

To obtain a ranking of objects on an attribute, the objects may be presented two at a time in all possible pairs and a judge required to make a choice on the presentation of each pair. Thus a choice is made between every object and every other object. This procedure is known as the *method of paired comparisons* and has been widely used in psychological work. The method is usually assumed to yield a more reliable ordering than that obtained by requiring a judge to order a whole group of objects directly. The number of possible pairs is the number of combinations of N things taken two at a time, or $N(N-1)/2$. As N increases, the number of comparisons increases very rapidly; consequently for large N the method is frequently impractical.

In the method of paired comparisons we may wish to ascertain the consistency of the choices made. Let A, B, and C be three objects. If A is preferred to B and B is preferred to C, consistency of judgment would require that A be preferred to C. If C is preferred to A, this latter choice is clearly inconsistent with the two previous choices. What meaning attaches to the presence of inconsistent choices? Let A, B, and C be red, blue, and yellow cards, each of a different saturation. A judge may prefer red to blue,

blue to yellow, and then may indicate a preference of yellow to red. This inconsistent choice may result because the judge may be unable to discriminate and may indicate preferences in a more or less haphazard fashion. Many inconsistent choices in the method of paired comparisons result because the task requires a refinement of discrimination which is beyond the capacity of the judge. Inconsistent responses may also arise because the dimension of judgment has changed. The red card may be preferred to the blue and the blue to the yellow on the basis of hue. The yellow may be preferred to the red on the basis of saturation. A different dimension is used as a basis of choice and leads to the presence of an inconsistency. To illustrate further, an orange may be preferred to a peach because of its color, a peach may be preferred to a pear because of its flavor, a pear may be preferred to an orange because of its shape, and thus an inconsistency arises. Where inconsistencies are numerous, a question may attach to the meaning of the rank ordering of objects obtained. It is convenient to represent a choice A in preference to B by the notation $A \rightarrow B$ and a choice of B to A by $B \rightarrow A$. The sequence $A \rightarrow B \rightarrow C \rightarrow A$ is an inconsistent triplet, or triad, of choices. For any set of paired comparisons between N objects the number of inconsistent triads may be counted and used to define a coefficient of consistency of response.

Responses obtained by the method of paired comparisons may be represented in tabular fashion in the form of a response pattern as shown in Table 21.4. This table shows paired comparisons between nine objects, A, $B, C, \ldots, H, I$. A is preferred to B, and a 1 is entered in the cell corresponding to row A and col. B above the main diagonal. A complementary 0

Table 21.4 **Response pattern for paired comparisons between nine objects and calculation of coefficient of consistence.**

	A	B	C	D	E	F	G	H	I	Row sum R	$(R - \bar{R})^2$
A	—	1	0	0	1	1	1	0	1	5	1
B	0	—	1	1	1	1	0	1	1	6	4
C	1	0	—	0	0	1	1	1	1	5	1
D	1	0	1	—	1	1	1	1	1	7	9
E	0	0	1	0	—	1	1	1	0	4	0
F	0	0	0	0	0	—	1	1	1	3	1
G	0	1	0	0	0	0	—	1	1	3	1
H	1	0	0	0	0	0	0	—	0	1	9
I	0	0	0	0	1	0	0	1	—	2	4

$\bar{R} = 4 \qquad \Sigma(R - \bar{R})^2 = 30$

$$K = \frac{12\Sigma(R - \bar{R})^2}{N(N^2 - 1)} = \frac{12 \times 30}{9(9^2 - 1)} = .500$$

is entered in col. *A* and row *B* below the main diagonal. All other choices may be similarly represented. We note that where no response inconsistencies are present, all entries on one side of the main diagonal are 1's and all entries on the other side 0's. In this table the presence of some 0's above the main diagonal and the complementary 1's below it indicate the presence of inconsistencies. Let us now sum the rows of Table 21.4. If no inconsistencies were present, the row sums would be the numbers 8, 7, 6, 5, 4, 3, 2, 1, 0. Because of the presence of inconsistencies, the actual obtained numbers are 7, 6, 5, 5, 4, 3, 3, 2, 1, although not in that order. The effect of inconsistencies is to reduce the variability of the numbers obtained by adding up the rows of the response pattern. Denote a row sum by R. The mean of the row sums is $\bar{R} = \Sigma R/N$, which may be shown equal to $(N-1)/2$. The sum of squares of row sums is:

[21.21] $\quad \Sigma(R - \bar{R})^2 = \Sigma R^2 - \dfrac{N(N-1)^2}{4}$

It is appropriate to consider the maximum and minimum values of this sum of squares. The maximum value of $\Sigma(R - \bar{R})^2$ occurs when no inconsistencies are present in the response pattern and is equal to $N(N^2 - 1)/12$. The minimum value of $\Sigma(R - \bar{R})^2$ depends on whether N is odd or even. If N is *odd*, the minimum value of $\Sigma(R - \bar{R})^2$ is 0. If N is *even*, it may be shown that the minimum value of $\Sigma(R - \bar{R})^2$ is not 0, but is $N/4$ (Kendall, 1943). We then define a *coefficient of consistence* of response K as follows:

[21.22] $\quad K = \dfrac{\text{observed sum of squares} - \text{minimum sum of squares}}{\text{maximum sum of squares} - \text{minimum sum of squares}}$

Simple substitution shows that if N is *odd*

[21.23] $\quad K = \dfrac{12\Sigma(R - \bar{R})^2}{N(N^2 - 1)}$

and if N is *even*

[21.24] $\quad K = \dfrac{12\Sigma(R - \bar{R})^2 - 3N}{N(N^2 - 4)}$

This is Kendall's coefficient of consistence. It has an expected value of 0 when responses are assigned at random, the case of maximal inconsistency, and 1 when no inconsistencies are present.

The calculation of K is illustrated in Table 21.4. In this example N is odd, $\Sigma(R - \bar{R})^2 = 30$, and $K = .500$.

How may the coefficient K be interpreted? The number of inconsistent triads of the kind $A \to B \to C \to A$ may be denoted by d, which is related to the coefficient K. It may be shown that when N is odd

[21.25] $\quad d = \dfrac{N(N^2 - 1)(1 - K)}{24}$

and when N is even

[21.26] $d = \dfrac{N(N^2 - 4)(1 - K)}{24}$

In the example of Table 21.4, the number of inconsistencies d is found to be 15. The maximum possible number of inconsistencies is 30. Thus one-half the triadic relations are inconsistent, the other half consistent, and $K = .50$. A K of .20 would mean that four-fifths of the relations were inconsistent and one-fifth consistent.

21.16   THE SIGNIFICANCE OF THE COEFFICIENT OF CONSISTENCE

The significance of the coefficient of consistence may be approached by considering the distribution of the number of triadic relations where choices are made at random. Kendall (1955) provides a table of probabilities that particular values of d will be attained or exceeded for $N = 2$ to 7. For $N > 7$, Kendall has shown that a χ^2 test may be used which provides approximate probabilities. The quantity

[21.27] $\chi^2 = \dfrac{8}{N - 4}\left(\dfrac{1}{4}\,C_3{}^N - d + \dfrac{1}{2}\right) + df$

has an approximate χ^2 distribution with degrees of freedom given by

[21.28] $df = \dfrac{N(N - 1)(N - 2)}{(N - 4)^2}$

The term $C_3{}^N$ in the expression for χ^2 is the number of combinations of N things taken three at a time, or $N!/3!(N - 3)!$. In using this test the required probability that a value of d equal to or greater than that obtained will result where choices are allotted at random is the *complement* of the probability of χ^2.

For the data of Table 21.4, $N = 9$ and $d = 15$. We have

$$df = \dfrac{9 \times 8 \times 7}{(9 - 4)^2} = 20.16$$

$$\chi^2 = \dfrac{8}{9 - 4}\left(\dfrac{1}{4} \times \dfrac{9!}{3!6!} - 15 + \dfrac{1}{2}\right) + 20.16 = 28.96$$

The probability associated with this value of χ^2 is greater than .99. This means that the significance level for d is less than .01, the complement of .99. We conclude that the consistency represented in the data is greater than we could reasonably expect on the assignment of choices at random. The coefficient of consistence $K = .50$ may be said to be significantly different from zero at better than the .01, or 1 per cent, level.

EXERCISES 1 Calculate Spearman's rank correlation coefficient for the following paired ranks:

X	1	2	3	4	5	6	7	8
Y	2	4	5	1	6	3	8	7

Does the coefficient obtained differ significantly from zero?

2 Convert the following measurements to ranks:

X	4	4	7	7	7	9	16	17	21	25
Y	8	16	8	8	16	20	12	15	25	20

Calculate Spearman's rank correlation coefficient. Does the coefficient obtained differ significantly from zero?

3 Are the following values of ρ significantly different from zero? (a) $\rho = .30$ for $N = 25$, (b) $\rho = .60$ for $N = 15$, (c) $\rho = .70$ for $N = 10$.

4 Calculate the statistics S for the following sets of paired ranks:

a	X	1	2	3	4	5	6
	Y	6	4	2	1	3	5

b	X	1	2	3	4	5	6
	Y	6	3.5	1.5	1.5	3.5	5

c	X	2	2	2	3	4	5
	Y	6	3.5	1.5	1.5	3.5	5

d	X	2	2	2	5	5	5
	Y	6	4	2	1	3	5

e	X	2	2	2	5	5	5
	Y	6	3.5	1.5	1.5	3.5	5

f	X	2	2	2	5	5	5
	Y	5	5	2	2	2	5

5 Calculate the sampling variances for the statistic S obtained for the sets of paired ranks in Exercise 4 above. In each case obtain the normal deviate z with a continuity correction.

6 Three judges rank order a group of seven students on an examination as follows:

Judge	Student						
	a	b	c	d	e	f	g
A	1	2	3	4	5	6	7
B	2	3	4	5	1	7	6
C	5	4	1	2	3	6	7

Compute the Spearman rank coefficients between judges and the coefficient of concordance.

7   A supervisor ranks six employees A, B, C, D, E, and F on job performance using the method of paired comparisons. The data are as follows: $A \rightarrow B$, $A \rightarrow C$, $A \rightarrow D$, $E \rightarrow A$, $F \rightarrow A$, $B \rightarrow C$, $D \rightarrow B$, $B \rightarrow E$, $B \rightarrow F$, $C \rightarrow D$, $C \rightarrow E$, $F \rightarrow C$, $D \rightarrow E$, $D \rightarrow F$, $E \rightarrow F$. Calculate the coefficient of consistence for these data. How may this coefficient be interpreted?

8   Calculate a coefficient of consistence for the table composed of the first five rows and columns of Table 21.4.

NONPARAMETRIC
TESTS OF SIGNIFICANCE

22.1 INTRODUCTION

Many tests of significance involve assumptions about the nature of the distributions of the variables in the populations from which the samples are drawn. The t test and the analysis of variance, for example, assume normality of the parent distribution. In experimental work situations arise where either little is known about the population distribution of the dependent variable or this distribution is known to depart appreciably from the normal form. In such situations *nonparametric tests* may be appropriately used. Nonparametric tests make few assumptions about the properties of the parent distribution. Assumptions about the parent distribution are involved in nonparametric tests, but these are usually fewer in number, weaker, and easier to satisfy in data situations. Nonparametric tests are frequently spoken of as *distribution-free* tests. The implication is that they are free, or independent, of some characteristics of the population distribution.

The reader will recall the distinction between nominal, ordinal, interval, and ratio variables. Nonparametric methods are appropriate for nominal and ordinal data; parametric methods for interval and ratio data. In practice, nonparametric methods are frequently used with data of this latter type. The data are reduced to a form such that a nominal, or an ordinal, statistical procedure may be applied to them. An important class of nonparametric tests employs only the sign properties of the data. All observations above a fixed value, such as the median, may be assigned a plus, and all below, a minus. The original variable is replaced by, or transformed to, another variable which takes the sign values plus or minus. Another class of nonparametric test employs the rank properties of the data. The original observations are replaced by the numbers 1, 2, 3, . . . , N. Subsequent statistical manipulation and inferences are based on ranks.

In applying a conventional test of significance, such as a two-sample t test, estimates $\overline{X}_1$ and $\overline{X}_2$ of the population means μ_1 and μ_2 are used to test the null hypotheses $H_0:\mu_1 = \mu_2$ under the assumption that the distributions are normal and the variances are equal, $\sigma_1{}^2 = \sigma_2{}^2$. In many nonparametric procedures the null hypothesis under test is not formulated in terms of the parameters of parent populations, nor are estimates of population parameters calculated. A number of commonly used two-sample procedures test the null hypothesis in its most general form. This null hypothesis is that the samples come from populations with the same distribution. This is tested against the alternative that samples come from populations with different distributions. Of course, if we are willing to assume in a two-sample test that the distributions are normal and $\sigma_1{}^2 = \sigma_2{}^2$, then if the distributions are found to differ, the differences must be due to differences in location. Some nonparametric tests are tests of central position in a rank order and are in effect tests of medians. If the assumption is made that the parent populations are symmetrical, such tests become also tests of means.

The question arises as to the criteria to be applied in comparing alternate procedures for testing the same hypothesis. The procedures to be compared may be two nonparametric tests, or a nonparametric test and a test of the conventional type. Alternate tests of significance are usually compared in terms of efficiency. The reader will recall that in the discussion on estimation in Chap. 10 the concept of relative efficiency was introduced. This was defined as the ratio of two sampling variances and could be related to sample size. A median, for example, calculated on a sample of 100 cases has a sampling variance equal to a mean calculated on 64 cases. Thus alternate procedures for estimating population parameters have the same efficiency when their sampling variances are equal. In comparing alternate procedures for testing hypotheses a different approach is used. This approach uses the concept of the power of a test.

The power of a statistical test is the probability of rejecting the null hypothesis when that hypothesis is false. It is one minus the Type II error, or $1 - \beta$. The power of a statistical test depends on the level of significance, the alternate hypothesis H_1, and the sample size. Two tests, A and B, may be compared by considering the relative sample size required to make them equally powerful. The *relative efficiency* of the two tests is given by N_a/N_b, where N_b is the number of observations required to make test B as powerful as test A with N_a observations. Because this definition of relative efficiency depends, of course, also on the level of significance, the alternate hypothesis H_1, and the sample size, it is too complex for most practical purposes. This has led to the use of the *asymptotic relative efficiency*, which is the limiting value of the ratio N_a/N_b as N approaches infinity and H_1 approaches the null hypothesis H_0. Questions can be raised regarding the practical usefulness of the asymptotic relative efficiency, since in most cases interest does not reside in large sample size or in alternate hypotheses close to the null hypothesis. The asymptotic relative efficiency of many nonparametric tests is known relative to the most powerful test available, which for two-sample tests is frequently the t test. Comparisons with the t test, for example, are made for normal distributions where both a conventional and a nonparametric procedure may be applied.

For a comprehensive treatment of nonparametric tests the reader is referred to Siegel (1956) and to Bradley (1968). Both books contain useful tables.

22.2 A SIGN TEST FOR TWO INDEPENDENT SAMPLES

This test is known as the *median test*. It compares the medians of two independent samples. The null hypothesis is that no difference exists between the medians of the populations from which the samples are drawn. The corresponding parametric test is a t test for comparing the means of independent samples. The median test is based on the idea that in two samples drawn from the same population the expectation is that as many observations in each sample will fall above as below the joint median.

The data consist of two independent samples of N_1 and N_2 observations. To apply the median test the median of the combined $N_1 + N_2$ observations is calculated. In each sample, observations above the joint median are assigned a $+$ and those at or below it a $-$. The number of $+$ and $-$ signs for each sample is ascertained. A χ^2 test is used to determine whether the observed frequencies of $+$ and $-$ signs depart significantly from expectation under the null hypothesis.

The following are observations for two independent samples:

Sample I	10	10	10	12	15	17	17	19	20	22	25	26
Sample II	6	7	8	8	12	16	19	19	22			

The median of the $N_1 + N_2$ observations is 16. Assigning a $+$ to values above the median and a $-$ to values at or below it, we obtain

Sample I	$-$	$-$	$-$	$-$	$-$	$+$	$+$	$+$	$+$	$+$	$+$	$+$
Sample II	$-$	$-$	$-$	$-$	$-$	$-$	$+$	$+$	$+$			

These data may be tabulated in the form of a 2×2 table as follows:

	$+$	$-$	
Sample I	7	5	12
Sample II	3	6	9
	10	11	

The value of χ^2 for this table with Yates's correction for continuity is .51. The value of χ^2 required for significance at the 5 per cent level is 3.84. Obviously, in this case we have no grounds for rejecting the null hypothesis that the samples came from populations with the same median. This is a two-tailed test.

The asymptotic relative efficiency of the sign test for independent samples when compared to the t test under the assumptions of normality and equal variance is $2/\pi = .637$.

22.3 A SIGN TEST FOR TWO CORRELATED SAMPLES

This test compares two correlated samples, and is applicable to data composed of N paired observations. The difference between each pair of observations is obtained. The null hypothesis is that the median difference between the pairs is zero. The test is based on the idea that under the null hypothesis the expectation is that half the differences between the paired observations will be positive and the other half negative. The symmetrical binomial $(\frac{1}{2} + \frac{1}{2})^N$ is used to obtain the probabilities required for a one-tailed or a two-tailed test.

The following are paired observations, X and Y, for a sample of 10 individuals together with the sign of the difference between X and Y:

X	15	19	31	36	10	11	19	15	10	16
Y	19	30	26	8	10	6	17	13	22	8
Sign of $X - Y$	$-$	$-$	$+$	$+$	0	$+$	$+$	$+$	$-$	$+$

Under the null hypothesis the probability that X is greater than Y is equal to the probability that Y is greater than X, which in turn is equal to $\frac{1}{2}$. The expected numbers of $+$ and $-$ signs are equal. In this example we have six plus signs, three minus signs, and one zero difference. The zero difference is discarded. From the binomial expansion $(\frac{1}{2} + \frac{1}{2})^9$ we can ascertain the exact probability of obtaining six or more plus signs under the null hypothesis. This probability is .254. This is a one-tailed test. The probability of obtaining either six or more plus signs or six or more minus signs is .508. This is a two-tailed test. Clearly here we have no grounds for rejecting the null hypothesis.

Where N is not too small, the normal approximation to the binomial or χ^2 may be used, preferably with Yates's correction. In this case the expected values are $N/2$. In the above example the observed values are 6 and 3, the expected values are 4.5 and 4.5, the corrected observed values are 5.5 and 3.5, and $\chi^2 = .44$. The probability of obtaining a χ^2 equal to or greater than .44 under the null hypothesis is .507. Although N is small, this is in close agreement with the exact probability of .508 obtained from the binomial. The reader will recall that χ^2 provides the probability for a two-tailed test.

Instead of the χ^2 procedure described above, a computationally simpler method may be used. Obtain the difference between the number of $+$ and $-$ signs. Denote this difference by D. It may be shown that

$$z = \frac{|D| - 1}{\sqrt{N}}$$

approaches the normal form as N increases in size. This formula incorporates a continuity correction. Values of 1.96 and 2.58 are required for significance at the .05 and .01 levels of significance, respectively, for a nondirectional test. In the above example we have six plus signs and three minus signs. $D = 6 - 3 = 3$, and

$$z = \frac{|3| - 1}{\sqrt{9}} = .67$$

Quite clearly this falls short of significance. The reader should recall that for $df = 1$ the quantity $z^2 = \chi^2$. In the above example, if the calculation is carried beyond two decimal places, $z^2 = .6667^2 = .4444$ and $\chi^2 = .4444$.

The asymptotic relative efficiency of this test when compared to the corresponding t test, and under the required assumptions, is $2/\pi = .637$.

22.4 A SIGN FOR k INDEPENDENT SAMPLES

This is an obvious extension of the median test for two independent samples. The data are comprised of k samples of $n_1, n_2, \ldots, n_k$ observations. As before, the null hypothesis is that no difference exists in the medians of the populations from which the samples are drawn. The median of the combined $n_1 + n_2 + \cdots + n_k$ observations is calculated. For each sample, observations above the joint median are assigned a $+$ and those either at or below the joint median a $-$. The data are arranged in a $2 \times k$ contingency table, and a χ^2 test applied.

The following are data for four samples.

Sample I	3	6	11	14	17	18	21	33
Sample II	3	3	4	5	5	8	9	14
Sample III	18	18	25	26	29	31		
Sample IV	14	18	20	22	22	25	27	35

The total number of observations is 30. The median is 18. Assigning a $+$ to values above the median and a $-$ to values at or below, we obtain

Sample I	$-$	$-$	$-$	$-$	$-$	$-$	$+$	$+$
Sample II	$-$	$-$	$-$	$-$	$-$	$-$	$-$	$-$
Sample III	$-$	$-$	$+$	$+$	$+$	$+$		
Sample IV	$-$	$-$	$+$	$+$	$+$	$+$	$+$	$+$

These data may be arranged in a 2 × 4 table as follows:

	+	−	
Sample I	2	6	8
Sample II	0	8	8
Sample III	4	2	6
Sample IV	6	2	8
	12	18	30

The value of χ^2 calculated on this table is 7.56. The number of degrees of freedom is $(4-1)(2-1) = 3$. The value of χ^2 required for significance at the 5 per cent level is 7.82. The observed value falls just below this.

22.5 A RANK TEST FOR TWO INDEPENDENT SAMPLES

The most commonly used rank test for comparing two independent samples is the Wilcoxon rank sum test. Tests which are equivalent to the rank sum test have been developed by Mann and Whitney, and others. The hypothesis under test here is that the two samples come from populations with the same distribution. If assumptions are made regarding the equivalence of shapes and variances of the two distributions, the Wilcoxon test becomes a test of central location.

In applying the Wilcoxon test the N_1 and N_2 observations are combined. The $N_1 + N_2$ observations are then arranged in order. A rank 1 is assigned to the smallest value, a rank 2 to the next smallest, and so on. The sum of ranks, R_1, is obtained for the *smaller* of the two samples, if the samples are unequal in size. If the samples are equal in size, either rank sum may be used. R_1 is then evaluated in relation to its distribution, which is discussed below.

The model distribution against which particular values of R are evaluated is obtained by considering a finite population of integers comprised of $N_1 + N_2 = N$ members. These integers are 1, 2, 3, . . . , N. The problem of drawing samples of size N_1 from this population of $N_1 + N_2 = N$ members may be considered. The number of possible equiprobable samples is $C_{N_1}{}^N$. A value R_1 may be obtained for each sample and a frequency distribution made of the values of R_1. This distribution can be used to evaluate particular values of R_1. If a particular value of R_1 has a small probability in relation to this distribution, we reject the hypothesis that the two samples come from the same population. The reader should note that this distribution is closely related to the sampling distribution of means from a finite population discussed in Sec. 9.5. Here the sampling distribution of sums, and not means, is under consideration, and the population from which samples are drawn is one of integers extending from 1 to N. The mean of this distribution, $\bar{R}_1$, is N_1 times the mean of the $N_1 + N_2$ ranks and is

[22.1] $\text{Mean} = \bar{R}_1 = \dfrac{N_1(N_1 + N_2 + 1)}{2}$

The variance of the distribution of R_1 can be shown to be

[22.2] $\text{Variance} = \sigma_{R_1}{}^2 = \dfrac{N_1 N_2(N_1 + N_2 + 1)}{12}$

The exact distributions of R_1 are known for N_1 and N_2 up to 25. The distribution approaches the normal form fairly rapidly. When both N_1 and N_2 are equal to or greater than 8 or 10, a large-sample procedure using the normal approximation and a continuity correction will lead to estimates of the required probabilities which do not differ much from those obtained from the exact distributions. The normal deviate z with a continuity correction is given by:

[22.3] $z = \dfrac{|R_1 - \bar{R}_1| - 1}{\sqrt{\dfrac{N_1 N_2(N_1 + N_2 + 1)}{12}}}$

If this value is equal to or greater than 1.96 or 2.58, we reject the null hypothesis for a nondirectional test at the .05 or .01 level and accept the alternative hypothesis that the samples are from different populations. For a directional test the .05 and .01 levels are 1.64 and 2.33.

Consider the following observations:

Sample I	27	33	37	52	53	57	69	70	71	77		
Sample II	6	9	14	16	29	43	45	47	50	55	63	72

Assigning ranks, proceeding from the smallest to the largest values, we obtain

Sample I	5	7	8	13	14	16	18	19	20	22		
Sample II	1	2	3	4	6	9	10	11	12	15	17	21

The sum of ranks R_1 for sample I is 142. The mean of the distribution of R_1, that is, $\bar{R}_1$, is 115. The normal deviate is

$$z = \frac{|142 - 115| - 1}{\sqrt{\dfrac{10 \times 12 \ (10 + 12 + 1)}{12}}} = 1.71$$

Since this falls below 1.96, we have no grounds for rejecting the null hypothesis for a two-tailed test. The result is, however, significant at the 5 per cent level for a one-tailed test.

When ties occur, the tied observations may be assigned the average of the ranks they would occupy if no ties had occurred. If ties are fairly numerous, a correction may be applied to the standard deviation in the denominator of the z ratio. Corrected for ties, that ratio becomes

[22.4] $$z = \frac{|R_1 - \bar{R}_1| - 1}{\sqrt{\left[\dfrac{N_1 N_2}{N(N-1)}\right] \left(\dfrac{N^3 - N}{12} - \Sigma T\right)}}$$

where $N = N_1 + N_2$ and $T = (t^3 - t)/12$, where t is the number of values tied at a particular rank. The summation of T extends over all groups of ties.

The above procedures use the normal approximation to the distribution of R_1. If an exact test is required, Table K may be used. This table shows the exact lower-tail critical values of R_1 for N_1 and N_2 up to 25 at probability levels equal to or less than .10, .05, .025, .01, .005, and .001. For example, the table shows a $p = .05$ for $R_1 = 19$ where $N_1 = 5$ and $N_2 = 5$. This means that the probability of obtaining a value equal to or less than R_1 in samples of this size is equal to or less than .05. This is a directional test which uses the lower tail of the distribution. Since the distribution is symmetrical, the corresponding upper-tail values are given by noting that a lower-tail value R_1 is $\bar{R}_1 - R_1$ points below the mean. The corresponding value above the mean is $\bar{R}_1 + (\bar{R}_1 - R_1) = 2\bar{R}_1 - R_1$. If R_1 is an upper-tail value, that is, if it is above the mean, Table K is entered with $2\bar{R}_1 - R_1$. The null hypothesis is rejected if it is smaller than the critical value. To assist calculation Table K shows values of $2\bar{R}_1$. In the illustrative example above, R_1 is 142 for $N_1 = 10$ and $N_2 = 12$. The mean value $\bar{R}_1$ is 115. The lower-tail value corresponding to R_1 is $2\bar{R}_1 - R_1 = 230 - 142 = 88$. Entering Table K we note that p is slightly less than .05. Thus we may assert significance at this level for a directional test. For a nondirectional or two-tailed test, either R_1 or $2\bar{R}_1 - R_1$, whichever is appropriate, is referred to Table K and the probabilities in the table are doubled. The appropriate choice between R_1 and $2\bar{R}_1 - R_1$ is, of course, always the smaller value.

The above discussion sounds rather complex. In practice the procedure is simple. *First*, calculate R_1. Calculate $\bar{R}_1$ using formula [22.1]. *Second*, if R_1 is less than $\bar{R}_1$ refer R_1 to Table K to obtain the required probability. *Third*, if R_1 is greater than $\bar{R}_1$ calculate $2\bar{R}_1 - R_1$, and refer this quantity to Table K to obtain the required probability.

The Wilcoxon rank sum test is in effect the same as the Mann-Whitney U test. Mann and Whitney (1947) studied the distribution of a statistic U, which is related in a simple way to R_1.

[22.5] $\quad U_1 = N_1N_2 + \dfrac{N_1(N_1 + 1)}{2} - R_1$

[22.6] $\quad U_2 = N_1N_2 + \dfrac{N_2(N_2 + 1)}{2} - R_2$

U is the smaller of these two values. Tables showing critical values of U are reproduced in Siegel (1956).

The Wilcoxon rank sum test has an asymptotic relative efficiency when compared with the t test for independent samples of $3/\pi = .955$. This comparison assumes that the distributions are normal. If the distributions are rectangular, the asymptotic relative efficiency is 1.00. For certain other types of distributions the asymptotic relative efficiency is greater than 1.00. All the available evidence indicates that the Wilcoxon rank sum test is an excellent alternative to the t test.

22.6 A RANK TEST FOR TWO CORRELATED SAMPLES

The rank test described here is due to Wilcoxon and is usually called the *Wilcoxon matched-pairs signed-ranks test*. The data are a set of N paired observations on X and Y. The difference, d, between each pair is calculated. If the two observations in a pair are the same, then $d = 0$ and the pair is deleted from the analysis. Values of d may be either positive or negative. The d's are then ranked without regard to sign, that is, the absolute values $|X_i - Y_i|$ are ranked. A rank of 1 is assigned to the smallest d, of 2 to the next smallest, and so on. If two or more d's are tied, the practice usually adopted is to assign to the tied ranks the average of the ranks they would have been assigned if they had differed. The sign of the difference d is attached to each rank. If d is positive, the rank is positive; if d is negative, the rank is negative. Denote the sum of the positive ranks by W_+ and the sum of the negative ranks by W_-.

The model null distribution against which W_+ is evaluated is obtained in the following way. If the two samples X and Y are samples from the same population, then the probability that $X_i - Y_i$ is either plus or minus is one-half. Also the probability that the rank corresponding to $|X_i - Y_i|$ is either plus or minus is one-half. All possible arrangements of plus or minus in relation to the N ranks may be considered. There are 2^N such arrangements. These are viewed as equiprobable. W_+ is calculated for each arrangement and the frequency distribution of W_+ becomes the null distribution against which particular values of W_+ are evaluated. To illustrate, for $N = 3$ the number of arrangements of plus and minus is $2^3 = 8$. These are as follows:

1	2	3	W_+
+	+	+	6
−	+	+	5
+	−	+	4
+	+	−	3
−	−	+	3
−	+	−	2
+	−	−	1
−	−	−	0

Eight different values of W_+ occur, ranging from 0 to 6. In general W_+ will range from 0 to $N(N+1)/2$. It can be shown that the mean and variance of the distribution of W_+ are as follows:

[22.7] $$\text{Mean} = \overline{W}_+ = \frac{N(N+1)}{4}$$

[22.8] $$\text{Variance} = \sigma_{W_+}{}^2 = \frac{N(N+1)(2N+1)}{24}$$

The distribution of W_+ approaches the normal form fairly rapidly. Consequently for samples of reasonable size the normal approximation may be used. The normal deviate z is given by

[22.9] $$z = \frac{W_+ - \dfrac{N(N+1)}{4}}{\sqrt{\dfrac{N(N+1)(2N+1)}{24}}}$$

Values of 1.96 and 2.58 are, as usual, required for significance at the 5 per cent and 1 per cent levels for a nondirectional or two-tailed test.

For N up to 40, values of W_+ may be referred to Table I of the Appendix. This table provides the lower-tail cumulative probabilities for values of W_+ that lie close to the selected significance levels, .05, .025, .01, and .005. To illustrate, for $N = 10$ and a significance level of .05, two values of W_+ are shown: 10 with an associated probability of .0420 and 11 with a probability of .0527. This means that the probability of obtaining a value of W_+ equal to or less than 10, when the null hypothesis is true, is .0420, and the probability of obtaining a value of W_+ equal to or less than 11 is .0527. This is a directional test. It uses the lower tail of the distribution. A directional test that uses the upper tail of the distribution may be required. Here we recall that the maximum value of W_+ is $N(N+1)/2$. Since the distribution is symmetrical, the corresponding upper-tail value is $[N(N+1)/2] - W_+$. This quantity is equal in absolute magnitude to the sum of the negative ranks $|W_-|$. To illustrate, suppose $N = 10$ and $W_+ = 50$. What is the probability of obtaining a value of W_+ equal to or greater than 50 when the null hypothesis is true? Here we calculate $[N(N+1)/2] - W_+ = 10 \times 11/2 - 50 = 5$. The value 5 is referred to Table I and a probability of .0098 is obtained. Thus under the null hypothesis the probability of obtaining a value of W_+ equal to or greater than 50 is in the neighborhood of .01. For a directional or two-tailed test we refer either W_+ or $[N(N+1)/2] - W_+$, whichever is the smaller, to the table and double the probabilities.

The discussion above is somewhat involved. In practice a simple procedure may be used. *First*, calculate W_+ and $|W_-|$. *Second*, take the smaller of these two quantities and refer this to Table I with the appropriate value of N.

The following are paired observations, X and Y, for a sample of 10 individuals:

X	15	19	31	36	10	11	19	15	10	16
Y	19	30	26	8	10	6	17	13	22	8
d	-4	-11	5	28	0	5	2	2	-12	8
Rank	-3	-7	4.5	9		4.5	1.5	1.5	-8	6

Values of d are calculated. One pair of observations is tied and is deleted from subsequent consideration. The d's are rank-ordered by absolute magnitude. The lowest values are a pair of 2's. These are assigned rank values of 1.5. The sum W_+, the sum of positive ranks, is 27. The value $[N(N+1)/2] - W_+ = 9 \times 10/2 - 27 = 18$. Note that $|W_-|$, the absolute value of the negative ranks, is also 18. This value 18 is referred to Table I. No basis exists in this case for rejecting the null hypothesis for either a directional or a nondirectional test.

The asymptotic relative efficiency of the Wilcoxon signed-rank test relative to the t test is .955. This test can be viewed as a useful and satisfactory alternative to the t test.

22.7 A RANK TEST FOR k INDEPENDENT SAMPLES

A rank test for k independent samples is the Kruskal-Wallis (1952) one-way analysis of variance by ranks. This is a generalization of the Wilcoxon rank sum test to k groups. The null hypothesis is that the k independent samples of $n_1, n_2, \ldots, n_k$ members are from the same population. To apply the test all the observations for the k samples are ranked. The lowest value is assigned a rank of 1, the next lowest 2, and so on. The sum of the ranks, R_i, for each of the k samples is obtained. If all k samples are from the same population, the expectation is that the mean rank sums $\bar{R}_i$ will be equal for the k groups, and equal to the mean of the N ranks, which is $(N+1)/2$.

The null distribution here involves a consideration of the $N!$ arrangements of ranks in k groups. Each arrangement is regarded as equiprobable.

For each arrangement the following statistic could be calculated:

$$[22.10] \quad S = \sum_{i}^{k} n_i \left(\overline{R}_i - \frac{N+1}{2} \right)^2$$

The quantity in parentheses in the above expression is simply the squared difference between the means of the ith group and its expected value under the null hypothesis. Since the groups may be unequal in size, each squared difference is weighted according to group size, n_i, to obtain a final sum of squares, S. S could be calculated for each of the $N!$ arrangements of ranks and its frequency distribution examined. Some advantages attach to the study of a statistic H, which is closely related to S and is given by:

$$[22.11] \quad H = \frac{12S}{N(N+1)} = \frac{12}{N(N+1)} \sum_{i}^{k} n_i \left[\overline{R}_i - \frac{N(N+1)}{2} \right]^2$$

The distribution of H approximates the distribution of chi square with $k-1$ degrees of freedom. For $k=3$ and $n_i \leqslant 5$, tables of critical values of H with exact probabilities have been prepared by Kruskal and Wallis. For larger values of k and n_i the chi-square approximation must be used. For computational purposes it is convenient to write H in the form

$$[22.12] \quad H = \frac{12}{N(N+1)} \sum^{k} \left(\frac{R_i^2}{n_i} \right) - 3(N+1)$$

where R_i is the sum of ranks for the kth group.

When ties occur, the usual convention is adopted of assigning to the tied observations the average of the ranks they would otherwise occupy. The value of H is then divided by

$$1 - \frac{\Sigma T}{N^3 - N}$$

where $T = t^3 - t$, and t is the number of tied observations in a group. The quantity H corrected for ties is

$$[22.13] \quad H = \frac{\dfrac{12}{N(N+1)} \sum_{i=1}^{k} \left(\dfrac{R_i^2}{n_i} \right) - 3(N+1)}{1 - \dfrac{\Sigma T}{N^3 - N}}$$

The correction for ties will increase the value of H.

The following are data for three samples:

Sample I	3	7	11	16	22	29	31	36	
Sample II	3	4	7	18	19	32			
Sample III	22	38	46	47	47	50	53	54	56

In this example $n_1 = 8$, $n_2 = 6$, $n_3 = 9$, and $N = 8 + 6 + 9 = 23$. All 23 observations are ranked to obtain

Sample I	1.5	4.5	6	7	10.5	12	13	15	
Sample II	1.5	3	4.5	8	9	14			
Sample III	10.5	16	17	18.5	18.5	20	21	22	23

The sums of ranks are calculated. These are $R_1 = 69.5$, $R_2 = 40$, and $R_3 = 166.5$. We note that we have four sets of ties of two observations each. $T = 2^3 - 2 = 6$, and for the four sets $\Sigma T = 24$. The value of H is then

$$H = \frac{\dfrac{12}{23(23+1)} \left(\dfrac{69.5^2}{8} + \dfrac{40^2}{6} + \dfrac{166.5^2}{9} \right) - 3(23+1)}{1 - \dfrac{24}{23^3 - 23}} = 13.88$$

In this example the effect of the correction for ties is negligible and may for all practical purposes be ignored. On reference to a table of χ^2 with $df = 2$, we note that an H of 13.88 is significant at better than the 1 per cent level. We may then reject the hypothesis that the samples are from the same population.

The Kruskal-Wallis test has an asymptotic relative efficiency of .955 when compared with the F test resulting from the analysis of variance applied to independent samples. For $k = 2$ this test is equivalent to the Wilcoxon rank sum test.

22.8 A RANK TEST FOR k CORRELATED SAMPLES

A rank test for k correlated samples is the Friedman two-way analysis of variance by ranks (1937). The data are a set of k observations for a sample of N individuals. Such data arise in many experiments where subjects are tested under a number of different experimental conditions. The corresponding parametric test is an analysis of variance for two-way classification where observations are made on each of a group of individuals under more than two conditions. If there is reason to believe that the assumptions underlying the analysis of variance are not satisfied by the data, the Friedman rank method is appropriate.

The data are arranged in a table containing N rows and k columns. The rows correspond to individuals, or groups, and the columns to experimental conditions. Table 22.1 shows such an arrangement of data for eight subjects tested under four experimental conditions. The observations in the rows are ordered as shown in Table 22.2. For example, the four observations in the top row are 4, 5, 9, and 3. These are replaced by the ranks 2, 3, 4, and 1. The ranks in each column are summed. If the samples are from the same population, the ranks in each column will be a random arrangement of the numbers 1, 2, 3, and 4. Under these circumstances the sums of

Table 22.1   Material recalled after four time intervals for a group of eight subjects

Subject	Time interval			
	I	II	III	IV
1	4	5	9	3
2	8	9	14	7
3	7	13	14	6
4	16	12	14	10
5	2	4	7	6
6	1	4	5	3
7	2	6	7	9
8	5	7	8	9

ranks for columns will tend to be the same. If these sums differ significantly, the hypothesis that they are from the same population may be rejected.

The null distribution here involves a consideration of the $k!$ arrangements of ranks in any row. These are considered equiprobable. Given N rows, the number of possible equiprobable arrangements of ranks is $(k!)^N$. For each of these arrangements a statistic S may be calculated, where

[22.14] $$S = \sum_{i}^{k} (R_i - \bar{R})^2$$

R_i is the sum of ranks for the ith column, and $\bar{R}$ is the mean rank sum. S is simply the sum of squares of rank sums about the mean rank sum. If the

Table 22.2   Ranks assigned by rows for the data of Table 22.1

Subject	Time interval			
	I	II	III	IV
1	2	3	4	1
2	2	3	4	1
3	2	3	4	1
4	4	2	3	1
5	1	2	4	3
6	1	3	4	2
7	1	2	3	4
8	1	2	3	4
R_i	14	20	29	17

samples are for the same population, the expectation is that the R_i's are equal and the expected value of S is zero. At least in theory, for any N and k a frequency distribution may be made of the $(k!)^N$ values of S. This distribution may be used to evaluate particular values of S. If the probability associated with a particular value of S is small, the null hypothesis is rejected. For small values of k and N the exact distributions of S are known. Bradley (1968) provides a table of exact critical values of S for $k = 3$ and N up to 15, and for $k = 4$ and N up to 8.

For values that lie outside the tabled values of S it is customary to use a statistic which is a function of S. This statistic is given by

[22.15] $$\chi_r^2 = \frac{12S}{Nk(k+1)}$$

This statistic has an approximate chi-square distribution with $k-1$ degrees of freedom.

For computational purposes a more convenient way of writing χ_r^2 is

[22.16] $$\chi_r^2 = \frac{12}{Nk(k+1)} \sum_{i}^{k} R_i^2 - 3N(k+1)$$

For the data of Table 22.2 we have

$$\chi_r^2 = \frac{12}{8 \times 4(4+1)} (14^2 + 20^2 + 29^2 + 17^2) - 3 \times 8(4+1) = 9.45$$

This result for $df = 4 - 1 = 3$ falls between the .05 and .01 levels of significance. Actually it is a little above the 2 per cent level. If this level of confidence is acceptable, we may conclude that the samples are not drawn from the same population and that a difference in the experimental conditions is exerting an effect. In this example $S = 126$. If this is referred to a table of exact critical values of S, as given in Bradley (1968), the associated probability is found to fall between .01 and .05, not far from the .02 level. The chi-square approximation is in close agreement with the more exact test.

The asymptotic relative efficiency of the Friedman test relative to the F test resulting from a two-way analysis of variance with one observation per cell is

$$\left(\frac{3}{\pi}\right)\left(\frac{k}{k+1}\right)$$

The efficiency of the test increases as k increases, and extends from .637 for $k = 2$ to a maximum of .955 for $k = \infty$. For $k = 2$ this test is the same as the sign test for correlated samples.

22.9 MONOTONIC TREND TEST FOR INDEPENDENT SAMPLES

In Chap. 19 methods of trend analysis for parametric data were described. Methods of trend analysis using ranks may be used. These are the nonparametric analogues of the tests described in Chap. 19. A test of monotonic trend is the nonparametric analogue of a test of linear trend. This section describes a simple test of monotonic trend for independent samples, which employs the statistic S as used in the definition of Kendall's tau (see Chap. 21).

Comment on the concept of monotonicity is appropriate here. A function $Y = f(X)$ is said to be a *monotonic increasing function* if any increment in X is associated with an increment in Y. Also, if any increment in X is associated with a decrement in Y, the function is said to be a *monotonic decreasing function*. The magnitude of the increment or decrement in X associated with the increment or decrement in Y is irrelevant to the concept of a monotonic function. Monotonicity is an order concept.

A common type of experiment is one in which k treatments are applied to k independent groups of $n_1, n_2, \ldots, n_k$ members, and a measurement obtained for each member. Such data are frequently analyzed using the analysis of variance for one-way classification. The nonparametric analogue of this is the Kruskal-Wallis one-way analysis of variance by ranks, described in Sec. 22.7. If, however, the k treatments exhibit an order, the question of monotonic trend may be raised. In effect, this question is answered by testing for significance the correlation between the ranks corresponding to the $n_1 + n_2 + \cdots + n_k = N$ measurements and the ranks for the k treatments. The ranks for treatments consist of k sets of tied members with $n_1, n_2, \ldots, n_k$ members respectively, in each of the k sets. Thus for $k = 3$, and $n_1 = n_2 = n_3 = 10$, the ranks for treatments consist of three sets of 10 tied values.

To illustrate, the following are measurements for three independent samples obtained under three ordered treatments.

Sample I	3	7	11	16	22	29	31	36	
Sample II	3	4	7	18	19	32			
Sample III	22	38	46	47	47	50	53	54	56

These values are ranked as in the Kruskal-Wallis one-way analysis of variance by ranks. The values 1, 2, and 3 are assigned to indicate membership in the three samples. Thus the data are represented by a set of paired ranks as follows:

Sample I	X	1	1	1	1	1	1	1	1
	Y	1.5	4.5	6	7	10.5	12	13	15

Sample II	X	2	2	2	2	2	2		
	Y	1.5	3	4.5	8	9	14		

Sample III	X	3	3	3	3	3	3	3	3	3
	Y	10.5	16	17	18.5	18.5	20	21	22	23

The average rank procedure for tied ranks could, of course, be applied to the X variable, in which case the values 4.5, 11.5, and 19 would replace the values 1, 2, and 3. No advantage attaches to this. In practical computation the X variable need not be recorded at all. The X variable is shown here merely to illustrate the fact that the problem is a simple correlational one. A value of S as used in the definition of Kendall's tau, and described in Chap. 21 of this book, is calculated. The value of S in this example is as follows:

$$S = 14 + 10 + 9 + 9 + 4 + 3 + 3 + 1 + 9 + 9 + 9 + 9 + 9 + 7 = 105$$

The value 14 is obtained by comparing the initial Y value of 1.5 for sample I with all Y values for samples II and III. The Y value of 1.5 for sample I need not be compared with other Y values for that sample because all values in sample I are tied on X.

The sampling variance of S is obtained from formula [21.14]. In this example $N = 23$. The X variable contains three groups of ties—one group of eight ties, one of six ties, and one of nine ties. The Y variable contains four groups of tied pairs. The variance of S from formula [21.14] is 1,245.25, and the standard error is 35.28. Subtracting unity as a continuity correction and assuming normality of the sampling distribution, we obtain the normal deviate $z = 104/35.28 = 2.95$, which warrants rejection of the null hypothesis.

The steps involved in this procedure may be stated in summary as follows:

1 Rank all the observations from 1 to N on the Y variable, as in the Kruskal-Wallis one-way analysis of variance by ranks, substituting average ranks for tied values. Use the ranks for the order of treatments as the X variable, which will consist of as many sets of ties as there are treatments.

2 Calculate S in the usual way.

3 Calculate the sampling variance of S by applying formula [21.14].

4 Subtract unity from S as a continuity correction, and divide this by its standard error to obtain the normal deviate z.

22.10 MONOTONIC TREND TEST FOR CORRELATED SAMPLES

This test may be applied to data obtained by making measurements on N subjects, each under k ordered conditions. We proceed by replacing the original measurements by ranks as in the Friedman two-way analysis of variance by ranks described in Sec. 22.8. The statistic S, as used in the definition of Kendall's tau, is calculated for each of the N subjects. The values of S are summed over the N subjects to obtain ΣS. The quantity ΣS is a measure of monotonic trend. It is descriptive of the increase, or decrease, in the experimental variable with increase in the treatment variable.

In the absence of ties the sampling variance of S for any subject is given by formula [21.11] and is the same for all subjects. The variance of ΣS is the sum of the separate variances. Thus $\sigma_{\Sigma S}^2 = \Sigma \sigma_S^2 = N \sigma_S^2$. Assuming the normality of the distribution of ΣS, the normal deviate z is

[22.17] $z = \dfrac{\Sigma S}{\sigma_{\Sigma S}}$

The normal deviate z takes the usual critical values 1.96 and 2.58 at the .05 and .01 levels, respectively, for a nondirectional test. In many experimental situations in which trend tests are used, some prior basis exists for predicting the direction of the trend; consequently a directional test will frequently be appropriate.

Estimates of probabilities obtained from the normal approximation to the distribution of ΣS will be improved by using a correction for continuity. To apply this correction, we subtract unity from ΣS if it is positive and add unity if it is negative. Thus the absolute value of ΣS is reduced by unity.

The method described above is illustrated with reference to the data of Table 22.1, which show hypothetical measurements for eight subjects under four treatments. These measurements have been ranked for each of the N subjects, as shown in Table 22.2. A value of S is calculated for each of the N subjects. The values of S summed over subjects are ΣS as follows:

$$S = 0 + 0 + 0 - 4 + 4 + 2 + 6 + 6 = 14$$

For $k = 4$ from formula [21.11] the sampling variance $\sigma_S^2 = 8.67$. Note that in this case in using formula [21.11] we write $N = k$, the number of treatment groups. The variance of $\sigma_{\Sigma S}^2 = 8 \times 8.67 = 69.33$, and $\sigma_{\Sigma S} = 8.33$. Reducing ΣS by unity as a continuity correction results in $z = 13/8.33 = 1.56$, which falls short of significance at the .05 level. Thus we cannot argue for a significant monotonic trend from these data.

It is of incidental interest to note that for $k = 2$ the quantity $(|\Sigma S| - 1)^2/N$ is distributed approximately as χ^2 with $df = 1$, and the quantity $(|\Sigma S| - 1)/\sqrt{N}$ has an approximately normal distribution. In this case the present test is the same as the usual sign test for two correlated samples.

The steps involved in the above procedure may be summarized as follows:

1 Rank the scores for each subject from 1 to k.

2 Calculate S for each subject.

3 Sum S for all subjects to obtain ΣS.

4 Calculate the sampling variance of S, σ_S^2, using formula [21.11], and multiply this by N to obtain the sampling variance of ΣS, $\sigma_{\Sigma S}^2$. The square root of this quantity is the standard error.

5 Divide $|\Sigma S| - 1$ by the standard error $\sigma_{\Sigma S}$ to obtain the normal deviate z.

6 Reject the null hypothesis for a nondirectional test at the .05 level if $z \geq 1.96$, and at the .01 level if $z \geq 2.58$. The corresponding values of z for a directional test are 1.64 and 2.33.

Ties may occur in the measurements obtained for each subject. Ties will presumably not ordinarily occur in the treatment variable, the values of that variable being controlled by the experimenter. Under these circumstances the value σ_S^2 may be calculated separately for each subject using formula [21.13], and the values of σ_S^2 summed to obtain $\sigma_{\Sigma S}^2$. A more convenient procedure is to calculate $\sigma_{\Sigma S}^2$ as in the untied case, and then subtract unity from this variance for each tied pair, 3.67 for each triplet of ties, and so on.

Illustrative data are shown in Table 22.3. This table shows measurements obtained for eight subjects under four conditions. The quantity ΣS has been calculated, and is found to be -25. The data contain five tied

Table 22.3 Data illustrating trend test for correlated data with ties

| Subject | Measurements under four conditions | | | |
	I	II	III	IV
1	7	7	4	2
2	6	1	5	5
3	3	2	2	2
4	4	2	2	2
5	8	6	5	2
6	4	7	2	2
7	9	4	3	1
8	10	4	2	2

| Subject | Ranks under four conditions | | | | |
	I	II	III	IV	S
1	3.5	3.5	2	1	−5
2	4	1	2.5	2.5	−1
3	2	1	3.5	3.5	+3
4	4	2	2	2	−3
5	4	3	2	1	−6
6	2	4	1.5	1.5	−3
7	4	3	2	1	−6
8	4	3	1.5	1.5	−4

$$z = \frac{|\Sigma S| - 1}{\sigma_{\Sigma s}} = \frac{24}{7.79} = 3.08 \qquad p < .01 \qquad \Sigma S = -25$$

pairs, and one triplet of ties. For $k = 4$ without ties the sampling variance $\sigma_S^2 = 8.67$, and the variance of $\sigma_{\Sigma S}^2 = 8 \times 8.67 = 69.33$. We subtract from this 5.00 for the five tied pairs and 3.67 for the triplet of ties to obtain a corrected value of the variance of 60.67 and a standard error of 7.79. Reducing the absolute value of ΣS by unity as a continuity correction results in $z = 24/7.79 = 3.08$, which is significant at better than the .01 level for a nondirectional test.

For a more detailed discussion of nonparametric trend tests see Ferguson (1965).

22.11 EXACT TEST OF SIGNIFICANCE FOR A 2 × 2 TABLE

An exact test of significance for a 2 × 2 table has been developed by R. A. Fisher. This test enables the calculation of exact probabilities and avoids the use of the continuous chi-square distribution to obtain approximate probabilities. It may be used appropriately where the expected cell frequencies are small. The principal objection to its use is the laborious calculation required.

In tossing a number of coins a finite number of events may result. In tossing six coins, seven outcomes are possible. We may toss 0, 1, 2, 3, 4, 5, or 6 heads. The binomial distribution may be used to determine the exact probabilities associated with these seven outcomes. Similarly, for any 2 × 2 table, given the restrictions imposed by the marginal totals, a finite number of arrangements of the cell frequencies may result. For example, for the table

2	1	3
3	5	8
5	6	

only four arrangements of the cell frequencies are possible. These are as follows:

1				2				3				4		
0	3	3		1	2	3		2	1	3		3	0	3
5	3	8		4	4	8		3	5	8		2	6	8
5	6			5	6			5	6			5	6	

The exact probability associated with each arrangement may be calculated. To conceptualize the situation here, consider an urn containing three black and eight white balls. Withdraw the balls one at a time and assign five of them at random to a black box and six to a white box. Count the number of black balls in the black box. Repeat the experiment many times and calculate the relative frequencies of the four possible outcomes. These relative frequencies are experimentally determined estimates of the probabilities of occurrence of the four possible 2×2 tables. The required probabilities may be calculated without this laborious experimental procedure. The probability of any arrangement of cell frequencies, given the marginal restrictions, is obtained by

[22.18] $$p = \frac{(A + B)!(C + D)!(A + C)!(B + D)!}{N!A!B!C!D!}$$

The numerator is the product of the factorials of the marginal totals. The denominator is $N!$ times the product of the factorials of the cell frequencies. The factorial of any number, say, 5, is $5 \times 4 \times 3 \times 2 \times 1 = 120$; also $0! = 1$. The probabilities associated with the four tables above are

1 $\quad p_1 = \dfrac{3!8!5!6}{11!0!3!5!3!} = \dfrac{4}{33} = .1212$

2 $\quad p_2 = \dfrac{3!8!5!6!}{11!1!2!4!4!} = \dfrac{15}{33} = .4545$

3 $\quad p_3 = \dfrac{3!8!5!6!}{11!2!1!3!5!} = \dfrac{12}{33} = .3636$

4 $\quad p_4 = \dfrac{3!8!5!6!}{11!3!0!2!6!} = \dfrac{2}{33} = .0606$

$\quad$ Total $\qquad\qquad\qquad = .9999$

Clearly, in this case we have no grounds for rejecting the hypothesis that the two variables are independent. The probability of obtaining a degree of association equal to or better than the one observed, and in the same direction, is obtained by summing the probabilities of arrangements 3 and 4. This probability is $.3636 + .0606 = .4242$. Thus in about 42 samples in 100, a result equal to or better than the one observed would occur by chance. With the present data no arrangement of the 2×2 table can lead to a statistically significant result.

Usually the probabilities associated with all possible arrangements of the 2×2 table need not be calculated. We need only calculate the probabilities associated with the observed table and those that represent more extreme departures from expectation in the same direction. Let Table 1 below represent the observed data. Tables 2 and 3 are the two more extreme tables in the same direction.

1				2				3		
4	2	6		5	1	6		6	0	6
3	5	8		2	6	8		1	7	8
7	7	14		7	7	14		7	7	14

The probabilities associated with these three tables are .2448, .0490, and .0023. The sum of these probabilities is .2961. This falls far short of significance, and we conclude that the evidence is insufficient to warrant rejection of the hypothesis of independence. The sum of the probabilities associated with Tables 2 and 3 is .0513. Thus the arrangement of Table 2 above, if it did occur, would fall short of significance at the 5 per cent level for a one-tailed test. The only arrangement of the three shown which could lead to a conclusion of significance, given the marginal restrictions, is that shown in Table 3.

Tables to assist the application of exact tests of significance to 2×2 tables have been prepared by Finney (1948). An adaptation of Finney's tables is given by Siegel (1956) and Bradley (1968).

EXERCISES **1** The following are data for two groups of experimental animals:

Group I	104	109	127	143	187	204	209	266	277
Group II	62	82	89	90	101	106	109	109	205

Apply a sign test to test the hypothesis that the two samples come from populations with the same median.

2 The following are data for a sample of nine animals tested under control and experimental conditions:

Control	21	24	26	32	55	82	46	55	88
Experimental	18	9	23	26	82	199	42	30	62

Test the significance of the difference between the two medians using a sign test.

3 The following data are for four groups of experimental animals:

Group I	5	7	16	14	19
Group II	8	15	18	20	24
Group III	17	21	22	25	29
Group IV	23	27	28	31	32

Apply a sign test to test the hypothesis that the four samples come from populations with the same median.

4 Apply the Wilcoxon rank sum test to the data of Exercise 1 above.

5 Apply the Wilcoxon matched-pairs signed-ranks test to the data of Exercise 2 above.

6 Apply a Kruskal-Wallis one-way analysis of variance by ranks to the data of Exercise 3 above.

7 Apply a Friedman two-way analysis of variance by ranks to the following data:

	Treatment			
Subject	*I*	*II*	*III*	*IV*
1	5	9	4	1
2	6	8	7	3
3	9	10	8	7
4	5	10	4	2
5	8	6	4	1
6	10	8	7	5
7	14	12	13	10

8 Apply a monotonic trend test to the data of Exercise 3 above.

9 Apply a monotonic trend test to the data of Exercise 7 above.

10 Obtain the exact probabilities associated with all possible arrangements of cell frequencies for the following 2×2 tables:

a		
3	1	4
2	4	6
5	5	10

b		
1	4	5
7	3	10
8	7	15

In either case would any arrangement of cell frequencies justify a rejection of the hypothesis of independence?

PSYCHOLOGICAL TEST AND MULTIVARIATE STATISTICS

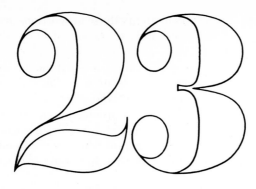

TEST CONSTRUCTION STATISTICS

23.1 INTRODUCTION

Many commonly used psychological tests of ability, achievement, and aptitude are constructed of items which permit two categories of response only, either a *pass* or a *fail*. With such items a weight of 1 is commonly assigned for a pass and a weight of 0 for a failure. Such items are ordinarily spoken of as *dichotomous* items. A person's score, X, on a test of n items is simply the number of items done correctly, although, as will be subsequently shown, certain corrections or transformations may be applied to this score. A substantial body of theory, and statistical technique, has been developed which is concerned with the construction, standardization, accuracy of measurement, and validation of such tests. This body of theory and technique is commonly spoken of as *psychological test*, or *mental test*, theory. For a thorough and comprehensive discussion of this subject the reader is referred to Gulliksen (1950), Magnusson (1967), and Lord and Novick (1968). A few elementary aspects only of the subject are discussed here.

23.2 THE MEAN AND VARIANCE OF TEST ITEMS

Consider a test of n items administered to a sample of N individuals. The number of items passed by each individual may be obtained. These are the test scores for the N individuals denoted by $X_1, X_2, \ldots, X_N$. In addition to this the number of individuals passing each of the n items may be obtained. These numbers, which may be denoted by $P_1, P_2, \ldots, P_n$, are ordinarily divided by N, the number of individuals, to obtain the proportions $p_1, p_2, \ldots, p_n$. With tests of ability these proportions are spoken of as the *difficulty values* of the items, and are presumed to describe the difficulty of the item. Clearly, if few people in the sample under consideration pass the item, say $p_i = .05$ or $p_j = .10$, the item is a difficult one. If a high proportion in the sample of people pass the item, say $p_i = .80$ or $p_j = .90$, the item may be viewed as an easy item. Note that the proportion is inversely related to difficulty: the higher the proportion, the easier the item. Note further that the proportion p_i is obtained by adding together all the scores of 1 and 0 for a particular item, and dividing by the number of individuals in the sample. The quantity p_i is, therefore, the *arithmetic mean* of the individual item.

Each of N individuals obtains a score of 1 or 0 on a particular item, say item i, whose mean is p_i. The variance of the individual item may be considered, where in this case the variance is defined as $s^2 = \Sigma(X - \bar{X})^2/N$. If we substitute 1's or 0's for the X's, and p_i for $\bar{X}$, it can be shown that the individual-item variance is given by $s_i^2 = p_i q_i$, where $q_i = 1 - p_i$. The item standard deviation is given by $s_i = \sqrt{p_i q_i}$.

The difficulty value of an item, the item mean, is obviously not independent of the item variance. The variance is a maximum when $p_i = .50$, and $s_i^2 = .50 \times .50 = .25$ and departs from this maximum as p_i departs from .50. The item variance approaches zero as p_i approaches 1.00 or zero.

23.3 CORRELATION BETWEEN TWO TEST ITEMS: THE PHI COEFFICIENT

In psychological test work the correlation between two test items is the usual product-moment correlation between two variables, where the variables are restricted to the integers 1 or 0. This statistic is the phi coefficient and is applicable to 2×2 tables only. It is related to χ^2. Although it may be used to describe the relation between any two dichotomous variables, its most common application is in describing the relation between two test items.

One formula for calculating the phi coefficient, or ϕ, is

[23.1] $$\phi = \frac{BC - AD}{\sqrt{(A + B)(C + D)(A + C)(B + D)}}$$

where A, B, C, and D are the four cell frequencies. The term in the denominator of the above expression is the square root of the product of the four marginal totals.

Table 23.1 shows a 2×2 table illustrating the relationship between two psychological test items. The value of ϕ based on this table is .376. The reader will note that in this example the two underlying variables may be regarded as continuous. The categories "pass" and "fail" may be considered a dichotomy of an underlying continuous ability variable. Individuals above a certain threshold value on the ability variable pass the item; those below it fail the item.

Table 23.1   Computation of phi coefficient of correlation between two test items

		Frequency Item 2					Proportion Item 2		
		Fail	Pass				Fail	Pass	
Item 1	Pass	11 (A)	19 (B)	30	Item 1	Pass	.22 (a)	.38 (b)	.60 (p_1)
	Fail	15 (C)	5 (D)	20		Fail	.30 (c)	.10 (d)	.40 (q_1)
		26	24	50			.52 (q_2)	.48 (p_2)	

$$\phi = \frac{19 \times 15 - 11 \times 5}{\sqrt{30 \times 20 \times 24 \times 26}} = .376$$

The phi coefficient is related to χ^2 calculated on a 2×2 table by the expression

$$\phi = \sqrt{\frac{\chi^2}{N}}$$

or

[23.2] $\chi^2 = N\phi^2$

Any formula for calculating χ^2 for a 2×2 table may with minor modification be used for calculating ϕ.

Alternative formulas for computing ϕ may be stated. Let us represent the proportion passing item i by p_i and those failing by q_i, where $p_i = 1 - q_i$. Similarly, the proportion passing item j is p_j and the proportion failing q_j. The proportion passing both items i and j is represented by p_{ij}. The ϕ coefficient of correlation between two test items may then be written as

[23.3] $\phi = \dfrac{p_{ij} - p_i p_j}{\sqrt{p_i p_j q_i q_j}}$

For the example of Table 23.1, $p_{ij} = .38$ and the phi coefficient is

$$\phi = \frac{.38 - .60 \times .48}{\sqrt{.60 \times .52 \times .40 \times .48}} = .376$$

which checks with the result previously obtained. When one of the variables is evenly divided, $p_i = q_i = .50$, the formula for ϕ simplifies to

[23.4] $\phi = \dfrac{2p_{ij} - p_j}{\sqrt{p_j q_j}}$

When both variables are evenly divided and $p_i = q_i = p_j = q_j = .50$, the formula becomes

[23.5] $\phi = 4p_{ij} - 1$

The phi coefficient has been widely used in statistical work associated with psychological tests. Usually when investigators speak of the correlation between dichotomously scored test items, the reference is to the phi coefficient.

The phi coefficient is a particular case of the product-moment correlation coefficient. If we assign integers, say, 1 and 0, to represent the two categories of each variable and calculate the product-moment correlation coefficient in the usual way, the result will be identical with ϕ.

The phi coefficient has a minimum value of -1 in the case of perfect negative and a maximum value of $+1$ in the case of perfect positive association. These limits, however, can be attained only when the two variables are evenly divided; that is, $p_i = q_i = p_j = q_j = .50$. When the variables are the same shape, $p_i = p_j$ and $q_i = q_j$, but are asymmetrical, $p_i \neq q_i$ and $p_j \neq q_j$, one or the other of the limits -1 or $+1$ may be attained but not both. The maximum and minimum values of phi are clearly influenced by the marginal totals. Consider the following 2×2 tables:

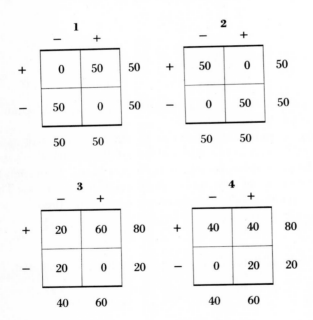

In Tables 1 and 2 both variables are evenly divided and coefficients of $+1$ and -1 are possible. Table 3 represents the maximum positive association possible, given the restriction of the marginal totals. The phi coefficient is .613. Table 4 shows the most extreme negative association possible with the same marginal totals. The phi coefficient is $-.403$. For this particular set of marginal totals phi can extend from a minimum of $-.403$ to a maximum of .613.

While the influence of the marginal totals on the range of values of phi may in some of its applications prove to be a disadvantage, this effect is in no way inconsistent with correlation theory. If a correlation coefficient is viewed as a measure of the efficacy of prediction, then perfect prediction in both a positive and a negative direction is possible only when the two distributions have the same shape and are symmetrical. If one variable is normally distributed and the other is rectangular, perfect prediction of the one from the other is not possible and the correlation coefficient reflects this fact. Perfect prediction in one direction requires symmetry also. The phi coefficient, although affected by the marginal totals, is a measure of the efficacy of prediction. From this viewpoint it quite rightly reflects the loss in degree of prediction resulting from the lack of concordance of the two marginal distributions.

Because $\chi^2 = N\phi^2$, we can readily test the significance of ϕ by referring $N\phi^2$ to a chi-square table with 1 degree of freedom. When $df = 1$, χ^2 is a normal deviate and we may refer $\phi\sqrt{N}$ to tables of the normal curve. In sampling from a population where no association exists, the distribution of ϕ should be approximately normal with a standard error of $1/\sqrt{N}$. Of course, all considerations pertaining to small frequencies (Sec. 13.6) apply here. N should, clearly, not be too small.

23.4 RESPONSE PATTERNS

The responses of N subjects on a test of n items may be represented in the form of a table containing n rows and N columns. The elements in this table are 1's and 0's. Such an arrangement of the data is spoken of as a *response* or *answer pattern*. Table 23.2 shows a hypothetical response pattern for a test of five items administered to a sample of ten individuals.

Difficulty values are shown in the column to the right of Table 23.2 and scores are shown in the bottom row. The mean score, $\overline{X}$, is 2.80. Note that $\overline{X} = \Sigma X/N = \Sigma p_i = 2.80$. The mean score on the test is the sum of the dif-

Table 23.2 Response pattern for a test of five items administered to a sample of ten subjects

					Individuals							
		1	2	3	4	5	6	7	8	9	10	p_i
Items	1	1	1	1	.	1	1	.	1	1	1	.80
	2	1	1	1	1	.	1	1	.	.	.	.60
	3	1	1	.	1	1	1	1	.	.	.	.60
	4	1	1	1	1	.	.	.	.	.	.	.40
	5	1	1	1	.	1	.	.	.	.	.	.40
	X_i	5	5	4	3	3	3	2	1	1	1	$\overline{X} = 2.80$

ficulty values. The variance of scores on the test, where the variance is defined as $s_x^2 = \Sigma(X - \bar{X})^2/N$, is 2.16.

23.5 RELATION BETWEEN THE VARIANCE OF TEST SCORES AND THE PROPERTIES OF ITEMS

In test construction a knowledge of the relations between the variance of test scores and the characteristics of test items is of interest. To ascertain the nature of these relations each item on a psychological test is viewed as being interrelated with every other item, the nature of these interrelations being most usefully represented by the interitem covariances. The interitem covariance between two items i and j is given by $r_{ij}s_is_j = p_{ij} - p_ip_j$. The table, or matrix, of item variances and interitem covariances may be represented as follows:

$$
\begin{matrix}
s_1^2 & r_{12}s_1s_2 & \cdots & r_{1n}s_1s_n \\
r_{12}s_1s_2 & s_2^2 & \cdots & r_{2n}s_2s_n \\
\cdot & \cdot & \cdots & \cdot \\
\cdot & \cdot & \cdots & \cdot \\
\cdot & \cdot & \cdots & \cdot \\
r_{1n}s_1s_n & r_{2n}s_2s_n & \cdots & s_n^2
\end{matrix}
$$

Such a table of variances and covariances is called a *covariance matrix*. This matrix is symmetrical with item variances along the diagonal and interitem covariances along both sides of the diagonal.

The reader may recall from Sec. 7.6 that the variance of the sum of a set of variables is the sum of all the elements in the covariance matrix. In the present context this means that the sum of all the elements in the interitem covariance matrix is the variance, s_x^2, of scores on the test as a whole. This relation may be written as

$$[23.6] \quad s_x^2 = \sum_i^n s_i^2 + 2 \sum_{i,j}^{n(n-1)/2} r_{ij}s_is_j$$

This formula describes the relation between the variance of the test and the properties of the individual test items of which the test is comprised. It expresses the test variance as a function of the item variances and the interitem covariances. For the illustrative response pattern in Table 23.2 for a test of five items administered to a sample of 10 subjects the interitem covariance matrix is:

.16	−.08	−.08	−.02	.08
−.08	.24	.14	.16	.06
−.08	.14	.24	.06	.06
−.02	.16	.06	.24	.14
.08	.06	.06	.14	.24

The elements in the diagonal are the item variances, $s_i^2 = p_i q_i$. The elements on either side of the diagonal are the item covariances, $r_{ij} s_i s_j = p_{ij} - p_i p_j$. The sum of all the elements in this matrix is 2.16, which is the variance of test scores, s_{ij}^2.

The relation shown in formula [23.6] between test variance and the variances and covariances of the items of which the test is comprised has important implications for test construction. The purpose of any test is the differentiation of individuals or the description of individual differences. In the construction of a test an important consideration is to ensure that the test as a whole discriminates, or shows variation, between those individuals to whom it can be appropriately administered. Clearly a test on which every individual made a zero score, a perfect score, or the same score, and for which $s_x^2 = 0$, would serve no useful purpose. Test makers, therefore, in the development of tests attempt to ensure that the test scores have a fairly large variance in relation to the number of items the test contains. Formula [23.6] shows that the larger the variance of a particular item, the greater the contribution tends to be of that item to the variance of the test as a whole. This has led some test makers to suggest that only items whose difficulty values were not distantly removed from .50, say between .30 and .70, be included in a test. Formula [23.6] shows also that the greater the interitem covariance between two items, the greater the contribution of those two items to the test variance. This has led test makers in their test construction procedures to include those items which had the higher interitem covariances and correlations.

23.6 INTERNAL CONSISTENCY

Inspection of the response pattern in Table 23.2 shows that individual 4 failed item 1, which had a difficulty value of .80. On the other hand the same individual passed items 2, 3, and 4 with difficulty values of .60, .60, and .40, respectively. His performance on item 1 is clearly inconsistent with his performance on items 2, 3, and 4. Likewise individual 5 failed item 2 with a difficulty value of .60 and passed item 5, a more difficult item, with a difficulty value of .40. This again is an inconsistency of response. If an individual who obtains a score X_i on a test obtains this score by passing the X_i easier items on the test and failing the $n - X_i$ more difficult items, his performance contains no inconsistencies. If all N individuals taking a test obtain their score X_i in this way, and no inconsistencies are present in the response pattern, that response pattern may be spoken of as an *internally consistent* pattern. With real data, response patterns that exhibit no inconsistencies do not ordinarily occur. Response patterns for different tests exhibit varying numbers of inconsistencies and varying degrees of internal consistency.

Although the term *internal consistency* has been used here to refer to a particular property of a response pattern, a variety of other terms is quite commonly used. Ferguson (1941) used the term *uniqueness* to refer to a

Table 23.3　　Internally consistent response pattern for a test of five items administered to a sample of ten subjects.

		Individuals										
		1	**2**	**3**	**4**	**5**	**6**	**7**	**8**	**9**	**10**	p_i
	1	1	1	1	1	1	1	1	1	.	.	.80
	2	1	1	1	1	1	1	.	.	.	.	.60
Items	**3**	1	1	1	1	1	1	.	.	.	.	.60
	4	1	1	1	1	.	.	.	.	.	.	.40
	5	1	1	1	1	.	.	.	.	.	.	.40
	X_i	5	5	5	5	3	3	1	1	.	.	$\bar{X} = 2.80$

response pattern with no inconsistencies. Loevinger (1947) used the term *homogeneity*. DuBois (1970) draws distinctions between different types of test homogeneity.

Table 23.3 shows an example of an internally consistent response pattern. The difficulty values of the items are the same as those in Table 23.2. Table 23.3 shows what the response pattern for the data of Table 23.2 would have been had no inconsistencies been present. Although the item difficulties are the same for Table 23.3 as for Table 23.2, the test scores are different. The test scores for internally consistent pattern of Table 23.3 are more variable. The variance of scores for Table 23.3 is $s_x^2 = 3.90$, as compared with a variance $s_x^2 = 2.16$ of Table 23.2. The effect of inconsistencies in the response pattern is to reduce the variance of test scores.

What would the variance of test scores be if all responses were assigned at random, given the restriction of the difficulty values? Under these conditions we would expect, since the responses are random, that all the interitem correlations and covariances would be zero. The expected value of the variance under these conditions is $s_x^2 = \Sigma s_i^2 = \Sigma p_i q_i$. Thus the test variance is the sum of the elements in the diagonal of the covariance matrix. All elements on either side of the diagonal are zero. For the data of Table 23.2 the expected variance, if all responses are assigned at random, is $s_x^2 = 1.12$. For any given set of difficulty values the test variance may vary from a minimum of $s_x^2 = \Sigma s_i^2$ to a maximum value which is the value of the variance which would result if the response pattern contained no inconsistencies.

23.7　THE ITEM SELECTION PROBLEM

Test makers in the construction of psychological tests have devised methods for pretesting test items. Not infrequently a preliminary form is prepared which contains a much larger number of items than will be included in the final test. If a test of 50 items is required, the test maker may begin with an initial collection of 100 items. These are administered on a trial basis to a sample of the population on which the test will ultimately be used. How does the test maker decide which are the better 50 items of the 100 items available? What considerations lead to the selection of one item and the rejection of another?

In some cases external criteria are available. Such criteria may be measures of job performance, average grades following one year at the university, or other indices of performance. Each test item may be correlated with the criterion variable, and the correlation coefficients used in the selection of items. Presumably, if the ultimate objective is to construct a test to predict the criterion variable, those items would be included in the final test that have a high correlation with that criterion. Items with a low correlation with the criterion would be rejected. At times criterion groups are available. The criterion may be membership in a top, middle, or low group on job performance; a group of psychotic patients and a group of normal subjects; groups of individuals who complete a four-year university course and those who do not. Here again correlation coefficients, or other statistics, may be used to describe how the test items discriminate between groups.

In many situations no external criterion against which the test items can be validated is available. Under these circumstances a criterion of internal consistency is commonly used; that is, the test maker by one means or another selects a subset of items that exhibits fewer inconsistencies in its response pattern than many of the other subsets that might have been selected. The problem may be stated in this way. For any group of n items a subset k may be selected in C_k^n possible ways. By what method may a set of k items be selected from the C_k^n possible sets such that the internal consistency for the set of k items selected is greater than the internal consistency for any of the $C_k^n - 1$ remaining sets? Many methods of item selection in common use, which are applied in the absence of an external criterion, provide *approximate* solutions to this problem. The subset of items selected is ordinarily not the most internally consistent subset, but is one of the more internally consistent subsets. Although an exact solution to the item selection problem as stated above is possible, its application would be very laborious except for quite small sets of items.

Although a great many methods of item selection exist, the most commonly used involves calculating the correlations between the test items and scores on the whole test. Since the items are dichotomously scored and scores on the whole test may be viewed for all practical purposes as continuous, a form of correlation is required which describes the relation between a dichotomous, or two-categoried, variable and a continuous variable. The forms of correlation used for this purpose are the point biserial correlation and the biserial correlation.

23.8 POINT BISERIAL CORRELATION

Point biserial correlation provides a measure of the relation between a continuous variable, such as scores on a test, and a two-categoried, or dichotomous, variable, such as "pass" or "fail" on a psychological test item. The data when arranged in the form of a frequency distribution compose a table comprised of R rows and 2 columns. Although commonly used as a measure of the correlation between test scores and test items, point biserial correlation may be used in other situations as well. For example, the continuous variable may be scores on a psychological test, and the dichotomous variable may be male or female, or high school graduates and university graduates, or a group of normal persons and a group of neurotics.

Point biserial correlation is a product-moment correlation. If we assign a 1 to individuals in one category and a 0 to individuals in the other and calculate the product-moment correlation, the result is a point biserial coefficient. Weights other than 1 and 0 may be assigned to the categories. The coefficient is not dependent on the weights assigned.

The formula for point biserial r is

$$[23.7] \quad r_{pbi} = \frac{\overline{X}_p - \overline{X}_q}{s_x} \sqrt{pq}$$

In this formula s_x is the standard deviation of scores on the continuous variable, defined as $\Sigma(X - \overline{X})^2/N$. If the continuous variable is a test, s_x is the standard deviation of test scores. The quantities p and q are the proportions of individuals in the two categories of the dichotomous variable. If the dichotomous variable is a test item, p is the proportion of individuals who pass the item and q is the proportion who fail. $\overline{X}_p$ and $\overline{X}_q$ are the mean scores on the continuous variable of individuals within the two categories. Again, if the continuous variable is a set of test scores, $\overline{X}_p$ is the mean score of those who pass the item and $\overline{X}_q$ is the mean score of those who fail.

The calculation of point biserial correlation from ungrouped data is illustrated in Table 23.4. This table shows hypothetical scores on a test, and on a test item, for a group of 14 individuals. The mean score, $\overline{X}_p$, on the test for the six individuals who pass the item is 38.17, and the mean $\overline{X}_q$ for the eight individuals who fail the item is 23.88. The standard deviation of test scores is 18.19. The proportion of individuals who pass and fail the item, p and q, are, respectively, .43 and .57. The point biserial correlation is .39. In this example the point biserial correlation coefficient is a measure of the capacity of the test to discriminate between two groups. This statistic can always be interpreted as a measure of the degree to which the continuous variable differentiates, or discriminates, between the two categories of the dichotomous variable. The reader should note in Table 23.4 that if the eight individuals who fail the item were the eight individuals making the eight lowest scores on the test, and the six individuals who pass the item were the six individuals making the six highest scores on the test, then the point biserial correlation would assume a maximum value for the data. Also, if "pass"

Table 23.4   Calculation of point biserial correlation from ungrouped data

Individual	Inventory score	Item score
1	6	0
2	8	1
3	8	0
4	11	0
5	16	1
6	25	0
7	27	0
8	31	0
9	31	1
10	39	0
11	44	0
12	50	1
13	56	1
14	68	1

Mean score for those who pass: $\overline{X}_p = 38.17$

Mean score for those who fail: $\overline{X}_q = 23.88$

$$s_x = 18.19 \qquad p = \frac{6}{14} = .43 \qquad q = \frac{8}{14} = .57$$

$$r_{pbi} = \frac{38.17 - 23.88}{18.19} \sqrt{.43 \times .57} = .39$$

and "fail" were arranged at random in relation to test scores, the point biserial correlation would have an expected value of zero.

An alternate method of calculating point biserial correlation is the formula

[23.8] $$r_{pbi} = \frac{\overline{X}_p - \overline{X}}{s_x} \sqrt{\frac{p}{q}}$$

where $\overline{X}$ is the mean of all scores on the continuous variable, and s_x, $\overline{X}_p$, p, and q are as defined in formula [23.7].

Point biserial correlation is not independent of the proportions in the two categories. When $p = q = .50$, its maximum and minimum values will differ from those which would be obtained when, say, $p = .20$ and $q = .80$. The maximum value of r_{pbi} never reaches $+1$; the minimum value never reaches -1. In predicting a two-categoried variable from a continuous variable, perfect prediction is possible and occurs when the two frequency distributions do not overlap. Perfect prediction of a continuous variable from a two-categoried variable is obviously impossible. Some error in prediction must always occur in predicting a variable which may take a wide range of values from a variable which may take two values only. The point biserial correlation coefficient reflects this fact. It is worth noting here that the regression line obtained by calculating the means of the two columns is of necessity linear, there being only two points. The regression line obtained by calculating the means of the rows cannot be linear except under certain special circumstances.

To test the significance of r_{pbi} from zero the situation may be treated as one requiring a comparison of the two means X_p and X_q. The appropriate value of t may be written

[23.9] $$t = r_{pbi} \sqrt{\frac{N-2}{1 - r_{pbi}^2}}$$

The number of degrees of freedom is $N - 2$. This is a two-tailed test. For large N the quantity $1/\sqrt{N}$ may be used as the standard error of r_{pbi} in testing the significance of the difference from zero.

23.9 BISERIAL CORRELATION

Biserial correlation is an estimate of the relation between a continuous variable and a dichotomous variable, it being assumed that the variable underlying the dichotomy is continuous and normal. If a bivariate table comprised of R rows and C columns is dichotomized and reduced to a table of R rows and 2 columns, biserial correlation will provide a more accurate estimate of the correlation based on the $R \times C$ table than point biserial correlation. Biserial correlation is sometimes used in place of point biserial correlation in correlating test scores with individual test items. The assumption here is that "pass" and "fail" on a test item represent the dichotomy of an underlying normally distributed ability variable.

The formula for biserial correlation is

[23.10] $$r_{bi} = \frac{\overline{X}_p - \overline{X}_q}{s_x} \frac{pq}{y}$$

Here s_x is the standard deviation of test scores, defined as for point biserial correlation; $\overline{X}_p$ and $\overline{X}_q$ are the mean test scores for those who, respectively, pass and fail the item; and p and q are the proportions who pass and fail the item. The quantity y is the height of the ordinate of the unit normal curve at the point of division between the p and q proportions of cases. Thus if $p = .30$ and $q = .70$, by consulting the table of areas and ordinates of the normal curve, Table A of the Appendix, we can ascertain that the height of the ordinate y at the point of dichotomy is .348.

For the data of Table 23.4 we may, for illustrative purposes, assume that the pass-fail dichotomy is a division of a normally distributed continuous variable. For these data $p = .43$, $q = .57$. The height of the ordinate of the unit normal curve at the point of dichotomy is $y = .393$, $\overline{X}_p = 38.17$, $\overline{X}_q = 23.88$, $s_x = 18.19$, and

$$r_{bi} = \frac{38.17 - 23.88}{18.19} \times \frac{.43 \times .57}{.393} = .49$$

An alternate formula for biserial correlation is

[23.11] $r_{bi} = \dfrac{\overline{X}_p - \overline{X}}{s_x} \dfrac{p}{y}$

where $\overline{X}$ is the mean score on the test for the total sample. Applying this formula to the data of Table 23.4 we obtain, as before, $r_{bi} = .49$.

Theoretically, the maximum and minimum values of r_{bi} are independent of the point of dichotomy and are -1 and $+1$. An implicit assumption underlying this statistic is that the continuous many-valued variable is normal, as well as the variable underlying the dichotomy. Values of r_{bi} greater than unity can occur under gross departures from normality.

Some difficulties surround the sampling distribution of r_{bi}. The standard error of r_{bi} in sampling from a population where the correlation is zero is roughly

[23.12] $s_{r_{bi}} = \dfrac{1}{y} \sqrt{\dfrac{pq}{N}}$

When N is large, this formula may be used with reference to the normal curve to test the significance of r_{bi}. It should, however, be used with caution, because the probabilities thereby obtained are somewhat inaccurate. The standard error tends to increase with the extremeness of the dichotomies. The reader may wish to compare the standard error of r_{bi} with the corresponding large-sample formula for the ordinary product-moment correlation, $s_r = 1/\sqrt{N}$. The standard error of r_{bi} is always larger than the standard error of the ordinary product-moment correlation. Where $p = q = .5$, the standard error of r_{bi} is 1.25 times as large as the standard error of r. Where $p = .90$ and $q = .10$, the standard error of r_{bi} is 1.71 times that of r.

The relation between biserial and point biserial correlation is given by the expression

[23.13] $r_{bi} = r_{pbi} \dfrac{\sqrt{pq}}{y}$

The factor $\sqrt{pq}/y$ varies from 1.25 where $p = q = .5$ to 3.73 where $p = .99$ and $q = .01$. Thus r_{bi} is always greater than r_{pbi}, and the difference increases with extremeness of the dichotomies.

23.10 CONTRIBUTION OF ITEMS TO THE TEST VARIANCE

The proportion of individuals passing an item, p_i, has been discussed as a measure of item difficulty. Point biserial and biserial correlation have been discussed as measures of the discriminatory capacity of test items. Difficulty and discriminatory capacity are dual criteria in item selection. Because we have two criteria, a problem arises. Is item i with $p_i = .50$ and $r_{pbi} = .40$ a better item than item j with $p_j = .40$ and $r_{pbi} = .50$? The relative importance of item difficulty and item discriminatory capacity in the selection of items has been viewed as a problem by some test makers.

This problem has a simple solution which involves combining difficulty and discriminatory capacity into a single index. This index is the contribution of the individual item to the variance of test scores. The presumption here is that in the construction of tests we wish, as it were, to acquire or capture variance. An item that contributes more to the total variance is presumed to be a better item than one that contributes less.

In Sec. 23.5 the total test variance was shown to be the sum of all the elements in the interitem covariance matrix. The sum of a column, or row, in the covariance matrix may be viewed as the contribution of the individual item to the total variance. This is one way only of defining the contribution of the item to the total variance. For the illustrative numerical example of a covariance matrix in Sec. 23.5 the column sums for the five items are as follows:

Item	1	2	3	4	5
Sum	.06	.52	.42	.58	.58

The total variance is $s_x^2 = 2.16$. Item 1 is viewed as contributing .06 to this; item 2, .52; item 3, .42; and so on.

It can be readily shown that the contribution of item i to the total test variance is the item-test covariance, $r_{ix}s_is_x$, where r_{ix} is the point biserial correlation coefficient. The total variance is the sum of the covariances for the n items, that is,

[23.14] $$s_x^2 = \sum_{i=1}^{n} r_{ix}s_is_x$$

If we multiply the formula for the point biserial correlation coefficient by s_is_x, we obtain the item-test covariance, which is

[23.15] $$r_{ix}s_is_x = (\bar{X}_p - \bar{X}_q)pq$$

Thus the contribution of the individual item to the total variance is, very simply, the difference between the means of those who pass and those who fail the item multiplied by pq.

In the covariance term $r_{ix}s_is_x$, s_x is the same for all items. This circumstance has led to the use of $r_{ix}s_i$, instead of $r_{ix}s_is_x$, as an index of item selection. The quantity $r_{ix}s_i$ has been referred to by some writers as the *reliability index*. If we divide formula [23.14] by s_x, we note that

[23.16] $$s_x = \sum_{i=1}^{n} r_{ix}s_i$$

Whereas the sum of the n terms $\sum^{n} r_{ix}s_is_x$ is the test variance, the sum of the n terms $\sum^{n} r_{ix}s_i$ is the test standard deviation.

The values of $r_{ix}s_i$ are slightly more difficult to compute than those of $r_{ix}s_is_x$, since $r_{ix}s_i = (\overline{X}_p - \overline{X}_q)pq/s_x$. The statistic $r_{ix}s_is_x$, or its simple modification $r_{ix}s_i$, is probably the most useful single index for the selection of items in common use.

EXERCISE

1 The following is the response pattern for a test of five items administered to a sample of 12 subjects.

							Subjects							
		1	**2**	**3**	**4**	**5**	**6**	**7**	**8**	**9**	**10**	**11**	**12**	P_i
Items	**1**	1	1	.	1	.	1	1	1	1	1	.	1	9
	2	1	1	1	.	1	.	.	1	1	1	1	.	8
	3	1	.	1	1	1	1	1	.	.	.	.	.	6
	4	1	1	1	1	.	.	.	.	.	.	.	.	4
	5	1	1	.	.	1	.	.	.	.	.	.	.	3
	X_i	5	4	3	3	3	2	2	2	2	2	1	1	$\overline{X} = 2.50$

Test scores, X_i, are shown along the bottom of the table. The number of individuals passing each item, P_i, are shown to the right. Calculate (*a*) the item means and (*b*) the item variances.

2 Calculate (*a*) the interitem covariances and (*b*) the intercorrelations for the data of Exercise 1.

3 If the response pattern of Exercise 1 had contained no inconsistencies, what would the variance of scores have been?

4 If the responses in the rows of the response pattern of Exercise 1 had been assigned at random, what would be the expected variance of scores?

5 Calculate the point biserial correlation between the individual items and the total test scores for the data of Exercise 1.

6 For the data of Exercise 1 calculate the biserial correlation between item 1 and scores on the whole test.

7 For the data of Exercise 1 calculate the contribution of each item to the total test variance.

8 For the data of Exercise 1 calculate the contribution of each item to the total test standard deviation.

ERRORS OF MEASUREMENT

24.1 THE NATURE OF ERROR

The measurements obtained in the conduct of experiments are subject to error of greater or less degree. In measuring the activity of a rat, the intelligence of a child, or the response latency of an experimental subject, we may assume that the individual measurements are subject to some error. In general, the concept of error always implies a true, fixed, standard, or parametric value which we wish to estimate and from which an observed measurement may differ by some amount. The difference between a true value and an observed value is an error. If we represent a particular observation by X_i, the true value which it purports to estimate by T_i, and an error by e_i, we may write

[24.1] $e_i = X_i - T_i$

where e_i may take either positive or negative values.

A distinction may be made between *systematic* and *random* error. Observations which consistently overestimate or underestimate the true value are subject to systematic error. A stopwatch which underestimates time intervals will yield observations with systematic errors. A random error exhibits no systematic tendency to be either positive or negative and is assumed to average to zero over a large number of subjects or trials. Random errors are also assumed to be uncorrelated both with true scores and with each other. The discussion in this chapter is concerned exclusively with random errors.

Any definition of error as the difference between an observed and true value is meaningless unless a precise definition is attached to the concept of true value. In theory a true value is sometimes conceptualized as the mean of an indefinitely large number of measurements of an attribute made under conditions such that the true value remains constant, and the procedures used in making the measurements do not change from trial to trial in any known systematic fashion. In mathematical language the true value may be defined as

$$T_i = \lim_{K \to \infty} \frac{\sum_{j=1}^{K} X_j}{K}$$

where X_j refers to the jth measurement. Thus the true value is the limit approached by the arithmetic mean as the number of repeated observations K is increased indefinitely. This concept of true value is appropriate for the measurements of physical quantities. For example, a yardstick may be used to measure the length of a desk. The measurement procedure may be repeated many times, and the variation in the observations attributed to error. It may be assumed that a considerable number of repeated observations may be made under fairly constant conditions, neither the desk nor the yardstick changing in any systematic way. By increasing the number of observations and taking their mean, the error in estimating the true value may be reduced. Theoretically, this error may be made as small as we like by increasing the number of observations. As the number of observations becomes indefinitely large, the mean approaches the true value as a limit.

Questions may be raised about the appropriateness of this concept of true value in the measurement of psychological quantities. Clearly, in the measurement of human behavior the making of a large number of repeated observations is usually not possible. The attribute being measured may fluctuate or change markedly with time, or the process of repeated measurement may modify the attribute under study. For example, in measuring the intelligence of a child, it is obviously out of the question to administer the same intelligence test 100 times to obtain an estimate of error. Quite apart from the labor involved in such estimation, the results obtained would be invalidated by practice, fatigue, and other effects. This circumstance has given rise in psychological work to a variety of procedures for estimating error other than by a series of repeated measurements. Despite the operational impracticality in psychology of estimating error by making a large number of repeated measurements, the concept of true score as the mean of an indefinitely large number of such measurements is still an important concept in the study of errors of measurement. Here we note that the role of true score is analogous to that of population parameter in sampling statistics. The difference between the sample statistic and the population parameter is a sampling error. By increasing sample size the magnitude of sampling error is reduced. For an infinite population an unbiased sample statistic will approach the population parameter as a limit as the sample becomes indefinitely large. A sampling error is an error associated with a statistic based on a sample of observations. An error of measurement is usually construed to be an error associated with a particular observation which is an estimate of a true value. In most instances both population parameters and true values cannot be known but can only be estimated from fallible data. This circumstance does not detract from the meaningfulness of these concepts, nor does it prevent the making of meaningful statements about the magnitude of error.

24.2 EFFECT OF MEASUREMENT ERROR ON THE MEAN AND VARIANCE

Consider a population of measurements. Each measurement is subject to error and may be written as $X_i = T_i + e_i$, where X_i is the observed and T_i the true measurement. By summation over all members of the population we obtain $\Sigma X_i = \Sigma T_i + \Sigma e_i$. If we assume that measurement error is random, and as often positive as negative, we may write $\Sigma e_i = 0$. Consequently, the sum of measurements subject to error is equal to the sum of true measurements. It follows also that the means of the observed and true values are equal, both being equal to the population mean μ. We conclude that measurement error exerts no systematic effect on the arithmetic mean. A mean based on a sample of N measurements will exhibit no tendency to be either greater than or less than the mean of true measurements. The expectations of the mean of observed and true scores are equal to the population mean μ; that is,

[24.2] $\quad E(\overline{X}) = E(\overline{T}) = \mu$

Measurement error exerts an effect on the sampling variance of the arithmetic mean. This point is discussed in Sec. 24.6.

Measurement error exerts a systematic effect on the variance. We may write $(X_i - \mu) = (T_i - \mu) + e_i$. If we square this identity, sum over all members of the population, and divide by N_p, where N_p is the number of members in the population, we obtain

$$\frac{\Sigma(X_i - \mu)^2}{N_p} = \frac{\Sigma(T_i - \mu)^2}{N_p} + \frac{\Sigma e_i^2}{N_p} + \frac{2\Sigma(T_i - \mu)e_i}{N_p}$$

On the assumption that measurement errors are random and uncorrelated with true scores, the third term to the right is equal to zero, and we may write

[24.3] $\quad \sigma_x^2 = \sigma_T^2 + \sigma_e^2$

Thus the variance of observed scores is equal to the variance of true scores plus the variance of the errors of measurement. For a fixed σ_T^2, the more inaccurate the measurements the greater the value of σ_e^2 and the greater the variance σ_x^2.

24.3 THE RELIABILITY COEFFICIENT

Consider a situation where each member of a population has been measured on two separate occasions. Two observations are available for each member. Both are presumed to be measures of the same attribute, and both are subject to error. We may write

$$X_{i1} = T_i + e_{i1}$$
$$X_{i2} = T_i + e_{i2}$$

In deviation form these become

$$(X_{i1} - \mu) = (T_i - \mu) + e_{i1}$$
$$(X_{i2} - \mu) = (T_i - \mu) + e_{i2}$$

By multiplying these two equations, summing over a population of N_p members, and dividing by $N_p \sigma_1 \sigma_2$, we obtain

$$\rho_{xx} = \frac{\Sigma(X_{i1} - \mu)(X_{i2} - \mu)}{N_p \sigma_1 \sigma_2}$$

$$= \frac{\Sigma(T_i - \mu)^2 + \Sigma e_{i1} e_{i2} + \Sigma e_{i1}(T_i - \mu) + \Sigma e_{i2}(T_i - \mu)}{N_p \sigma_1 \sigma_2}$$

On the assumption that errors are random and uncorrelated with each other and with true scores, the three terms on the right in the numerator are equal to zero. Because the paired observations are measures of the same attribute, $\sigma_1 = \sigma_2$. Also $\Sigma(T_i - \mu)^2 = N_p \sigma_T^2$. Hence, writing $\sigma_1 = \sigma_2 = \sigma_x$,

[24.4] $$\rho_{xx} = \frac{\sigma_T^2}{\sigma_x^2}$$

where ρ_{xx} is the reliability coefficient. The reliability coefficient is a simple proportion. It is the proportion of obtained variance that is true variance. If $\sigma_x^2 = 400$ and $\sigma_T^2 = 360$, the reliability coefficient $\rho_{xx} = .90$. This means that 90 per cent of the variation in the measurements is attributable to variation in true score, the remaining 10 per cent being attributable to error. Where sample estimates are used we may write

[24.5] $$r_{xx} = \frac{s_T^2}{s_x^2}$$

where r_{xx} is the sample estimate of the reliability coefficient.

24.4 METHODS FOR ESTIMATING RELIABILITY

Above, the reliability coefficient has been discussed without reference to methods for obtaining such coefficients in practice. A number of different practical methods for estimating reliability are used. These methods are as follows:

1 **Test-retest method**   The same measuring instrument is applied on two occasions to the same sample of individuals. When the instrument is a psy-

chological test, the test is administered twice to a sample of individuals and the scores correlated.

2 **Parallel-forms method** Parallel or equivalent forms of a test may be administered to the same group of subjects, and the paired observations correlated. Criteria of parallelism are required.

3 **Split-half method** This method is appropriate where the testing procedure may in some fashion be divided into two halves and two scores obtained. These may be correlated. With psychological tests a common procedure is to obtain scores on the odd and even items.

4 **Internal-consistency methods** These are used with psychological tests comprised of a series of items, usually dichotomously scored, a 1 being assigned for a pass and a 0 for a failure. These methods require a knowledge of certain test-item statistics.

The interpretation of a reliability coefficient depends on the method used to obtain it. When the same test is administered twice to the same group with a time interval separating the two administrations, some variation, fluctuation, or change in the ability or function measured may occur. The departure of r_{xx} from unity may be construed to result in part from error and in part from changes in the ability or function measured. With many psychological tests the value of r_{xx} will show a systematic decrease with increase in the time interval separating the two administrations. When the time interval is short, memory effects may operate. The subject may recall many of his previous responses and proceed to reproduce them. A spuriously high correspondence between measurements obtained at the two testings may thereby result. Regardless of the time interval separating the two testings, varying environmental conditions such as noise, temperature, and other factors may affect the result obtained. Likewise, varying physiological factors, fatigue and the like, may exert an influence.

In estimating reliability by the administration of parallel or equivalent forms of a test, criteria of parallelism are required. Test content, type of item, instructions for administering, and the like, should be similar for the different forms. Also the parallel forms should have approximately equal means and standard deviations. In addition, the intercorrelations should be equal. Thus with three parallel tests the intercorrelations should be such that $r_{12} = r_{13} = r_{23}$. A discussion of criteria for parallel tests is given by Gulliksen (1950). Situations arise where a large pool or population of test items is available. Samples of items may be drawn at random. Each sample of items is a randomly parallel form. This approach to the development of parallel tests has been studied at length by Lord (1955a, 1955b).

In many situations a single administration only of a test may be possible. The test is divided into two halves. A not uncommon procedure is to divide a test into odd and even items. Scores are obtained on the two halves, and these are correlated. The result is a reliability coefficient for a half test. Given a reliability coefficient for a half test, the reliability coefficient for a whole test may be estimated using the Spearman-Brown formula. This formula is

[24.6] $$r_{xx} = \frac{2r_{hh}}{1 + r_{hh}}$$

where r_{hh} is the reliability of a half test. If, for example, $r_{hh} = .80$, then $r_{xx} = .89$. The Spearman-Brown formula provides an estimate of the reliability of the whole test. It estimates what the reliability would be if each test half were made twice as long.

The split-half method should not be used with highly speeded test material. Obviously, if a test is comprised of easy items, and a subject is required to complete as many items as possible within a limited time interval, and all or nearly all items are correct, the scores on the two halves would be about the same and the correlation would be close to $+1.00$.

A method of obtaining reliability coefficients using test-item statistics has been developed by Kuder and Richardson (1937). Many psychological tests are constructed of dichotomously scored items. An individual either passes or fails the item. A 1 is assigned for a pass, and 0 for a failure. The score is the number of items done correctly. The proportion of individuals passing item i is denoted by the symbol p_i, and the proportion failing, by q_i, where $q_i = 1 - p_i$. An estimate of reliability is given by

[24.7] $$r_{xx} = \frac{n}{n - 1} \frac{s_x^2 - \sum_{i=1}^{n} p_i q_i}{s_x^2}$$

where n = number of items
s_x^2 = variance of scores on test defined as $\Sigma (X - \overline{X})^2 / N$
$p_i q_i$ = product of proportion of passes and fails for item i

$\sum_{i=1}^{n} p_i q_i$ = sum of these products for n items

This formula is frequently referred to as Kuder-Richardson formula 20. The coefficient r_{xx} computed by this formula will take values ranging from zero to unity. If the responses of individuals to the test items are assigned at random, the expectation of s_x^2 is equal to $\sum_{i=1}^{n} p_i q_i$ and the expectation of r_{xx} is zero. If all items are perfectly correlated, a situation which can only arise when all have the same difficulty, $r_{xx} = 1$. The correlation between items is the phi coefficient.

If all assumptions implicit in the split-half method of estimating reliability coefficients are satisfied, the split-half and Kuder-Richardson formula

20 will yield identical results (Ferguson, 1951). Because these assumptions are rarely, if ever, satisfied in practice, differences in the coefficients obtained will result. One difficulty with the split-half method is that a test may be split in a great many ways, yielding many different values of r_{xx}. It may be shown that if a test is split in all possible ways, the average of all the split-half reliability coefficients with the Spearman-Brown correction is the Kuder-Richardson formula 20. This coefficient has a simple unique value for any particular test.

The Kuder-Richardson formula 20 is a measure of the internal consistency, or homogeneity, or scalability, of the test material. In this context these three terms may be considered synonymous. If the items on a test have high intercorrelations with each other and are measures of much the same attribute, then the reliability coefficient will be high. If the intercorrelations are low, either because the items measure different attributes or because of the presence of error, then the reliability coefficient will be low.

The Kuder-Richardson formula 20 may be applied to tests comprised of items which elicit more than two categories of response. Personality and interest inventories and attitude scales frequently permit three or more response categories. For a dichotomously scored item we note that $p_i q_i$ is the item variance s_i^2 and $\sum_{i=1}^{n} p_i q_i = \sum_{i=1}^{n} s_i^2$, the sum of the item variances.

For an item with more than two response categories, where each category has been assigned a weight, the individual item variances may be calculated and their sum may be substituted in Kuder-Richardson formula 20 for $\sum_{i=1}^{n} p_i q_i$. Consider a test comprised of statements which elicit the possible responses "agree," "undecided," "disagree." Let p_1, p_2, and p_3 be the proportion of individuals responding in the three categories. If weights 3, 2, 1 or +1, 0, −1, or any other system of weights, are assigned to the categories, the item variance may be calculated. These may be summed, and the sum substituted for $\sum_{i=1}^{n} p_i q_i$. The quantity s_x^2 is, of course, the variance of scores obtained by summing items with the assigned weights. For further discussion see Ferguson (1951).

On the assumption that all test items are of equal difficulty, a simplified form of the Kuder-Richardson formula may be obtained for use with dichotomously scored test items. This formula may be written as

$$[24.8] \quad r_{xx} = \frac{n}{n-1} \left[1 - \frac{\bar{X}(n - \bar{X})}{n s_x^2} \right]$$

where $\bar{X}$ is the mean test score and s_x^2 is the variance. This formula is referred to as Kuder-Richardson formula 21. The formula may be derived using the assumptions implicit in the concept of randomly parallel tests (Sec. 24.9).

24.5 EFFECT OF TEST LENGTH ON THE RELIABILITY COEFFICIENT

In discussing split-half reliability, a formula was given for estimating the reliability of a whole test from the reliability of a half test. This formula is a particular case of a more general Spearman-Brown formula for estimating increased reliability with increased test length. The more general formula is

$$[24.9] \quad r_{kk} = \frac{k r_{xx}}{1 + (k-1) r_{xx}}$$

where $r_{xx} =$ an estimate of reliability of a test of unit length
$r_{kk} =$ reliability of test made k times as long

If $r_{xx} = .60$ and the test is made four times as long, the reliability coefficient r_{kk} for the lengthened test is estimated as .86. From a theoretical point of view a test may be made as reliable as we like by increasing its length. Practical considerations, of course, restrict test length.

Because reliability is a function of test length, reliability coefficients calculated on tests of different lengths are, for certain purposes, not directly comparable. If, for example, we wish to compare the reliability of different types of test material, we presumably should require measures which were independent of the differing lengths of the tests. One procedure here is to use the Spearman-Brown formula and calculate reliability coefficients for a standard test of 100 items. If a test has 40 items, then a value of $k = \frac{100}{40} = 2.50$ would be used in estimating the reliability of the standard test. If another test has 150 items, then $k = \frac{100}{150} = .67$, and so on. Thus a comparison of the reliabilities of different tests may be made which is independent of differing test lengths.

24.6 EFFECT OF MEASUREMENT ERROR ON THE SAMPLING VARIANCE OF THE MEAN

Because measurement error affects the variance of a set of measurements it will also affect the sampling variance of the mean. The sampling variance of the arithmetic mean may be written as

$$[24.10] \quad \sigma_{\bar{x}}^2 = \frac{\sigma_x^2}{N} = \frac{\sigma_T^2}{N} + \frac{\sigma_e^2}{N}$$

The component σ_T^2/N is the sampling variance of the means of samples of true measurements, and σ_e^2/N is the component of the sampling variance attributable to measurement error. While measurement error exerts no systematic effect on the sample mean as an estimate of μ, such error increases the variation in sample means with repeated sampling. The increase in sampling variance over that with no measurement error present is σ_e^2/N.

The ratio of the sampling variance of the mean of true scores to the sampling variance of the mean of obtained scores is the reliability coefficient. Thus

[24.11] $\rho_{xx} = \dfrac{\sigma_T^2}{\sigma_x^2} = \dfrac{\sigma_T^2/N}{\sigma_x^2/N} = \dfrac{\sigma_{\bar{T}}^2}{\sigma_{\bar{x}}^2}$

This means that the reliability coefficient may be interpreted as descriptive of the loss in efficiency of estimation resulting from measurement error. To illustrate, a mean calculated on a sample of 100 cases, where $\rho_{xx} = .80$, has a sampling variance equal to that of a mean calculated on a sample of 80 cases where $\rho_{xx} = 1.00$. The loss in efficiency of estimation resulting from measurement error amounts to 20 cases in 100.

24.7   EFFECT OF ERRORS OF MEASUREMENT ON THE CORRELATION COEFFICIENT

Errors of measurement tend to reduce the size of the correlation coefficient. The correlation between true scores will tend to be greater than the correlation between obtained scores. If ρ_{xy} is the correlation between X and Y in the population, the relation between the correlation of true and obtained scores is given by

[24.12] $\rho_{T_x T} = \dfrac{\rho_{xy}}{\sqrt{\rho_{xx}\rho_{yy}}}$

where $\rho_{T_x T_y}$ = correlation between true scores
ρ_{xx} = reliability of X
ρ_{yy} = reliability of Y

This formula is known as the *correction for attenuation*. Errors tend to attenuate the correlation coefficient between obtained scores from the correlation between true scores. For a derivation of this formula and a discussion of the simplifying assumptions involved, see Walker and Lev (1953). The corresponding sample form of the correction for attenuation is

[24.13] $r_{T T} = \dfrac{r_{xy}}{\sqrt{r_{xx}r_{yy}}}$

To illustrate, let $r_{xy} = .60$, $r_{xx} = .80$, and $r_{yy} = .90$. The correlation between true scores on X and Y, estimated by the above formula, is .707. The correlation may be viewed as attenuated from .707 to .60 because of errors of measurement. The squares of these coefficients yield a better appreciation of the loss in predictive capacity due to errors of measurement. The squares of .707 and .60 are .50 and .36. We conclude that the presence of errors of measurement results in 14 per cent loss in predictive capacity. If the correlation between two variables is low, the correlation will not be markedly increased by improvements in reliability. If the correlation is high, improving reliability may result in substantial gains in the prediction of one variable from another.

Because the correlation between true scores can never exceed unity, the maximum correlation between two variables arises where $r_{T_x T_y} = 1$. Under this circumstance $r_{xy} = \sqrt{r_{xx} r_{yy}}$. This is an estimate of the maximum correlation between X and Y. If $r_{xx} = .80$ and $r_{yy} = .90$, the maximum possible correlation between X and Y is estimated as $\sqrt{.80 \times .90} = .85$

24.8 RELIABILITY OF DIFFERENCE SCORES

Situations arise where the difference between two sets of measurements is defined as a score. The two measurements may be initial, or prestimulus values, and values obtained in the presence of a stimulus factor. If differences are obtained between standard scores on X and Y, that is, between z_x and z_y, the reliability of the differences may be estimated by

[24.14] $$r_{dd} = \frac{r_{xx} + r_{yy} - 2r_{xy}}{2 - 2r_{xy}}$$

where r_{xx} and r_{yy} = reliability coefficients for X and Y
r_{dd} = reliability of difference $z_x - z_y$

For fixed values of r_{xx} and r_{yy} the reliability of the difference will decrease with increase in r_{xy} from zero. If $r_{xx} = .90$ and $r_{yy} = .80$, for $r_{xy} = .80$ the reliability of differences $r_{dd} = .25$. For $r_{xy} = 0$, $r_{dd} = .85$. As r_{xy} departs in a positive direction from zero, the error variance accounts for an increasing proportion of the total variance of differences, with a resulting decrease in reliability. The point to note here is that difference scores may be grossly unreliable and should be used only after careful scrutiny of the data. When the correlation between the two variables is reasonably high, it is probable that with many sets of data most of the variance of differences is error variance.

24.9 THE STANDARD ERROR OF MEASUREMENT

Because $\rho_{xx} = \sigma_T^2/\sigma_x^2$ and $\sigma_x^2 = \sigma_T^2 + \sigma_e^2$, we may write

[24.15] $$\rho_{xx} = 1 - \frac{\sigma_e^2}{\sigma_x^2}$$

and

[24.16] $$\sigma_e = \sigma_x \sqrt{1 - \rho_{xx}}$$

This latter formula is the standard error of measurement. Where s_x and r_{xx} are used as estimates of σ_x and ρ_{xx}, we obtain

[24.17] $$s_e = s_x \sqrt{1 - r_{xx}}$$

as the corresponding sample estimate. If it may be assumed that errors of measurement are independent of the magnitude of test score, then s_e may be used as the standard error associated with a single score and interpreted in the same way as the standard error of any statistic. On the assumption of a normal-curve approximation, the 95 and 99 per cent confidence intervals of an individual's score X_i are estimated by $X_i \pm 1.96s_e$ and $X_i \pm 2.58s_e$, respectively. With most psychological tests, however, errors of measurement are not independent of the magnitude of test score. The standard error is higher in the middle-score range and diminishes in size as the score departs from the average. Because of this the use of s_e to estimate confidence intervals for particular scores may yield misleading results. The variance s_e^2 is a sort of average value, and s_e when applied to particular scores has meaning only in relation to scores near the average.

The problem of the standard error of measurement associated with psychological test scores has been investigated by Lord (1955a, 1955b, 1957). Lord defines the standard error of measurement as the standard deviation of scores an individual might be expected to obtain on a large number of randomly parallel test forms. The assumption is that the ability of the individual remains unchanged and is not affected by practice, fatigue, and the like. Randomly parallel forms are viewed as composed of items drawn at random from a large pool or population of items. The items are scored 1 for a pass and 0 for a failure, a score on a test being the sum of item scores. The proportion of items in the population which individual i can do correctly is θ_i. The true score of individual i for a test of n items is $T_i = n\theta_i$. The number of items done correctly by individual i for a random sample of n items is X_i. The standard deviation of the sampling distribution of the X_i's is the standard error. This is obtained from the standard deviation of the binomial and is given by

$$\sigma_e(X_i) = \sqrt{n\theta_i(1 - \theta_i)}$$

[24.18]

$$= \sqrt{\frac{1}{n} T_i(m - T_i)}$$

An individual's score X_i may be used as an estimate of T_i. Introducing the factor $n/(n - 1)$ to obtain an unbiased estimate yields

[24.19] $$s_e(X_i) = \sqrt{\frac{X_i(n - X_i)}{n - 1}}$$

This formula may be used for estimating the standard error of a test score X_i. Where $n = 100$ and $X_i = 50$, $s_e(X_i) = 5.02$. Where $X_i = 80$ and $n = 100$, $s_e(X_i) = 4.02$. The standard error diminishes in size as the more extreme values are approached.

Lord (1955a) has shown that if s_e^2 is taken as the average of $s_e^2(X_i)$ and substituted in $r_{tt} = 1 - s_e^2/s_t^2$, unbiased variance estimates being used throughout, Kuder-Richardson formula 21, described in Sec. 24.4, is obtained.

In most practical situations where parallel tests are used, the tests are not randomly parallel in the strictest sense. The items are matched to some extent. The standard error for such tests will be less than that estimated by

$s_e(X_i)$. Thus $s_e(X_i)$ in most situations will tend to be a moderate overestimate. It is of interest to note that $s_e(X_i)$ is independent of the characteristics of the items of which a test is comprised, provided, of course, that these are scored 1 for a pass and 0 for a failure.

24.10 CONCLUDING OBSERVATIONS

The theory and method associated with the study of measurement error in psychology have been developed in relation to psychological testing. Much of this theory and method is generally applicable to measurements of all kinds. Little attention has been directed to the study of measurement error by experimental psychologists. It is probable that in much work in the field of human and animal learning, fairly gross error attaches to many of the measurements made. Reliability coefficients less than .50 are not uncommon, and coefficients of zero are perhaps not isolated curiosities. The errors which attach to measurements in the field of animal experimentation are known quite often to be substantial. Low reliability does not necessarily invalidate a technique as a device for drawing valid inferences. Low reliability may be compensated for by increase in sample size. An unreliable technique used with a small sample is, however, capable of detecting gross differences only, and the probability of not rejecting the null hypothesis when it is false may be high. When significant results are reported with an unreliable technique on a small sample, the treatment applied is usually exerting a gross effect.

A common type of experimental design requires the making of measurements on an experimental group in the presence of a treatment and on a control group in the absence of the treatment. Although substantive evidence is lacking, it is probable that in many experiments the measurements are less reliable under the experimental than under the control conditions, one of the effects of the treatment being to increase measurement error. It seems probable that this effect is more likely to occur when the treatment is in the nature of a gross assault on the normal functioning of the organism, as is the case with certain drugs, stress agents, and operative procedures. Experimental situations may be found where the treatment may increase rather than decrease the reliability of the measurements. This author can recall one experiment where the important effect of the treatment was to stabilize and make more reliable the responses of the experimental animals.

The discussion of measurement error given in this chapter is of necessity brief and incomplete. The most comprehensive, and theoretically most advanced, discussion available on measurement error as applied to psychological tests is found in Lord and Novick (1968). A straightforward and more elementary discussion of this topic is given by Magnusson (1967). For a consideration of the analysis of variance as applied to test reliability and other specialized topics, including the Kuder-Richardson formulas, the reader is referred to the monograph by Jackson and Ferguson (1941).

EXERCISES 1 For $r_{xx} = .90$ and $s_x = 15$, estimate the variance of true scores and the error variance. What percentage of the obtained variance is due to error?

2 The following are correlations between half tests: .30, .50, .72, .80, .96. Find reliability coefficients for the whole tests.

3 The following are difficulty values for a test of 20 items:

1	.97	6	.53	11	.50	16	.04
2	.95	7	.75	12	.55	17	.35
3	.76	8	.40	13	.42	18	.27
4	.80	9	.82	14	.30	19	.15
5	.60	10	.20	15	.15	20	.09

The standard deviation of test scores is 6.5. Calculate reliability coefficients using both Kuder-Richardson formulas 20 and 21. Explain the difference between the two coefficients.

4 For a particular test, $r_{xx} = .50$. What is the effect on the reliability coefficient of making the test five times as long?

5 The sampling variance of an arithmetic mean of a test is 6.2 where $r_{xx} = .80$. What part of the sampling variance is due to sampling error, and what part to measurement error? If the test were made three times as long, what proportion of the sampling variance of the mean of the lengthened test would be due to measurement error?

6 Estimate the correlation between true scores on X and Y where $r_{xy} = .60$, $r_{xx} = .80$, and $r_{yy} = .90$. What is the maximum possible correlation between X and Y?

7 For the data of Exercise 6 above, calculate the reliability of difference scores in standard-score form between X and Y.

8 Estimate the standard error associated with the individual scores 7, 26, and 44 for a test of 50 items.

SCORE TRANSFORMATIONS: NORMS

25.1 INTRODUCTION

Many varieties of transformations are used in the interpretation and analysis of statistical data. A transformation is any systematic alteration in a set of observations whereby certain characteristics of the set are changed and other characteristics remain unchanged. The representation of a set of observations X as deviations from the mean $X - \overline{X} = x$ is a simple transformation. The mean of the transformed values is zero. All other characteristics of the transformed values are the same as those of the original values. The variability, skewness, and kurtosis remain unchanged. The ordinal properties of the data are preserved. The rank ordering of the observations is the same as before. The transformation of a variable X to standard-score form $(X - \overline{X})/s = z$ results in a change both in mean and standard deviation. The mean of the transformed values is zero, and the standard deviation is unity. Skewness, kurtosis, and rank order are unchanged.

Certain commonly used transformations change the shape of the frequency distribution of the variable. The variable may, for example, be transformed to the normal form. This may involve not only a change in mean and standard deviation, but also a change in skewness and kurtosis. The original observations may be negatively skewed and leptokurtic. The transformed values may be normally distributed, or approximately so. This type of transformation does not change the rank order of the observations. The transformations most commonly used by psychologists that alter the shape of the frequency distribution are to the normal and rectangular forms. The conversion of a set of observations to percentile ranks is a transformation to a rectangular distribution.

The conversion of a set of frequencies $f_1, f_2, f_3, \ldots, f_k$ to proportions by dividing each frequency by N, or to percentages by dividing by N and multiplying by 100, is a simple transformation. The ordering of the transformed values is the same as the ordering of the original frequencies. If each frequency is divided by different values of N, say $N_1, N_2, N_3, \ldots, N_k$, then the transformed values will quite probably have an order different from the original values. The conversion of a mental age to an intelligence quotient by dividing by chronological age and multiplying by 100 is a transformation which changes the ordinal properties of the data. In converting mental ages to intelligence quotients, not only is the order changed, but also the mean, standard deviation, skewness, and kurtosis. The transformed values have a mean of 100 in the standardization group and are approximately normally distributed with a known standard deviation.

Raw scores on psychological tests are usually highly arbitrary. The values of the mean, standard deviation, and possible range of scores reside in large measure in the predilection of the test constructor. Unless the mean, standard deviation, and something about the shape of the score distribution are known, no proper interpretation can be attached to the original, or raw, scores. Such scores are frequently transformed to normal distributions with an agreed mean and standard deviation. For example, a psychological test when administered to a representative sample of individuals from the population for which the test is intended may have a mean of 37 and a standard deviation of 9.6 and be positively skewed. Scores may be transformed to a normally distributed variable with a mean of 100 and a standard deviation of 16. Scores thus transformed to the normal form immediately take on meaning. If an individual has a score of 116, we know that he is one standard deviation unit above the average. Because the scores are normally distributed we know that his performance is better than that of about 84 per cent of the population and below the performance of about 16 per cent of the population. The procedure for developing such a transformation is known as *standardization*. A psychological test is said to be standardized when transformed scores are available, based on a reference group of acceptable size. The transformed scores themselves are called *norms*. An individual's score takes on meaning in relation to a standard, or normative, group. Tests are frequently standardized to permit age allowances. This means in effect that separate norms have been prepared for each age group. The average child in each age group may have a mean transformed score of, say, 100. The standard deviation of scores for each age group may be 16. Thus a younger child may make a lower raw score than an older child but have a considerably higher transformed score. Intelligence quotients are transformed scores which make adjustments for the differing chronological ages of children taking the test. Intelligence quotients are presumed to be independent of chronological age within an accepted age range. Most published tests are accompanied by manuals containing conversion tables which permit the transformation of raw scores to standardized scores. Both normal and rectangular transformations are used in test standardization.

25.2 TRANSFORMATIONS TO STANDARD MEASURE

A standard score is a deviation from the mean divided by the standard deviation; thus $z = (X - \overline{X})/s$. The mean is the origin, and the standard deviation is the unit of measurement. Thus a particular value is z standard deviation units above or below the mean. The mean of z scores is zero, and the standard deviation is unity. The skewness and kurtosis of the distribution are unchanged. The distribution of z scores has the same shape as the distribution of X. Standard scores on two or more variables are directly comparable only in the sense that they have the same mean and standard deviation.

A standard-score transformation does not change the proportionality of scale intervals. If X_1, X_2, and X_3 are three measurements in raw-score form and z_1, z_2, and z_3 are the same three measurements in standard-score form, then

$$\frac{X_1 - X_2}{X_2 - X_3} = \frac{z_1 - z_2}{z_2 - z_3}$$

This means that the *relative* distances between the variate values remain unchanged under a standard-score transformation. Let X_1, X_2, and X_3 be 20, 30, and 50. If $\overline{X} = 40$ and $s = 15$, then z_1, z_2, and z_3 become -1.33, $-.67$, and $.67$. We note that

$$\frac{20 - 30}{30 - 50} = \frac{-1.33 + .67}{-.67 - .67} = .50$$

Standard scores involve the use of decimals and plus and minus signs. This is sometimes inconvenient. Also the range of values will seldom exceed the limits -3 and $+3$. It is not uncommon to select an arbitrary origin and standard deviation to ensure that all, or nearly all, the measures have a plus sign and that decimals are eliminated. For this purpose a mean of 50 and a standard deviation of 15 are sometimes used. If z' denotes this type of score, then

$$z' = 50 + 15 \left(\frac{X - \overline{X}}{s} \right)$$
$$= 50 + 15z$$

To change the standard deviation we multiply every standard score by 15. To change the origin we merely add 50. Values of z' are rounded to the nearest integer. In comparing performance on a series of tests, standard-score values z' are more convenient than z. Of course, any other mean and standard deviation could be selected.

25.3 PERCENTILE POINTS AND PERCENTILE RANKS

In the standardization of psychological tests transformations to percentile ranks have frequent application. Such transformations are rectangular.

Each percentile rank has the same frequency of occurrence. The frequency distribution is flat.

A clear distinction must be made between *percentile points* and *percentile ranks*. If k per cent of the members of a sample have scores less than a particular value, that value is the kth percentile point. It is a value of the variable below which k per cent of individuals lie. On an examination, if 85 per cent of individuals score less than 60, then 60 is the 85th percentile point. If a frequency distribution is represented graphically and ordinates raised at all percentile points, the total area under the frequency distribution is divided into 100 equal parts.

Percentile points may be represented by the symbols $P_0, P_1, P_2, \ldots, P_{100}$. The points P_0 and P_{100} are limits which include all members of the sample. A percentile rank, as distinct from a percentile point, is a value on the transformed scale corresponding to the percentile point. If 60 is a score below which 85 per cent of individuals fall, then 85 is the corresponding percentile rank. As in all transformations, values on the original scale correspond to certain values on the transformed scale. In the present context the values on the original scale are percentile points, the corresponding values on the transformed scale are percentile ranks.

The reader will recall that the median is a value of the variable above and below which 50 per cent of cases lie. The median is the 50th percentile point, P_{50}. The upper quartile is a value of the variable above which 25 per cent of cases and below which 75 per cent of cases lie; conversely for the lower quartile. The upper quartile is the 75th percentile point, or P_{75}, and the lower quartile is the 25th percentile point, or P_{25}. Decile points are sometimes used. These, as the name implies, involve dividing into tenths. A *decile point* is a value of the variable below which a certain percentage of individuals fall, the percentage being taken in units of 10. Decile ranks are transformed values corresponding to the decile points and taking the integer values 1 to 10. The median is the 5th decile. An ordinate at the median divides the area under the frequency distribution into 2 equal parts; ordinates at the upper quartile, the median, and the lower quartile divide the area into 4 equal parts; ordinates at the decile points divide the area into 10 equal parts; ordinates at the percentile points divide the area into 100 equal parts.

For small N the computation of percentile points and percentile ranks is not a very meaningful procedure. Given the scores 8, 17, 23, 42, 61, and 63, obviously little meaning could possibly attach to P_{30} or P_{80}. The conversion of these scores to percentile ranks would be a somewhat spurious procedure, with no advantage over ordinary ranks.

25.4 COMPUTATION OF PERCENTILE POINTS AND RANKS—UNGROUPED DATA

To illustrate the computation of percentile points and ranks for ungrouped data, consider the psychological-test scores tabulated in Table 25.1. We adopt the convention that any score value X has exact limits given by

Table 25.1   Psychological-test scores for a group of 60 children arranged in order

Individual	Score	Individual	Score	Individual	Score
1	83	21	110	41	123
2	88	22	110	42	124
3	88	23	110	43	124
4	91	24	110	44	125
5	91	25	111	45	125
6	93	26	112	46	125
7	93	27	114	47	126
8	93	28	115	48	126
9	97	29	116	49	127
10	98	30	116	50	128
11	98	31	116	51	130
12	98	32	117	52	130
13	100	33	118	53	131
14	101	34	119	54	132
15	103	35	120	55	135
16	107	36	121	56	135
17	107	37	122	57	136
18	108	38	123	58	136
19	109	39	123	59	136
20	110	40	123	60	139

$X - .5$ and $X + .5$. The variable is presumed to be continuous. Thus the score 116 has exact limits 115.5 and 116.5. This convention is the same as that used in determining the exact limits of class intervals. Let us now calculate P_{40}, the 40th percentile point, the point below which 40 per cent of individuals lie. $N = 60$, and 40 per cent of this is 24. The 24th individual has a score of 110, the exact upper limit is 110.5, and this is taken as the point on the test scale below which 24 individuals lie. Thus $P_{40} = 110.5$. Note in this case that the 25th individual has a score of 111. The exact lower limit of this score is also 110.5. Consider now the calculation of P_{20}. We require a point on the test scale below which 12 and above which 48 individuals lie. The score of the 12th individual is 98 with an upper limit of 98.5. We note also that the score of the 13th individual is 100 with a lower limit of 99.5. Presumably the median falls somewhere between 98.5 and 99.5. It is indeterminate. As an arbitrary working procedure the percentile P_{20} may be taken halfway between these two values. Thus $P_{20} = 99.0$. To illustrate the handling of ties in the computation of percentile points let us calculate P_{10}. A score is required below which 6 and above which 54 individuals fall. We note that individuals 6, 7, and 8 have the same score, 93. Thus three individuals have scores within the exact limits 92.5 and 93.5. Since we require a point below which 6 individuals fall, we interpolate one-third of the way into this interval. One-third of this interval is .33, and $P_{10} = 92.50 + .33 = 92.83$. With the above data P_0 may be taken as the lower exact limit of the lowest score, or 82.5. Similarly, P_{100} may be taken as the upper exact limit of the highest score, or 139.5.

The calculation of percentile ranks as distinct from percentile points is the reverse of the above process. Above we calculated scores corresponding to particular ranks. We may now attend to the calculation of ranks corresponding to particular scores. To illustrate, consider individual 32 in Table 25.1. This individual is 32d from the bottom. His test score is 117. The number of individuals scoring below 117 is 31. The percentage below is $\frac{31}{60} \times 100 = 51.67$. The number scoring above 117 is 28. The percentage is $\frac{28}{60} \times 100 = 46.67$. These two percentages do not add to 100. Individual 32 occupies $\frac{1}{60} \times 100 = 1.67$ per cent of the total scale. His percentile rank falls between 51.67 and $51.67 + 1.67 = 53.33$. We may take the mid-point of this interval as the required percentile rank. Thus the percentile rank corresponding to score 117 is

$$51.67 + \frac{1.67}{2} = 52.50$$

This method assumes that any rank R covers the interval $R - .5$ and $R + .5$.

Consider the question of ties. We note that five individuals score 110. The number of individuals scoring below 110 is 19, or $\frac{19}{60} \times 100 = 31.67$ per cent of the total. The number scoring above 110 is 36, or $\frac{36}{60} \times 100 = 60.00$ per cent. The number occupying the score position 110 is 5, or $\frac{5}{60} \times 100 = 8.33$ per cent. The required percentile rank may be taken as the mid-point of the interval 31.67, and $31.67 + 8.33 = 40.00$. Thus the percentile rank of the score 110 is $31.67 + 8.33/2 = 35.83$.

Percentile ranks may be obtained by using the simple formula

[25.1] $$PR = 100\,\frac{R - .5}{N}$$

where R = rank of individual, counting from the bottom
N = total number of cases

Where ties occur, R is taken as the average rank which the tied observations occupy. The average rank of the five individuals who score 110 is 22, and the corresponding percentile rank is, as before,

$$100\,\frac{22 - .5}{60} = 35.83$$

Percentile ranks are ordinarily rounded to the nearest whole number. Thus the rank 35.83 becomes 36.

25.5 CALCULATION OF PERCENTILE POINTS AND RANKS—GROUPED DATA

The calculation of percentile points and ranks for grouped data will be discussed with reference to the data of Table 25.2. Cumulative frequencies are recorded in col. 3, and cumulative percentages in col. 4.

Let us calculate P_{25}. $N = 200$, and 25 per cent of N is 50. We observe that the 50th case falls within the interval 65 to 69. The exact limits of this interval are 64.5 and 69.5. We must now interpolate within the interval to locate a point below which 50 cases fall. We note that 36 cases fall below and 26 cases within the interval containing P_{25}. To arrive at the 50th case we require 14 of the cases within the interval. Thus we take $\frac{14}{26}$ of the interval 64.5 to 69.5. This is $\frac{14}{26} \times 5 = 2.69$. We add this to the lower limit of the interval to obtain P_{25}, which is $64.5 + 2.69 = 67.19$.

The following formula may be used to calculate percentile points:

$$[25.2] \quad P_i = L + \frac{pN - F}{f_i} \times h$$

where $P_i = k$th percentile point

p = proportion corresponding to ith percentile point; thus if $i = 62$, $p = .62$

L = exact lower limit of interval containing P_i

F = sum of all frequencies below L

f_i = frequency of interval containing P_i

h = class interval

For P_{25} in Table 25.2 we have $L = 64.5$, $p_i = .25$, $F = 36$, $f_i = 26$, and $h = 5$. Thus

$$P_{25} = 64.5 + \frac{.25 \times 200 - 36}{26} \times 5 = 67.19$$

This result is identical with that obtained previously for P_{25}. The reader will observe that for P_{50} this formula is the same as that given previously for calculating the median from grouped data.

The calculation of percentile ranks is the reverse of the above procedure. The cumulative percentages shown in col. 4 of Table 25.2 are the percentile ranks corresponding to the exact top limits of the intervals. Thus 56.0 is the percentile rank corresponding to the percentile point 74.5, the exact top limit of the interval 70 to 74. Likewise 11.0 is the percentile rank corresponding to the percentile point 59.5, the exact top limit of the interval 55 to 59. The percentile rank of any score may be obtained by interpolation. What is the percentile rank corresponding to the score 81? The score 81 falls within an interval with exact limits 79.5 and 84.5. It is 1.5 score units above the bottom of this interval. The lower limit has a percentile rank of 76.0 and the upper limit 92.5. Thus we have two points on the score scale corresponding to two points on the percentile-rank scale. Five units on the score scale is equal to $92.5 - 76.0 = 16.5$ units on the percentile-rank scale, and 1.5 units on the score scale is equal to $(92.5 -$

Table 25.2   Cumulative frequencies and percentages of test scores

1 Class interval	2 Frequency	3 Cumulative frequency	4 Cumulative percentage
95–99	1	200	100.0
90–94	6	199	99.5
85–89	8	193	96.5
80–84	33	185	92.5
75–79	40	152	76.0
70–74	50	112	56.0
65–69	26	62	31.0
60–64	14	36	18.0
55–59	10	22	11.0
50–54	6	12	6.0
45–49	4	6	3.0
40–44	2	2	1.0
Total	200		

76.0)1.5/5 = 4.95 units on the rank scale. We now take $76.0 + 4.95 =$ 80.95 as the percentile rank of the score 81. Rounding this to the nearest integer we obtain a rank of 81. It is pure coincidence that in this case the percentile rank is numerically equal to the score.

The steps involved in finding percentile ranks from grouped data may be summarized as follows:

1   Find the exact lower limit of the interval containing the score X whose percentile rank is required.

2   Find the difference between X and the lower limit of the interval containing it.

3   Divide this by the class interval and multiply by the percentage within the interval.

4   Add this to the percentile rank corresponding to the bottom of the interval.

Usually, where percentile ranks are calculated we are interested in preparing a table for converting any score value to percentile ranks. Thus for every possible score, we require the corresponding percentile rank. This may be done by systematically computing all percentile-rank values in the manner described above. A somewhat easier procedure is to make a graphical plotting on suitable graph paper of cumulative percentages against the corresponding upper limits of the class intervals. Score values are plotted on the horizontal axis, and cumulative percentages on the vertical axis. The points may be joined by straight lines. Percentile ranks corresponding to scores may then be read directly from the graph. If the points are joined by straight lines, these rank values will be the same, within limits of error, as those obtained by linear interpolation directly on the numerical values. If the sample is small, the points when plotted may show considerable irregularity and it may be advisable to fit a smoothed curve to the data. The fitting of a smoothed curve by freehand methods is accurate enough for most practical purposes. A procedure related to the method described above is to calculate certain selected percentile points and then interpolate either numerically or graphically between these points. The percentile points P_{10}, P_{20}, P_{30}, . . . , P_{90} may be calculated. To achieve greater accuracy at the tails of the distribution it may be desirable to calculate P_1 and P_5; also P_{95} and P_{99}.

25.6 NORMAL TRANSFORMATIONS

The transformation of a variable to the normal form is a frequent procedure in test standardization and correlational analysis. Not uncommonly, test norms are normal transformations of the original raw scores with arbitrarily selected means and standard deviations. A type of normal transformation used by educationists is a T score. T scores are normally distributed, usually with a mean of 50 and a standard deviation of 10. A normal transformation with a mean of 100 and a standard deviation of 15, or thereabouts, resembles an IQ scale.

Transforming a set of scores to the normal form is a relatively simple procedure. Every percentile rank corresponds to a point on the base line of the unit normal curve measured from a mean of zero in standard deviation units. A percentile rank of 50 corresponds to the zero point. A rank of 60 is .25 standard deviation units above the mean. A rank of 70 is .52 standard deviation units above the mean. Table 25.3 shows points on the base line of the unit normal curve corresponding to selected percentile ranks. These and other points are readily obtained from any table of areas under the normal curve (Table A of the Appendix).

In summary, the steps used in transforming a variable to the normal form are as follows. Percentile ranks corresponding to certain points on the score scale may be calculated. A table of areas under the normal curve is used to find the points on the base line of the unit normal curve corresponding to these percentile ranks. These points correspond to the percentile points on the original score scale. Thus a correspondence is established between a set of points on the original score scale and points on a normal distribution of zero mean and unit standard deviation. Percentile ranks are stepping-stones in establishing this correspondence. The normal standard scores are multiplied by a constant to obtain any desired standard devia-

Table 25.3   Points on the base line of the unit normal curve corresponding to selected percentile ranks

Percentile rank	Standard deviation
99	+2.33
95	+1.65
90	+1.28
80	+0.84
70	+0.52
60	+0.25
50	0.00
40	−0.25
30	−0.52
20	−0.84
10	−1.28
5	−1.65
1	−2.33

tion of the transformed values. A constant is usually added to produce a change in means, thus eliminating negative signs. A transformed value corresponding to any score value on the original scale may be obtained by interpolation.

Some freedom of choice is possible in the selection of a set of points on the score scale with associated percentile ranks. *First*, we may use the exact top limits of the intervals and obtain the corresponding percentile ranks from the cumulative-percentage frequencies. *Second*, we may take the mid-points of the class intervals and obtain percentile ranks corresponding to these. *Third*, we may use a selected set of percentile points with associated percentile ranks. Thus P_{10}, P_{20}, P_{30}, . . . , P_{90} may be used. P_1, P_5, and P_{95}, P_{99} may be added at the tails as a refinement. *Fourth*, we may select certain equally spaced points on the normal standard-score scale and ascertain their percentile ranks and the corresponding percentile-point scores. These equally spaced points may, for example, be −2.5, −2.0, −1.5, . . . , +1.5, +2.0, +2.5. The difference between the four alternatives outlined above is a matter of units. The first uses units of class interval of the original variable, a unit extending from the top of one interval to the top of the next. The second also uses units of class interval of the original variable, a unit extending from the mid-point of one interval to the mid-point of the next. The third, excluding the tails, uses equal units

'able 25.4 Illustration of the transformation of scores to a normal distribution—data of Table 25.2

Class interval 1	Mid-point 2	Frequency 3	Cumulative frequency to mid-point 4	Cumulative percentage to mid-point 5	Normal standard deviation unit z 6	T score z × 10 7	z × 10 +50 8
95–99	97	1	199.5	99.75	2.81	28.1	78.1
90–94	92	6	196.0	98.00	2.06	20.6	70.6
85–89	87	8	189.0	89.50	1.25	12.5	62.5
80–84	82	33	168.5	84.25	1.00	10.0	60.0
75–79	77	40	132.0	66.00	.41	4.1	54.1
70–74	72	50	87.0	43.50	−.16	−1.6	48.4
65–69	67	26	49.0	24.50	−.69	−6.9	43.1
60–64	62	14	29.0	14.50	−1.06	−10.6	39.4
55–59	57	10	17.0	8.50	−1.37	−13.7	36.3
50–54	52	6	9.0	4.50	−1.70	−17.0	33.0
45–49	47	4	4.0	2.00	−2.05	−20.5	29.5
40–44	42	2	1.0	.50	−2.58	−25.8	24.2
Total	. . .	200					

on the percentile-rank scale. The fourth alternative uses equal units on the normal standard-score scale. While minor advantages may be claimed for one procedure in preference to another, the differences, where N is fairly large, are trivial. Any one of the four procedures is satisfactory enough for most practical purposes.

To illustrate the transformation of a set of scores to the normal form we shall use the second alternative and take the mid-points of the class intervals with their corresponding percentile ranks. Table 25.4 shows a frequency distribution of test scores. Column 2 shows the exact midpoints of the class intervals. Column 3 shows the frequencies. Column 4 shows the cumulative frequencies to the mid-points. These are the cumulative frequencies to the bottom of the interval plus half the frequencies within the interval. The number of cases below the interval 60 to 64 is 22. The number within the interval is 14. Half this number is 7. The cumulative frequency to the mid-point is $22 + 7 = 29$. Column 5 shows the cumulative percentage frequencies to the mid-points. These cumulative percentage frequencies are percentile ranks corresponding to the mid-points of the intervals. The numbers in col. 6 are points on the base line of the unit normal curve in standard deviation units from a zero mean. The percentage of the area of the unit normal curve falling below a standard score of 2.81 is 99.75, the percentage below a standard score of 2.06 is 98.00, and so on. These values are normalized standard scores corresponding to the mid-points of the original score intervals. Thus we have a set of values on the original scale paired with a set of values on a normal transformed scale. Transformed values corresponding to any score on the original scale may be obtained by either arithmetical or graphical interpolation.

Table 25.4 shows a T-score transformation. In col. 7 the standard scores of col. 6 are multiplied by 10, thus yielding transformed scores with a standard deviation of 10. In col. 8 a constant value 50 is added to the values of col. 7, thus changing the origin from zero to 50 and eliminating negative values. If we had multiplied by 15 and added 100, the transformed values would have a standard deviation of 15 and a mean of 100. Any other standard deviation and mean could be used.

25.7 THE STANINE SCALE

During World War II the United States Army Air Force Aviation Psychology Program used a stanine scale. Scores on psychological tests were converted to stanines. A stanine scale is an approximately normal transformation. A coarse grouping is used, only nine score categories being allowed. The transformed values are assigned the integers 1 to 9. The mean of a stanine scale is 5, and the standard deviation is 1.96. The percentage of cases in the stanine-score categories from 1 to 9 are 4, 7, 12, 17, 20, 17, 12, 7, and 4. Thus 4 per cent have a stanine score 1, 7 per cent a score 2, 12 per cent a score 3, and so on. If a set of scores is ordered from the lowest to the highest, the lowest 4 per cent assigned a score 1, the next lowest a score 2, the next lowest a score 3, and the process continued until the top 4 per cent receives a score of 9, the transformed scores are roughly normal and form a stanine scale. Stanine scores correspond to equal intervals in standard deviation units on the base line of the unit normal curve. A stanine of 5 covers the interval from $-.25$ to $+.25$ in standard deviation units. Roughly 20 per cent of the area of the unit normal curve falls within this interval. A stanine of 6 covers the interval $+.25$ to $+.75$ in standard deviation units. Roughly 17 per cent of the area of the unit normal curve falls within this interval. The interval used is one-half a standard deviation unit; a stanine of 9 includes all cases above $+2.25$, and a stanine of 1 all cases below -2.25 standard deviation units. Test scores can rapidly be converted to stanines. A stanine transformation is a simple method of converting scores to an approximate normal form. The grouping, although coarse, is sufficiently refined for many practical purposes.

25.8 REGRESSION TRANSFORMATIONS

The data resulting from certain psychological experiments are comprised of a set of initial measurements, obtained in the absence of an experimental treatment, and a set of subsequent measurements obtained on the same subjects in the presence of an experimental treatment. These latter measurements are a function both of the initial measurements and the effects of the experimental treatment. The investigator may wish to transform the measurements obtained under the treatment to a new variable which is independent of the initial measurements, the transformed variable being the object of further analysis. To illustrate, measures of motor performance may be obtained both in the absence and the presence of a stress agent. The scores obtained under stress conditions are not independent of the initial scores. A person may have a low score under stress because his initial level of motor performance is low, or he may have a high score because his initial level is high, quite apart from the effects of the stress agent. We require a transformation that removes the effect of the initial values. The variation in the transformed measurements is presumably the result of the stress agent, the effects of initial level of performance being removed.

Various approaches to this problem have been used. Some investigators have employed difference scores, the presumption here being that the increase or decrease in score over the initial value must result from the experimental treatment. Other investigators have used ratio scores. These methods do not achieve independence with respect to initial values. A straightforward approach to this problem is to remove the effects of initial values by simple linear regression, assuming of course that a linear-regression model is appropriate to the data.

Let X_0 and X_1 be scores obtained under the two conditions. Let z_0 and z_1 be the corresponding standard scores. The regression equation for predicting z_1 from z_0 is $z_1' = r_{01}z_0$, where r_{01} is the correlation between measures obtained under the two conditions, and z_1' is a standard score predicted from the initial values. The values z_1' are points on the regression line used in predicting z_1 from z_0. The difference between z_1 and z_1' is a deviation from the regression line and may be written as $z_1 - r_{01}z_0$. These deviations are transformed values which are quite independent of the initial values. The effect of initial performance level has been removed. The variation in the transformed values results from the experimental condition plus error. Of course, in any practical situation the data may be contaminated by other factors unless adequate controls are exercised.

The scores $z_1 - r_{01}z_0$ are errors of estimation with zero mean and a standard deviation given by $\sqrt{1 - r_{01}^2}$. They may be expressed in standard-score form by writing

[25.3] $$\delta = \frac{z_1 - r_{01}z_0}{\sqrt{1 - r_{01}^2}}$$

In this form they may be referred to as δ scores, or delta scores. These transformed scores have a mean of zero and a standard deviation of unity. Their skewness and kurtosis are not a simple matter. Such scores may be multiplied by a constant to obtain any desired standard deviation. Any constant may be added to change the mean.

This type of simple regression transformation is quite general and is applicable in many situations where we wish to remove the effects of one variable on another.

25.9　TRANSFORMATIONS WITH AGE ALLOWANCES

Any detailed consideration of a score transformation with age allowances is beyond the scope of this book. A few comments may, however, be appropriate. This transformation is a variant of the regression transformation described in the previous section. Its purpose is to achieve comparability between children of different ages by transforming to a variable which is independent of chronological age. An older child A may have greater ability than a younger child B. Relative to his age group, however, his ability may be appreciably less. We require an answer to the question, how would child A compare with child B if both were the same age? This question is answered by a transformation to a variable which is independent of chronological age. Age transformations usually incorporate a normalizing process. The transformed scores are normally distributed with a fixed mean and standard deviation.

Such a transformation may be effected in a variety of ways. One method involves the following general steps. Obtain the frequency distributions of scores for each age group. If the age group is restricted to 1 year, say 11, a frequency distribution may be prepared for each month of age, 11 years 0 months, 11 years 1 month, 11 years 2 months, and so on. For a group covering a wider age range, 3- or 6-month intervals may be used. The next step is to compute certain selected percentile points for each frequency distribution. Let us calculate the 5th, 16th, 50th, 84th, and 95th percentile points. These percentiles correspond roughly to the points on the base line of the unit normal curve of -1.65, -1.00, 0.00, $+1.00$, and $+1.65$. If greater accuracy is required, additional percentile points may be calculated. We thus have 5 percentile points for each age group. We may now fit lines to these percentile points, using either mathematical or graphical methods. Thus we fit a line to all the 5th-percentile points, another line to the 16th-percentile points, another line to the 50th-percentile points, and so on. These lines describe the increase in score with increase in age at each percentile-rank level. For a fairly narrow age range a straight line may prove an adequate fit to the data. Such a line may be fitted by the method of least squares. For a broad age range the lines may exhibit certain curvilinear properties and it may be advisable to fit a smooth curve to the points using graphical methods. These percentile lines smooth out irregularities in the data. The original percentile points are replaced by points on these lines. Let us now assume that we require a transformed variable with a mean of 100 and a standard deviation of 15. All percentile points on the 50th-percentile line correspond to a score of 100 on the transformed variable. All percentile points on the 84th-percentile line correspond to a score of 115 on the transformed variable. All points on the 95th-percentile line correspond to a score of 125 on the transformed variable. Points on the 5th- and 16th-percentile lines correspond to scores of 75 and 85 on the transformed variable. Thus for each age group we have a set of percentile points, points on a fitted line, and a corresponding set of transformed values. By interpolation and extrapolation a transformed value corresponding to each original score value may be obtained and a conversion table prepared.

Transformed scores obtained by this general method will be approximately normal with a mean of 100 and a standard deviation of 15. Any other appropriate mean and standard deviation may be used. The transformed scores are independent of age. The correlation between chronological age and transformed score is about zero.

Many variants and refinements of this general method may be applied. Many investigators may prefer to use a large number of percentile lines and equal standard-score units.

EXERCISES

1 State the difference between percentile points and percentile ranks.

2 For the data of Table 25.1, compute (a) percentile points P_{25}, P_{50}, and P_{75} and (b) percentile ranks for scores 103, 123, and 136.

3 For the data of Table 25.2, compute (a) percentile points P_{10}, P_{30}, and P_{80} and (b) percentile ranks for the scores 59, 74, and 82.

4 Develop a T-score transformation for the data of Table 3.1.

5 Develop a stanine transformation for the data of Table 3.1.

6 The following are measures of motor skill under initial nonstress conditions and subsequent stress conditions for a sample of 12 individuals.

Nonstress	26	33	41	53	28	36	44	28	52	47	59	37
Stress	18	29	52	40	25	30	38	50	41	39	50	45

Apply a regression transformation to these data. What purpose would such a transformation serve?

PARTIAL AND MULTIPLE
CORRELATION

26.1 INTRODUCTION

Previous discussion of correlation has been concerned with the relationship between two variables. In many investigations data on more than two variables are gathered and forms of multivariate analysis are required. Two forms of correlational analysis which may be applied to multivariate data are *partial* and *multiple* correlation. Partial correlation deals with the residual relationship between two variables where the common influence of one or more other variables has been removed. Multiple correlation deals with the calculation of weights which produce the maximum possible correlation between a criterion variable and the weighted sum of two or more predictor variables. Its purpose is to maximize the efficiency of prediction. Other forms of multivariate analysis exist, but these are beyond the scope of the present elementary discussion.

26.2 PARTIAL CORRELATION

Let us assume that a test of intelligence and a test of psychomotor ability have been administered to a group of children showing considerable variation in age. Both intelligence and psychomotor ability increase with age. Ten-year-old children are on the average more intelligent than six-year-old children. They also have more highly developed psychomotor abilities. Scores on the two tests will correlate with each other because both are correlated with age. Partial correlation may be used with such data to obtain a measure of correlation with the effect of age eliminated or removed.

What is meant by *eliminating*, or *removing*, the effect of a third variable? These terms in the present context have a precise statistical meaning. Let X_1, X_2, and X_3 be three variables. All or part of the correlation between X_1 and X_2 may result because both are correlated with X_3. The reader will recall from previous discussion on correlation that a score on X_1 may be divided into two parts. One part is a score predicted from X_3. The other part is the residual, or error of estimate, in predicting X_1 from X_3. These two parts are independent, or uncorrelated. Similarly, a score on X_2 may be divided into two parts, a part predictable from X_3 and a residual, or error of estimate, in predicting X_2 from X_3. The correlation between the two sets of residuals, or errors of estimate, in predicting X_1 from X_3 and X_2 from X_3 is the partial correlation coefficient. It is the part of the correlation which remains when the effect of the third variable is eliminated, or removed.

The formula for calculating the partial correlation coefficient to eliminate a third variable is

[26.1] $$r_{12.3} = \frac{r_{12} - r_{13}r_{23}}{\sqrt{(1 - r_{13}{}^2)(1 - r_{23}{}^2)}}$$

The notation $r_{12.3}$ means the correlation between residuals when X_3 has been removed from both X_1 and X_2. This is sometimes called a first-order partial correlation coefficient.

Let X_1 and X_2 be scores on an intelligence and a psychomotor test for a group of school children. Let X_3 be age. Let the correlation between the three variables be as follows: $r_{12} = .55$, $r_{13} = .60$, and $r_{23} = .50$. The partial correlation coefficient is

$$r_{12.3} = \frac{.55 - .60 \times .50}{\sqrt{(1 - .60^2)(1 - .50^2)}} = .36$$

Using a variance interpretation, the proportion overlap between X_1 and X_2 is $r_{12}{}^2 = .55^2 = .303$. The proportion overlap with X_3 eliminated is $r_{12.3}{}^2 = .36^2 = .127$. The proportion overlap which results from the effects of age is $.303 - .127 = .176$. It would also be appropriate to state that the percentage of the total association present resulting from the effect of age is $(.176/.303)100 = 58$ per cent. The remaining 42 per cent of the association results from other factors.

Partial correlation may be used to remove the effect of more than one variable. The partial correlation between X_1 and X_2 with the effects of both X_3 and X_4 removed is

[26.2] $$r_{12.34} = \frac{r_{12.4} - r_{13.4}r_{23.4}}{\sqrt{(1 - r_{13.4}{}^2)(1 - r_{23.4}{}^2)}}$$

This is a *second-order* partial correlation coefficient. Because of difficulties of interpretation, partial correlation coefficients involving the elimination of more than one variable are infrequently calculated.

A t test may be used to test whether a partial correlation coefficient is significantly different from zero. The required t is

[26.3] $t = \dfrac{r_{12.3}}{\sqrt{(1 - r_{12.3}^2)/(N - 3)}}$

This may be referred to a table of t with $N - 3$ degrees of freedom.

26.3   MULTIPLE REGRESSION AND CORRELATION

The correlation coefficient may be used to predict or estimate a score on an unknown variable from knowledge of a score on a known variable. The regression equation in standard-score form is $z_1' = r_{12}z_2$, where z_1' is a predicted or estimated standard score. In this situation we have one dependent and one independent variable. If $z_2 = 1.2$ and $r_{12} = .80$, the best estimate of an individual's standard score on variable 1 is $z_1' = .80 \times 1.2 = .96$. The estimate is that the individual is .96 standard deviation units above the average.

We may consider a situation where we have one dependent and two independent variables. The dependent variable may be a measure of scholastic success. The independent variables may be two psychological tests used at university entrance. The dependent variable is spoken of as the *criterion*. The two independent variables are *predictors*. How may scores on the two predictors be combined to predict scholastic success? The correlation between the three variables may be arranged in a small table. Let these correlations be as follows:

	1	2	3
1	1.0	.8	.3
2	.8	1.0	.5
3	.3	.5	1.0

Variable 1 is the criterion, and variables 2 and 3 are the predictors. Note that 1.0's have been entered along the main diagonal. In estimating standard scores on 1 from standard scores on 2 and 3 separately, the two regression equations are $z_1' = .8z_2$ and $z_1' = .3z_3$. Variable 2 is a much better predictor than variable 3. Presumably, by employing a knowledge of both 2 and 3, a better estimate of the criterion may be obtained.

Consider the straight sum of standard scores on 2 and 3. If the *sums* of the values in the four quadrants of the correlation table are represented by

A	C
C	B

the correlation between a standard score on 1 with the sum of standard scores on 2 and 3 is given by

$$\frac{C}{\sqrt{AB}}$$

In our example this becomes

$$r_{z_1(z_2+z_3)} = \frac{.8 + .3}{\sqrt{1.0 + 1.0 + .5 + .5}} = \frac{1.1}{\sqrt{3}} = .635$$

If we express variables 2 and 3 in standard measure, add them together, and correlate the sum with standard scores on the criterion, the correlation will be .635. This is not as good as the prediction obtained with variable 2 taken alone. The straight sum of standard scores assigns equal weight to the two variables. When variables are added together directly, they are weighted in a manner proportional to their standard deviations. The standard deviation of standard scores is 1. Consequently, on adding together standard scores, the variables are equally weighted.

Let us select some arbitrary set of weights and observe the result. Let us assign weights of 4 and 1 to the two predictors. Thus one predictor will receive four times the weight of the other. Write these weights along the top and to the side of the correlation table as follows:

		4	1
	1.0	.8	.3
4	.8	1.0	.5
1	.3	.5	1.0

We now multiply the rows and columns by these weights to obtain

1.0	3.2	.3
3.2	16.0	2.0
.3	2.0	1.0

The correlation of the criterion with the sum $4z_2 + z_3$ is again given by $C/\sqrt{AB}$ and is:

$$r_{z_1(4z_2+z_3)} = \frac{3.2 + .3}{\sqrt{16.0 + 1.0 + 2.0 + 2.0}} = .765$$

This particular arrangement of weights, 4 and 1, results in a correlation which is substantially better than that obtained with equal weights. Obviously, these are not the best possible weights. The correlation of the weighted sum with the criterion is less than that obtained with variable 2 taken separately.

How may a set of weights be obtained which will maximize the correlation between the criterion and the sum of scores on the dependent variables? Let us represent weights by the symbols β_2 and β_3. An estimated standard score on 1 is then given by $z_1' = \beta_2 z_2 + \beta_3 z_3$. We wish to calculate weights β_2 and β_3 such that the correlation between z_1 and z_1' is a maximum. Mathematically, the problem reduces to the calculation of weights which will minimize the average sum of squares of differences between the criterion score z_1 and the estimated criterion score z_1'. We require values of β_2 and β_3 such that

$$\frac{1}{N} \sum (z_1 - z_1')^2 = \text{a minimum}$$

The values of β_2 and β_3 are multiple regression weights for standard scores. They are sometimes called *beta* coefficients.

With three variables the values of β_2 and β_3 are given by

[26.4] $\beta_2 = \dfrac{r_{12} - r_{13} r_{23}}{1 - r_{23}^2}$

[26.5] $\beta_3 = \dfrac{r_{13} - r_{12} r_{23}}{1 - r_{23}^2}$

In the above example

$\beta_2 = \dfrac{.8 - .3 \times .5}{1 - .5^2} = .867$

$\beta_3 = \dfrac{.3 - .8 \times .5}{1 - .5^2} = -.133$

Let us write these weights above and to the side of the correlation table and multiply the rows and columns as follows:

.867   −1.33

1.0	.8	.3
.8	1.0	.5
.3	.5	1.0

=

1.000	.694	−.040
.694	.752	−.058
−.040	−.058	.018

.867
−.133

The sums of the elements in the four quadrants are

1.000	.654
.654	.654

The correlation between the criterion and the weighted sum is

$$C/\sqrt{AB} = .654/\sqrt{.654} = \sqrt{.654} = .809$$

This is a multiple correlation coefficient and may be denoted by R. No other system of weights will yield a higher correlation between the criterion and the weighted sum of predictors.

Note that the sum of elements in the top right quadrant of the weighted correlation table is equal to the sum in the lower right, or $C = B$. This circumstance will occur if the weights used are multiple regression weights. It provides a check on the calculation. We note also that $R^2 = C$ and $R = \sqrt{C}$. Thus the multiple correlation coefficient may be obtained by the formula

[26.6] $\quad R = \sqrt{\beta_2 r_{12} + \beta_3 r_{13}}$

This is the commonly used formula for calculating a multiple correlation coefficient.

In our example the multiple correlation is .809. The correlation of variable 2 with the criterion is .8. The addition of the third variable increases prediction very slightly. In a practical situation the third variable could safely be discarded as contributing a negligible amount to the efficacy of prediction.

26.4 THE REGRESSION EQUATION FOR RAW SCORES

The equation $z_1' = \beta_2 z_2 + \beta_3 z_3$ is a regression equation in standard-score form. It will yield the best possible linear prediction of a standard score on 1 from standard scores on 2 and 3. In practice, we usually require a regression equation for predicting a raw score on 1 from a raw score on 2 and 3. Let X_1' be a predicted raw score on 1, and X_2 and X_3 the obtained raw score on 2 and 3. The estimated standard score z_1' and the observed standard scores z_2 and z_3 may be written as

$$z_1' = \frac{X_1' - \bar{X}_1}{s_1}$$

$$z_2 = \frac{X_2 - \bar{X}_2}{s_2}$$

$$z_3 = \frac{X_3 - \bar{X}_3}{s_3}$$

By substituting these values in the regression equation in standard-score form we obtain

$$\frac{X_1' - \bar{X}_1}{s_1} = \beta_2 \frac{X_2 - \bar{X}_2}{s_2} + \beta_3 \frac{X_3 - \bar{X}_3}{s_3}$$

Rearranging terms and writing the expression explicit for X_1' yields

[26.7] $\quad X_1' = \beta_2 \frac{s_1}{s_2} X_2 + \beta_3 \frac{s_1}{s_3} X_3 + \left(\bar{X}_1 - \beta_2 \frac{s_1}{s_2} \bar{X}_2 - \beta_3 \frac{s_1}{s_3} \bar{X}_3 \right)$

This is a regression equation in raw-score form. It may be used to predict a raw score on 1 from a raw score on 2 and 3. The values $\beta_2 s_1/s_2$ and $\beta_3 s_1/s_3$ act as weights. The quantity to the right in parentheses is a constant.

In the example of the previous section $\beta_2 = .867$ and $\beta_3 = -.133$. Let us assume that $s_1 = 5$, $s_2 = 10$, $s_3 = 20$; also $\overline{X}_1 = 20$, $\overline{X}_2 = 40$, and $\overline{X}_3 = 60$. The regression equation in raw-score form is written as

$$X'_1 = (.867) \tfrac{5}{10} \ X_2 + (-.133) \tfrac{5}{20} \ X_3$$
$$+ [20 - (.867) \tfrac{5}{10} (40) - (-.133) \tfrac{5}{20} (60)]$$
$$= .434X_2 - .033X_3 + 4.62$$

26.5 THE GEOMETRY OF MULTIPLE REGRESSION

Given two variables X_1 and X_2, each pair of observations may be plotted as a point on a plane. If interest resides in predicting one variable from a knowledge of another, a straight regression line may be fitted to the points and this line used for prediction purposes.

Given three variables X_1, X_2, and X_3, each triplet of observations may be plotted as a point in a space of three dimensions as shown in Fig. 26.1. Instead of two axes at right angles to each other, we now have three. All triplets of observations may be plotted as points. If the correlations between the three variables are all positive, the assembly of points will show some tendency to cluster along the diagonal of the space of three dimensions extending from the origin O to Y. A plane may be fitted to the assembly of points. With two variables a regression line is fitted to points in a two-dimensional space. With three variables a regression plane is fitted to points in a three-dimensional space. In Fig. 26.1 this plane is represented

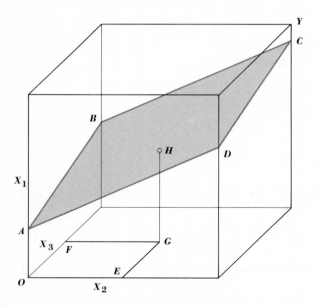

Fig. 26.1 Geometrical representation of multiple regression. *ABCD* is a multiple regression plane.

by *ABCD*. With two variables the regression equation is the equation for a straight line and is of the type $X_1' = b_2 X_2 + a$, where b_2 is the slope of the line and a is the point where the line intercepts the X_1 axis. With three variables the regression equation is the equation for a plane and is of the type $X_1' = b_2 X_2 + b_3 X_3 + a$. Here b_2 is the slope of the line *AD* in Fig. 26.1 and b_3 is the slope of the line *AB*. The constant a is the point where the plane intercepts the X_1 axis. In Fig. 26.1 it is the distance *AO*.

Consider now a particular individual. Represent his score on X_2 by *OE* and on X_3 by *OF*. We locate the point *G* in the plane of X_2 and X_3 and proceed upward until we reach the point *H* in the regression plane *ABCD*. The distance *GH* is the best estimate of the individual's score on X_1 given his scores on X_2 and X_3. It is the best estimate in the sense that the regression plane is so located as to minimize the sums of squares of deviations from it parallel to the X_1 axis.

The reader will observe that the three-variable case is a simple extension of the two-variable case. A plane is used instead of a straight line. With four or more variables the idea is essentially the same. With four variables, in effect, we plot points in a space of four dimensions and fit a three-dimensional hyperplane to these points. By increasing the number of variables we may complicate the arithmetic. We do not complicate the idea.

26.6 MORE THAN THREE VARIABLES

In the discussion above we have considered the multiple regression case with three variables only, one criterion and two predictors. With k variables the multiple regression equation in standard-score form is

[26.8] $z_1' = \beta_2 z_2 + \beta_3 z_3 + \cdots + \beta_k z_k$

The raw-score form of this equation may be obtained, as previously, by substituting for the values of z_i the values $(X_i - \overline{X}_i)/s_i$ and rearranging terms. We thereby obtain

[26.9] $X_1' = \beta_2 \dfrac{s_1}{s_2} X_2 + \beta_3 \dfrac{s_1}{s_3} X_3 + \cdots + \beta_k \dfrac{s_1}{s_k} X_k + A$

where A is given by

[26.10] $A = \overline{X}_1 - \beta_2 \dfrac{s_1}{s_2} \overline{X}_2 - \beta_3 \dfrac{s_1}{s_3} \overline{X}_3 - \cdots - \beta_k \dfrac{s_1}{s_k} \overline{X}_k$

The multiple correlation coefficient is given by

26.11] $R = \sqrt{\beta_2 r_{12} + \beta_3 r_{13} + \cdots + \beta_k r_{1k}}$

Thus to calculate this coefficient we multiply each correlation of a predictor with the criterion by its corresponding regression coefficient, sum these products, and take the square root.

A number of computational procedures exist for calculating the required regression weights with more than three variables. The method described here originates with Aitken (1937) and has been called the method of pivotal condensation. It is described in detail in Thomson (1951). A modification of the Aitken method which reduces the amount of calculation has been developed by DuBois (1965).

26.7 AITKEN'S NUMERICAL SOLUTION

To illustrate the application of Aitken's method let us consider a problem with five variables, one criterion and four predictors. Denote the criterion by X_1 and the predictors by X_2, X_3, X_4, and X_5. The criterion may be regarded as a measure of success in an occupation, and the predictors may be psychological tests used to predict performance in the occupation.

The intercorrelations between the five variables are shown in Table 26.1. The means and standard deviations of the five variables are shown in Table 26.2. Table 26.3 shows the procedure for calculating the multiple regression weights. This procedure requires the successive calculation of differences between cross products. If the four cell values are

$a \quad b$
$c \quad d$

the difference between cross products is $ad - cb$. In this case the cell value a is the pivotal element.

Table 26.1 Correlation coefficients between a criterion and four predictors

	X_1	X_2	X_3	X_4	X_5
X_1	1.00	.72	.58	.41	.63
X_2	.72	1.00	.69	.49	.39
X_3	.58	.69	1.00	.38	.19
X_4	.41	.49	.38	1.00	.27
X_5	.63	.39	.19	.27	1.00

Table 26.2 Means and standard deviations for criterion and four predictors

	$\overline{X}$	s
X_1	8.72	5.68
X_2	104.65	15.71
X_3	43.22	9.92
X_4	14.98	6.32
X_5	87.22	14.09

Table 26.3 Aitken's method for computing regression coefficients*

									Check
	(1)	.69	.49	.39	−1	·	·	·	1.57
	.69	1	.38	.19	·	−1	·	·	1.26
A	.49	.38	1	.27	·	·	−1	·	1.14
	.39	.19	.27	1	·	·	·	−1	.85
	.72	.58	.41	.63	·	·	·	·	2.34
	(1.908)	(.524)	.042	−.079	.690	−1	·	·	.177
		1.000	.080	−.151	1.317	−1.908	·	·	.338
B		.042	.760	.079	.490	·	−1	·	.371
		−.079	.079	.848	.390	·	·	−1	.238
		.083	.057	.349	.720	·	·	·	1.210
	(1.321)		(.757)	.085	.435	.080	−1	·	.357
			1.000	.112	.575	.106	−1.321	·	.472
C			.085	.836	.494	−.151	·	−1	.265
			.050	.362	.611	.158	·	·	1.182
	(1.211)			(.826)	.445	−.160	.112	−1	.225
D				1.000	.539	−.194	.136	−1.211	.272
				.356	.582	.153	.066	·	1.158
E					.390	.222	.018	.431	1.061

Regression coefficients

* Example from Godfrey H. Thomson, *The factorial analysis of human ability*, 5th ed., University of London Press Ltd., London, 1951.

The steps in the calculation are as follows:

1 Write down the matrix of intercorrelations between the predictors, that is, between variables X_2, X_3, X_4, and X_5. Insert 1's along the diagonal. Beneath this matrix write a row containing the correlations of the predictors with the criterion. The resulting matrix is shown to the left of slab A in Table 26.3.

2 To the right of the above matrix record another matrix with −1's down the diagonal. All other elements are zero, including those in the bottom row. In Table 26.3 a dot represents a zero.

3 Sum the rows to obtain the values in the *check* column.

4 Calculate the differences between cross products for the first two rows of slab A, using the 1 in the top left cell as the *pivotal* element. Thus the following product differences are formed:

$$1 \times 1 - .69 \times .69 = .524 \qquad 1 \times (-1) - 0 \times .69 = -1$$
$$1 \times .38 - .49 \times .69 = .042 \qquad 1 \times 0 - 0 \times .69 = 0$$
$$1 \times .19 - .39 \times .69 = .079 \qquad 1 \times 0 - 0 \times .69 = 0$$
$$1 \times 0 - (-1) \times .69 = .690$$

These values are recorded in the first row of slab B. The check value is obtained by forming the product difference $1 \times 1.26 - 1.57 \times .69 = .177$. If the calculation is correct to this point, the sum of elements in the first row of slab B will equal the product difference .177.

5 Beneath the first row of slab B write a second version of it obtained by dividing each element by the top left element, .524. The result is a row with unity as the pivot. This assists subsequent calculation. This part of the procedure is most readily accomplished by multiplying the elements in the row by the reciprocal of .524, or by 1.908.

6 The remaining elements in slab B are obtained by forming product differences using the first row of slab A with the third, fourth, and fifth rows of slab A, successively, always using the 1 in the top left cell as the pivotal element. Thus

$$1 \times .38 - .69 \times .49 = .042$$
$$1 \times 1 - .49 \times .49 = .760$$

and so on. Each row is summed to provide a check on the calculation. The result is a reduction of the original 5×4 matrix of slab A to the 4×3 matrix of slab B.

7 The procedure is now repeated to obtain slabs C, D, and E. At each stage, with the exception of the last, the top row in each slab is divided by the left-hand cell value, or multiplied by the reciprocal of that value, to obtain a second version of the top row. The appropriate reciprocal for row C is 1.321, and for row D it is 1.211.

8 By proceeding with the calculation, the original matrix is condensed to the cell values in slab E. These four values are the multiple regression coefficients for predicting a standard score on the criterion from standard scores on the four predictors.

In this example the regression equation for predicting the criterion from the predictors in standard-score form is $z_1' = .390z_2 + .222z_3 + .018z_4 + .431z_5$. No other system of weights will provide a better estimate of the criterion. The correlations of the four predictors with the criterion are

.72   .58   .41   .63

By multiplying these by the corresponding regression coefficients, summing the resulting products, and taking the square root, we obtain the multiple correlation coefficient as follows:

$$R = \sqrt{.390 \times .72 + .222 \times .58 + .018 \times .41 + .431 \times .63} = .83$$

A multiple correlation coefficient is amenable to the same general type of interpretation as any other correlation coefficient. It is the correlation between a criterion variable and the weighted sum of the predictors, the predictors being weighted in order to maximize that correlation.

To obtain a multiple regression equation in raw-score form we require the means and standard deviations of Table 26.2. We may write

$$X_1' = (.390) \frac{5.68}{15.71} X_2 + (.222) \frac{5.68}{9.92} X_3$$

$$+ (.018) \frac{5.68}{6.32} X_4 + (.431) \frac{5.68}{14.09} X_5 + A$$

The constant A is given by

$$A = 8.72 - (.390) \frac{5.68}{15.71} (104.65) - (.222) \frac{5.68}{9.92} (43.22)$$

$$- (.018) \frac{5.68}{6.32} (14.98) - (.431) \frac{5.68}{14.09} (87.22) = -26.81$$

With any substantial number of variables the calculation of multiple regression weights is clearly a laborious procedure and requires the use of modern computing devices.

26.8 THE SIGNIFICANCE OF A MULTIPLE CORRELATION COEFFICIENT

An F ratio may be used to test whether an observed multiple correlation coefficient is significantly different from zero. The required value of F is given by the formula

[26.12] $\quad F = \dfrac{R^2}{1 - R^2} \dfrac{N - k - 1}{k}$

where R = multiple correlation coefficient
$\quad N$ = number of observations
$\quad k$ = number of independent variables or predictors

The table of F is entered with $df_1 = k$ and $df_2 = N - k - 1$.

26.9 SHRINKAGE IN MULTIPLE CORRELATION

The multiple correlation coefficient is a measure of the efficacy of prediction for a particular sample. It is not, however, an unbiased estimate of the population correlation coefficient. Also, if the multiple regression weights calculated on one sample are applied to a second sample, the correlation between the weighted predictors and the criterion in the second sample will be less than the multiple correlation originally calculated on the first sample. The bias in the multiple correlation results because the process of determining regression weights, which minimize the average squared error, takes advantage, as it were, of the idiosyncrasies of the sample. The extent of the bias depends on the population values of the multiple correlation coefficient, the sample size, and the number of predictor variables. The phenomenon under discussion here is commonly referred to as *shrinkage*. One estimate of the population value of the squared multiple correlation coefficient in the population is given by the formula

$$[26.13] \quad R'^2 = 1 - \frac{N-1}{N-k-1}(1-R^2)$$

where R^2 is the squared multiple correlation in the sample, N is sample size, and k is the number of predictors. The extent of the bias in R^2, as an estimate of the corresponding population value, decreases with increase in sample size and increases with increase in the number of predictors. For $R = .707$, $R^2 = .500$, $N = 26$, and $k = 10$, $R'^2 = .167$ and $R' = .409$. In this example the shrinkage in the multiple correlation coefficient is substantial. Herzberg (1969) in a computer simulation study has investigated formula [26.13] for estimating the squared multiple correlation. His conclusion is that this formula provides quite a reasonable estimate of the squared population multiple correlation.

On occasion multiple regression weights calculated on one sample are applied to a second sample. This procedure is known as *cross-validation*. The correlation between the criterion and the weighted predictors is known as the *cross-validity* and is denoted by r_c. The quantity r_c is generally an underestimate of the population multiple correlation. The bias tends to decrease with sample size and increase with the number of predictors.

26.10 SOME OBSERVATIONS ON MULTIPLE CORRELATION

The techniques of multiple correlation have practical application in occupational and scholastic selection where it becomes necessary to combine a number of variables to provide the best possible estimate of a criterion measure. An appreciation of the relative contributions of the independent variables in predicting the criterion is not readily grasped by simple inspection of the multiple regression coefficients. With two predictors the square of the multiple correlation coefficient may be shown equal to $R^2 = \beta_2^2 + \beta_3^2 + 2\beta_2\beta_3 r_{23}$. Thus the predicted variance is comprised of three additive parts. β_2^2 represents a contribution by X_2, β_3^2 a contribution by X_3, and the term $2\beta_2\beta_3 r_{23}$ is a component which involves the correlation between X_2 and X_3. Thus the evaluation of the relative contributions of the different variables is not a simple matter of direct comparison of the relative magnitudes of the regression coefficients but requires also a consideration of the correlation terms.

Frequently, in practical work, the greater part of the prediction achieved can be attributed to a relatively small number of variables, perhaps four or five or six, and the inclusion of additional variables contributes only small and diminishing amounts to prediction. Tests of significance may be applied to decide whether or not the addition of one or more variables to a subset of variables will significantly improve prediction.

Investigators concerned with problems of prediction frequently attempt to identify independent variables which show a high correlation with the criterion and a low correlation with each other. If two variables have a fairly high correlation with the criterion and a low correlation with each other, both measure different aspects of the criterion and both will contribute substantially to prediction. If two variables have a high correlation with each other, they are measures of much the same thing, and the inclusion of both, instead of either one or the other, will contribute little to the prediction achieved.

EXERCISES

1 Given the correlations $r_{12} = .70$, $r_{13} = .50$, and $r_{23} = .60$, compute $r_{12.3}$. What percentage of the association between variables 1 and 2 results because of the effect of variable 3?

2 The mean and standard deviation of a criterion variable are $\overline{X}_1 = 24.56$ and $s_1 = 4.52$. The means and standard deviations for two predictor variables are $\overline{X}_2 = 36.48$, $\overline{X}_3 = 16.95$, and $s_2 = 5.49$, $s_3 = 3.66$. The correlations are $r_{12} = .70$, $r_{13} = .65$, and $r_{23} = .33$. Compute (a) the correlation between standard scores on the criterion and the sum of standard scores on the two predictors, (b) the correlation between raw scores on the criterion and the sum of raw scores on the two predictors, (c) the multiple regression equation in standard-score form, (d) the multiple regression equation in raw-score form, (e) the multiple correlation coefficient.

3 The following are intercorrelations between first-year university averages and five university entrance examinations. Means and standard deviations are also given:

	X_1	X_2	X_3	X_4	X_5	X_6	$\overline{X}_i$	s_i
X_1	1.00	.62	.55	.43	.38	.09	72.61	6.56
X_2	.62	1.00	.72	.55	.43	.49	62.50	4.28
X_3	.55	.72	1.00	.55	.36	.47	58.65	5.33
X_4	.43	.55	.55	1.00	.65	.50	65.80	5.77
X_5	.38	.43	.36	.65	1.00	.20	69.75	3.91
X_6	.09	.49	.47	.50	.20	1.00	71.80	4.45

Compute (a) the multiple regression equation in standard-score form, (b) the multiple regression equation in raw-score form, (c) the multiple correlation coefficient, (d) the multiple correlation coefficients obtained by successively dropping variables 6, 5, and 4.

AN INTRODUCTION TO FACTOR ANALYSIS

27.1 INTRODUCTION

Factor analysis is a multivariate statistical method which is used in the analysis of tables, or matrices, of correlation coefficients. These coefficients are usually, although not necessarily, product-moment correlation coefficients. In most applications of factor analysis the variables are psychological tests. The method is, however, quite general and can be applied to correlations between variables of any type, e.g., economic, physiological, meteorological, or physical. Direct inspection of any large matrix of correlation coefficients indicates immediately that no simple intuitive interpretation of the pattern of interrelations between the variables is possible. The method of factor analysis reduces the original set of variables to a smaller number of variables, called *factors*, which are amenable to interpretation. The information which a complex pattern of interrelations contains can, thereby, be understood. In multiple regression a distinction is made between a dependent variable, or variables, and a set of independent variables. Factor analysis is usually applied to data where no distinction between dependent and independent variables is possible. Concern is with the study of interdependencies and the discovery of structure among interdependencies.

Factor analysis as a method originated with C. Spearman. In a paper on the theory of intelligence, published in 1904, Spearman argued that "all branches of intellectual activity have in common one fundamental function (or group of functions), whereas the remaining or specific elements of the activity seem in every case to be wholly different from that in all others." Spearman analyzed tables of intercorrelation between psychological tests and purportedly was able to show that the intercorrelations could be accounted for in terms of one general factor, common to all tests, and factors which were specific, or unique to each test. This was known as the *theory of two factors*. The general factor was referred to as *g*. It is of interest to note that Spearman was primarily concerned with psychological theory and viewed the factor analysis of tables of intercorrelations as providing evidence to support his theory. Spearman's theory had neurophysiological aspects. He linked his general factor with energy, which was thought to serve the whole cortex or nervous system, and his specific factors with particular groups of neurons, which served particular kinds of operations.

Following Spearman, much work on factor analysis was done by Cyril Burt, Godfrey H. Thomson, Karl J. Holzinger, L. L. Thurstone, and others. The theory of two factors was the object of much criticism and was rapidly discarded, although the concept of a general factor common to many types of intellectual activity still persists. L. L. Thurstone in effect generalized the factor analytic method and invented multiple-factor analysis, which could be applied to data involving any number of factors.

Probably the most significant influence on the development of the factor analytic method, and its application, over the past fifteen years has been the electronic computer. Prior to the availability of the electronic computer the completion of a factor analysis was arithmetically very laborious. Many problems in factor analysis could not be properly investigated because of the enormous amount of computation required. The computer has not only extended the applications of factor analysis, but has led to the solution of problems which were hitherto arithmetically impractical.

The purpose of the present chapter is to explain in a simple way the nature of the ideas involved in factor analysis. No attempt is made to instruct the reader in computational procedures, since all such computation is now done by computer. The student who wishes to pursue the study of factor analysis in greater depth is referred to the comprehensive and up-to-date treatment of the subject by Harry H. Harman (1967). The present discussion follows Harman's statistical notation. This may assist the reader who wishes to pursue the topic beyond the elementary introduction presented here.

27.2 AN ILLUSTRATIVE ANALOGY

Some aspects of the nature of factor analysis can be readily understood by considering a simple analogy. Consider five urns, each one containing a certain proportion of red balls and a proportion of balls of another color. The proportions are as shown in Table 27.1. Most people would agree that Table 27.1 provides a convenient way of describing the composition of the balls in the five urns with respect to color.

Let us now suppose that samples consisting of pairs of balls are drawn from all possible pairs of urns in turn, and the number of agreements counted. For example, a ball is drawn from the first urn, another from the

Table 27.1   Proportion of balls of different colors in five urns

Urn	Red	Orange	Yellow	Green	Blue	Purple
1	.50	.50				
2	.60		.40			
3	.70			.30		
4	.80				.20	
5	.90					.10

second. If the two balls are the same with respect to color, this is an agreement. This process is continued until samples are drawn from all possible pairs of urns. We note, however, that the probability of drawing a red ball from the first urn is .50, the probability of drawing a red ball from the second urn is .60; consequently, using the multiplication theorem of probability, the expectation is that the proportion of agreements in drawing samples of pairs from the first and second urns is $.50 \times .60 = .30$. Similarly, the expected proportion of agreements in drawing samples of pairs from the first and third urns is $.50 \times .70 = .35$, from the first and fourth urns is $.50 \times .80 = .40$, and so on. The expected proportion of agreements in drawing samples of pairs from all pairs of urns is given in Table 27.2.

Let us now suppose that we are given only the information contained in Table 27.2. Is it possible to reconstruct the description of the urns contained in Table 27.1? This is directly analogous to the factor analytic problem. In factor analysis we are presented initially with a table of intercorrelations which is the analogue of the proportions of Table 27.2. These intercorrelations are measures, as it were, of what all possible pairs of variables have in common. Factor analysis enables us to derive from the table of intercorrelations a description of the variables which is analogous to the description of the composition of the urns shown in Table 27.1. How may the description as shown in Table 27.1 be obtained from the information presented in Table 27.2? The proportion of red balls in each of the five urns

Table 27.2   Expected proportion of agreements in drawing pairs of balls from all possible pairs of urns

	1	2	3	4	5
1					
2	.30				
3	.35	.42			
4	.40	.48	.56		
5	.45	.54	.63	.72	

may be denoted by p_1, p_2, p_3, p_4, and p_5. The proportion of orange, yellow, green, blue, and purple balls may be denoted by q_1, q_2, q_3, q_4, and q_5. Likewise the proportion of agreements in Table 27.2 may be represented by p_{12}, p_{13}, . . . , p_{45}. Also, we have the relations

$$p_1 p_2 = p_{12}$$
$$p_1 p_3 = p_{13}$$
$$p_2 p_3 = p_{23}$$

$$\cdot \quad \cdot$$
$$\cdot \quad \cdot$$
$$\cdot \quad \cdot$$

$$p_4 p_5 = p_{45}$$

If we multiply $p_1 p_2 = p_{12}$ by $p_1 p_3 = p_{13}$, divide by $p_2 p_3 = p_{23}$, and take the square root, we obtain

[27.1] $$p_1 = \sqrt{\frac{p_{12} p_{13}}{p_{23}}}$$

Similarly we can solve for p_2, p_3, p_4, and p_5. In general

[27.2] $$p_i = \sqrt{\frac{p_{ij} p_{ik}}{p_{jk}}}$$

This means that, given only the data in Table 27.2, the description of Table 27.1 can be obtained. For example,

$$p_1 = \sqrt{\frac{.30 \times .35}{.42}} = .50$$

Similarly $p_2 = .60$, $p_3 = .70$, $p_4 = .80$, and $p_5 = .90$.

Two points which relate to the present analogy should be noted. First, the statistical model used in obtaining the description of Table 27.1 from the information in Table 27.2 assumes that there is one factor only in common. While this model is clearly appropriate here, models involving other assumptions might be considered. Second, no information exists in Table 27.2 which would make possible an identification of factors. While the proportions in Table 27.1 can be obtained from Table 27.2, the procedure whereby this is done does not indicate that the common factor is red and the specific factors are orange, yellow, and so on. The procedure indicates only the presence of a factor common to all variables and factors specific to each variable. In any application of the factor analytic method the situation is the same. The analysis provides no information whereby the nature of the factor can be identified and named. The indentification of factors depends on considerations that are apart from the statistical analysis used.

27.3 BASIC EQUATIONS

The basic factor analytic model is that a score of individual i on variable j can be conceptualized as the weighted sum of scores on a smaller number of derived variables, called *factors*. This is a linear model, and may be expressed in standard-score form as follows:

[27.3] $z_{ji} = a_{j1}F_{1i} + a_{j2}F_{2i} + \cdots + a_{jm}F_{mi} + d_jU_{ji}$

The quantity z_{ji} is the standard score of individual i on variable j. F_{1i} is the standard score of individual i on the first common factor; F_{2i} is his standard score on the second common factor; and F_{mi} is his standard score on the mth common factor. The quantity U_{ji} is the standard score of individual i on what is called a *unique* factor; that is, a factor that is involved in a single variable only, in this case the variable j. The coefficients $a_{j1}, a_{j2}, \ldots, a_{jm}$ are factor loadings. These are weights which attach to the common-factor scores. The coefficient d_j is the weight which attaches to the unique-factor score. Factor analysis is concerned primarily with the determination of the coefficients, or loadings, $a_{j1}, a_{j2}, \ldots, a_{jm}$. It is not usually concerned with estimating the factor scores F_{ji}, although methods for such estimation exist.

The analogic relation between equation [27.3] and an ordinary multiple regression equation is of interest. A multiple regression equation in standard-score form may be written

[27.4] $z_1' = \beta_2 z_2 + \beta_3 z_3 + \cdots + \beta_m z_m$

where z_1' is a predicted standard score on the dependent variable; z_2, $z_3, \ldots, z_m$ are the standard scores on the independent variables; and $\beta_2, \beta_3, \ldots, \beta_m$ are weights so determined as to maximize the correlation between the dependent variable and the weighted sum of scores on the independent variables. What are the points of similarity and difference between a multiple regression equation in [27.4] and the factor analytic equation in [27.3]? First, the multiple regression equation states that a person's predicted score on a dependent variable may be conceptualized as the weighted sum of standard scores on a number of independent variables, or predictors. Similarly, the factor analytic equation states that a person's standard score on one variable may be conceptualized as the weighted sum of standard scores on a number of other variables, called *factors*. Second, both the regression coefficients, β_j, and the factor loadings, a_{ji}, are weights. In multiple regression, however, a distinction is made between a dependent variable and a set of independent variables. The values of β_j are determined in order to maximize the prediction of the former from the latter. In factor analysis no distinction is made between a dependent variable and a set of independent variables. This distinction is without meaning. The investigation begins with a set of interdependent variables which cannot be classified as dependent and independent. The values, a_{ji}, as will subsequently be seen, are determined by the structure, organization, or patterning that exists among the variables. Multiple regression is concerned with prediction, and equation [27.4] is a prediction equation. Factor analysis is concerned with the discovery and description of structure in complex patterns of relations. Third, it is worth noting that in a multiple regression equation the independent variables $z_2, z_3, \ldots, z_m$ are usually not independent of each other, but are correlated. In the factor analytic equation, with some sets of data the factors are defined in such a way as to be independent or uncorrelated. This means that the factor scores $F_{1i}, F_{2i}, \ldots, F_{mi}$ are uncorrelated with each other. With other sets of data correlated factors may be used. The F_{ji} are not independent but are correlated. When the factors are uncorrelated, they are said to be *orthogonal*. A method that leads to a set of uncorrelated factors is spoken of as an *orthogonal solution*. When the factors are correlated, they are said to be *oblique*. A method that leads to a set of correlated factors is spoken of as an *oblique solution*.

Equation [27.3] expresses a standard score z_{ji} as a linear function of weighted-factor scores. Alternatively, the model may be written in a form which expresses the variable z_j as a linear function of weighted factors, the factors being viewed as variables. Thus

[27.5] $z_j = a_{j1} F_1 + a_{j2}F_2 + \cdots + a_{jm}F_m + d_jU_j$

Here the $a_{j1}, a_{j2}, \ldots, a_{jm}$ are factor loadings and $F_1, F_2, \ldots, F_m$ are the factors as variables. For a set of n variables this model may be written as

[27.6]
$$
\begin{aligned}
z_1 &= a_{11}F_1 + a_{12}F_2 + \cdots + a_{1m}F_m + d_1U_1 \\
z_2 &= a_{21}F_1 + a_{22}F_2 + \cdots + a_{2m}F_m + d_2U_2 \\
&\quad \cdots \cdots \cdots \cdots \cdots \cdots \cdots \cdots \\
z_n &= a_{n1}F_1 + a_{n2}F_2 + \cdots + a_{nm}F_m + d_nU_n
\end{aligned}
$$

Such an arrangement is called a *factor pattern*. Usually a factor pattern is written with the factor loadings recorded in columns with the number of common factors shown at the top of the column. To illustrate, the following is a factor pattern for three factors and five variables. The value h_j^2 is

	Factors			
Variables	*I*	*II*	*III*	h_j^2
1	a_{11}	a_{12}	a_{13}	h_1^2
2	a_{21}	a_{22}	a_{23}	h_2^2
3	a_{31}	a_{32}	a_{33}	h_3^2
4	a_{41}	a_{42}	a_{43}	h_4^2
5	a_{51}	a_{52}	a_{53}	h_5^2

known as the *communality*, and is the sum of the squares of the common-factor loadings. The communality is discussed in the section to follow.

27.4 COMPONENTS OF VARIANCE

Equation [27.3] is in standard-score form. Thus the scores z_{ji} and the factor scores F_{ji} have zero mean and unit variance. By squaring both sides of equation [27.3], summing over N cases, dividing by N, and assuming that factor scores are uncorrelated, it can be readily shown that

[27.7] $\quad s_j^2 = 1 = a_{j1}^2 + a_{j2}^2 + \cdots + a_{jm}^2 + d_j^2$

This equation states that the total variance of a test j may be partitioned into additive parts. The component of variance due to the first factor is a_{j1}^2, that due to the second factor is a_{j2}^2, and so on. The a_{j1} are correlation coefficients between the variables and the factors. Using the variance interpretation of the correlation coefficient, the a_{ji}^2 may be interpreted as the proportion of the total variance that may be attributed to the different factors.

In the terminology of factor analysis different components of variance have been assigned different names. The *communality* of a variable, usually represented by the symbol h_j^2, is the sum of the squares of the common-factor loadings, a common factor being one that is common to more than one variable in the set. Thus

[27.8] $\quad h_j^2 = a_{j1}^2 + a_{j2}^2 + \cdots + a_{jm}^2$

The communality is that part of the variance that may be attributed to common factors. The part of the variance that is left over, and that cannot be attributed to common factors, is called the *uniqueness* and is represented by the symbol d_j^2. The uniqueness is sometimes partitioned into two components, *specificity*, b_j^2, and *error variance*, e_j^2. The specificity is that part of the total variance which is due to factors that are specific to the particular variable and not due to measurement error. Since all measurement involves error in some degree, in all applications of factor analysis part of the uniqueness will be due to measurement error. If the reliability coefficient for a particular variable is known, the error variance may be readily calculated from the formula $e_j^2 = 1 - r_{jj}$, where r_{jj} is a reliability coefficient for test j. The quantity e_j^2 is nothing other than the error variance associated with a single test score in standard-score form.

27.5 REPRODUCING THE CORRELATION COEFFICIENTS

A factor analysis begins with a table of correlation coefficients between variables. The analysis represents the variables as the weighted sum of factors. What relation exists between the factors and the original correlation coefficients? How may the correlation coefficient be accounted for, as it were, by the common-factor loadings? Or, stated in different terms, how good a fit is the linear model to the observed correlations?

The correlation between two variables in standard-score form is given by

[27.9] $\quad r_{jk} = \dfrac{\displaystyle\sum_{i}^{N} z_{ji} z_{ki}}{N - 1}$

We may write the following:

[27.10]
$$z_{ji} = a_{j1}F_{1i} + a_{j2}F_{2i} + \cdots + a_{jm}F_{mi} + d_jU_{ji}$$
$$z_{ki} = a_{k1}F_{1i} + a_{k2}F_{2i} + \cdots + a_{km}F_{mi} + d_kU_{ki}$$

These two equations, which are also in standard-score form, may be multiplied together, summed over N individuals, and divided by $N-1$. The common factors are uncorrelated with each other. The unique factors are uncorrelated with each other and with the common factors. Given these simplifying assumptions, the following result is obtained:

[27.11] $r'_{jk} = a_{j1}a_{k1} + a_{j2}a_{k2} + \cdots + a_{jm}a_{km}$

This equation, which applies only to uncorrelated factors, shows how the correlation coefficient between two variables may be reproduced from the factor loadings. The quantity $a_{j1}a_{k1}$ is viewed as the contribution of the first factor to the correlation, $a_{j2}a_{k2}$ as the contribution of the second factor to the correlation, and so on. We have used r'_{jk} to refer to the reproduced correlation coefficient to distinguish it from r_{jk}, the observed correlation. The reproduced correlation coefficient between two variables is the sum of the products of the common-factor loadings. With real data, because of sampling or other kinds of error, the correlation reproduced from the factor loadings will differ in some degree from the observed correlations. The differences between the observed and the reproduced correlations are known as *residual correlations*, $\bar{r}_{jk} = r_{jk} - r'_{jk}$. The magnitude of the residual correlations indicates the extent to which the factors account for the observed correlations, or how good a fit the linear model is to the observed correlations.

27.6 THE GEOMETRY OF FACTOR ANALYSIS

Many problems in factor analysis are illuminated by conceptualizing them in geometrical terms. We begin by considering the geometrical representation of the correlation coefficient. Variable j may be represented by a vector of length h_j, and variable k by a vector of length h_k. Denote the angle between the two vectors by ϕ_{jk}. It can be shown that the correlation coefficient between two variables is equal to the length of the two vectors multiplied by the angular separation between them; that is,

[27.12] $r_{jk} = h_jh_k \cos \phi_{jk}$

For a simple proof of this result see Thurstone (1947). If the vectors are of unit length, $h_j = h_k = 1$, and the correlation is simply the cosine of the angle between the vectors. Fig. 27.1 shows the geometrical representation of correlation coefficients of different magnitudes.

For vectors of unit length, a correlation of zero is represented by two vectors at right angles to each other, a correlation of .707 by two vectors at a 45° angle to each other, and a correlation of $-.500$ by two vectors with an angular separation of 120°. Since the cosine of a zero angle is 1.00, a correlation of 1.00 is represented by two vectors that coincide.

Fig. 27.1   **Geometrical representation of correlation coefficients.**

The geometrical representation of tables of three variables may be considered. The following are three tables of correlations between three variables.

	1	2	3
1			
2	.000		
3	.000	.000	

(a)

	1	2	3
1			
2	.707		
3	.707	.707	

(b)

	1	2	3
1			
2	.000		
3	.707	.500	

(c)

The geometrical representations of these three tables of correlations, assuming a vector of unit length, are shown in Fig. 27.2.

In Fig. 27.2a all correlations are zero and the three vectors are at right angles to each other. In Fig. 27.2b the three correlations are .707 and the three vectors are at a 45° angle to each other. In Fig. 27.2c the three vectors have different angular separations. The examples of Fig. 27.2 must be viewed as being in a space of three dimensions.

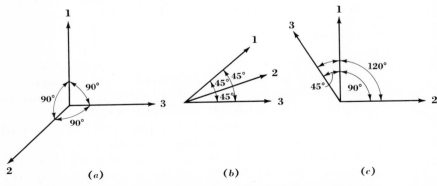

Fig. 27.2   **Geometrical representation of three tables of correlation coefficients.**

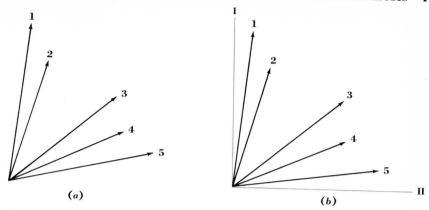

Fig. 27.3   Configuration of vectors shown without and with reference axes.

The geometrical representation of a table of correlations of more than three variables is a simple extension of the above. The vector model is a sheaf of vectors of varying lengths and angular separations. The lengths of the vectors, h_1, h_2, . . . , h_m, are the square roots of the communalities. The sheaf of vectors is commonly known as a *configuration*. A vector model for any substantial number of variables must ordinarily be conceptualized as existing in a multidimensional space.

Geometrically the factorization of a table of correlations involves inserting a set of references axes in the configuration of vectors and describing the terminal points of the vectors in relation to the reference axes. The projections of the terminal points of the vectors on the reference axes are the factor loadings. If the factors are uncorrelated, the reference axes are at right angles to each other. To illustrate, the following is a table of correlations between five variables.

	1	2	3	4	5
1					
2	.65				
3	.51	.47			
4	.33	.33	.51		
5	.17	.23	.53	.51	

The configuration of vectors corresponding to this table of correlations is shown in Fig. 27.3a. Figure 27.3b shows two orthogonal reference axes inserted in the configuration of vectors. The projections of the terminal points of the vectors on the reference axes are the factor loadings for the five variables. These are as follows:

	I	*II*	h^2
1	.90	.10	.82
2	.70	.20	.53
3	.50	.60	.61
4	.30	.60	.45
5	.10	.80	.65

The communalities, which are the sums of the squares of the factor loadings, are shown. The square roots of the communalities correspond to the lengths of the vectors shown in Fig. 27.3*a*.

A variety of methods exist for inserting reference axes in a configuration of vectors and thereby arriving at a factor matrix which reproduces approximately the correlation matrix. Quite clearly an indefinitely large number of solutions to this problem exist, since the reference axes could be located anywhere at all. In Fig. 27.3*b* the location of the reference axes is arbitrary. The reference axes could have been located elsewhere and a different factor matrix obtained. This circumstance has led to a two-stage process in factor analysis. The first stage involves locating a reference axis in the configuration of vectors by one of a number of methods which meet certain statistical criteria. A factor matrix is thereby obtained. Such solutions are referred to by Harman (1967) as *direct solutions*. The number of factors which might prove of interest is determined at the direct-solution stage. Although such solutions may have interesting statistical properties, the factors obtained are not viewed as amenable to interpretation. If the variables are psychological variables, no psychological meaning can ordinarily be attached to them. The second stage involves rotating the reference axes to a new position such that the factors can be interpreted. This has been known as the *rotational problem* in factor analysis. Such solutions are referred to by Harman as *derived solutions*. Although a variety of derived solutions exist, all such solutions involve locating the reference axes in a position which is determined by the structural properties of the configuration of vectors. This concept is discussed in detail in Sec. 27.9.

27.7 THE COMMUNALITIES

A factor analysis begins with a matrix of correlations. What are the appropriate values to insert as diagonal elements in the matrix of correlations? Most factor analytic methods assume that the appropriate diagonal elements are the communalities, h_j^2, the sums of the squares of the common-factor loadings. Initially these are, however, unknown and must be estimated from the data. A variety of methods for estimating communalities have been used. One very arbitrary method takes the largest correlation in each row or column of the correlation matrix as an estimate. Another method uses the average correlation of each variable with every other variable. One of the best estimates of the communality, which has much to commend it, is the square multiple correlation, R_j^2, of every variable with each of the remaining variables. The squared multiple correlations, R_1^2, R_2^2, . . . , R_n^2, of each variable with the $n - 1$ remaining variables can readily be obtained with an electronic computer.

27.8   PRINCIPAL-FACTOR SOLUTION

The most commonly used *direct* method of factoring in use at the present time is the principal-factor solution. This method was developed by Hotelling in 1933. Because of the laborious nature of the arithmetical calculations involved, it was not widely used until electronic computers became available.

The geometrical model for the principal-factor solution is of interest. Measurements on two variables for a sample of N members can be plotted as points in relation to two orthogonal reference axes. The result is the usual scatter diagram. If a correlation exists between the two variables, the arrangement of points will be elliposidal in form. Given more than two variables, say n, the points may be plotted with reference to n reference axes. The result is an elliposidal swarm of points. The principal factors correspond to the principal axes of this ellipsoid. Questions may be raised regarding the dimensionality of the ellipsoid. If 1's are inserted in the diagonals of the correlation matrix prior to factoring, the ellipsoid will have n principal axes, and n components will result. If estimates of the communalities are placed in the diagonals, m factors will result, where m is ordinarily less than n.

The identification of the principal axes of an ellipsoidal swarm of points can be shown to reduce algebraically to the identification of a set of factors in decreasing order of their contribition to the total communality, assuming that estimates of the communalities are used in the diagonal of the correlation matrix.

The brief discussion here attempts to present the algebraic rationale of a solution whereby the factors are obtained successively. In the computational procedures which are ordinarily employed in practice, the factors are obtained simultaneously. The solution amounts to the identification of a factor F, which maximizes the quantity

[27.13] $$V_1 = a_{11}^2 + a_{21}^2 + \cdots + a_{n1}^2$$

where V_1 is the contribution of F_1 to the total communality. The maximization of V_1 is subject to the condition that the sum of the products of the factor loadings will reproduce the correlation coefficients. Following the calculation of F_1 a matrix of first-factor residuals is obtained. The contribution of the first factor to r_{12} is $a_{11}a_{21}$, and the residual correlation is $r_{12} - a_{11}a_{21}$. In general, the first-factor residual correlations are given by $r_{jk} - a_{j1}a_{k1}$. A table of residual correlations results. This is a table of correlations with the influence of the first factor removed and with residual communalities in the diagonal. The second factor, F_2, is obtained by maximizing

[**27.14**] $V_2 = a_{12}^2 + a_{22}^2 + \cdots + a_{n2}^2$

where V_2 is the contribution of the second factor to the communality. Again a table of residuals is obtained which shows those parts of the correlation that remain after the second factor has been removed. This procedure is continued until the part of the total communality that remains can be viewed as trivial, whatever this might mean. No rigorous method exists for determining the number of factors. A rule commonly employed in the principal-factor solution is that the contribution of the factor to the total communality should not be less than 1. This rule is, of course, arbitrary.

In this section the principal-factor solution only has been briefly discussed. Another direct method of factoring which has been very widely used is the centroid solution, developed by L. L. Thurstone. Many examples of factor analysis reported in the literature use the centroid solution. The centroid solution involves placing the first reference axis through the centroid of the configuration of vectors; obtaining a table of residual correlations, which are subject to certain adjustments; placing the second factor through the centroid corresponding to the table of residual correlations; and continuing the process until the magnitude of the residuals can be considered inconsequential. The centroid solution is an approximation to the principal-axes solution, and was used prior to the general availability of computers because it involved less arithmetical labor than the principal-factor solution, which is to be preferred. Another method of factoring which will undoubtedly prove to be of increasing interest is the maximum-likelihood solution developed by D. N. Lawley (1940). See also Lawley and Maxwell (1963). This method has never been widely used because of difficulties of computation, although the electronic computer now makes its application practical.

27.9 ILLUSTRATIVE EXAMPLE OF PRINCIPAL-FACTOR SOLUTION

The principal-factor solution will be illustrated using data originally gathered by Holzinger and Swineford (1939). These data are reproduced from Harman (1967).

Twenty-four psychological tests were administered to a group of 145 children in grades seven and eight in a suburb of Chicago. The 24 psychological tests, together with means, standard deviations, and reliability coefficients, are shown in Table 27.3. The intercorrelations between these 24 psychological tests are shown in Table 27.4.

Note that correlation coefficients below the main diagonal are shown. The table is, of course, symmetric about the main diagonal.

The data of Table 27.4 were analyzed using the principal-factor solution. In applying this method squared multiple correlations (SMCs) have been

Table 27.3   Basic statistics for twenty-four psychological tests*

Test X_j	Mean $\overline{X}_j$	Standard deviation s_j	Reliability coefficient r_{ii}
1. Visual Perception	29.60	6.90	.756
2. Cubes	24.84	4.50	.568
3. Paper Form Board	15.65	3.07	.544
4. Flags	36.31	8.38	.922
5. General Information	44.92	11.75	.808
6. Paragraph Comprehension	9.95	3.36	.651
7. Sentence Completion	18.79	4.63	.754
8. Word Classification	28.18	5.34	.680
9. Word Meaning	17.24	7.89	.870
10. Addition	90.16	23.60	.952
11. Code	68.41	16.84	.712
12. Counting Dots	109.83	21.04	.937
13. Straight-Curved Capitals	191.81	37.03	.889
14. Word Recognition	176.14	10.72	.648
15. Number Recognition	89.45	7.57	.507
16. Figure Recognition	103.43	6.74	.600
17. Object-Number	7.15	4.57	.725
18. Number-Figure	9.44	4.49	.610
19. Figure-Word	15.24	3.58	.569
20. Deduction	30.38	19.76	.649
21. Numerical Puzzles	14.46	4.82	.784
22. Problem Reasoning	27.73	9.77	.787
23. Series Completion	18.82	9.35	.931
24. Arithmetic Problems	25.83	4.70	.836

* Reproduced from Harry H. Harman's *Modern factor analysis*, 2d ed., University of Chicago Press, Chicago, 1967, with kind permission of the author and publisher.

used as estimates of the communalities. These are the squared multiple correlations of each variable with the $n-1$ remaining variables. Table 27.5 shows the principal-factor solution obtained for these data. Five factors have been extracted. The body of the table shows the factor loadings for the 24 tests on the five factors. The rows at the bottom of the table show the contribution of each factor to the total communality. The contribution of each factor is simply the sum of the squares of the factor loadings. The per cent of the total communality contributed by each factor is also shown. To the right of Table 27.5 the original SMC communalities are shown together with the communalities obtained following the calculation by summing the squares of the factor loadings for the individual rows.

The reader should note that the contribution of the factors decreases from the first to the fifth factor. The first factor accounts for 64.2 per cent of

Table 27.4

Intercorrelations of twenty-four psychological tests for 145 children*

Test: j	1	2	3	4	5	6	7	8	9	10	11	12	13	14	15	16	17	18	19	20	21	22	23	24
1	—																							
2	.318	—																						
3	.403	.317	—																					
4	.468	.230	.305	—																				
5	.321	.285	.247	.227	—																			
6	.335	.234	.268	.327	.622	—																		
7	.304	.157	.223	.335	.656	.722	—																	
8	.332	.157	.382	.391	.578	.527	.619	—																
9	.326	.195	.184	.325	.723	.714	.685	.532	—															
10	.116	.057	-.075	.099	.311	.203	.246	.285	.170	—														
11	.308	.150	.091	.110	.344	.353	.232	.300	.280	.484	—													
12	.314	.145	.140	.160	.215	.095	.181	.271	.113	.585	.428	—												
13	.489	.239	.321	.327	.344	.309	.345	.395	.280	.408	.535	.512	—											
14	.125	.103	.177	.066	.280	.292	.236	.252	.260	.172	.350	.131	.195	—										
15	.238	.131	.065	.127	.229	.251	.172	.175	.248	.154	.240	.173	.139	.370	—									
16	.414	.272	.263	.322	.187	.291	.180	.296	.242	.124	.314	.119	.281	.412	.325	—								
17	.176	.005	.177	.187	.208	.273	.228	.255	.274	.289	.362	.278	.194	.341	.345	.324	—							
18	.368	.255	.211	.251	.263	.167	.159	.250	.208	.317	.350	.349	.323	.201	.334	.344	.448	—						
19	.270	.112	.312	.137	.190	.251	.226	.274	.274	.190	.290	.110	.263	.206	.192	.258	.324	.358	—					
20	.365	.292	.297	.339	.398	.435	.451	.427	.446	.173	.202	.246	.241	.302	.272	.388	.262	.301	.167	—				
21	.369	.306	.165	.349	.318	.263	.314	.362	.266	.405	.399	.355	.425	.183	.232	.348	.173	.357	.331	.413	—			
22	.413	.232	.250	.380	.441	.386	.396	.357	.483	.160	.304	.193	.279	.243	.246	.283	.273	.317	.342	.463	.374	—		
23	.474	.348	.383	.335	.435	.431	.405	.501	.504	.262	.251	.350	.382	.242	.256	.360	.287	.272	.303	.509	.451	.503	—	
24	.282	.211	.203	.248	.420	.433	.437	.388	.424	.531	.412	.414	.358	.304	.165	.262	.326	.405	.374	.366	.448	.375	.434	—

* Reproduced, with minor modification, from Harry H. Harman's *Modern factor analysis*, 2d ed., University of Chicago Press, Chicago, 1967, with kind permission of the author and publisher.

Table 27.5   Principal-factor solution for twenty-four psychological tests (communality estimates: SMCs)*

Test	Common factors					Communality	
	P_1	P_2	P_3	P_4	P_5	Original	Calculated
1	.595	−.039	.369	−.184	−.073	.511	.531
2	.374	.026	.270	−.147	.121	.300	.250
3	.433	.115	.396	−.112	−.276	.440	.446
4	.501	.108	.290	−.178	.044	.409	.380
5	.701	.312	−.273	−.050	.003	.673	.666
6	.683	.404	−.213	.067	−.102	.677	.690
7	.676	.412	−.284	−.082	−.046	.684	.716
8	.680	.206	−.088	−.115	−.118	.564	.540
9	.690	.446	−.212	.076	.036	.713	.727
10	.456	−.469	−.446	−.128	.105	.579	.654
11	.589	−.372	−.198	.076	−.185	.541	.565
12	.448	−.491	−.154	−.263	.044	.537	.537
13	.590	−.268	.018	−.300	−.255	.539	.575
14	.435	−.063	−.012	.418	−.057	.358	.371
15	.390	−.102	.055	.362	.101	.293	.307
16	.512	−.098	.325	.259	.006	.429	.444
17	.471	−.212	−.036	.388	−.087	.412	.426
18	.521	−.331	.118	.145	.028	.443	.417
19	.450	−.115	.110	.167	−.178	.367	.287
20	.623	.135	.142	.049	.252	.464	.492
21	.596	−.220	.076	−.140	.204	.473	.471
22	.600	.103	.138	.053	.142	.449	.413
23	.685	.063	.160	−.096	.154	.561	.532
24	.635	−.169	−.192	−.009	.079	.527	.475
V_p	7.665	1.672	1.208	.920	.447	11.943	11.912
$100V_p/11.943$	64.2	14.0	10.1	7.7	3.7	100.	99.7

* Reproduced from Harry H. Harman's *Modern factor analysis*, 2d ed., University of Chicago Press, Chicago, 1967, with kind permission of the author and publisher.

the total communality, the second for 14.0 per cent, and the fifth for only 3.7 per cent. Note also that all loadings for the first factor are positive, whereas subsequent factors contain both positive and negative loadings.

The data of Table 27.5 are not amenable to any psychological interpretation. Before any meaning can be assigned to factors, the factors obtained by the principal-factor solution must be subject to further analysis.

27.10   THE CONCEPT OF STRUCTURE

The factors obtained by the principal-factor or other direct solutions are not ordinarily viewed as amenable to interpretation. A rotation of the reference axes is made to a new position and derived factors are obtained. These can be assigned meaning. The need for the rotation of the reference axes has not always been fully understood although the reasons for this are quite straightforward.

As discussed in Sec. 27.6 a table of correlations can be represented geometrically as a configuration of vectors. It is possible to conceptualize a random configuration of vectors, that is, a configuration where the vectors are arranged at random in relation to each other. Such a configuration is represented in Fig. 27.4a. Just as a random arrangement of points in a space of two dimensions has no structure or organization, similarly a random arrangement of vectors in a space of m dimensions has no structure or organization. The configurations of vectors that correspond to most tables

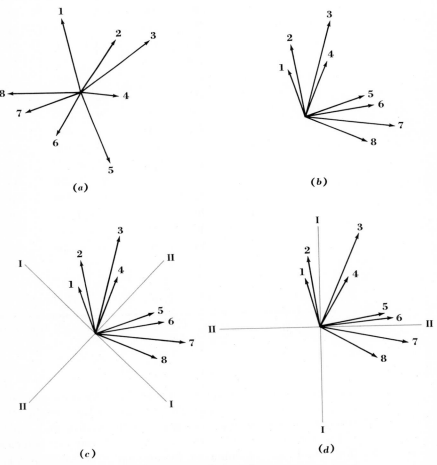

Fig. 27.4   Configurations of vectors with and without reference axes.

of correlations are not random, but they have certain structural properties. In Fig. 27.4b the vectors are clustered in two groups of four vectors each, the two clusters being roughly orthogonal to each other. The arrangement of vectors is not random. It has structure. Here structure, or organization, is viewed as progression from a state of randomness, and randomness as the obverse regression. The configuration of vectors in Fig. 27.4b can be described by inserting reference axes in an indefinitely large number of possible locations. One such location, obtained possibly by the principal-factor solution, is shown in Fig. 27.4c. While this location provides a description of the terminal points of the vectors, intuitively it does not communicate readily to our understanding the structural properties of the configuration. By rotating the reference axes to the position shown in Fig. 27.4d, a factor matrix is obtained which clearly shows the structure of the configuration. In Fig. 27.4d variables 1, 2, 3, and 4 have high loadings in factor I and small or zero loadings in factor II. Similarly variables 5, 6, 7, and 8 have high loadings in factor II and small or zero loadings in factor I. Most observers would agree that intuitively the reference axes in Fig. 27.4d provide a more meaningful description of the configuration than those in Fig. 27.4c.

The purpose of factor analysis is to discover, and satisfactorily describe, the structural properties of the matrix of correlations, or the corresponding configuration of vectors. This purpose is, it is hoped, achieved by a rotation of the reference axes. The reader should note that the new location of the reference axes is determined, or compelled, by the structural properties of the configuration. The situation here is directly analogous to fitting a straight line to a set of points by the method of least squares. The precise location of the line, and the fact that it is a straight line rather than a curved line, is determined by the structural properties of the set of points.

For many years the rotation of axes was done using laborious graphical methods. These methods have been replaced by analytical methods which maximize or minimize certain statistical criteria.

27.11 ANALYTICAL METHODS OF ROTATION

Analytical methods of rotation have been developed by Carrol (1953), Ferguson (1954), Neuhaus and Wrigley (1954), Saunders (1953), and Kaiser (1958). All of these methods involve, with modification, essentially the same idea. This idea, as stated by Ferguson (1954), is as follows:

Consider a point p, plotted in relation to two orthogonal reference axes. These axes may be rotated into an indefinitely large number of positions, yielding, thereby, an indefinitely large number of descriptions of the location of the point. Which position provides the most parsimonious description? Intuitively it appears that the most parsimonious description will result when one or other of the axes passes through the point, the point being, thereby, described as a distance measured from an origin. Likewise as one or other of the reference axes is rotated in the direction of the point the product of the two coordinates grows smaller. This product is a minimum and equal to zero when one of the axes passes through the point, and is a maximum when a line from the origin and passing through the point is at a 45° angle to both axes. These circumstances suggest that some function of the product of the two coordinates might be used as a measure of the amount of parsimony associated with the description of the point. Note, further, that this product is an area, and is a measure of the departure of the point from the reference frame. Consider now any set of points with both positive and negative coordinates, the usual factor case. Here the sum of products is inappropriate because a zero sum can result when the sums of negative and positive products are equal. Here following the usual statistical convention we may use the sum of squares of products of coordinates, or some related quantity, as a measure of parsimony.

The quantity under consideration for two factors and four tests is illustrated in Fig. 27.5. Here a measure of the departure of the terminal points of the vectors from the reference frame is the sum of the squares of the areas

$$(a_{11}a_{12})^2 + (a_{21}a_{22})^2 + (a_{31}a_{32})^2 + (a_{41}a_{42})^2$$

For m factors and n variables we consider the $m(m-1)/2$ sum of products of pairs of coordinates, which may be written as

[27.15]
$$\sum_{p<q=1}^{m} \sum_{j=1}^{n} (a_{jp}a_{jq})^2$$

The a_{jp} and a_{jq} are unrotated factor loadings. If b_{jp} and b_{jq} are now taken to denote factor loadings following rotation, the problem of rotation becomes one of locating the reference axes in such a position that the quantity

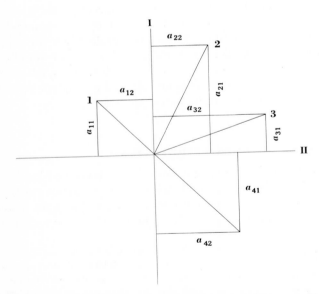

Fig. 27.5   Departure of points from reference frame. The sum of squares of the areas $a_{11}a_{12}$, $a_{21}a_{22}$, $a_{31}a_{32}$, and $a_{41}a_{42}$ is a measure of the departure of points from the reference frame.

[**27.16**] $\displaystyle\sum_{p<q=1}^{m}\sum_{j=1}^{n}(b_{jp}b_{jq})^2$

is a *minimum*. By some simple algebra it can be shown that *minimizing* this quantity is the same thing as maximizing

[**27.17**] $\displaystyle Q=\sum_{p=1}^{m}\sum_{j=1}^{n}b_{jp}{}^{4}$

Thus the problem becomes one of locating the reference axes in a position such that the sum of all factor loadings in the factor matrix raised to the fourth power is a maximum. The statistic Q is a measure of parsimony. It is a measure of the departure of the terminal points of the vectors from the reference frame. It is a measure in the least square sense of the goodness of fit of the reference frame to the configuration. One difficulty with the Q criterion relates to the weighting of variables. This method weights each variable in a manner proportional to the square of its communality. This means that a variable with a communality of .80 will exert four times the weight in determining the final solution that a test with a communality of .40 will exert. A possible solution here is to use "normalized factor" loadings, that is, factor loadings adjusted in such a way that the sum of their squares for each variable, the communality, is equal to 1. In the geometrical model this amounts to extending all vectors to unit length. The rotation of normalized-factor loadings assigns equal weight to each of the variables.

All analytical methods of rotation in common use involve, with modification, the essential ideas described above. Probably the method most widely accepted at present is a modification proposed by Kaiser (1958), who was led to consider the simplicity of a factor matrix. He defined the simplicity of any factor as the variance of its squared factor loadings and the simplicity of the factor matrix as the sum of these variances over all factors. This quantity, which is known as the *raw varimax criterion*, is as follows:

[**27.18**] $\displaystyle s^2=\frac{1}{n}\sum_{p=1}^{m}\sum_{q=1}^{n}b_{jp}{}^{4}-\frac{1}{n^2}\sum_{p=1}^{m}\left(\sum_{j=1}^{n}b_{jp}{}^{2}\right)^{2}$

This criterion weights each variable in a manner proportional to the square of the communality. Since empirically this weighting system did not yield a satisfactory result, Kaiser proposed a "normal" varimax criterion which may be written as follows:

[**27.19**] $\displaystyle V=n\sum_{p=1}^{m}\sum_{j=1}^{n}(b_{jp}/h_j)^4-\sum_{p=1}^{m}\left(\sum_{j=1}^{n}b_{jp}{}^{2}/h_j{}^{2}\right)^{2}$

Here all factors are normalized prior to rotation. All communalities are 1, all vectors are of unit length, and all variables receive equal weight. The normal varimax criterion is commonly used in factor analytic studies at the present time.

**27.12 ILLUSTRATIVE EXAMPLE OF THE NORMAL
VARIMAX METHOD OF ROTATION**

The normal varimax method of rotation will be illustrated with reference to
the principal-factor solution for 24 psychological tests as shown in Table
27.5. The first four factors of Table 27.5 were rotated in order to maximize
the normal varimax criterion of equation [27.19]. The fifth factor of Table
27.5 was excluded from this analysis because of its very small contribution
to the total communality.

Table 27.6 shows the rotated factors. All factors greater than .400 are

Table 27.6 Normal varimax solution for twenty-four psychological tests

Test	Verbal M_1	Speed M_2	Deduction M_3	Memory M_4	Calculated communality
1	.152	.204	**.658**	.165	.526
2	.111	.089	**.458**	.069	.235
3	.144	−.013	**.579**	.118	.370
4	.231	.090	**.557**	.080	.378
5	**.745**	.220	.199	.149	.666
6	**.764**	.078	.199	.225	.680
7	**.803**	.155	.196	.084	.714
8	**.577**	.233	.345	.137	.526
9	**.795**	.048	.203	.224	.726
10	.172	**.762**	−.090	.157	.643
11	.191	**.572**	.102	.396	.530
12	.018	**.704**	.178	.083	.535
13	.168	**.552**	.413	.078	.510
14	.207	.081	.060	**.561**	.368
15	.125	.082	.105	**.513**	.297
16	.073	.059	.414	**.514**	.444
17	.142	.222	.061	**.588**	.419
18	.026	.356	.285	**.455**	.416
19	.130	.168	.250	.385	.256
20	.384	.100	**.426**	.300	.429
21	.171	**.437**	.405	.212	.429
22	.351	.113	**.407**	.301	.392
23	.368	.225	**.521**	.225	.508
24	.351	**.479**	.176	.292	.469
V_p	3.613	2.616	2.934	2.300	11.462
$100V_p/11.462$	31.522	22.823	25.598	20.066	100.

shown in boldface type. The contributions of each factor to the total communality are shown at the bottom of the table. The communalities calculated for the four factors, instead of the original five, are shown to the right of the table.

The rotated factors of Table 27.6 are amenable to psychological interpretation. Inspection of the tests used in this study suggests that the first factor is a verbal factor. Tests 5, 6, 7, 8, and 9 have high loadings on this factor. These tests are very largely tests of verbal ability. Inspection of the tests suggests that the second factor is a speed factor. The tests with the highest loadings in this factor are tests 10, 11, 12, and 13. The third factor is a deduction factor, and the tests loading highest in this factor are tests 1, 3, 4, and 23. The fourth factor has its highest loadings in tests 14, 15, 16, 17, and 18, and may be identified as a memory factor.

EXERCISES 1 If there is one general factor only, the general factor loading for variable i may be calculated by the formula:

$$g_i = \sqrt{\frac{r_{ij}r_{ik}}{r_{jk}}}$$

This formula is seen to be directly analogous to formula [27.2]. Calculate the (a) general and (b) specific factor loadings from the following table of intercorrelations.

	1	2	3	4	5
1					
2	.72				
3	.56	.63			
4	.16	.18	.14		
5	.18	.18	.14	.04	

2 Two variables, z_1 and z_2, may be expressed as linear functions of weighted factors as follows:

$$z_1 = .40F_1 + .60F_2 - .05F_3 + .10F_4 + .78U_1$$
$$z_2 = .05F_1 + .70F_2 - .10F_3 + .60F_4 \cdot \cdot \cdot + .37U_2$$

Calculate the correlation between z_1 and z_2.

3   The following are factor loadings on two-factors for six variables:

Variable	F_1	F_2
1	.9	.3
2	.8	.3
3	.7	.3
4	.6	.6
5	.5	.6
6	.4	.6

Calculate (a) the communalities, (b) the uniqueness for each variable, (c) the contribution of each factor to the total communality, (d) the per cent contribution of each factor to the total communality.

4   For the data of Exercise 3 above, write down the table of reproduced correlation coefficients.

5   The reliability coefficient for variable 6 in Exercise 3 is .80. For this variable calculate that part of the variance that is error variance.

6   In a normalized configuration the vectors are of unit length. Normalized factor loadings are obtained by dividing the factor loadings for any variable by the square root of the communality. Calculate the normalized factor loadings for the data of Exercise 3.

7   The loadings for a particular variable on two factors, F_1 and F_2, are, respectively, .3 and .5. These loadings may be plotted as a point with reference to two orthogonal reference axes. Calculate the distance of the point from the origin.

8   The following are loadings for six variables on two factors:

Variable	F_1	F_2
1	.2	.2
2	.3	.3
3	.4	.4
4	.2	$-.2$
5	.3	$-.3$
6	.4	$-.4$

Rotate these factors to what appears to be the most parsimonious position. Calculate the rotated factor loadings.

ANSWERS TO EXERCISES

CHAPTER 1

2 **a.** ratio   **b.** ratio   **c.** ordinal   **d.** nominal
    e. nominal   **f.** interval   **g.** ratio   **h.** nominal
    i. interval   **j.** ordinal

3 **a.** $\displaystyle\sum_{i=1}^{15} X_i$ **b.** $\displaystyle\sum_{i=1}^{N} Y_i$ **c.** $\displaystyle\sum_{i=1}^{7} X_i + \sum_{i=1}^{7} Y_i$

    d. $\displaystyle\sum_{i=1}^{N} X_i Y_i$ **e.** $\displaystyle\sum_{i=1}^{N} X_i^3 Y_i$ **f.** $\displaystyle\sum_{i=1}^{8} X_i + 8c$

    g. $\displaystyle c\sum_{i=1}^{25} X_i$ **h.** $\displaystyle\frac{1}{c}\sum_{i=1}^{N} X_i$ **i.** $\displaystyle c\sum_{i=1}^{N} X_i^2 Y_i$

4 **a.** $X_1 + X_2$ **b.** $X_1 Y_1 + X_2 Y_2 + X_3 Y_3$
    c. $(X_1 + Y_1) + (X_2 + Y_2) + (X_3 + Y_3) + (X_4 + Y_4) + (X_5 + Y_5)$
    d. $cX_1^2 Y_1 + cX_2^2 Y_2 + cX_3^2 Y_3$
    e. $(X_1 + c) + (X_2 + c) + (X_3 + c) + (X_4 + c)$
    f. $X_1/c + X_2/c + X_3/c + X_4/c + X_5/c + cY_1 + cY_2 + cY_3 + cY_4 + cY_5$

5 $\displaystyle\sum_{i=1}^{N} (X_i + c)^2 = \sum_{i=1}^{N} (X_i^2 + 2X_i c + c^2) = \sum_{i=1}^{N} X_i^2 + 2c\sum_{i=1}^{N} X_i + Nc^2$

6 **a.** 139   **b.** 120   **c.** 14   **d.** 40   **e.** 86
    f. -14 **g.** 1,955   **h.** 94   **i.** 11.300

7 **a.** false   **b.** false   **c.** true   **d.** true

8 5,050

9 **a.** 30   **b.** 560

10 **a.** 24   **b.** Nc **c.** $2N$

CHAPTER 2

1

Class Interval	Frequency	Cumulative Frequency
95–99	2	40
90–94	2	38
85–89	4	36
80–84	6	32
75–79	5	26
70–74	4	21
65–69	5	17
60–64	2	12
55–59	2	10
50–54	4	8
45–49	1	4
40–44	2	3
35–39	1	1
Total	40	

2

Interval	Exact Limits	Mid-points
95–99	94.5–99.5	97
90–94	89.5–94.5	92
85–89	84.5–89.5	87
80–84	79.5–84.5	82
75–79	74.5–79.5	77
70–74	69.5–74.5	72
65–69	64.5–69.5	67
60–64	59.5–64.5	62
55–59	54.5–59.5	57
50–54	49.5–54.5	52
45–49	44.5–49.5	47
40–44	39.5–44.5	42
35–39	34.5–39.5	37

3

	Class Interval	Exact Limits	Mid-points
a.	85–89	84.5–89.5	87
	80–84	79.5–84.5	82
	75–79	74.5–79.5	77
	etc.	etc.	etc.
b.	135–137	134.5–137.5	136.0
	132–134	131.5–134.5	133.0
	129–131	128.5–131.5	130.0
	etc.	etc.	etc.
c.	850–899	849.5–899.5	874.50
	800–849	799.5–849.5	824.50
	750–799	749.5–799.5	774.50
	etc.	etc.	etc.
d.	.800–.849	.7995–.8495	.8245
	.750–.799	.7495–.7995	.7745
	.700–.749	.6995–.7495	.7245
	etc.	etc.	etc.

4 *Class Interval*	*Frequency*	*Cumulative Frequency*	*Cumulative % Frequency*
17–18	3	50	100
15–16	7	47	94
13–14	1	40	80
11–12	5	39	78
9–10	6	34	68
7–8	6	28	56
5–6	8	22	44
3–4	8	14	28
1–2	6	6	12

8 The distribution of intelligence quotients in a random sample of the population at large would ordinarily be approximately symmetrical with a measure of central location in the neighborhood of 100. The distribution of intelligence quotients in a sample of university students would in most cases be negatively skewed, with a measure of central location substantially greater than 100, possibly 115 or 120. Very few university students would be expected to have intelligence quotients below 100.

9 Marks on mathematics examinations are not infrequently more variable than marks on history examinations, a higher proportion of individuals making very high and very low marks.

CHAPTER 3

1 3.46

2 66.58

3 If a constant c is added to all observations in a sample with mean $\bar{X}$, the new mean is $\bar{X} + c$. Multiplying by c results in a mean $c\bar{X}$.

4 **a.** 52.00 **b.** 40.00

5 1,225

6 **a.** 15 **b.** 23.5 **c.** 32.5 **d.** 24 **e.** 5.0

7 **a.** 5.5 **b.** 4

8 Ex. 1 3.55, 4.00
Ex. 2 67.36, 67.00

CHAPTER 4

1 **a.** 17 **b.** 4.67 **c.** 39.20 **d.** 6.26

2 1,485

3 12.50

4 **a.** 500 **b.** .80

5 $\Sigma(X - \bar{X})^2 = \Sigma(X^2 + \bar{X}^2 - 2X\bar{X}) = \Sigma X^2 + N\bar{X}^2 - 2N\bar{X}^2 = \Sigma X^2 - N\bar{X}^2$

6 **a.** 67.78, 8.23 **b.** 477.63, 21.85

7 $-1.342, -.447, .000, +.447, +1.342; N - 1 = 4$

8 $-1.2, 1.3, 2.8$

9 $m_2 = 20.80, m_3 = 0, m_4 = 620.80; g_1 = 0, g_2 = -1.57$

CHAPTER 5

1 $\frac{1}{3}$

2 $\frac{2}{9}$

3 $\frac{31}{36}$

4 $\frac{10}{36}$

5 $\frac{1}{6}$

6 $\frac{4}{52} \times \frac{3}{51} \times \frac{2}{50} \times \frac{1}{49}$

7 $\frac{4}{7} \times \frac{3}{6} \times \frac{3}{5} \times \frac{2}{4} \times \frac{2}{3} \times \frac{1}{2}$

8 $8 \times 7 \times 6 \times 5 \times 4 \times 3 \times 2 \times 1 = 40{,}320$

9 12

10 $\frac{1}{2}$

11 20

12 10

13 $\frac{6}{100}$

14 20 minutes

15 $\frac{36}{1024}$

16 $(\frac{3}{4})^6$, $\frac{1}{4}(\frac{3}{4})^5$

17 The expected distribution of heads in tossing six coins 64 times is

H	0	1	2	3	4	5	6
f	1	6	15	20	15	6	1

18 The expected distribution of 6's in rolling six dice 64 times is:

6's	f
0	$15{,}625/46{,}656 \times 64$
1	$18{,}750/46{,}656 \times 64$
2	$9{,}375/46{,}656 \times 64$
3	$2{,}500/46{,}656 \times 64$
4	$375/46{,}656 \times 64$
5	$30/46{,}656 \times 64$
6	$1/46{,}656 \times 64$

19 598/4,096

20 $7/64 \times 31/46{,}656$

CHAPTER 6

1 .0395, .1238, .3980, .2444, .0088

2 3.50, 25.90, 79.40, 62.46, 34.28

3 **a.** .4319   **b.** .3962   **c.** .4013   **d.** .0668   **e.** .1038

   f. .8289   **g.** .3830   **h.** .7066   **i.** .1337   **j.** .0025

4 **a.** .675   **b.** 1.281   **c.** 1.281   **d.** ±1.281

5 **a.** .0099   **b.** .0918   **c.** .2514   **d.** .9050

6 68

7 An error score of 66 is more than four standard deviations above the mean. The occurrence of a score equal to or greater than 66 is very improbable in random samples.

8

Grade	Score Interval
A	62.8–
B	55.2–62.8
C	44.8–55.2
D	37.2–44.8
E	–37.2

9 **a.** 79 **b.** 201

CHAPTER 7

1 **a.** The correlation between the intelligence of parents and that of their offspring has frequently been reported to be in the neighborhood of .50. **b.** Probably low positive. **c.** Fairly high positive in samples of individuals covering a broad age range up to about the age of 17. **d.** Fairly high positive. **e.** Fairly high positive under relatively stable economic conditions. **f.** Probably about zero, although data have been reported to the contrary. **g.** Probably about zero, although arguments may be advanced for a possible negative correlation.

2 $-.063$

3 $s_x = \sqrt{\dfrac{\Sigma x^2}{N-1}}, \; s_y = \sqrt{\dfrac{\Sigma y^2}{N-1}}$

Hence by substitution

$$\frac{\Sigma xy}{(N-1)\,s_x s_y} = \sqrt{\frac{\Sigma xy}{\Sigma x^2 \Sigma y^2}}$$

4 $+1$ and -1

5. The paired measurements may be in exactly the same order, but the value of z_x may not be exactly equal to the value of z_y with which it is paired. Consequently $\Sigma z_x z_y$ will not necessarily be equal to Σz_x^2 and Σz_y^2. Under these circumstances $\Sigma z_x z_y$ will be less than either Σz_x^2 or Σz_y^2, and r will be less than 1.

6 .544

7 The mean of $X - Y$ is $\bar{X} - \bar{Y}$. Hence we may write

$$s_{X-Y}^2 = \frac{\Sigma[(X-Y)-(\bar{X}-\bar{Y})]^2}{N-1} = \frac{\Sigma[(X-\bar{X})-(Y-\bar{Y})]^2}{N-1}$$

$$= \frac{\Sigma(X-\bar{X})^2}{N-1} + \frac{\Sigma(Y-\bar{Y})^2}{N-1} - \frac{2\Sigma(X-\bar{X})(Y-\bar{Y})}{N-1}$$

$$= s_X^2 + s_Y^2 - 2rs_X s_Y$$

8 The variance of sums is equal to the variance of the differences when the variables are uncorrelated.

9 No

10 $\dfrac{\Sigma z_x z_y}{N}$

CHAPTER 8

1 **a.** $r_{XY} = .809$ **b.** $b_{XY} = 1.200$ **c.** $b_{YX} = .545$
   d. $z'_X = .809Z_Y$ **e.** $X' = 1.200Y + .400$
   f. $s_{Y.X}^2 = .865$

2 **a.** .468   **b.** $Y' = .482X + 38.360$ **c.** 72.751   **d.** 103.033

3 $-1.00, -.84, .09, .58, .866$

4 $Y' = .171X + 5.593, X' = 2.100Y + 14.050$

5 From [8.8],

$$b_{YX} = r_{XY}\frac{s_Y}{s_X}, \; b_{XY} = r_{XY}\frac{s_X}{s_Y}$$

Hence $r_{XY} = \sqrt{b_{YX}b_{XY}}$

6 This statement is correct when the regression lines are not linear. Despite the possibly poor fit of the linear regression model, the fact remains that 50 per cent of the variance of one variable is predictable from the other variable. It is of course the case that, had a nonlinear regression model been used, more than 50 per cent of the variance of one variable might have been predictable from the other variable.

7 **a.** 300   **b.** 100

8 .866

9 ± 9.80

CHAPTER 9

1 **a.** A random sample is such that every member of the population is viewed as having an equal probability of being included in it. In a stratified random sample the members in each stratum are viewed as having an equal probability of being included in the subsample drawn from that stratum.   **b.** If the proportions in the various strata in the sample are the same as the proportions in the population, the sample is a proportional stratified sample.   **c.** An experimental sampling distribution is obtained by the actual drawing of samples from a population. It is the distribution of the values of the statistic thus obtained. A theoretical sampling distribution, on the other hand, is obtained using a theoretical method based on probability considerations. No samples are actually drawn.

2 A variety of methods suggest themselves. Students may be assigned numbers, using a directory. Numbers may be drawn using a lottery method, or a table of random numbers may be used.

3 Student population may be stratified by, for example, faculty. Let us suppose that a university has 5,000 students in four faculties, A, B, C, and D, the proportion in these four faculties being .50, .25, .15, and .10, respectively. In forming a sample of 100, 50 may be drawn from faculty A, 25 from faculty B, and so on. Faculty A contains 2,500 students. Using a directory, or faculty index, of students, every fiftieth name may be chosen. Faculty B contains 1,250 students. Here again every fiftieth name may be chosen, and so on. The resulting sample is a systematic proportional stratified sample.

4 A random sample of names selected from a telephone directory may not be considered an appropriate sample of voting behavior. Owning a telephone may be correlated with income, and possibly other variables, which may in turn be correlated with particular political predilections. Bias might result. Such a sample would clearly exhibit less bias in 1966 than, say, in 1931.

5   **a.**

$\overline{X}$	f
2.00	1
2.33	1
2.67	2
3.00	3
3.33	4
3.67	5
4.00	6
4.33	6
4.67	6
5.00	6
5.33	5
5.67	4
6.00	3
6.33	2
6.67	1
7.00	1
	56

b. 1.12   **c.** .0179

6 $\sigma \sqrt{N}$

7   With replacement $= 8.00$.   Without replacement $= 7.53$.

9   **a.** 200   **b.** 400

10   **a.**

p	f
.50	15
.75	40
1.00	15

b. $\sigma_p = .1637$

11   Increase

CHAPTER 10

1   95: 103.04–106.96
99: 102.42–107.58
75: 103.85–106.15
85: 103.56–106.44

2   95: 48.24–51.76
99: 47.68–52.32

3   By tripling sample size the standard error is multiplied by .5774, or divided by $\sqrt{3}$.

4

	95%	**99%**
a.	23.55–28.85	22.71–29.69
b.	55.35–61.25	54.42–62.18
c.	43.18–49.42	42.19–50.41
d.	5.73–11.07	4.89–11.91

5   **a.** 2.086   **b.** -3.850 **c.** 1.725   **d.** ± 1.725

6   **a.** 2.571   **b.** -6.859 **c.** 2.015   **d.** ± 2.015

7   **a.** .005   **b.** .050   **c.** .999   **d.** .015   **e.** .965

8

Number	f
0	.0007
1	.0079
2	.0413
3	.1239
4	.2322
5	.2787
6	.2090
7	.0896
8	.0168

$\sigma_p = .1732$

9 .685–.815, .664–.836

CHAPTER 11

1 $t = 1.74$; $p > .05$

2 $t = 1.62$; $p > .05$

3 $t = 2.81$; $p < .05$

4 $t = 1.22$

5 $t = 2.96$

6 Groups are ordinarily matched to reduce sampling error. If matching leads to a positive correlation between the values of the independent variables for the matched pairs, the standard error of the difference between means, for example, will be less than that which would have been obtained for independent samples without matching.

7 Cochran & Cox: $t_{.05} = 2.17$; $p > .05$
Welch: $t_{.05} = 2.05$, $df = 27$; $p > .05$

CHAPTER 12

1 $z = 3.086$; $p < .01$

2 $z = 1.414$; $p > .05$

3 $z = 1.704$; $p > .05$

4 $z = 3.620$; $p < .001$

5 $F = 1.563$; $p > .10$

6 $t = 1.350$; $p > .05$

7 **a.** .867 **b.** .050 **c.** $-.693$ **d.** -2.647

8 **a.** $t = 1.414$ **b.** $t = 1.939$ **c.** $t = 2.309$ **d.** $t = 2.450$

9 $z = 2.064$; $p < .05$

10 $t = 3.222$; $p < .01$

CHAPTER 13

1   $\chi^2 = 11.333$, $df = 5$; $p < .05 > .02$

2

Class Interval	Observed Frequency	Theoretical Frequency
90–99	1 ⎫ 6	
80–89	5 ⎭	8
70–79	17	17
60–69	30	31
50–59	50	39
40–49	35	34
30–39	10	20
20–29	6	8
10–19	4 ⎫ 6	
0–9	2 ⎭	3
	160	160

$\chi^2 = 9.498$, $df = 5$; $p > .05$

3   a. 1   b. 2   c. 8

4   $\chi^2 = 3.163$; $p > .05$

5   $\chi^2 = 14.287$; $p < .001$

6   $\chi^2 = 5.922$; $p < .02$

7   a. $\chi^2 = .633$; $p > .05$ b. $\chi^2 = 5.206$; $p > .05$

CHAPTER 14

1   a. In the simplest type of experiment two variables are involved, an independent and a dependent variable. The value of the dependent variables is thought to depend in some way on the value of the independent variable. To illustrate, an experiment might be conducted to investigate the effects of different amounts of practice on intelligence test items on intelligence test performance. Here intelligence test performance is the dependent variable and amount of practice constitutes the independent variable. In an experiment designed to investigate the relation between size of lesion in a particular brain locus and maze performance, measures of maze performance constitute the dependent variable and size of lesion the independent variable. The investigator wishes to ascertain how maze performance depends on lesion size.   b. A treatment variable is in effect the creation of the investigator, who decides the particular values which the variable will assume and the frequency of occurrence of these values. Different methods of learning a language, different dosages of a drug, and different amounts of practice in learning a task are examples of treatment variables. A classification variable, on the other hand, is one whose values, and the frequency of their occurrence, are not within the control of the investigator but exist, as it were, in nature. Height, weight, College Board test scores, and the like, are examples of classification variables.

2   Consider a simple experimental situation involving an experimental and a control group. Subjects in the two groups are matched on a variable X; consequently the two groups are equivalent with respect to X. If X is positively correlated with Y, a possibility exists that the paired values of Y for the experimental and control groups will also be positively correlated. If this is so, the sampling variance of the difference between means, and other statistics, will be less than had the two groups been independent.

3   In the above question matching of experimental subjects on a single variable X was discussed. Clearly the subjects do not have a single attribute which may be correlated with Y, but possibly a large number of such attributes. Randomization tends to ensure that

groups of subjects will not differ significantly on a large number of such attributes. Randomization also tends to ensure that groups of subjects will not differ significantly on a large number of attributes, some of which may be correlated with Y.

4   One very simple procedure is to number subjects from 1 to 100, write the numbers on slips of paper, and draw the slips of paper from a well-shuffled container, allocating a subject in turn to each of the five groups. Tables of random numbers could be used.

5   Subjects may be assigned to treatments at random, as described in Sec. 14.7. Since $N \neq k^2$, in this case no systematic arrangement of the Latin square type is possible.

6   Unless either equal or proportional groups are used, the factors are not independent of each other. When the groups are either equal or proportional, conclusions can be reached about the effect of each of the factors on the dependent variable, quite apart from, and independently of, the other factors used in the experiment.

7   A randomized block experiment is one in which a basis exists for arranging subjects into subgroups or blocks. Subjects may be assigned at random to treatments, a restriction being that each treatment occurs once only in each block.

8   A   B   C   D   E
     B   C   D   E   A
     C   D   E   A   B
     D   E   A   B   C
     E   A   B   C   D

9   See Sec. 1.8.

CHAPTER 15

1   **a.** $df_1 = 1$, $df_2 = 38$ **b.** $df_1 = 3$, $df_2 = 52$ **c.** $df_1 = 3$, $df_2 = 51$

2 $F = 23.31$, $df_1 = 4$, $df_2 = 20$; $p < .01$

3 $F = 2.06$, $df_1 = 3$, $df_2 = 21$; $p > .05$

4 $F = 2.27$, $df_1 = 2$, $df_2 = 27$; $p > .05$

5 $s_b^2 = 59.36$, $s_w^2 = 68.49$, $F = 0.87$, $df_1 = 1$, $df_2 = 11$; $p > .05$

6   The assumption of normality, homogeneity of variance, and additivity

7 $F = 3.73$, $df_1 = 2$, $df_2 = 12$; $p > .05$

CHAPTER 16

1   **a.**

	C_1	C_2	C_3	
R_1	10.2	3.7	14.8	9.6
R_2	10.5	4.0	15.1	9.9
	10.4	3.9	15.0	9.8

     b. 212.3

2

Source	Sum of Squares	df	Variance Estimate
Row	754.08	11	68.55
Column	142.17	2	71.09
Interaction	598.50	22	27.20
Total	1494.75	35	

$F = 2.61$, $df_1 = 2$, $df_2 = 22$; $p > .05$

3 $F_r = 1.23$; $p > .05$ $F_c = .23$; $p > .05$ $F_i = .31$; $p > .05$

4 $F_r = 3.85$; $p > .05$ $F_c = 8.19$; $p < .01$ $F_i = .08$; $p > .05$

CHAPTER 17

1

Source	df
R	4
C	3
L	2
$R \times C$	12
$R \times L$	8
$C \times L$	6
$R \times C \times L$	24
Within cells	540
Total	599

2

Source	SS
$R \times C$	562.50
$R \times L$	.00
$C \times L$	.00
$R \times C \times L$	.00

3 **a.** $R \times C \times L$ interaction   **b.** $R \times C$ interaction   **c.** within cells

4

Source	SS	df	Mean Square	F
Rows	12.00	1	12.00	.21
Columns	216.75	1	216.75	3.80
Layers	1800.75	1	1800.75	31.57*
$R \times C$	280.33	1	280.33	4.92*
$R \times L$	8.33	1	8.33	.15
$C \times L$	114.08	1	114.08	2.00
$R \times C \times L$	208.33	1	208.33	3.65
Within cells	2281.30	40	57.03	

* Significant at the .05 level.

5

Source	SS	df	Mean Square	F
Row	1763.57	2	881.78	28.17*
Column	91.01	2	45.51	1.45
Layer	3.56	1	3.56	.11
$R \times C$	162.32	4	40.58	1.30
$R \times L$	36.78	2	18.39	.59
$C \times L$	2507.70	2	1253.85	40.06*
$R \times C \times L$	101.62	4	25.41	.81
Within cells	4507.65	144	31.30	

* Significant at the .01 level.

6

Source	Sum of Squares	df	Mean Square	F
Subjects	258.69	3	86.23	
Rows	105.06	1	105.06	1.42 (SR)
Columns	189.06	1	189.06	26.15* (SC)
$S \times R$	220.69	3	73.56	
$S \times C$	21.69	3	7.23	
$R \times C$	68.06	1	68.06	16.10* (SRC)
$S \times R \times C$	12.69	3	4.23	

Note: Error term shown in parentheses following F ratio.

* Significant at the .05 level.

CHAPTER 18

1 $t = 4.23$, $df = 20$; $p < .01$

2

Comparison	F
I, II	.006
I, III	19.91
I, IV	65.01
I, V	14.71
II, III	20.61
II, IV	66.27
II, V	15.32
III, IV	12.97
III, V	.39
IV, V	17.87

$df_1 = 4$, $df_2 = 20$
$F' = 17.72$
Significant comparison ($p < .01$)
I, III; I, IV; II, III;
II, IV; IV, V

3

Table of Q

	II	I	V	III	IV
II		.111	5.535*	6.421*	11.513*
I			5.424*	6.310*	5.978*
V				.886	5.978*
III					5.092

$Q_2 = 4.02$, $Q_3 = 4.64$, $Q_4 = 5.02$, $Q_5 = 5.29$

* Significant at the .01 level.

4

Comparison	F
I and II	.26
I and III	.68
I and IV	2.06
II and III	2.14
II and IV	1.04
III and IV	5.90

$df = 3$, $df_2 = 21$
$F'_{01} = 14.61$
No comparisons are significant
at .01 level.

CHAPTER 19

1   **a.** 565.35, 275.63, 3.00

   b. 9.92, 275.63, 3.00

   c. 27.79, .30

2   2.62

3 $k = 6$

c_{1j}: $-5, -3, -1, 1, 3, 5$

c_{2j}: $5, -1, -4, -4, -1, 5$

c_{3j}: $-5, 7, 4, -4, -7, 5$

4 Linear   17.72

   Quadratic   13.64

   Cubic   134.09

   Deviation   79.17

   Within-groups   3.62

   $F_1 = 4.90$ $(p < .05)$, $F_2 = 3.77$ $(p > .05)$, $F_3 = 37.04$ $(p < .01)$

5 See Sec. 19.10.

CHAPTER 20

1 Adjusted total sums of squares on $X = 319.13$

   Adjusted within-groups sums of squares on $X = 128.73$

   Adjusted between-groups sums of squares on $X = 190.41$

   $F = 8.14$, $df_1 = 2$, $df_2 = 11$; $p < .01$

2 **a.** 1.39   **b.** .72   **c.** 1.75

3 Gp I, 15.14; Gp II, 22.51; Gp III, 29.95

4 $A = 125.54$, $B = 128.73$, $F = .15$; $p > .05$

CHAPTER 21

1 $\rho = .643$, $p = .05$ for one-tailed test

2 $\rho = .612$, $p < .05$ for one-tailed test

3 **a.** $p > .05$ for one-tailed test   **b.** $p < .05$ for one-tailed test   **c.** $p < .05$ for one-tailed test

4 **a.** $S = -3$ **b.** $S = -1$ **c.** $S = 2$ **d.** $S = -3$ **e.** $S = -1$ **f.** $S = -3$

5 **a.** $\sigma_s^2 = 28.33$ **b.** $\sigma_s^2 = 26.33$ **c.** $\sigma_s^2 = 23.07$ **d.** $\sigma_s^2 = 21.00$ **e.** $\sigma_s^2 = 19.80$

       $z = .376$ $z = .000$ $z = .208$ $z = .437$ $z = .000$

   f. $\sigma_s^2 = 16.20$

       $z = .498$

6 $\rho_{ab} = .607$, $\rho_{ac} = .429$, $\rho_{bc} = .393$, $W = .651$

7 $K = .000$

8 $K = .400$

CHAPTER 22

1 $\chi^2 = 5.63$, $p < .02 > .01$

2 $p = .18$

3 $\chi^2 = 13.60$, $p < .01$

4 $z = 2.69$, $p < .01$

5 $W_+ = 28$, $p > .05$

6 $H = 12.87$, $p < .01$

7 $\chi_r^2 = 15.86$, $p < .01$

8 $z = 3.87$, $p < .01$

9 $z = -3.73$

10 **a.** $p_1 = .0238$ **b.** $p_1 = .0070$
$p_2 = .2381$ $p_2 = .0932$
$p_3 = .4762$ $p_3 = .3263$
$p_4 = .2381$ $p_4 = .3916$
$p_5 = .0238$ $p_5 = .1632$
$p_6 = .0186$

For **a** above, two arrangements would prove to be significant with $p < .05$. For **b**, one arrangement is significant with $p < .01$ and two with $p < .05$.

CHAPTER 23

1 **a.** .750, .667, .500, .333, .250

b. .188, .222, .250, .222, .188

2 **a.**

	1	2	3	4	5
1					
2	−.083				
3	−.042	−.083			
4	.000	.030	.083		
5	−.021	.083	.042	.084	

b.

	1	2	3	4	5
1					
2	−.405				
3	−.194	−.350			
4	.000	.136	.350		
5	−.112	.405	.194	.410	

3 3.583

4 1.070

5 .087, .316, .447, .790, .776

6 .118

7 .042, .169, .250, .419, .376

8 .038, .151, .224, .375, .336

CHAPTER 24

1 $s_t^2 = 202.50$, $s_e^2 = 22.50$; 10 per cent

2 .462, .667, .837, .889, .980

3 .968, .928

4 $r_{kk} = .833$

5 $s_t^2/N = 4.96$, $s_e^2/N = 1.24$. The proportion due to error will be .077.

6 .707, .849

7 .625

8 $s_e\ (7) = 2.75$, $s_e\ (26) = 3.57$, $s_e\ (44) = 2.32$

CHAPTER 25

1 The kth percentile point is a value such that k per cent of the members of a sample have less than that value. A percentile rank is a value on a transformed rectangular scale which extends from zero to 100.

2 **a.** $p_{25} = 105.00$ **b.** $PR\ (103) = 24.17$
$p_{50} = 116.17$ $PR\ (123) = 65.00$
$p_{75} = 125.17$ $PR\ (136) = 95.83$

3 **a.** $p_{10} = 56.32$ **b.** $PR\ (59) = 10.50$
$p_{20} = 65.71$ $PR\ (74) = 53.50$
$p_{80} = 80.71$ $PR\ (82) = 84.25$

4

Class Interval	Mid-point	Frequency	T Score
45–49	47	1	74.8
40–44	42	2	69.4
35–39	37	3	65.6
30–34	32	6	61.8
25–29	27	8	58.1
20–24	22	17	53.2
15–19	17	26	45.9
10–14	12	11	37.1
5–9	7	2	27.8
0–4	2	0	
		76	

5

Stanine	Exact Limits of Interval
9	39.43–
8	32.53–39.43
7	26.08–32.53
6	21.44–26.08
5	17.85–21.44
4	15.36–17.85
3	12.39–15.36
2	9.97–12.39
1	– 9.97

6 δ-scores: -1.451, $-.605$, 1.560, $-.537$, $-.767$, $-.669$, $-.229$, 2.106, $-.362$, $-.292$, $.254$, $.994$

CHAPTER 26

1 31.4 per cent

2 **a.** $.828$ **b.** $.826$ **c.** $z'_1 = .545z_2 + 470z_3$ **d.** $x'_1 = .449x_2 + .580x_3 - 1.65$ **e.** $.830$

3 **a.** $z'_1 = .439z_2 + .268z_3 - .071z_4 + 400z_5 - 296z_6$
b. $x'_1 = 673x_2 + .330x_3 - .081x_4 + 672x_5 - .436x_6 - 61.59$
c. $.771$ **d.** $.734, .642, .638$

CHAPTER 27

1

Variable	g	s
1	.8	.6000
2	.9	.4359
3	.7	.7141
4	.2	.9798
5	.2	.9798

2 .505

3 **a.** .90, .73, .58, .72, .61, .52

    b. .10, .27, .42, .36, .39, .48

    c. 2.71, 1.35   **d.** 66.75, 33.25

4

	1	2	3	4	5	6
1						
2	.81					
3	.72	.65				
4	.72	.66	.60			
5	.63	.58	.53	.66		
6	.54	.50	.46	.60	.56	

5 .20

6 F_1: .949, .936, .919, .707, .640, .555
F_2: .316, .351, .394, .707, .768, .832

7 .5831

8 Rotated loadings are:
F_1: .2828, .4243, .5657, . . . , ... , . . .
F_2: . . . , ... , .2828, .4243, .5657

GLOSSARY OF SYMBOLS

For most commonly used statistics, Roman letters denote sample values and Greek letters denote parameters. Exceptions to this are made either for convenience or in conformity with common usage. For example, ρ denotes both the population value of the product-moment correlation coefficient and the sample value of Spearman's rank-order correlation coefficient. η denotes the sample value of the correlation ratio. T, not τ, denotes a true measurement.

A bar above a symbol always indicates the arithmetic mean of a sample of observations. A few symbols are used with double or triple meanings. Homonyms are permissible in the language of mathematics as in any other.

Some symbols with idiosyncratic use in a restricted context are not listed.

a	Constant in a regression equation; used with subscripts as a_{yx} and a_{xy}; first subscript denotes the predicted variable, second the observed variable. Geometrically, a_{yx} and a_{xy} are distances from the origin where the regression lines intercept the Y and X axes.
b	Regression weight applied to an independent variable, or predictor, in original units; used with subscripts as b_{yx} and b_{xy} to distinguish predicted from observed variable. Geometrically, b_{yx} and b_{xy} are the slopes of regression lines.
c	(1) A constant.
	(2) Denotes cth column in a set of C columns.
C	(1) Number of columns.
	(2) Contingency coefficient, measure of association between nominal variables.
C_r^N	Number of combinations of N things taken r at a time.
d	(1) Difference between paired ranks.
	(2) Difference between the mean of a subgroup and the mean of combined groups, $\overline{X}_i - \overline{X} = d_i$.
D	Difference between paired measurements.
df	Degrees of freedom.
e	Base of Napierian logarithms, 2.7183.
e_i	Sampling or measurement error associated with the ith value.
E	(1) Expected frequency in the calculation of χ^2.
	(2) The expectation of or expected value, as $E(\overline{X})$ or $E(X - \mu)^2$.
f	Frequency in a distribution; used with subscript to denote interval or subclass, f_i.
f_x, f_y	Marginal frequencies in bivariate distribution.

f_{xy}	Cell frequency in bivariate distribution.
F	Ratio of two sample variances.
g_1	Measure of skewness.
g_2	Measure of kurtosis.
h	Size of a class interval.
H_0	Null hypothesis, as in H_0: $\mu_1 - \mu_2 = 0$.
i, j, k	Subscripts used to identify particular observations in a group.
k	(1) Number of subclasses.
	(2) Number of times a test is lengthened.
K	Kendall's coefficient of consistence.
m_r	The rth moment about the arithmetic mean.
n	(1) Number of observations in a subclass; used with subscript to indicate subclass, n_i, n_j, n_{ij}, etc. Always used where the number of subclasses is greater than 2.
	(2) Number of test items.
N	Number of observations in a sample.
N_p	Number of members in a population.
O	Observed frequency in the calculation of χ^2.
p_i	(1) Sample proportion in the ith class; estimate of the probability of the occurrence of the ith event.
	(2) In relation to psychological tests denotes the proportion of individuals passing item i.
P_i	Percentile point; subscript denotes particular percentile point, as P_{20}, P_{50}, etc.
$P_r{}^N$	Number of permutations of N things taken r at a time.
p	Sample proportion or probability estimate in one of two mutually exclusive classes.
q	$1 - p$.
r	(1) Sample value of the correlation coefficient; used with subscripts to denote variables correlated, r_{xy}, r_{12}, etc.
	(2) Denotes rth row in a set of R rows.
r_{bi}	Biserial correlation coefficient.
r_{pbi}	Point biserial correlation coefficient.
r_t	Tetrachoric correlation coefficient.
r_{xx}	Reliability coefficient.
r_{hh}	Reliability coefficient for a half test.
r_{kk}	Reliability coefficient for a test lengthened k times.
$r_{xy.z}$, $r_{12.3}$	Partial correlation coefficient.
R	(1) Number of rows.
	(2) Sum of ranks; used with subscript to denote sum of ranks for the jth group, R_j.
	(3) Multiple correlation coefficient.
s	(1) Standard deviation of a sample; used with subscript to denote variable, s_x, s_y, etc.
	(2) Estimate of the standard error of a statistic; subscript denotes statistic, s_x, s_p, s_s, etc.
s^2	Variance estimate, the square of any standard deviation; used with subscripts as indicated under s.
s_c	Sample standard deviation corrected for grouping error.
$s_{y.x}$, $s_{x.y}$	Standard error of estimate.
s_y', s_x'	Standard deviation of predicted values of Y and X.
$s_e(X_i)$	Standard error associated with individual measurement, X_i.
S	Difference between number of agreements and disagreements in the calculation of Kendall's tau.
t	(1) Ratio of normally distributed variable to an estimate of the standard error of that variable. Deviation from the origin along the base line of distribution of t.

	(2) Number of values tied at a particular rank in a set of ranks.
T	(1) True value of an observation or measurement; used with subscript, T_i.
	(2) In the analysis of variance denotes the sum of observations; T_j is the sum of observations in the jth group.
	(3) Correction factor for ties in the calculation of Kendall's tau and coefficient of concordance; used with subscript to denote variable, T_x, T_y.
W	Kendall's coefficient of concordance.
x, y	Variable expressed as deviation from the arithmetic mean; subscripts denote particular values of the variable, x_i, y_i, etc.
x', y'	Variable expressed as deviation from arbitrary origin, sometimes involving a change in unit. Computation variable.
X, Y	Variable in original units; subscripts used to denote particular values of variable, X_i, Y_j, etc.
$\overline{X}, \overline{Y}$	Arithmetic mean of a sample. A bar above a symbol always denotes a sample mean.
X_0, Y_0	Arbitrary origin.
y	Ordinate of unit normal curve.
z	(1) Variable expressed in standard-score form, $z = (X - \overline{X})/s_x$. Subscript used to denote variable, z_x, z_y.
	(2) Deviation from origin along base line of normal curve of unit area and unit standard deviation.
z_r	Transformation of the correlation coefficient to approximate normal form; used in tests of significance on r.
β	Regression weight in a multiple regression equation applied to an independent variable or predictor in standard-score form.
η_{yx}, η_{xy}	Correlation ratio.
θ	Population value of a proportion.
μ	Population mean; used with subscript to indicate variable, μ_x.
π	Ratio of the circumference of a circle to the diameter, 3.1416.
ρ	(1) Correlation coefficient in a population.
	(2) Sample value of the rank-order correlation coefficient.
$\hat{\rho}$	Maximum likelihood estimate of ρ.
σ	(1) Standard deviation of a population; used with subscript to denote variable, σ_x, σ_y, etc.
	(2) Standard error, standard deviation of sampling distribution; subscript denotes statistics, σ_x, σ_p, etc.
$\hat{\sigma}$	Maximum likelihood estimate of σ.
$\displaystyle\sum_{i=1}^{N}$	The sum of; the operation of adding a set of variate values. Symbols above and below define limits of the summation.
τ	Kendall's coefficient of rank correlation, tau.
ϕ	Phi coefficient, measure of fourfold point correlation.
χ^2	Chi square.
$a > b$	a is greater than b.
$a < b$	a is less than b.
$a \geqslant b$	a is greater than or equal to b.
$a \leqslant b$	a is less than or equal to b.
$a >> b$	a is very much greater than b.
$a << b$	a is very much less than b.
$=$	Is equal to.
$\neq$	Is not equal to.
$\lvert a \rvert$	Absolute value of a.
$\sqrt{a}$	The square root of a.
∞	Infinity.

APPENDIX TABLES

Table A Ordinates and areas of the normal curve*

(In terms of σ units)

$\dfrac{x}{\sigma}$	Area	Ordinate	$\dfrac{x}{\sigma}$	Area	Ordinate	$\dfrac{x}{\sigma}$	Area	Ordinate
00	.0000	.3989	.50	.1915	.3521	1.00	.3413	.2420
.01	.0040	.3989	.51	.1950	.3503	1.01	.3438	.2396
.02	.0080	.3989	.52	.1985	.3485	1.02	.3461	.2371
.03	.0120	.3988	.53	.2019	.3467	1.03	.3485	.2347
.04	.0160	.3986	.54	.2054	.3448	1.04	.3508	.2323
.05	.0199	.3984	.55	.2088	.3429	1.05	.3531	.2299
.06	.0239	.3982	.56	.2123	.3410	1.06	.3554	.2275
.07	.0279	.3980	.57	.2157	.3391	1.07	.3577	.2251
.08	.0319	.3977	.58	.2190	.3372	1.08	.3599	.2227
.09	.0359	.3973	.59	.2224	.3352	1.09	.3621	.2203
.10	.0398	.3970	.60	.2257	.3332	1.10	.3643	.2179
.11	.0438	.3965	.61	.2291	.3312	1.11	.3665	.2155
.12	.0478	.3961	.62	.2324	.3292	1.12	.3686	.2131
.13	.0517	.3956	.63	.2357	.3271	1.13	.3708	.2107
.14	.0557	.3951	.64	.2389	.3251	1.14	.3729	.2083
.15	.0596	.3945	.65	.2422	.3230	1.15	.3749	.2059
.16	.0636	.3939	.66	.2454	.3209	1.16	.3770	.2036
.17	.0675	.3932	.67	.2486	.3187	1.17	.3790	.2012
.18	.0714	.3925	.68	.2517	.3166	1.18	.3810	.1989
.19	.0753	.3918	.69	.2549	.3144	1.19	.3830	.1965
.20	.0793	.3910	.70	.2580	.3123	1.20	.3849	.1942
.21	.0832	.3902	.71	.2611	.3101	1.21	.3869	.1919
.22	.0871	.3894	.72	.2642	.3079	1.22	.3888	.1895
.23	.0910	.3885	.73	.2673	.3056	1.23	.3907	.1872
.24	.0948	.3876	.74	.2703	.3034	1.24	.3925	.1849
.25	.0987	.3867	.75	.2734	.3011	1.25	.3944	.1826
.26	.1026	.3857	.76	.2764	.2989	1.26	.3962	.1804
.27	.1064	.3847	.77	.2794	.2966	1.27	.3980	.1781
.28	.1103	.3836	.78	.2823	.2943	1.28	.3997	.1758
.29	.1141	.3825	.79	.2852	.2920	1.29	.4015	.1736
.30	.1179	.3814	.80	.2881	.2897	1.30	.4032	.1714
.31	.1217	.3802	.81	.2910	.2874	1.31	.4049	.1691
.32	.1255	.3790	.82	.2939	.2850	1.32	.4066	.1669
.33	.1293	.3778	.83	.2967	.2827	1.33	.4082	.1647
.34	.1331	.3765	.84	.2995	.2803	1.34	.4099	.1626
.35	.1368	.3752	.85	.3023	.2780	1.35	.4115	.1604
.36	.1406	.3739	.86	.3051	.2756	1.36	.4131	.1582
.37	.1443	.3725	.87	.3078	.2732	1.37	.4147	.1561
.38	.1480	.3712	.88	.3106	.2709	1.38	.4162	.1539
.39	.1517	.3697	.89	.3133	.2685	1.39	.4177	.1518
.40	.1554	.3683	.90	.3159	.2661	1.40	.4192	.1497
.41	.1591	.3668	.91	.3186	.2637	1.41	.4207	.1476
.42	.1628	.3653	.92	.3212	.2613	1.42	.4222	.1456
.43	.1664	.3637	.93	.3238	.2589	1.43	.4236	.1435
.44	.1700	.3621	.94	.3264	.2565	1.44	.4251	.1415
.45	.1736	.3605	.95	.3289	.2541	1.45	.4265	.1394
.46	.1772	.3589	.96	.3315	.2516	1.46	.4279	.1374
.47	.1808	.3572	.97	.3340	.2492	1.47	.4292	.1354
.48	.1844	.3555	.98	.3365	.2468	1.48	.4306	.1334
.49	.1879	.3538	.99	.3389	.2444	1.49	.4319	.1315
.50	.1915	.3521	1.00	.3413	.2420	1.50	.4332	.1295

* Reproduced from J. E. Wert, *Educational statistics*, by courtesy of McGraw-Hill Book Company, New York.

Table A (*Continued*)

$\frac{x}{\sigma}$	Area	Ordinate	$\frac{x}{\sigma}$	Area	Ordinate	$\frac{x}{\sigma}$	Area	Ordinate
1.50	.4332	.1295	2.00	.4772	.0540	2.50	.4938	.0175
1.51	.4345	.1276	2.01	.4778	.0529	2.51	.4940	.0171
1.52	.4357	.1257	2.02	.4783	.0519	2.52	.4941	.0167
1.53	.4370	.1238	2.03	.4788	.0508	2.53	.4943	.0163
1.54	.4382	.1219	2.04	.4793	.0498	2.54	.4945	.0158
1.55	.4394	.1200	2.05	.4798	.0488	2.55	.4946	.0154
1.56	.4406	.1182	2.06	.4803	.0478	2.56	.4948	.0151
1.57	.4418	.1163	2.07	.4808	.0468	2.57	.4949	.0147
1.58	.4429	.1145	2.08	.4812	.0459	2.58	.4951	.0143
1.59	.4441	.1127	2.09	.4817	.0449	2.59	.4952	.0139
1.60	.4452	.1109	2.10	.4821	.0440	2.60	.4953	.0136
1.61	.4463	.1092	2.11	.4826	.0431	2.61	.4955	.0132
1.62	.4474	.1074	2.12	.4830	.0422	2.62	.4956	.0129
1.63	.4484	.1057	2.13	.4834	.0413	2.63	.4957	.0126
1.64	.4495	.1040	2.14	.4838	.0404	2.64	.4959	.0122
1.65	.4505	.1023	2.15	.4842	.0395	2.65	.4960	.0119
1.66	.4515	.1006	2.16	.4846	.0387	2.66	.4961	.0116
1.67	.4525	.0989	2.17	.4850	.0379	2.67	.4962	.0113
1.68	.4535	.0973	2.18	.4854	.0371	2.68	.4963	.0110
1.69	.4545	.0957	2.19	.4857	.0363	2.69	.4964	.0107
1.70	.4554	.0940	2.20	.4861	.0355	2.70	.4965	.0104
1.71	.4564	.0925	2.21	.4864	.0347	2.71	.4966	.0101
1.72	.4573	.0909	2.22	.4868	.0339	2.72	.4967	.0099
1.73	.4582	.0893	2.23	.4871	.0332	2.73	.4968	.0096
1.74	.4591	.0878	2.24	.4875	.0325	2.74	.4969	.0093
1.75	.4599	.0863	2.25	.4878	.0317	2.75	.4970	.0091
1.76	.4608	.0848	2.26	.4881	.0310	2.76	.4971	.0088
1.77	.4616	.0833	2.27	.4884	.0303	2.77	.4972	.0086
1.78	.4625	.0818	2.28	.4887	.0297	2.78	.4973	.0084
1.79	.4633	.0804	2.29	.4890	.0290	2.79	.4974	.0081
1.80	.4641	.0790	2.30	.4893	.0283	2.80	.4974	.0079
1.81	.4649	.0775	2.31	.4896	.0277	2.81	.4975	.0077
1.82	.4656	.0761	2.32	.4898	.0270	2.82	.4976	.0075
1.83	.4664	.0748	2.33	.4901	.0264	2.83	.4977	.0073
1.84	.4671	.0734	2.34	.4904	.0258	2.84	.4977	.0071
1.85	.4678	.0721	2.35	.4906	.0252	2.85	.4978	.0069
1.86	.4686	.0707	2.36	.4909	.0246	2.86	.4979	.0067
1.87	.4693	.0694	2.37	.4911	.0241	2.87	.4979	.0065
1.88	.4699	.0681	2.38	.4913	.0235	2.88	.4980	.0063
1.89	.4706	.0669	2.39	.4916	.0229	2.89	.4981	.0061
1.90	.4713	.0656	2.40	.4918	.0224	2.90	.4981	.0060
1.91	.4719	.0644	2.41	.4920	.0219	2.91	.4982	.0058
1.92	.4726	.0632	2.42	.4922	.0213	2.92	.4982	.0056
1.93	.4732	.0620	2.43	.4925	.0208	2.93	.4983	.0055
1.94	.4738	.0608	2.44	.4927	.0203	2.94	.4984	.0053
1.95	.4744	.0596	2.45	.4929	.0198	2.95	.4984	.0051
1.96	.4750	.0584	2.46	.4931	.0194	2.96	.4985	.0050
1.97	.4756	.0573	2.47	.4932	.0189	2.97	.4985	.0048
1.98	.4761	.0562	2.48	.4934	.0184	2.98	.4986	.0047
1.99	.4767	.0551	2.49	.4936	.0180	2.99	.4986	.0046
2.00	.4772	.0540	2.50	.4938	.0175	3.00	.4987	.0044

Table B   Critical values of t^*

		Level of significance for one-tailed test				
	.10	.05	.025	.01	.005	.0005
		Level of significance for two-tailed test				
df	.20	.10	.05	.02	.01	.001
1	3.078	6.314	12.706	31.821	63.657	636.619
2	1.886	2.920	4.303	6.965	9.925	31.598
3	1.638	2.353	3.182	4.541	5.841	12.941
4	1.533	2.132	2.776	3.747	4.604	8.610
5	1.476	2.015	2.571	3.365	4.032	6.859
6	1.440	1.943	2.447	3.143	3.707	5.959
7	1.415	1.895	2.365	2.998	3.499	5.405
8	1.397	1.860	2.306	2.896	3.355	5.041
9	1.383	1.833	2.262	2.821	3.250	4.781
10	1.372	1.812	2.228	2.764	3.169	4.587
11	1.363	1.796	2.201	2.718	3.106	4.437
12	1.356	1.782	2.179	2.681	3.055	4.318
13	1.350	1.771	2.160	2.650	3.012	4.221
14	1.345	1.761	2.145	2.624	2.977	4.140
15	1.341	1.753	2.131	2.602	2.947	4.073
16	1.337	1.746	2.120	2.583	2.921	4.015
17	1.333	1.740	2.110	2.567	2.898	3.965
18	1.330	1.734	2.101	2.552	2.878	3.922
19	1.328	1.729	2.093	2.539	2.861	3.883
20	1.325	1.725	2.086	2.528	2.845	3.850
21	1.323	1.721	2.080	2.518	2.831	3.819
22	1.321	1.717	2.074	2.508	2.819	3.792
23	1.319	1.714	2.069	2.500	2.807	3.767
24	1.318	1.711	2.064	2.492	2.797	3.745
25	1.316	1.708	2.060	2.485	2.787	3.725
26	1.315	1.706	2.056	2.479	2.779	3.707
27	1.314	1.703	2.052	2.473	2.771	3.690
28	1.313	1.701	2.048	2.467	2.763	3.674
29	1.311	1.699	2.045	2.462	2.756	3.659
30	1.310	1.697	2.042	2.457	2.750	3.646
40	1.303	1.684	2.021	2.423	2.704	3.551
60	1.296	1.671	2.000	2.390	2.660	3.460
120	1.289	1.658	1.980	2.358	2.617	3.373
∞	1.282	1.645	1.960	2.326	2.576	3.291

* Abridged from Table III of R. A. Fisher and F. Yates, *Statistical tables for biological, agricultural, and medical research*, published by Oliver & Boyd, Ltd., Edinburgh, by permission of the authors and publishers.

Table C Critical values of chi square*

Probability under H_0 that $\chi^2 \geq$ chi square

df	.99	.98	.95	.90	.80	.70	.50	.30	.20	.10	.05	.02	.01	.001
1	.00016	.00063	.0039	.016	.064	.15	.46	1.07	1.64	2.71	3.84	5.41	6.64	10.83
2	.02	.04	.10	.21	.45	.71	1.39	2.41	3.22	4.60	5.99	7.82	9.21	13.82
3	.12	.18	.35	.58	1.00	1.42	2.37	3.66	4.64	6.25	7.82	9.84	11.34	16.27
4	.30	.43	.71	1.06	1.65	2.20	3.36	4.88	5.99	7.78	9.49	11.67	13.28	18.46
5	.55	.75	1.14	1.61	2.34	3.00	4.35	6.06	7.29	9.24	11.07	13.39	15.09	20.52
6	.87	1.13	1.64	2.20	3.07	3.83	5.35	7.23	8.56	10.64	12.59	15.03	16.81	22.46
7	1.24	1.56	2.17	2.83	3.82	4.67	6.35	8.38	9.80	12.02	14.07	16.62	18.48	24.32
8	1.65	2.03	2.73	3.49	4.59	5.53	7.34	9.52	11.03	13.36	15.51	18.17	20.09	26.12
9	2.09	2.53	3.32	4.17	5.38	6.39	8.34	10.66	12.24	14.68	16.92	19.68	21.67	27.88
10	2.56	3.06	3.94	4.86	6.18	7.27	9.34	11.78	13.44	15.99	18.31	21.16	23.21	29.59
11	3.05	3.61	4.58	5.58	6.99	8.15	10.34	12.90	14.63	17.28	19.68	22.62	24.72	31.26
12	3.57	4.18	5.23	6.30	7.81	9.03	11.34	14.01	15.81	18.55	21.03	24.05	26.22	32.91
13	4.11	4.76	5.89	7.04	8.63	9.93	12.34	15.12	16.98	19.81	22.36	25.47	27.69	34.53
14	4.66	5.37	6.57	7.79	9.47	10.82	13.34	16.22	18.15	21.06	23.68	26.87	29.14	36.12
15	5.23	5.98	7.26	8.55	10.31	11.72	14.34	17.32	19.31	22.31	25.00	28.26	30.58	37.70
16	5.81	6.61	7.96	9.31	11.15	12.62	15.34	18.42	20.46	23.54	26.30	29.33	32.00	39.29
17	6.41	7.26	8.67	10.08	12.00	13.53	16.34	19.51	21.62	24.77	27.59	31.00	33.41	40.75
18	7.02	7.91	9.39	10.86	12.86	14.44	17.34	20.60	22.76	25.99	28.87	32.35	34.80	42.31
19	7.63	8.57	10.12	11.65	13.72	15.35	18.34	21.69	23.90	27.20	30.14	33.69	36.19	43.82
20	8.26	9.24	10.85	12.44	14.58	16.27	19.34	22.78	25.04	28.41	31.41	35.02	37.57	45.32
21	8.90	9.92	11.59	13.24	15.44	17.18	20.34	23.86	26.17	29.62	32.67	36.34	38.93	46.80
22	9.54	10.60	12.34	14.04	16.31	18.10	21.34	24.94	27.30	30.81	33.92	37.66	40.29	48.27
23	10.20	11.29	13.09	14.85	17.19	19.02	22.34	26.02	28.43	32.01	35.17	38.97	41.64	49.73
24	10.86	11.99	13.85	15.66	18.06	19.94	23.34	27.10	29.55	33.20	36.42	40.27	42.98	51.18
25	11.52	12.70	14.61	16.47	18.94	20.87	24.34	28.17	30.68	34.38	37.65	41.57	44.31	52.62
26	12.20	13.41	15.38	17.29	19.82	21.79	25.34	29.25	31.80	35.56	38.88	42.86	45.64	54.05
27	12.88	14.12	16.15	18.11	20.70	22.72	26.34	30.32	32.91	36.74	40.11	44.14	46.96	55.48
28	13.56	14.85	16.93	18.94	21.59	23.65	27.34	31.39	34.03	37.92	41.34	45.42	48.28	56.89
29	14.26	15.57	17.71	19.77	22.48	24.58	28.34	32.46	35.14	39.09	42.56	46.69	49.59	58.30
30	14.95	16.31	18.49	20.60	23.36	25.51	29.34	33.53	36.25	40.26	43.77	47.96	50.89	59.70

* Abridged from Table IV of R. A. Fisher and F. Yates: *Statistical tables for biological, agricultural, and medical research*, published by Oliver & Boyd, Ltd., Edinburgh, by permission of the authors and publishers.

Table D   Critical values of F^*

5 per cent (roman type) and 1 per cent (bold-face type) points for the distribution of F^*

Degrees of freedom for greater mean square

Degrees of freedom for lesser mean square	1	2	3	4	5	6	7	8	9	10	11	12	14	16	20	24	30	40	50	75	100	200	500	∞
1	161 **4052**	200 **4999**	216 **5403**	225 **5625**	230 **5764**	234 **5859**	237 **5928**	239 **5981**	241 **6022**	242 **6056**	243 **6082**	244 **6106**	245 **6142**	246 **6169**	248 **6208**	249 **6234**	250 **6258**	251 **6286**	252 **6302**	253 **6323**	253 **6334**	254 **6352**	254 **6361**	254 **6366**
2	18.51 **98.49**	19.00 **99.01**	19.16 **99.17**	19.25 **99.25**	19.30 **99.30**	19.33 **99.33**	19.36 **99.34**	19.37 **99.36**	19.38 **99.38**	19.39 **99.40**	19.40 **99.41**	19.41 **99.42**	19.42 **99.43**	19.43 **99.44**	19.44 **99.45**	19.45 **99.46**	19.46 **99.47**	19.47 **99.48**	19.47 **99.48**	19.48 **99.49**	19.49 **99.49**	19.49 **99.49**	19.50 **99.50**	19.50 **99.50**
3	10.13 **34.12**	9.55 **30.81**	9.28 **29.46**	9.12 **28.71**	9.01 **28.24**	8.94 **27.91**	8.88 **27.67**	8.84 **27.49**	8.81 **27.34**	8.78 **27.23**	8.76 **27.13**	8.74 **27.05**	8.71 **26.92**	8.69 **26.83**	8.66 **26.69**	8.64 **26.60**	8.62 **26.50**	8.60 **26.41**	8.58 **26.35**	8.57 **26.27**	8.56 **26.23**	8.54 **26.18**	8.54 **26.14**	8.53 **26.12**
4	7.71 **21.20**	6.94 **18.00**	6.59 **16.69**	6.39 **15.98**	6.26 **15.52**	6.16 **15.21**	6.09 **14.98**	6.04 **14.80**	6.00 **14.66**	5.96 **14.54**	5.93 **14.45**	5.91 **14.37**	5.87 **14.24**	5.84 **14.15**	5.80 **14.02**	5.77 **13.93**	5.74 **13.83**	5.71 **13.74**	5.70 **13.69**	5.68 **13.61**	5.66 **13.57**	5.65 **13.52**	5.64 **13.48**	5.63 **13.46**
5	6.61 **16.26**	5.79 **13.27**	5.41 **12.06**	5.19 **11.39**	5.05 **10.97**	4.95 **10.67**	4.88 **10.45**	4.82 **10.27**	4.78 **10.15**	4.74 **10.05**	4.70 **9.96**	4.68 **9.89**	4.64 **9.77**	4.60 **9.68**	4.56 **9.55**	4.53 **9.47**	4.50 **9.38**	4.46 **9.29**	4.44 **9.24**	4.42 **9.17**	4.40 **9.13**	4.38 **9.07**	4.37 **9.04**	4.36 **9.02**
6	5.99 **13.74**	5.14 **10.92**	4.76 **9.78**	4.53 **9.15**	4.39 **8.75**	4.28 **8.47**	4.21 **8.26**	4.15 **8.10**	4.10 **7.98**	4.06 **7.87**	4.03 **7.79**	4.00 **7.72**	3.96 **7.60**	3.92 **7.52**	3.87 **7.39**	3.84 **7.31**	3.81 **7.23**	3.77 **7.14**	3.75 **7.09**	3.72 **7.02**	3.71 **6.99**	3.69 **6.94**	3.68 **6.90**	3.67 **6.88**
7	5.59 **12.25**	4.74 **9.55**	4.35 **8.45**	4.12 **7.85**	3.97 **7.46**	3.87 **7.19**	3.79 **7.00**	3.73 **6.84**	3.68 **6.71**	3.63 **6.62**	3.60 **6.54**	3.57 **6.47**	3.52 **6.35**	3.49 **6.27**	3.44 **6.15**	3.41 **6.07**	3.38 **5.98**	3.34 **5.90**	3.32 **5.85**	3.29 **5.78**	3.28 **5.75**	3.25 **5.70**	3.24 **5.67**	3.23 **5.65**
8	5.32 **11.26**	4.46 **8.65**	4.07 **7.59**	3.84 **7.01**	3.69 **6.63**	3.58 **6.37**	3.50 **6.19**	3.44 **6.03**	3.39 **5.91**	3.34 **5.82**	3.31 **5.74**	3.28 **5.67**	3.23 **5.56**	3.20 **5.48**	3.15 **5.36**	3.12 **5.28**	3.08 **5.20**	3.05 **5.11**	3.03 **5.06**	3.00 **5.00**	2.98 **4.96**	2.96 **4.91**	2.94 **4.88**	2.93 **4.86**
9	5.12 **10.56**	4.26 **8.02**	3.86 **6.99**	3.63 **6.42**	3.48 **6.06**	3.37 **5.80**	3.29 **5.62**	3.23 **5.47**	3.18 **5.35**	3.13 **5.26**	3.10 **5.18**	3.07 **5.11**	3.02 **5.00**	2.98 **4.92**	2.93 **4.80**	2.90 **4.73**	2.86 **4.64**	2.82 **4.56**	2.80 **4.51**	2.77 **4.45**	2.76 **4.41**	2.73 **4.36**	2.72 **4.33**	2.71 **4.31**
10	4.96 **10.04**	4.10 **7.56**	3.71 **6.55**	3.48 **5.99**	3.33 **5.64**	3.22 **5.39**	3.14 **5.21**	3.07 **5.06**	3.02 **4.95**	2.97 **4.85**	2.94 **4.78**	2.91 **4.71**	2.86 **4.60**	2.82 **4.52**	2.77 **4.41**	2.74 **4.33**	2.70 **4.25**	2.67 **4.17**	2.64 **4.12**	2.61 **4.05**	2.59 **4.01**	2.56 **3.96**	2.55 **3.93**	2.54 **3.91**
11	4.84 **9.65**	3.98 **7.20**	3.59 **6.22**	3.36 **5.67**	3.20 **5.32**	3.09 **5.07**	3.01 **4.88**	2.95 **4.74**	2.90 **4.63**	2.86 **4.54**	2.82 **4.46**	2.79 **4.40**	2.74 **4.29**	2.70 **4.21**	2.65 **4.10**	2.61 **4.02**	2.57 **3.94**	2.53 **3.86**	2.50 **3.80**	2.47 **3.74**	2.45 **3.70**	2.42 **3.66**	2.41 **3.62**	2.40 **3.60**

Table D *(Continued)*

df																								
12	2.30 / 3.36	2.31 / 3.38	2.32 / 3.41	2.35 / 3.46	2.36 / 3.49	2.40 / 3.56	2.42 / 3.61	2.46 / 3.70	2.50 / 3.78	2.54 / 3.86	2.60 / 3.98	2.64 / 4.05	2.69 / 4.16	2.72 / 4.22	2.76 / 4.30	2.80 / 4.39	2.85 / 4.50	2.92 / 4.65	3.00 / 4.82	3.11 / 5.06	3.26 / 5.41	3.49 / 5.95	3.88 / 6.93	4.75 / 9.33
13	2.21 / 3.16	2.22 / 3.18	2.24 / 3.21	2.26 / 3.27	2.28 / 3.30	2.32 / 3.37	2.34 / 3.42	2.38 / 3.51	2.42 / 3.59	2.46 / 3.67	2.51 / 3.78	2.55 / 3.85	2.60 / 3.96	2.63 / 4.02	2.67 / 4.10	2.72 / 4.19	2.77 / 4.30	2.84 / 4.44	2.92 / 4.62	3.02 / 4.86	3.18 / 5.20	3.41 / 5.74	3.80 / 6.70	4.67 / 9.07
14	2.13 / 3.00	2.14 / 3.02	2.16 / 3.06	2.19 / 3.11	2.21 / 3.14	2.24 / 3.21	2.27 / 3.26	2.31 / 3.34	2.35 / 3.43	2.39 / 3.51	2.44 / 3.62	2.48 / 3.70	2.53 / 3.80	2.56 / 3.86	2.60 / 3.94	2.65 / 4.03	2.70 / 4.14	2.77 / 4.28	2.85 / 4.46	2.96 / 4.69	3.11 / 5.03	3.34 / 5.56	3.74 / 6.51	4.60 / 8.86
15	2.07 / 2.87	2.08 / 2.89	2.10 / 2.92	2.12 / 2.97	2.15 / 3.00	2.18 / 3.07	2.21 / 3.12	2.25 / 3.20	2.29 / 3.29	2.33 / 3.36	2.39 / 3.48	2.43 / 3.56	2.48 / 3.67	2.51 / 3.73	2.55 / 3.80	2.59 / 3.89	2.64 / 4.00	2.70 / 4.14	2.79 / 4.32	2.90 / 4.56	3.06 / 4.89	3.29 / 5.42	3.68 / 6.36	4.54 / 8.68
16	2.01 / 2.75	2.02 / 2.77	2.04 / 2.80	2.07 / 2.86	2.09 / 2.89	2.13 / 2.96	2.16 / 3.01	2.20 / 3.10	2.24 / 3.18	2.28 / 3.25	2.33 / 3.37	2.37 / 3.45	2.42 / 3.55	2.45 / 3.61	2.49 / 3.69	2.54 / 3.78	2.59 / 3.89	2.66 / 4.03	2.74 / 4.20	2.85 / 4.44	3.01 / 4.77	3.24 / 5.29	3.63 / 6.23	4.49 / 8.53
17	1.96 / 2.65	1.97 / 2.67	1.99 / 2.70	2.02 / 2.76	2.04 / 2.79	2.08 / 2.86	2.11 / 2.92	2.15 / 3.00	2.19 / 3.08	2.23 / 3.16	2.29 / 3.27	2.33 / 3.35	2.38 / 3.45	2.41 / 3.52	2.45 / 3.59	2.50 / 3.68	2.55 / 3.79	2.62 / 3.93	2.70 / 4.10	2.81 / 4.34	2.96 / 4.67	3.20 / 5.18	3.59 / 6.11	4.45 / 8.40
18	1.92 / 2.57	1.93 / 2.59	1.95 / 2.62	1.98 / 2.68	2.00 / 2.71	2.04 / 2.78	2.07 / 2.83	2.11 / 2.91	2.15 / 3.00	2.19 / 3.07	2.25 / 3.19	2.29 / 3.27	2.34 / 3.37	2.37 / 3.44	2.41 / 3.51	2.46 / 3.60	2.51 / 3.71	2.58 / 3.85	2.66 / 4.01	2.77 / 4.25	2.93 / 4.58	3.16 / 5.09	3.55 / 6.01	4.41 / 8.28
19	1.88 / 2.49	1.90 / 2.51	1.91 / 2.54	1.94 / 2.60	1.96 / 2.63	2.00 / 2.70	2.02 / 2.76	2.07 / 2.84	2.11 / 2.92	2.15 / 3.00	2.21 / 3.12	2.26 / 3.19	2.31 / 3.30	2.34 / 3.36	2.38 / 3.43	2.43 / 3.52	2.48 / 3.63	2.55 / 3.77	2.63 / 3.94	2.74 / 4.17	2.90 / 4.50	3.13 / 5.01	3.52 / 5.93	4.38 / 8.18
20	1.84 / 2.42	1.85 / 2.44	1.87 / 2.47	1.90 / 2.53	1.92 / 2.56	1.96 / 2.63	1.99 / 2.69	2.04 / 2.77	2.08 / 2.86	2.12 / 2.94	2.18 / 3.05	2.23 / 3.13	2.28 / 3.23	2.31 / 3.30	2.35 / 3.37	2.40 / 3.45	2.45 / 3.56	2.52 / 3.71	2.60 / 3.87	2.71 / 4.10	2.87 / 4.43	3.10 / 4.94	3.49 / 5.85	4.35 / 8.10
21	1.81 / 2.36	1.82 / 2.38	1.84 / 2.42	1.87 / 2.47	1.89 / 2.51	1.93 / 2.58	1.96 / 2.63	2.00 / 2.72	2.05 / 2.80	2.09 / 2.88	2.15 / 2.99	2.20 / 3.07	2.25 / 3.17	2.28 / 3.24	2.32 / 3.31	2.37 / 3.40	2.42 / 3.51	2.49 / 3.65	2.57 / 3.81	2.68 / 4.04	2.84 / 4.37	3.07 / 4.87	3.47 / 5.78	4.32 / 8.02
22	1.78 / 2.31	1.80 / 2.33	1.81 / 2.37	1.84 / 2.42	1.87 / 2.46	1.91 / 2.53	1.93 / 2.58	1.98 / 2.67	2.03 / 2.75	2.07 / 2.83	2.13 / 2.94	2.18 / 3.02	2.23 / 3.12	2.26 / 3.18	2.30 / 3.26	2.35 / 3.35	2.40 / 3.45	2.47 / 3.59	2.55 / 3.76	2.66 / 3.99	2.82 / 4.31	3.05 / 4.82	3.44 / 5.72	4.30 / 7.94
23	1.76 / 2.26	1.77 / 2.28	1.79 / 2.32	1.82 / 2.37	1.84 / 2.41	1.88 / 2.48	1.91 / 2.53	1.96 / 2.62	2.00 / 2.70	2.04 / 2.78	2.10 / 2.89	2.14 / 2.97	2.20 / 3.07	2.24 / 3.14	2.28 / 3.21	2.32 / 3.30	2.38 / 3.41	2.45 / 3.54	2.53 / 3.71	2.64 / 3.94	2.80 / 4.26	3.03 / 4.76	3.42 / 5.66	4.28 / 7.88
24	1.73 / 2.21	1.74 / 2.23	1.76 / 2.27	1.80 / 2.33	1.82 / 2.36	1.86 / 2.44	1.89 / 2.49	1.94 / 2.58	1.98 / 2.66	2.02 / 2.74	2.09 / 2.85	2.13 / 2.93	2.18 / 3.03	2.22 / 3.09	2.26 / 3.17	2.30 / 3.25	2.36 / 3.36	2.43 / 3.50	2.51 / 3.67	2.62 / 3.90	2.78 / 4.22	3.01 / 4.72	3.40 / 5.61	4.26 / 7.82

* Reprinted, by permission, from G. W. Snedecor and William G. Cochran, *Statistical methods*, 6th ed., Iowa State University Press, Ames, Iowa, 1967.

Table D (Continued)

Degrees of freedom for greater mean square

Degrees of freedom for lesser mean square	1	2	3	4	5	6	7	8	9	10	11	12	14	16	20	24	30	40	50	75	100	200	500	∞
25	4.24 **7.77**	3.38 **5.57**	2.99 **4.68**	2.76 **4.18**	2.60 **3.86**	2.49 **3.63**	2.41 **3.46**	2.34 **3.32**	2.28 **3.21**	2.24 **3.13**	2.20 **3.05**	2.16 **2.99**	2.11 **2.89**	2.06 **2.81**	2.00 **2.70**	1.96 **2.62**	1.92 **2.54**	1.87 **2.45**	1.84 **2.40**	1.80 **2.32**	1.77 **2.29**	1.74 **2.23**	1.72 **2.19**	1.71 **2.17**
26	4.22 **7.72**	3.37 **5.53**	2.98 **4.64**	2.74 **4.14**	2.59 **3.82**	2.47 **3.59**	2.39 **3.42**	2.32 **3.29**	2.27 **3.17**	2.22 **3.09**	2.18 **3.02**	2.15 **2.96**	2.10 **2.86**	2.05 **2.77**	1.99 **2.66**	1.95 **2.58**	1.90 **2.50**	1.85 **2.41**	1.82 **2.36**	1.78 **2.28**	1.76 **2.25**	1.72 **2.19**	1.70 **2.15**	1.69 **2.13**
27	4.21 **7.68**	3.35 **5.49**	2.96 **4.60**	2.73 **4.11**	2.57 **3.79**	2.46 **3.56**	2.37 **3.39**	2.30 **3.26**	2.25 **3.14**	2.20 **3.06**	2.16 **2.98**	2.13 **2.93**	2.08 **2.83**	2.03 **2.74**	1.97 **2.63**	1.93 **2.55**	1.88 **2.47**	1.84 **2.38**	1.80 **2.33**	1.76 **2.25**	1.74 **2.21**	1.71 **2.16**	1.68 **2.12**	1.67 **2.10**
28	4.20 **7.64**	3.34 **5.45**	2.95 **4.57**	2.71 **4.07**	2.56 **3.76**	2.44 **3.53**	2.36 **3.36**	2.29 **3.23**	2.24 **3.11**	2.19 **3.03**	2.15 **2.95**	2.12 **2.90**	2.06 **2.80**	2.02 **2.71**	1.96 **2.60**	1.91 **2.52**	1.87 **2.44**	1.81 **2.35**	1.78 **2.30**	1.75 **2.22**	1.72 **2.18**	1.69 **2.13**	1.67 **2.09**	1.65 **2.06**
29	4.18 **7.60**	3.33 **5.42**	2.93 **4.54**	2.70 **4.04**	2.54 **3.73**	2.43 **3.50**	2.35 **3.33**	2.28 **3.20**	2.22 **3.08**	2.18 **3.00**	2.14 **2.92**	2.10 **2.87**	2.05 **2.77**	2.00 **2.68**	1.94 **2.57**	1.90 **2.49**	1.85 **2.41**	1.80 **2.32**	1.77 **2.27**	1.73 **2.19**	1.71 **2.15**	1.68 **2.10**	1.65 **2.06**	1.64 **2.03**
30	4.17 **7.56**	3.32 **5.39**	2.92 **4.51**	2.69 **4.02**	2.53 **3.70**	2.42 **3.47**	2.34 **3.30**	2.27 **3.17**	2.21 **3.06**	2.16 **2.98**	2.12 **2.90**	2.09 **2.84**	2.04 **2.74**	1.99 **2.66**	1.93 **2.55**	1.89 **2.47**	1.84 **2.38**	1.79 **2.29**	1.76 **2.24**	1.72 **2.16**	1.69 **2.13**	1.66 **2.07**	1.64 **2.03**	1.62 **2.01**
32	4.15 **7.50**	3.30 **5.34**	2.90 **4.46**	2.67 **3.97**	2.51 **3.66**	2.40 **3.42**	2.32 **3.25**	2.25 **3.12**	2.19 **3.01**	2.14 **2.94**	2.10 **2.86**	2.07 **2.80**	2.02 **2.70**	1.97 **2.62**	1.91 **2.51**	1.86 **2.42**	1.82 **2.34**	1.76 **2.25**	1.74 **2.20**	1.69 **2.12**	1.67 **2.08**	1.64 **2.02**	1.61 **1.98**	1.59 **1.96**
34	4.13 **7.44**	3.28 **5.29**	2.88 **4.42**	2.65 **3.93**	2.49 **3.61**	2.38 **3.38**	2.30 **3.21**	2.23 **3.08**	2.17 **2.97**	2.12 **2.89**	2.08 **2.82**	2.05 **2.76**	2.00 **2.66**	1.95 **2.58**	1.89 **2.47**	1.84 **2.38**	1.80 **2.30**	1.74 **2.21**	1.71 **2.15**	1.67 **2.08**	1.64 **2.04**	1.61 **1.98**	1.59 **1.94**	1.57 **1.91**
36	4.11 **7.39**	3.26 **5.25**	2.86 **4.38**	2.63 **3.89**	2.48 **3.58**	2.36 **3.35**	2.28 **3.18**	2.21 **3.04**	2.15 **2.94**	2.10 **2.86**	2.06 **2.78**	2.03 **2.72**	1.98 **2.62**	1.93 **2.54**	1.87 **2.43**	1.82 **2.35**	1.78 **2.26**	1.72 **2.17**	1.69 **2.12**	1.65 **2.04**	1.62 **2.00**	1.59 **1.94**	1.56 **1.90**	1.55 **1.87**
38	4.10 **7.35**	3.25 **5.21**	2.85 **4.34**	2.62 **3.86**	2.46 **3.54**	2.35 **3.32**	2.26 **3.15**	2.19 **3.02**	2.14 **2.91**	2.09 **2.82**	2.05 **2.75**	2.02 **2.69**	1.96 **2.59**	1.92 **2.51**	1.85 **2.40**	1.80 **2.32**	1.76 **2.22**	1.71 **2.14**	1.67 **2.08**	1.63 **2.00**	1.60 **1.97**	1.57 **1.90**	1.54 **1.86**	1.53 **1.84**
40	4.08 **7.31**	3.23 **5.18**	2.84 **4.31**	2.61 **3.83**	2.45 **3.51**	2.34 **3.29**	2.25 **3.12**	2.18 **2.99**	2.12 **2.88**	2.07 **2.80**	2.04 **2.73**	2.00 **2.66**	1.95 **2.56**	1.90 **2.49**	1.84 **2.37**	1.79 **2.29**	1.74 **2.20**	1.69 **2.11**	1.66 **2.05**	1.61 **1.97**	1.59 **1.94**	1.55 **1.88**	1.53 **1.84**	1.51 **1.81**
42	4.07 **7.27**	3.22 **5.15**	2.83 **4.29**	2.59 **3.80**	2.44 **3.49**	2.32 **3.26**	2.24 **3.10**	2.17 **2.96**	2.11 **2.86**	2.06 **2.77**	2.02 **2.70**	1.99 **2.64**	1.94 **2.54**	1.89 **2.46**	1.82 **2.35**	1.78 **2.26**	1.73 **2.17**	1.68 **2.08**	1.64 **2.02**	1.60 **1.94**	1.57 **1.91**	1.54 **1.85**	1.51 **1.80**	1.49 **1.78**
44	4.06 **7.24**	3.21 **5.12**	2.82 **4.26**	2.58 **3.78**	2.43 **3.46**	2.31 **3.24**	2.23 **3.07**	2.16 **2.94**	2.10 **2.84**	2.05 **2.75**	2.01 **2.68**	1.98 **2.62**	1.92 **2.52**	1.88 **2.44**	1.81 **2.32**	1.76 **2.24**	1.72 **2.15**	1.66 **2.06**	1.63 **2.00**	1.58 **1.92**	1.56 **1.88**	1.52 **1.82**	1.50 **1.78**	1.48 **1.75**

Table D (*Continued*)

df	1	2	3	4	5	6	7	8	9	10	11	12	13	14	15	16	17	18	19	20	21	22	23	24
46	4.05 / 7.21	3.20 / 5.10	2.81 / 4.24	2.57 / 3.76	2.42 / 3.44	2.30 / 3.22	2.22 / 3.05	2.14 / 2.92	2.09 / 2.82	2.04 / 2.73	2.00 / 2.66	1.97 / 2.60	1.91 / 2.50	1.87 / 2.42	1.80 / 2.30	1.75 / 2.22	1.71 / 2.13	1.65 / 2.04	1.62 / 1.98	1.57 / 1.90	1.54 / 1.86	1.51 / 1.80	1.48 / 1.76	1.46 / 1.72
48	4.04 / 7.19	3.19 / 5.08	2.80 / 4.22	2.56 / 3.74	2.41 / 3.42	2.30 / 3.20	2.21 / 3.04	2.14 / 2.90	2.08 / 2.80	2.03 / 2.71	1.99 / 2.64	1.96 / 2.58	1.90 / 2.48	1.86 / 2.40	1.79 / 2.28	1.74 / 2.20	1.70 / 2.11	1.64 / 2.02	1.61 / 1.96	1.56 / 1.88	1.53 / 1.84	1.50 / 1.78	1.47 / 1.73	1.45 / 1.70
50	4.03 / 7.17	3.18 / 5.06	2.79 / 4.20	2.56 / 3.72	2.40 / 3.41	2.29 / 3.18	2.20 / 3.02	2.13 / 2.88	2.07 / 2.78	2.02 / 2.70	1.98 / 2.62	1.95 / 2.56	1.90 / 2.46	1.85 / 2.39	1.78 / 2.26	1.74 / 2.18	1.69 / 2.10	1.63 / 2.00	1.60 / 1.94	1.55 / 1.86	1.52 / 1.82	1.48 / 1.76	1.46 / 1.71	1.45 / 1.70
55	4.02 / 7.12	3.17 / 5.01	2.78 / 4.16	2.54 / 3.68	2.38 / 3.37	2.27 / 3.15	2.18 / 2.98	2.11 / 2.85	2.05 / 2.75	2.00 / 2.66	1.97 / 2.59	1.93 / 2.53	1.88 / 2.43	1.83 / 2.35	1.76 / 2.23	1.72 / 2.15	1.67 / 2.06	1.61 / 1.96	1.58 / 1.90	1.52 / 1.82	1.50 / 1.78	1.46 / 1.71	1.43 / 1.66	1.41 / 1.64
60	4.00 / 7.08	3.15 / 4.98	2.76 / 4.13	2.52 / 3.65	2.37 / 3.34	2.25 / 3.12	2.17 / 2.95	2.10 / 2.82	2.04 / 2.72	1.99 / 2.63	1.95 / 2.56	1.92 / 2.50	1.86 / 2.40	1.81 / 2.32	1.75 / 2.20	1.70 / 2.12	1.65 / 2.03	1.59 / 1.93	1.56 / 1.87	1.50 / 1.79	1.48 / 1.74	1.44 / 1.68	1.41 / 1.63	1.39 / 1.60
65	3.99 / 7.04	3.14 / 4.95	2.75 / 4.10	2.51 / 3.62	2.36 / 3.31	2.24 / 3.09	2.15 / 2.93	2.08 / 2.79	2.02 / 2.70	1.98 / 2.61	1.94 / 2.54	1.90 / 2.47	1.85 / 2.37	1.80 / 2.30	1.73 / 2.18	1.68 / 2.09	1.63 / 2.00	1.57 / 1.90	1.54 / 1.84	1.49 / 1.76	1.46 / 1.71	1.42 / 1.64	1.39 / 1.60	1.37 / 1.56
70	3.98 / 7.01	3.13 / 4.92	2.74 / 4.08	2.50 / 3.60	2.35 / 3.29	2.23 / 3.07	2.14 / 2.91	2.07 / 2.77	2.01 / 2.67	1.97 / 2.59	1.93 / 2.51	1.89 / 2.45	1.84 / 2.35	1.79 / 2.28	1.72 / 2.15	1.67 / 2.07	1.62 / 1.98	1.56 / 1.88	1.53 / 1.82	1.47 / 1.74	1.45 / 1.69	1.40 / 1.62	1.37 / 1.56	1.35 / 1.53
80	3.96 / 6.96	3.11 / 4.88	2.72 / 4.04	2.48 / 3.56	2.33 / 3.25	2.21 / 3.04	2.12 / 2.87	2.05 / 2.74	1.99 / 2.64	1.95 / 2.55	1.91 / 2.48	1.88 / 2.41	1.82 / 2.32	1.77 / 2.24	1.70 / 2.11	1.65 / 2.03	1.60 / 1.94	1.54 / 1.84	1.51 / 1.78	1.45 / 1.70	1.42 / 1.65	1.38 / 1.57	1.35 / 1.52	1.32 / 1.49
100	3.94 / 6.90	3.09 / 4.82	2.70 / 3.98	2.46 / 3.51	2.30 / 3.20	2.19 / 2.99	2.10 / 2.82	2.03 / 2.69	1.97 / 2.59	1.92 / 2.51	1.88 / 2.43	1.85 / 2.36	1.79 / 2.26	1.75 / 2.19	1.68 / 2.06	1.63 / 1.98	1.57 / 1.89	1.51 / 1.79	1.48 / 1.73	1.42 / 1.64	1.39 / 1.59	1.34 / 1.51	1.30 / 1.46	1.28 / 1.43
125	3.92 / 6.84	3.07 / 4.78	2.68 / 3.94	2.44 / 3.47	2.29 / 3.17	2.17 / 2.95	2.08 / 2.79	2.01 / 2.65	1.95 / 2.56	1.90 / 2.47	1.86 / 2.40	1.83 / 2.33	1.77 / 2.23	1.72 / 2.15	1.65 / 2.03	1.60 / 1.94	1.55 / 1.85	1.49 / 1.75	1.45 / 1.68	1.39 / 1.59	1.36 / 1.54	1.31 / 1.46	1.27 / 1.40	1.25 / 1.37
150	3.91 / 6.81	3.06 / 4.75	2.67 / 3.91	2.43 / 3.44	2.27 / 3.14	2.16 / 2.92	2.07 / 2.76	2.00 / 2.62	1.94 / 2.53	1.89 / 2.44	1.85 / 2.37	1.82 / 2.30	1.76 / 2.20	1.71 / 2.12	1.64 / 2.00	1.59 / 1.91	1.54 / 1.83	1.47 / 1.72	1.44 / 1.66	1.37 / 1.56	1.34 / 1.51	1.29 / 1.43	1.25 / 1.37	1.22 / 1.33
200	3.89 / 6.76	3.04 / 4.71	2.65 / 3.88	2.41 / 3.41	2.26 / 3.11	2.14 / 2.90	2.05 / 2.73	1.98 / 2.60	1.92 / 2.50	1.87 / 2.41	1.83 / 2.34	1.80 / 2.28	1.74 / 2.17	1.69 / 2.09	1.62 / 1.97	1.57 / 1.88	1.52 / 1.79	1.45 / 1.69	1.42 / 1.62	1.35 / 1.53	1.32 / 1.48	1.26 / 1.39	1.22 / 1.33	1.19 / 1.28
400	3.86 / 6.70	3.02 / 4.66	2.62 / 3.83	2.39 / 3.36	2.23 / 3.06	2.12 / 2.85	2.03 / 2.69	1.96 / 2.55	1.89 / 2.46	1.85 / 2.37	1.81 / 2.29	1.78 / 2.23	1.72 / 2.12	1.67 / 2.04	1.60 / 1.92	1.54 / 1.84	1.49 / 1.74	1.42 / 1.64	1.38 / 1.57	1.32 / 1.47	1.28 / 1.42	1.22 / 1.32	1.16 / 1.24	1.13 / 1.19
1000	3.85 / 6.66	3.00 / 4.62	2.61 / 3.80	2.38 / 3.34	2.22 / 3.04	2.10 / 2.82	2.02 / 2.66	1.95 / 2.53	1.89 / 2.43	1.84 / 2.34	1.80 / 2.26	1.76 / 2.20	1.70 / 2.09	1.65 / 2.01	1.58 / 1.89	1.53 / 1.81	1.47 / 1.71	1.41 / 1.61	1.36 / 1.54	1.30 / 1.44	1.26 / 1.38	1.19 / 1.28	1.13 / 1.19	1.08 / 1.11
∞	3.84 / 6.64	2.99 / 4.60	2.60 / 3.78	2.37 / 3.32	2.21 / 3.02	2.09 / 2.80	2.01 / 2.64	1.94 / 2.51	1.88 / 2.41	1.83 / 2.32	1.79 / 2.24	1.75 / 2.18	1.69 / 2.07	1.64 / 1.99	1.57 / 1.87	1.52 / 1.79	1.46 / 1.69	1.40 / 1.59	1.35 / 1.52	1.28 / 1.41	1.24 / 1.36	1.17 / 1.25	1.11 / 1.15	1.00 / 1.00

Table E   Transformation of r to z_r*

r	z_r	r	z_r	r	z_r	r	z_r	r	z_r
.000	.000	.200	.203	.400	.424	.600	.693	.800	1.099
.005	.005	.205	.208	.405	.430	.605	.701	.805	1.113
.010	.010	.210	.213	.410	.436	.610	.709	.810	1.127
.015	.015	.215	.218	.415	.442	.615	.717	.815	1.142
.020	.020	.220	.224	.420	.448	.620	.725	.820	1.157
.025	.025	.225	.229	.425	.454	.625	.733	.825	1.172
.030	.030	.230	.234	.430	.460	.630	.741	.830	1.188
.035	.035	.235	.239	.435	.466	.635	.750	.835	1.204
.040	.040	.240	.245	.440	.472	.640	.758	.840	1.221
.045	.045	.245	.250	.445	.478	.645	.767	.845	1.238
.050	.050	.250	.255	.450	.485	.650	.775	.850	1.256
.055	.055	.255	.261	.455	.491	.655	.784	.855	1.274
.060	.060	.260	.266	.460	.497	.660	.793	.860	1.293
.065	.065	.265	.271	.465	.504	.665	.802	.865	1.313
.070	.070	.270	.277	.470	.510	.670	.811	.870	1.333
.075	.075	.275	.282	.475	.517	.675	.820	.875	1.354
.080	.080	.280	.288	.480	.523	.680	.829	.880	1.376
.085	.085	.285	.293	.485	.530	.685	.838	.885	1.398
.090	.090	.290	.299	.490	.536	.690	.848	.890	1.422
.095	.095	.295	.304	.495	.543	.695	.858	.895	1.447
.100	.100	.300	.310	.500	.549	.700	.867	.900	1.472
.105	.105	.305	.315	.505	.556	.705	.877	.905	1.499
.110	.110	.310	.321	.510	.563	.710	.887	.910	1.528
.115	.116	.315	.326	.515	.570	.715	.897	.915	1.557
.120	.121	.320	.332	.520	.576	.720	.908	.920	1.589
.125	.126	.325	.337	.525	.583	.725	.918	.925	1.623
.130	.131	.330	.343	.530	.590	.730	.929	.930	1.658
.135	.136	.335	.348	.535	.597	.735	.940	.935	1.697
.140	.141	.340	.354	.540	.604	.740	.950	.940	1.738
.145	.146	.345	.360	.545	.611	.745	.962	.945	1.783
.150	.151	.350	.365	.550	.618	.750	.973	.950	1.832
.155	.156	.355	.371	.555	.626	.755	.984	.955	1.886
.160	.161	.360	.377	.560	.633	.760	.996	.960	1.946
.165	.167	.365	.383	.565	.640	.765	1.008	.965	2.014
.170	.172	.370	.388	.570	.648	.770	1.020	.970	2.092
.175	.177	.375	.394	.575	.655	.775	1.033	.975	2.185
.180	.182	.380	.400	.580	.662	.780	1.045	.980	2.298
.185	.187	.385	.406	.585	.670	.785	1.058	.985	2.443
.190	.192	.390	.412	.590	.678	.790	1.071	.990	2.647
.195	.198	.395	.418	.595	.685	.795	1.085	.995	2.994

* Reprinted, by permission, from Allen L. Edwards, *Statistical methods*, 2nd ed., Holt, Rinehart, and Winston, Inc., New York, 1967.

Table F   Critical values of the correlation coefficient*

Level of significance for one-tailed test

.05	.025	.01	.005

Level of significance for two-tailed test

df = N-2	.10	.05	.02	.01
1	.988	.997	.9995	.9999
2	.900	.950	.980	.990
3	.805	.878	.934	.959
4	.729	.811	.882	.917
5	.669	.754	.833	.874
6	.622	.707	.789	.834
7	.582	.666	.750	.798
8	.549	.632	.716	.765
9	.521	.602	.685	.735
10	.497	.576	.658	.708
11	.476	.553	.634	.684
12	.458	.532	.612	.661
13	.441	.514	.592	.641
14	.426	.497	.574	.623
15	.412	.482	.558	.606
16	.400	.468	.542	.590
17	.389	.456	.528	.575
18	.378	.444	.516	.561
19	.369	.433	.503	.549
20	.360	.423	.492	.537
21	.352	.413	.482	.526
22	.344	.404	.472	.515
23	.337	.396	.462	.505
24	.330	.388	.453	.496
25	.323	.381	.445	.487
26	.317	.374	.437	.479
27	.311	.367	.430	.471
28	.306	.361	.423	.463
29	.301	.355	.416	.456
30	.296	.349	.409	.449
35	.275	.325	.381	.418
40	.257	.304	.358	.393
45	.243	.288	.338	.372
50	.231	.273	.322	.354
60	.211	.250	.295	.325
70	.195	.232	.274	.303
80	.183	.217	.256	.283
90	.173	.205	.242	.267
100	.164	.195	.230	.254

* Abridged from R. A. Fisher and F. Yates, *Statistical tables for biological, agricultural, and medical research*, Oliver & Boyd, Ltd., Edinburgh, by permission of the authors and publishers.

Table G   Critical values of ρ, the Spearman rank correlation coefficient*

Significance level (one-tailed test)

N	.05	.01
4	1.000	
5	.900	1.000
6	.829	.943
7	.714	.893
8	.643	.833
9	.600	.783
10	.564	.746
12	.506	.712
14	.456	.645
16	.425	.601
18	.399	.564
20	.377	.534
22	.359	.508
24	.343	.485
26	.329	.465
28	.317	.448
30	.306	.432

* Adapted from E. G. Olds, Distributions of sums of squares of rank differences for small numbers of individuals, *Annals of Mathematical Statistics*, 9, 133–148, 1938; The 5% significance levels for sums of squares of rank differences and a correction, *Annals of Mathematical Statistics*, 20, 117–118, 1949; with the kind permission of the author and the publisher.

Table H Probabilities associated with values as large as observed values of S in the Kendall rank correlation coefficient*

	Values of N					Values of N		
S	4	5	8	9	S	6	7	10
0	.625	.592	.548	.540	1	.500	.500	.500
2	.375	.408	.452	.460	3	.360	.386	.431
4	.167	.242	.360	.381	5	.235	.281	.364
6	.042	.117	.274	.306	7	.136	.191	.300
8		.042	.199	.238	9	.068	.119	.242
10		.0083	.138	.179	11	.028	.068	.190
12			.089	.130	13	.0083	.035	.146
14			.054	.090	15	.0014	.015	.108
16			.031	.060	17		.0054	.078
18			.016	.038	19		.0014	.054
20			.0071	.022	21		.00020	.036
22			.0028	.012	23			.023
24			.00087	.0063	25			.014
26			.00019	.0029	27			.0083
28			.000025	.0012	29			.0046
30				.00043	31			.0023
32				.00012	33			.0011
34				.000025	35			.00047
36				.0000028	37			.00018
					39			.000058
					41			.000015
					43			.0000028
					45			.00000028

* Adapted by permission from M. G. Kendall, *Rank correlation methods*, 2d ed., Charles Griffin & Co., Ltd., London, 1955.

Table I Critical and quasi-critical lower-tail values of W_+ (and their probability levels) for Wilcoxon's signed-rank test*

N	$\alpha = .05$		$\alpha = .025$		$\alpha = .01$		$\alpha = .005$	
5	0	.0313						
	1	.0625						
6	2	.0469	0	.0156				
	3	.0781	1	.0313				
7	3	.0391	2	.0234	0	.0078		
	4	.0547	3	.0391	1	.0156		
8	5	.0391	3	.0195	1	.0078	0	.0039
	6	.0547	4	.0273	2	.0117	1	.0078
9	8	.0488	5	.0195	3	.0098	1	.0039
	9	.0645	6	.0273	4	.0137	2	.0059
10	10	.0420	8	.0244	5	.0098	3	.0049
	11	.0527	9	.0322	6	.0137	4	.0068
11	13	.0415	10	.0210	7	.0093	5	.0049
	14	.0508	11	.0269	8	.0122	6	.0068
12	17	.0461	13	.0212	9	.0081	7	.0046
	18	.0549	14	.0261	10	.0105	8	.0061
13	21	.0471	17	.0239	12	.0085	9	.0040
	22	.0549	18	.0287	13	.0107	10	.0052
14	25	.0453	21	.0247	15	.0083	12	.0043
	26	.0520	22	.0290	16	.0101	13	.0054
15	30	.0473	25	.0240	19	.0090	15	.0042
	31	.0535	26	.0277	20	.0108	16	.0051
16	35	.0467	29	.0222	23	.0091	19	.0046
	36	.0523	30	.0253	24	.0107	20	.0055
17	41	.0492	34	.0224	27	.0087	23	.0047
	42	.0544	35	.0253	28	.0101	24	.0055
18	47	.0494	40	.0241	32	.0091	27	.0045
	48	.0542	41	.0269	33	.0104	28	.0052
19	53	.0478	46	.0247	37	.0090	32	.0047
	54	.0521	47	.0273	38	.0102	33	.0054
20	60	.0487	52	.0242	43	.0096	37	.0047
	61	.0527	53	.0266	44	.0107	38	.0053

* Body of table is reproduced, with changes only in notation, from Table II in F. Wilcoxon, S. K. Katti, and Roberta A. Wilcox, *Critical values and probability levels for the Wilcoxon rank sum test and the Wilcoxon signed rank test*, American Cyanamid Company (Lederle Laboratories Division, Pearl River, N.Y.) and The Florida State University (Department of Statistics, Tallahassee, Fla.), August, 1963, with permission of the authors and publishers.

Table I (*Continued*)

N	$\alpha = .05$		$\alpha = .025$		$\alpha = .01$		$\alpha = .005$	
21	67	.0479	58	.0230	49	.0097	42	.0045
	68	.0516	59	.0251	50	.0108	43	.0051
22	75	.0492	65	.0231	55	.0095	48	.0046
	76	.0527	66	.0250	56	.0104	49	.0052
23	83	.0490	73	.0242	62	.0098	54	.0046
	84	.0523	74	.0261	63	.0107	55	.0051
24	91	.0475	81	.0245	69	.0097	61	.0048
	92	.0505	82	.0263	70	.0106	62	.0053
25	100	.0479	89	.0241	76	.0094	68	.0048
	101	.0507	90	.0258	77	.0101	69	.0053
26	110	.0497	98	.0247	84	.0095	75	.0047
	111	.0524	99	.0263	85	.0102	76	.0051
27	119	.0477	107	.0246	92	.0093	83	.0048
	120	.0502	108	.0260	93	.0100	84	.0052
28	130	.0496	116	.0239	101	.0096	91	.0048
	131	.0521	117	.0252	102	.0102	92	.0051
29	140	.0482	126	.0240	110	.0095	100	.0049
	141	.0504	127	.0253	111	.0101	101	.0053
30	151	.0481	137	.0249	120	.0098	109	.0050
	152	.0502	138	.0261	121	.0104	110	.0053
31	163	.0491	147	.0239	130	.0099	118	.0049
	164	.0512	148	.0251	131	.0105	119	.0052
32	175	.0492	159	.0249	140	.0097	128	.0050
	176	.0512	160	.0260	141	.0103	129	.0053
33	187	.0485	170	.0242	151	.0099	138	.0049
	188	.0503	171	.0253	152	.0104	139	.0052
34	200	.0488	182	.0242	162	.0098	148	.0048
	201	.0506	183	.0252	163	.0103	149	.0051
35	213	.0484	195	.0247	173	.0096	159	.0048
	214	.0501	196	.0257	174	.0100	160	.0051
36	227	.0489	208	.0248	185	.0096	171	.0050
	228	.0505	209	.0258	186	.0100	172	.0052
37	241	.0487	221	.0245	198	.0099	182	.0048
	242	.0503	222	.0254	199	.0103	183	.0050
38	256	.0493	235	.0247	211	.0099	194	.0048
	257	.0509	236	.0256	212	.0104	195	.0050
39	271	.0493	249	.0246	224	.0099	207	.0049
	272	.0507	250	.0254	225	.0103	208	.0051
40	286	.0486	264	.0249	238	.0100	220	.0049
	287	.0500	265	.0257	239	.0104	221	.0051

Table I (*Continued*)

N	$\alpha = .05$		$\alpha = .025$		$\alpha = .01$		$\alpha = .005$	
41	302	.0488	279	.0248	252	.0100	233	.0048
	303	.0501	280	.0256	253	.0103	234	.0050
42	319	.0496	294	.0245	266	.0098	247	.0049
	320	.0509	295	.0252	267	.0102	248	.0051
43	336	.0498	310	.0245	281	.0098	261	.0048
	337	.0511	311	.0252	282	.0102	262	.0050
44	353	.0495	327	.0250	296	.0097	276	.0049
	354	.0507	328	.0257	297	.0101	277	.0051
45	371	.0498	343	.0244	312	.0098	291	.0049
	372	.0510	344	.0251	313	.0101	292	.0051
46	389	.0497	361	.0249	328	.0098	307	.0050
	390	.0508	362	.0256	329	.0101	308	.0052
47	407	.0490	378	.0245	345	.0099	322	.0048
	408	.0501	379	.0251	346	.0102	323	.0050
48	426	.0490	396	.0244	362	.0099	339	.0050
	427	.0500	397	.0251	363	.0102	340	.0051
49	446	.0495	415	.0247	379	.0098	355	.0049
	447	.0505	416	.0253	380	.0100	356	.0050
50	466	.0495	434	.0247	397	.0098	373	.0050
	467	.0506	435	.0253	398	.0101	374	.0051

Table J   Coefficients of orthogonal polynomials

k	Polynomial	Coefficients										$\Sigma c_{ij}{}^2$
3	Linear	−1	0	1								2
	Quadratic	1	−2	1								6
	Linear	−3	−1	1	3							20
4	Quadratic	1	−1	−1	1							4
	Cubic	−1	3	−3	1							20
	Linear	−2	−1	0	1	2						10
5	Quadratic	2	−1	−2	−1	2						14
	Cubic	−1	2	0	−2	1						10
	Quartic	1	−4	6	−4	1						70
	Linear	−5	−3	−1	1	3	5					70
6	Quadratic	5	−1	−4	−4	−1	5					84
	Cubic	−5	7	4	−4	−7	5					180
	Quartic	1	−3	2	2	−3	1					28
	Linear	−3	−2	−1	0	1	2	3				28
7	Quadratic	5	0	−3	−4	−3	0	5				84
	Cubic	−1	1	1	0	−1	−1	1				6
	Quartic	3	−7	1	6	1	−7	3				154
	Linear	−7	−5	−3	−1	1	3	5	7			168
	Quadratic	7	1	−3	−5	−5	−3	1	7			168
8	Cubic	−7	5	7	3	−3	−7	−5	7			264
	Quartic	7	−13	−3	9	9	−3	−13	7			616
	Quintic	−7	23	−17	−15	15	17	−23	7			2184
	Linear	−4	−3	−2	−1	0	1	2	3	4		60
	Quadratic	28	7	−8	−17	−20	−17	−8	7	28		2772
9	Cubic	−14	7	13	9	0	−9	−13	−7	14		990
	Quartic	14	−21	−11	9	18	9	−11	−21	14		2002
	Quintic	−4	11	−4	−9	0	9	4	−11	4		468
	Linear	−9	−7	−5	−3	−1	1	3	5	7	9	330
	Quadratic	6	2	−1	−3	−4	−4	−3	−1	2	6	132
10	Cubic	−42	14	35	31	12	−12	−31	−35	−14	42	8580
	Quartic	18	−22	−17	3	18	18	3	−17	−22	18	2860
	Quintic	−6	14	−1	−11	−6	6	11	1	−14	6	780

Table K Critical lower-tail values of R_1 for rank test for two independent samples ($N_1 < N_2$)*

N_2	$N_1 = 1$							$N_1 = 2$							N_2
	0.001	0.005	0.010	0.025	0.05	0.10	$2\bar{R}_1$	0.001	0.005	0.010	0.025	0.05	0.10	$2\bar{R}_1$	
2							4						—	10	2
3							5						3	12	3
4							6					—	3	14	4
5							7					3	4	16	5
6							8					3	4	18	6
7							9				—	3	4	20	7
8						—	10				3	4	5	22	8
9					1		11				3	4	5	24	9
10					1		12				3	4	6	26	10
11					1		13				3	4	6	28	11
12					1		14			—	4	5	7	30	12
13					1		15			3	4	5	7	32	13
14					1		16			3	4	6	8	34	14
15					1		17			3	4	6	8	36	15
16					1		18			3	4	6	8	38	16
17					1		19			3	5	6	9	40	17
18				—	1		20		—	3	5	7	9	42	18
19			1	2			21		3	4	5	7	10	44	19
20			1	2			22		3	4	5	7	10	46	20
21			1	2			23		3	4	6	8	11	48	21
22			1	2			24		3	4	6	8	11	50	22
23			1	2			25		3	4	6	8	12	52	23
24			1	2			26		3	4	6	9	12	54	24
25	—	—	—	—	1	2	27	—	3	4	6	9	12	56	25

N_2	$N_1 = 3$							$N_1 = 4$							N_2
	0.001	0.005	0.010	0.025	0.05	0.10	$2\bar{R}_1$	0.001	0.005	0.010	0.025	0.05	0.10	$2\bar{R}_1$	
3					6	7	21								3
4				—	6	7	24			—	10	11	13	36	4
5				6	7	8	27		—	10	11	12	14	40	5
6			—	7	8	9	30		10	11	12	13	15	44	6
7			6	7	8	10	33		10	11	13	14	16	48	7
8		—	6	8	9	11	36		11	12	14	15	17	52	8
9		6	7	8	10	11	39	—	11	13	14	16	19	56	9
10		6	7	9	10	12	42	10	12	13	15	17	20	60	10
11		6	7	9	11	13	45	10	12	14	16	18	21	64	11
12		7	8	10	11	14	48	10	13	15	17	19	22	68	12
13		7	8	10	12	15	51	11	13	15	18	20	23	72	13
14		7	8	11	13	16	54	11	14	16	19	21	25	76	14
15		8	9	11	13	16	57	11	15	17	20	22	26	80	15
16	—	8	9	12	14	17	60	12	15	17	21	24	27	84	16
17	6	8	10	12	15	18	63	12	16	18	21	25	28	88	17
18	6	8	10	13	15	19	66	13	16	19	22	26	30	92	18
19	6	9	10	13	16	20	69	13	17	19	23	27	31	96	19
20	6	9	11	14	17	21	72	13	18	20	24	28	32	100	20
21	7	9	11	14	17	21	75	14	18	21	25	29	33	104	21
22	7	10	12	15	18	22	78	14	19	21	26	30	35	108	22
23	7	10	12	15	19	23	81	14	19	22	27	31	36	112	23
24	7	10	12	16	19	24	84	15	20	23	27	32	38	116	24
25	7	11	13	16	20	25	87	15	20	23	28	33	38	120	25

* This table is reproduced, with changes in notation, from Table I in L. R. Verdooren, Extended tables of critical values for Wilcoxon's test statistic, *Biometrika*, 50, 177–186, 1963, with permission of the author and editor.

The table shows (i) values, R_1, which are just significant at the probability level quoted at the head of column, (ii) twice the mean $2\bar{R}_1$. A blank space indicates that no value is significant at that level. The upper critical value is $2\bar{R}_1 - R_1$.

Table K (Continued)

$N_1 = 5$

N_2	0.001	0.005	0.010	0.025	0.05	0.10	$2\bar{R}_1$
5		15	16	17	19	20	55
6		16	17	18	20	22	60
7	—	16	18	20	21	23	65
8	15	17	19	21	23	25	70
9	16	18	20	22	24	27	75
10	16	19	21	23	26	28	80
11	17	20	22	24	27	30	85
12	17	21	23	26	28	32	90
13	18	22	24	27	30	33	95
14	18	22	25	28	31	35	100
15	19	23	26	29	33	37	105
16	20	24	27	30	34	38	110
17	20	25	28	32	35	40	115
18	21	26	29	33	37	42	120
19	22	27	30	34	38	43	125
20	22	28	31	35	40	45	130
21	23	29	32	37	41	47	135
22	23	29	33	38	43	48	140
23	24	30	34	39	44	50	145
24	25	31	35	40	45	51	150
25	25	32	36	42	47	53	155

$N_1 = 6$

0.001	0.005	0.010	0.025	0.05	0.10	$2\bar{R}_1$	N_2
—	23	24	26	28	30	78	6
21	24	25	27	29	32	84	7
22	25	27	29	31	34	90	8
23	26	28	31	33	36	96	9
24	27	29	32	35	38	102	10
25	28	30	34	37	40	108	11
25	30	32	35	38	42	114	12
26	31	33	37	40	44	120	13
27	32	34	38	42	46	126	14
28	33	36	40	44	48	132	15
29	34	37	42	46	50	138	16
30	36	39	43	47	52	144	17
31	37	40	45	49	55	150	18
32	38	41	46	51	57	156	19
33	39	43	48	53	59	162	20
33	40	44	50	55	61	168	21
34	42	45	51	57	63	174	22
35	43	47	53	58	65	180	23
36	44	48	54	60	67	186	24
37	45	50	56	62	69	192	25

$N_1 = 7$

N_2	0.001	0.005	0.010	0.025	0.05	0.10	$2\bar{R}_1$
7	29	32	34	36	39	41	105
8	30	34	35	38	41	44	112
9	31	35	37	40	43	46	119
10	33	37	39	42	45	49	126
11	34	38	40	44	47	51	133
12	35	40	42	46	49	54	140
13	36	41	44	48	52	56	147
14	37	43	45	50	54	59	154
15	38	44	47	52	56	61	161
16	39	46	49	54	58	64	168
17	41	47	51	56	61	66	175
18	42	49	52	58	63	69	182
19	43	50	54	60	65	71	189
20	44	52	56	62	67	74	196
21	46	53	58	64	69	76	203
22	47	55	59	66	72	79	210
23	48	57	61	68	74	81	217
24	49	58	63	70	76	84	224
25	50	60	64	72	78	86	231

$N_1 = 8$

0.001	0.005	0.010	0.025	0.05	0.10	$2\bar{R}_1$	N_2
40	43	45	49	51	55	136	8
41	45	47	51	54	58	144	9
42	47	49	53	56	60	152	10
44	49	51	55	59	63	160	11
45	51	53	58	62	66	168	12
47	53	56	60	64	69	176	13
48	54	58	62	67	72	184	14
50	56	60	65	69	75	192	15
51	58	62	67	72	78	200	16
53	60	64	70	75	81	208	17
54	62	66	72	77	84	216	18
56	64	68	74	80	87	224	19
57	66	70	77	83	90	232	20
59	68	72	79	85	92	240	21
60	70	74	81	88	95	248	22
62	71	76	84	90	98	256	23
64	73	78	86	93	101	264	24
65	75	81	89	96	104	272	25

$N_1 = 9$

N_2	0.001	0.005	0.010	0.025	0.05	0.10	$2\bar{R}_1$
9	52	56	59	62	66	70	171
10	53	58	61	65	69	73	180
11	55	61	63	68	72	76	189
12	57	63	66	71	75	80	198
13	59	65	68	73	78	83	207
14	60	67	71	76	81	86	216
15	62	69	73	79	84	90	225
16	64	72	76	82	87	93	234
17	66	74	78	84	90	97	243
18	68	76	81	87	93	100	252
19	70	78	83	90	96	103	261
20	71	81	85	93	99	107	270
21	73	83	88	95	102	110	279
22	75	85	90	98	105	113	288
23	77	88	93	101	108	117	297
24	79	90	95	104	111	120	306
25	81	92	98	107	114	123	315

$N_1 = 10$

0.001	0.005	0.010	0.025	0.05	0.10	$2\bar{R}_1$	N_2
65	71	74	78	82	87	210	10
67	73	77	81	86	91	220	11
69	76	79	84	89	94	230	12
72	79	82	88	92	98	240	13
74	81	85	91	96	102	250	14
76	84	88	94	99	106	260	15
78	86	91	97	103	109	270	16
80	89	93	100	107	113	280	17
82	92	96	103	110	117	290	18
84	94	99	107	113	121	300	19
87	97	102	110	117	125	310	20
89	99	105	113	120	128	320	21
91	102	108	116	123	132	330	22
93	105	110	119	127	136	340	23
95	107	113	122	130	140	350	24
98	110	116	126	134	144	360	25

Table K *(Continued)*

| | | $N_1 = 11$ | | | | | | | | $N_1 = 12$ | | | | | |
N_2	0.001	0.005	0.010	0.025	0.05	0.10	$2\bar{R}_1$	0.001	0.005	0.010	0.025	0.05	0.10	$2\bar{R}_1$	N_2
11	81	87	91	96	100	106	253								
12	83	90	94	99	104	110	264	98	105	109	115	120	127	300	12
13	86	93	97	103	108	114	275	101	109	113	119	125	131	312	13
14	88	96	100	106	112	118	286	103	112	116	123	129	136	324	14
15	90	99	103	110	116	123	297	106	115	120	127	133	141	336	15
16	93	102	107	113	120	127	308	109	119	124	131	138	145	348	16
17	95	105	110	117	123	131	319	112	122	127	135	142	150	360	17
18	98	108	113	121	127	135	330	115	125	131	139	146	155	372	18
19	100	111	116	124	131	139	341	118	129	134	143	150	159	384	19
20	103	114	119	128	135	144	352	120	132	138	147	155	164	396	20
21	106	117	123	131	139	148	363	123	136	142	151	159	169	408	21
22	108	120	126	135	143	152	374	126	139	145	155	163	173	420	22
23	111	123	129	139	147	156	385	129	142	149	159	168	178	432	23
24	113	126	132	142	151	161	396	132	146	153	163	172	183	444	24
25	116	129	136	146	155	165	407	135	149	156	167	176	187	456	25

| | | $N_1 = 13$ | | | | | | | | $N_1 = 14$ | | | | | |
N_2	0.001	0.005	0.010	0.025	0.05	0.10	$2\bar{R}_1$	0.001	0.005	0.010	0.025	0.05	0.10	$2\bar{R}_1$	N_2
13	117	125	130	136	142	149	351								
14	120	129	134	141	147	154	364	137	147	152	160	166	174	406	14
15	123	133	138	145	152	159	377	141	151	156	164	171	179	420	15
16	126	136	142	150	156	165	390	144	155	161	169	176	185	434	16
17	129	140	146	154	161	170	403	148	159	165	174	182	190	448	17
18	133	144	150	158	166	175	416	151	163	170	179	187	196	462	18
19	136	148	154	163	171	180	429	155	168	174	183	192	202	476	19
20	139	151	158	167	175	185	442	159	172	178	188	197	207	490	20
21	142	155	162	171	180	190	455	162	176	183	193	202	213	504	21
22	145	159	166	176	185	195	468	166	180	187	198	207	218	518	22
23	149	163	170	180	189	200	481	169	184	192	203	212	224	532	23
24	152	166	174	185	194	205	494	173	188	196	207	218	229	546	24
25	155	170	178	189	199	211	507	177	192	200	212	223	235	560	25

| | | $N_1 = 15$ | | | | | | | | $N_1 = 16$ | | | | | |
N_2	0.001	0.005	0.010	0.025	0.05	0.10	$2\bar{R}_1$	0.001	0.005	0.010	0.025	0.05	0.10	$2\bar{R}_1$	N_2
15	160	171	176	184	192	200	465								
16	163	175	181	190	197	206	480	184	196	202	211	219	229	528	16
17	167	180	186	195	203	212	495	188	201	207	217	225	235	544	17
18	171	184	190	200	208	218	510	192	206	212	222	231	242	560	18
19	175	189	195	205	214	224	525	196	210	218	228	237	248	576	19
20	179	193	200	210	220	230	540	201	215	223	234	243	255	592	20
21	183	198	205	216	225	236	555	205	220	228	239	249	261	608	21
22	187	202	210	221	231	242	570	209	225	233	245	255	267	624	22
23	191	207	214	226	236	248	585	214	230	238	251	261	274	640	23
24	195	211	219	231	242	254	600	218	235	244	256	267	280	656	24
25	199	216	224	237	248	260	615	222	240	249	262	273	287	672	25

| | | $N_1 = 17$ | | | | | | | | $N_1 = 18$ | | | | | |
N_2	0.001	0.005	0.010	0.025	0.05	0.10	$2\bar{R}_1$	0.001	0.005	0.010	0.025	0.05	0.10	$2\bar{R}_1$	N_2
17	210	223	230	240	249	259	595								
18	214	228	235	246	255	266	612	237	252	259	270	280	291	666	18
19	219	234	241	252	262	273	629	242	258	265	277	287	299	684	19
20	223	239	246	258	268	280	646	247	263	271	283	294	306	702	20
21	228	244	252	264	274	287	663	252	269	277	290	301	313	720	21
22	233	249	258	270	281	294	680	257	275	283	296	307	321	738	22
23	238	255	263	276	287	300	697	262	280	289	303	314	328	756	23
24	242	260	269	282	294	307	714	267	286	295	309	321	335	774	24
25	247	265	275	288	300	314	731	273	292	301	316	328	343	792	25

Table K   (*Continued*)

N_2	0.001	0.005	$N_1 = 19$ 0.010	0.025	0.05	0.10	$2\bar{R}_1$	0.001	0.005	$N_1 = 20$ 0.010	0.025	0.05	0.10	$2\bar{R}_1$	N_2
19	267	283	291	303	313	325	741								
20	272	289	297	309	320	333	760	298	315	324	337	348	361	820	20
21	277	295	303	316	328	341	779	304	322	331	344	356	370	840	21
22	283	301	310	323	335	349	798	309	328	337	351	364	378	860	22
23	288	307	316	330	342	357	817	315	335	344	359	371	386	880	23
24	294	313	323	337	350	364	836	321	341	351	366	379	394	900	24
25	299	319	329	344	357	372	855	327	348	358	373	387	403	920	25

N_2	0.001	0.005	$N_1 = 21$ 0.010	0.025	0.05	0.10	$2\bar{R}_1$	0.001	0.005	$N_1 = 22$ 0.010	0.025	0.05	0.10	$2\bar{R}_1$	N_2
21	331	349	359	373	385	399	903								
22	337	356	366	381	393	408	924	365	386	396	411	424	439	990	22
23	343	363	373	388	401	417	945	372	393	403	419	432	448	1012	23
24	349	370	381	396	410	425	966	379	400	411	427	441	457	1034	24
25	356	377	388	404	418	434	987	385	408	419	435	450	467	1056	25

N_2	0.001	0.005	$N_1 = 23$ 0.010	0.025	0.05	0.10	$2\bar{R}_1$	0.001	0.005	$N_1 = 24$ 0.010	0.025	0.05	0.10	$2\bar{R}_1$	N_2
23	402	424	434	451	465	481	1081								
24	409	431	443	459	474	491	1104	440	464	475	492	507	525	1176	24
25	416	439	451	468	483	500	1127	448	472	484	501	517	535	1200	25

N_2	0.001	0.005	$N_1 = 25$ 0.010	0.025	0.05	0.10	$2\bar{R}_1$
25	480	505	517	536	552	570	1275

Table L Critical values of the Studentized range statistic*

df for S_w^2	$1-\alpha$	\multicolumn{9}{c}{k = number of means or steps between ordered means}
		2	3	4	5	6	7	8	9	10
1	.95	18.0	27.0	32.8	37.1	40.4	43.1	45.4	47.4	49.1
	.99	90.0	135	164	186	202	216	227	237	246
2	.95	6.09	8.3	9.8	10.9	11.7	12.4	13.0	13.5	14.0
	.99	14.0	19.0	22.3	24.7	26.6	28.2	29.5	30.7	31.7
3	.95	4.50	5.91	6.82	7.50	8.04	8.48	8.85	9.18	9.46
	.99	8.26	10.6	12.2	13.3	14.2	15.0	15.6	16.2	16.7
4	.95	3.93	5.04	5.76	6.29	6.71	7.05	7.35	7.60	7.83
	.99	6.51	8.12	9.17	9.96	10.6	11.1	11.5	11.9	12.3
5	.95	3.64	4.60	5.22	5.67	6.03	6.33	6.58	6.80	6.99
	.99	5.70	6.97	7.80	8.42	8.91	9.32	9.67	9.97	10.2
6	.95	3.46	4.34	4.90	5.31	5.63	5.89	6.12	6.32	6.49
	.99	5.24	6.33	7.03	7.56	7.97	8.32	8.61	8.87	9.10
7	.95	3.34	4.16	4.69	5.06	5.36	5.61	5.82	6.00	6.16
	.99	4.95	5.92	6.54	7.01	7.37	7.68	7.94	8.17	8.37
8	.95	3.26	4.04	4.53	4.89	5.17	5.40	5.60	5.77	5.92
	.99	4.74	5.63	6.20	6.63	6.96	7.24	7.47	7.68	7.87
9	.95	3.20	3.95	4.42	4.76	5.02	5.24	5.43	5.60	5.74
	.99	4.60	5.43	5.96	6.35	6.66	6.91	7.13	7.32	7.49
10	.95	3.15	3.88	4.33	4.65	4.91	5.12	5.30	5.46	5.60
	.99	4.48	5.27	5.77	6.14	6.43	6.67	6.87	7.05	7.21
11	.95	3.11	3.82	4.26	4.57	4.82	5.03	5.20	5.35	5.49
	.99	4.39	5.14	5.62	5.97	6.25	6.48	6.67	6.84	6.99
12	.95	3.08	3.77	4.20	4.51	4.75	4.95	5.12	5.27	5.40
	.99	4.32	5.04	5.50	5.84	6.10	6.32	6.51	6.67	6.81
13	.95	3.06	3.73	4.15	4.45	4.69	4.88	5.05	5.19	5.32
	.99	4.26	4.96	5.40	5.73	5.98	6.19	6.37	6.53	6.67
14	.95	3.03	3.70	4.11	4.41	4.64	4.83	4.99	5.13	5.25
	.99	4.21	4.89	5.32	5.63	5.88	6.08	6.26	6.41	6.54
16	.95	3.00	3.65	4.05	4.33	4.56	4.74	4.90	5.03	5.15
	.99	4.13	4.78	5.19	5.49	5.72	5.92	6.08	6.22	6.35
18	.95	2.97	3.61	4.00	4.28	4.49	4.67	4.82	4.96	5.07
	.99	4.07	4.70	5.09	5.38	5.60	5.79	5.94	6.08	6.20
20	.95	2.95	3.58	3.96	4.23	4.45	4.62	4.77	4.90	5.01
	.99	4.02	4.64	5.02	5.29	5.51	5.69	5.84	5.97	6.09
24	.95	2.92	3.53	3.90	4.17	4.37	4.54	4.68	4.81	4.92
	.99	3.96	4.54	4.91	5.17	5.37	5.54	5.69	5.81	5.92
30	.95	2.89	3.49	3.84	4.10	4.30	4.46	4.60	4.72	4.83
	.99	3.89	4.45	4.80	5.05	5.24	5.40	5.54	5.56	5.76
40	.95	2.86	3.44	3.79	4.04	4.23	4.39	4.52	4.63	4.74
	.99	3.82	4.37	4.70	4.93	5.11	5.27	5.39	5.50	5.60
60	.95	2.83	3.40	3.74	3.98	4.16	4.31	4.44	4.55	4.65
	.99	3.76	4.28	4.60	4.82	4.99	5.13	5.25	5.36	5.45
120	.95	2.80	3.36	3.69	3.92	4.10	4.24	4.36	4.48	4.56
	.99	3.70	4.20	4.50	4.71	4.87	5.01	5.12	5.21	5.30
∞	.95	2.77	3.31	3.63	3.86	4.03	4.17	4.29	4.39	4.47
	.99	3.64	4.12	4.40	4.60	4.76	4.88	4.99	5.08	5.16

* This table is abridged from Table II.2 in *The probability integrals of the range and of the Studentized range*, prepared by H. Leon Harter, Donald S. Clemm, and Eugene H. Guthrie. These tables are published in WADC tech. Rep. 58–484, vol. 2, 1959, Wright Air Development Center, and are reproduced with the kind permission of the authors.

Table M Squares and square roots of numbers from 1 to 1,000*

N	N^2	$\sqrt{N}$	$\sqrt{10N}$	N	N^2	$\sqrt{N}$	$\sqrt{10N}$
1	1	1.0000	3.1623	51	26 01	7.1414	22.5832
2	4	1.4142	4.4721	52	27 04	7.2111	22.8035
3	9	1.7321	5.4772	53	28 09	7.2801	23.0217
4	16	2.0000	6.3246	54	29 16	7.3485	23.2379
5	25	2.2361	7.0711	55	30 25	7.4162	23.4521
6	36	2.4495	7.7460	56	31 36	7.4833	23.6643
7	49	2.6458	8.3666	57	32 49	7.5498	23.8747
8	64	2.8284	8.9443	58	33 64	7.6158	24.0832
9	81	3.0000	9.4868	59	34 81	7.6811	24.2899
10	1 00	3.1623	10.0000	60	36 00	7.7460	24.4949
11	1 21	3.3166	10.4881	61	37 21	7.8102	24.6982
12	1 44	3.4641	10.9545	62	38 44	7.8740	24.8998
13	1 69	3.6056	11.4018	63	39 69	7.9373	25.0998
14	1 96	3.7417	11.8322	64	40 96	8.0000	25.2982
15	2 25	3.8730	12.2474	65	42 25	8.0623	25.4951
16	2 56	4.0000	12.6491	66	43 56	8.1240	25.6905
17	2 89	4.1231	13.0384	67	44 89	8.1854	25.8844
18	3 24	4.2426	13.4164	68	46 24	8.2462	26.0768
19	3 61	4.3589	13.7840	69	47 61	8.3066	26.2679
20	4 00	4.4721	14.1421	70	49 00	8.3666	26.4575
21	4 41	4.5826	14.4914	71	50 41	8.4261	26.6458
22	4 84	4.6904	14.8324	72	51 84	8.4853	26.8328
23	5 29	4.7958	15.1658	73	53 29	8.5440	27.0185
24	5 76	4.8990	15.4919	74	54 76	8.6023	27.2029
25	6 25	5.0000	15.8114	75	56 25	8.6603	27.3861
26	6 76	5.0990	16.1245	76	57 76	8.7178	27.5681
27	7 29	5.1962	16.4317	77	59 29	8.7750	27.7489
28	7 84	5.2915	16.7332	78	60 84	8.8318	27.9285
29	8 41	5.3852	17.0294	79	62 41	8.8882	28.1069
30	9 00	5.4772	17.3205	80	64 00	8.9443	28.2843
31	9 61	5.5678	17.6068	81	65 61	9.0000	28.4605
32	10 24	5.6569	17.8885	82	67 24	9.0554	28.6356
33	10 89	5.7446	18.1659	83	68 89	9.1104	28.8097
34	11 56	5.8310	18.4391	84	70 56	9.1652	28.9828
35	12 25	5.9161	18.7083	85	72 25	9.2195	29.1548
36	12 96	6.0000	18.9737	86	73 96	9.2736	29.3258
37	13 69	6.0828	19.2354	87	75 69	9.3274	29.4958
38	14 44	6.1644	19.4936	88	77 44	9.3808	29.6648
39	15 21	6.2450	19.7484	89	79 21	9.4340	29.8329
40	16 00	6.3246	20.0000	90	81 00	9.4868	30.0000
41	16 81	6.4031	20.2485	91	82 81	9.5394	30.1662
42	17 64	6.4807	20.4939	92	84 64	9.5917	30.3315
43	18 49	6.5574	20.7364	93	86 49	9.6437	30.4959
44	19 36	6.6332	20.9762	94	88 36	9.6954	30.6594
45	20 25	6.7082	21.2132	95	90 25	9.7468	30.8221
46	21 16	6.7823	21.4476	96	92 16	9.7980	30.9839
47	22 09	6.8557	21.6795	97	94 09	9.8489	31.1448
48	23 04	6.9282	21.9089	98	96 04	9.8995	31.3050
49	24 01	7.0000	22.1359	99	98 01	9.9499	31.4643
50	25 00	7.0711	22.3607	100	1 00 00	10.0000	31.6228

With permission of the author and publisher abridged from J. P. Guilford, *Fundamental statistics in psychology and education*, 4th ed., 1965, McGraw-Hill Book Company, New York.

Table M (*Continued*)

N	N^2	$\sqrt{N}$	$\sqrt{10N}$	N	N^2	$\sqrt{N}$	$\sqrt{10N}$
101	1 02 01	10.0499	31.7805	151	2 28 01	12.2882	38.8587
102	1 04 04	10.0995	31.9374	152	2 31 04	12.3288	38.9872
103	1 06 09	10.1489	32.0936	153	2 34 09	12.3693	39.1152
104	1 08 16	10.1980	32.2490	154	2 37 16	12.4097	39.2428
105	1 10 25	10.2470	32.4037	155	2 40 25	12.4499	39.3700
106	1 12 36	10.2956	32.5576	156	2 43 36	12.4900	39.4968
107	1 14 49	10.3441	32.7109	157	2 46 49	12.5300	39.6232
108	1 16 64	10.3923	32.8634	158	2 49 64	12.5698	39.7492
109	1 18 81	10.4403	33.0151	159	2 52 81	12.6095	39.8748
110	1 21 00	10.4881	33.1662	160	2 56 00	12.6491	40.0000
111	1 23 21	10.5357	33.3167	161	2 59 21	12.6886	40.1248
112	1 25 44	10.5830	33.4664	162	2 62 44	12.7279	40.2492
113	1 27 69	10.6301	33.6155	163	2 65 69	12.7671	40.3733
114	1 29 96	10.6771	33.7639	164	2 68 96	12.8062	40.4969
115	1 32 25	10.7238	33.9116	165	2 72 25	12.8452	40.6202
116	1 34 56	10.7703	34.0588	166	2 75 56	12.8841	40.7431
117	1 36 89	10.8167	34.2053	167	2 78 89	12.9228	40.8656
118	1 39 24	10.8628	34.3511	168	2 82 24	12.9615	40.9878
119	1 41 61	10.9087	34.4964	169	2 85 61	13.0000	41.1096
120	1 44 00	10.9545	34.6410	170	2 89 00	13.0384	41.2311
121	1 46 41	11.0000	34.7851	171	2 92 41	13.0767	41.3521
122	1 48 84	11.0454	34.9285	172	2 95 84	13.1149	41.4729
123	1 51 29	11.0905	35.0714	173	2 99 29	13.1529	41.5933
124	1 53 76	11.1355	35.2136	174	3 02 76	13.1909	41.7133
125	1 56 25	11.1803	35.3553	175	3 06 25	13.2288	41.8330
126	1 58 76	11.2250	35.4965	176	3 09 76	13.2665	41.9524
127	1 61 29	11.2694	35.6371	177	3 13 29	13.3041	42.0714
128	1 63 84	11.3137	35.7771	178	3 16 84	13.3417	42.1900
129	1 66 41	11.3578	35.9166	179	3 20 41	13.3791	42.3084
130	1 69 00	11.4018	36.0555	180	3 24 00	13.4164	42.4264
131	1 71 61	11.4455	36.1939	181	3 27 61	13.4536	42.5441
132	1 74 24	11.4891	36.3318	182	3 31 24	13.4907	42.6615
133	1 76 89	11.5326	36.4692	183	3 34 89	13.5277	42.7785
134	1 79 56	11.5758	36.6060	184	3 38 56	13.5647	42.8952
135	1 82 25	11.6190	36.7423	185	3 42 25	13.6015	43.0116
136	1 84 69	11.6619	36.8782	186	3 45 96	13.6382	43.1277
137	1 87 69	11.7047	37.0135	187	3 49 69	13.6748	43.2435
138	1 90 44	11.7473	37.1484	188	3 53 44	13.7113	43.3590
139	1 93 21	11.7898	37.2827	189	3 57 21	13.7477	43.4741
140	1 96 00	11.8322	37.4166	190	3 61 00	13.7840	43.5890
141	1 98 81	11.8743	37.5500	191	3 64 81	13.8203	43.7035
142	2 01 64	11.9164	37.6829	192	3 68 64	13.8564	43.8178
143	2 04 49	11.9583	37.8153	193	3 72 49	13.8924	43.9318
144	2 07 36	12.0000	37.9473	194	3 76 36	13.9284	44.0454
145	2 10 25	12.0416	38.0789	195	3 80 25	13.9642	44.1588
146	2 13 16	12.0830	38.2099	196	3 84 16	14.0000	44.2719
147	2 16 09	12.1244	38.3406	197	3 88 09	14.0357	44.3847
148	2 19 04	12.1655	38.4708	198	3 92 04	14.0712	44.4972
149	2 22 01	12.2066	38.6005	199	3 96 01	14.1067	44.6094
150	2 25 00	12.2474	38.7298	200	4 00 00	14.1421	44.7214

Table M   (*Continued*)

N	N²	√N	√10N	N	N²	√N	√10N
201	4 04 01	14.1774	44.8330	251	6 30 01	15.8430	50.0999
202	4 08 04	14.2127	44.9444	252	6 35 04	15.8745	50.1996
203	4 12 09	14.2478	45.0555	253	6 40 09	15.9060	50.2991
204	4 16 16	14.2829	45.1664	254	6 45 16	15.9374	50.3984
205	4 20 25	14.3178	45.2769	255	6 50 25	15.9687	50.4975
206	4 24 36	14.3527	45.3872	256	6 55 36	16.0000	50.5964
207	4 28 49	14.3875	45.4973	257	6 60 49	16.0312	50.6952
208	4 32 64	14.4222	45.6070	258	6 65 64	16.0624	50.7937
209	4 36 81	14.4568	45.7165	259	6 70 81	16.0935	50.8920
210	4 41 00	14.4914	45.8258	260	6 76 00	16.1245	50.9902
211	4 45 21	14.5258	45.9347	261	6 81 21	16.1555	51.0882
212	4 49 44	14.5602	46.0435	262	6 86 44	16.1864	51.1859
213	4 53 69	14.5945	46.1519	263	6 91 69	16.2173	51.2835
214	4 57 96	14.6287	46.2601	264	6 96 96	16.2481	51.3809
215	4 62 25	14.6629	46.3681	265	7 02 25	16.2788	51.4782
216	4 66 56	14.6969	46.4758	266	7 07 56	16.3095	51.5752
217	4′ 70 89	14.7309	46.5833	267	7 12 89	16.3401	51.6720
218	4 75 24	14.7648	46.6905	268	7 18 24	16.3707	51.7687
219	4 79 61	14.7986	46.7974	269	7 23 61	16.4012	51.8652
220	4 84 00	14.8324	46.9042	270	7 29 00	16.4317	51.9615
221	4 88 41	14.8661	47.0106	271	7 34 41	16.4621	52.0577
222	4 92 84	14.8997	47.1169	272	7 39 84	16.4924	52.1536
223	4 97 29	14.9332	47.2229	273	7 45 29	16.5227	52.2494
224	5 01 76	14.9666	47.3286	274	7 50 76	16.5529	52.3450
225	5 06 25	15.0000	47.4342	275	7 56 25	16.5831	52.4404
226	5 10 76	15.0333	47.5395	276	7 61 76	16.6132	52.5357
227	5 15 29	15.0665	47.6445	277	7 67 29	16.6433	52.6308
228	5 19 84	15.0997	47.7493	278	7 72 84	16.6733	52.7257
229	5 24 41	15.1327	47.8539	279	7 78 41	16.7033	52.8205
230	5 29 00	15.1658	47.9583	280	7 84 00	16.7332	52.9150
231	5 33 61	15.1987	48.0625	281	7 89 61	16.7631	53.0094
232	5 38 24	15.2315	48.1664	282	7 95 24	16.7929	53.1037
233	5 42 89	15.2643	48.2701	283	8 00 89	16.8226	53.1977
234	5 47 56	15.2971	48.3735	284	8 06 56	16.8523	53.2917
235	5 52 25	15.3297	48.4768	285	8 12 25	16.8819	53.3854
236	5 56 96	15.3623	48.5798	286	8 17 96	16.9115	53.4790
237	5 61 69	15.3948	48.6826	287	8 23 69	16.9411	53.5724
238	5 66 44	15.4272	48.7852	288	8 29 44	16.9706	53.6656
239	5 71 21	15.4596	48.8876	289	8 35 21	17.0000	53.7587
240	5 76 00	15.4919	48.9898	290	8 41 00	17.0294	53.8516
241	5 80 81	15.5242	49.0918	291	8 46 81	17.0587	53.9444
242	5 85 64	15.5563	49.1935	292	8 52 64	17.0880	54.0370
243	5 90 49	15.5885	49.2950	293	8 58 49	17.1172	54.1295
244	5 95 36	15.6205	49.3964	294	8 64 36	17.1464	54.2218
245	6 00 25	15.6525	49.4975	295	8 70 25	17.1756	54.3139
246	6 05 16	15.6844	49.5984	296	8 76 16	17.2047	54.4059
247	6 10 09	15.7162	49.6991	297	8 82 09	17.2337	54.4977
248	6 15 04	15.7480	49.7996	298	8 88 04	17.2627	54.5894
249	6 20 01	15.7797	49.8999	299	8 94 01	17.2916	54.6809
250	6 25 00	15.8114	50.0000	300	9 00 00	17.3205	54.7723

Table M   (*Continued*)

N	N²	√N	√10N	N	N²	√N	√10N
301	9 06 01	17.3494	54.8635	351	12 32 01	18.7350	59.2453
302	9 12 04	17.3781	54.9545	352	12 39 04	18.7617	59.3296
303	9 18 09	17.4069	55.0454	353	12 46 09	18.7883	59.4138
304	9 24 16	17.4356	55.1362	354	12 53 16	18.8149	59.4979
305	9 30 25	17.4642	55.2268	355	12 60 25	18.8414	59.5819
306	9 36 36	17.4929	55.3173	356	12 67 36	18.8680	59.6657
307	9 42 49	17.5214	55.4076	357	12 74 49	18.8944	59.7495
308	9 48 64	17.5499	55.4977	358	12 81 64	18.9209	59.8331
309	9 54 81	17.5784	55.5878	359	12 88 81	18.9473	59.9166
310	9 61 00	17.6068	55.6776	360	12 96 00	18.9737	60.0000
311	9 67 21	17.6352	55.7674	361	13 03 21	19.0000	60.0833
312	9 73 44	17.6635	55.8570	362	13 10 44	19.0263	60.1664
313	9 79 69	17.6918	55.9464	363	13 17 69	19.0526	60.2495
314	9 85 96	17.7200	56.0357	364	13 24 96	19.0788	60.3324
315	9 92 25	17.7482	56.1249	365	13 32 25	19.1050	60.4152
316	9 98 56	17.7764	56.2139	366	13 39 56	19.1311	60.4979
317	10 04 89	17.8045	56.3028	367	13 46 89	19.1572	60.5805
318	10 11 24	17.8326	56.3915	368	13 54 24	19.1833	60.6630
319	10 17 61	17.8606	56.4801	369	13 61 61	19.2094	60.7454
320	10 24 00	17.8885	56.5685	370	13 69 00	19.2354	60.8276
321	10 30 41	17.9165	56.6569	371	13 76 41	19.2614	60.9098
322	10 36 84	17.9444	56.7450	372	13 83 84	19.2873	60.9918
323	10 43 29	17.9722	56.8331	373	13 91 29	19.3132	61.0737
324	10 49 76	18.0000	56.9210	374	13 98 76	19.3391	61.1555
325	10 56 25	18.0278	57.0088	375	14 06 25	19.3649	61.2372
326	10 62 76	18.0555	57.0964	376	14 13 76	19.3907	61.3188
327	10 69 29	18.0831	57.1839	377	14 21 29	19.4165	61.4003
328	10 75 84	18.1108	57.2713	378	14 28 84	19.4422	61.4817
329	10 82 41	18.1384	57.3585	379	14 36 41	19.4679	61.5630
330	10 89 00	18.1659	57.4456	380	14 44 00	19.4936	61.6441
331	10 95 61	18.1934	57.5326	381	14 51 61	19.5192	61.7252
332	11 02 24	18.2209	57.6194	382	14 59 24	19.5448	61.8061
333	11 08 89	18.2483	57.7062	383	14 66 89	19.5704	61.8870
334	11 15 56	18.2757	57.7927	384	14 74 56	19.5959	61.9677
335	11 22 25	18.3030	57.8792	385	14 82 25	19.6214	62.0484
336	11 28 96	18.3303	57.9655	386	14 89 96	19.6469	62.1289
337	11 35 69	18.3576	58.0517	387	14 97 69	19.6723	62.2093
338	11 42 44	18.3848	58.1378	388	15 05 44	19.6977	62.2896
339	11 49 21	18.4120	58.2237	389	15 13 21	19.7231	62.3699
340	11 56 00	18.4391	58.3095	390	15 21 00	19.7484	62.4500
341	11 62 81	18.4662	58.3952	391	15 28 81	19.7737	62.5300
342	11 69 64	18.4932	58.4808	392	15 36 64	19.7990	62.6099
343	11 76 49	18.5203	58.5662	393	15 44 49	19.8242	62.6897
344	11 83 36	18.5472	58.6515	394	15 52 36	19.8494	62.7694
345	11 90 25	18.5742	58.7367	395	15 60 25	19.8746	62.8490
346	11 97 16	18.6011	58.8218	396	15 68 16	19.8997	62.9285
347	12 04 09	18.6279	58.9067	397	15 76 09	19.9249	63.0079
348	12 11 04	18.6548	58.9915	398	15 84 04	19.9499	63.0872
349	12 18 01	18.6815	59.0762	399	15 92 01	19.9750	63.1664
350	12 25 00	18.7083	59.1608	400	16 00 00	20.0000	63.2456

Table M (*Continued*)

N	N²	√N	√10N	N	N²	√N	√10N
401	16 08 01	20.0250	63.3246	451	20 34 01	21.2368	67.1565
402	16 16 04	20.0499	63.4035	452	20 43 04	21.2603	67.2309
403	16 24 09	20.0749	63.4823	453	20 52 09	21.2838	67.3053
404	16 32 16	20.0998	63.5610	454	20 61 16	21.3073	67.3795
405	16 40 25	20.1246	63.6396	455	20 70 25	21.3307	67.4537
406	16 48 36	20.1494	63.7181	456	20 79 36	21.3542	67.5278
407	16 56 49	20.1742	63.7966	457	20 88 49	21.3776	67.6018
408	16 64 64	20.1990	63.8749	458	20 97 64	21.4009	67.6757
409	16 72 81	20.2237	63.9531	459	21 06 81	21.4243	67.7495
410	16 81 00	20.2485	64.0312	460	21 16 00	21.4476	67.8233
411	16 89 21	20.2731	64.1093	461	21 25 21	21.4709	67.8970
412	16 97 44	20.2978	64.1872	462	21 34 44	21.4942	67.9706
413	17 05 69	20.3224	64.2651	463	21 43 69	21.5174	68.0441
414	17 13 96	20.3470	64.3428	464	21 52 96	21.5407	68.1175
415	17 22 25	20.3715	64.4205	465	21 62 25	21.5639	68.1909
416	17 30 56	20.3961	64.4981	466	21 71 56	21.5870	68.2642
417	17 38 89	20.4206	64.5755	467	21 80 89	21.6102	68.3374
418	17 47 24	20.4450	64.6529	468	21 90 24	21.6333	68.4105
419	17 55 61	20.4695	64.7302	469	21 99 61	21.6564	68.4836
420	17 64 00	20.4939	64.8074	470	22 09 00	21.6795	68.5565
421	17 72 41	20.5183	64.8845	471	22 18 41	21.7025	68.6294
422	17 80 84	20.5426	64.9615	472	22 27 84	21.7256	68.7023
423	17 89 29	20.5670	65.0385	473	22 37 29	21.7486	68.7750
424	17 97 76	20.5913	65.1153	474	22 46 76	21.7715	68.8477
425	18 06 25	20.6155	65.1920	475	22 56 25	21.7945	68.9202
426	18 14 76	20.6398	65.2687	476	22 65 76	21.8174	68.9928
427	18 23 29	20.6640	65.3452	477	22 75 29	21.8403	69.0652
428	18 31 84	20.6882	65.4217	478	22 84 84	21.8632	69.1375
429	18 40 41	20.7123	65.4981	479	22 94 41	21.8861	69.2098
430	18 49 00	20.7364	65.5744	480	23 04 00	21.9089	69.2820
431	18 57 61	20.7605	65.6506	481	23 13 61	21.9317	69.3542
432	18 66 24	20.7846	65.7267	482	23 23 24	21.9545	69.4262
433	18 74 89	20.8087	65.8027	483	23 32 89	21.9773	69.4982
434	18 83 56	20.8327	65.8787	484	23 42 56	22.0000	69.5701
435	18 92 25	20.8567	65.9545	485	23 52 25	22.0227	69.6419
436	19 00 06	20.8806	66.0303	486	23 61 96	22.0454	69.7137
437	19 09 69	20.9045	66.1060	487	23 71 69	22.0681	69.7854
438	19 18 44	20.9284	66.1816	488	23 81 44	22.0907	69.8570
439	19 27 21	20.9523	66.2571	489	23 91 21	22.1133	69.9285
440	19 36 00	20.9762	66.3325	490	24 01 00	22.1359	70.0000
441	19 44 81	21.0000	66.4078	491	24 10 81	22.1585	70.0714
442	19 53 64	21.0238	66.4831	492	24 20 64	22.1811	70.1427
443	19 62 49	21.0476	66.5582	493	24 30 49	22.2036	70.2140
444	19 71 36	21.0713	66.6333	494	24 40 36	22.2261	70.2851
445	19 80 25	21.0950	66.7083	495	24 50 25	22.2486	70.3562
446	19 89 16	21.1187	66.7832	496	24 60 16	22.2711	70.4273
447	19 98 09	21.1424	66.8581	497	24 70 09	22.2935	70.4982
448	20 07 04	21.1660	66.9328	498	24 80 04	22.3159	70.5691
449	20 16 01	21.1896	67.0075	499	24 90 01	22.3383	70.6399
450	20 25 00	21.2132	67.0820	500	25 00 00	22.3607	70.7107

Table M   (*Continued*)

N	N²	√N	√10N	N	N²	√N	√10N
501	25 10 01	22.3830	70.7814	551	30 36 01	23.4734	74.2294
502	25 20 04	22.4054	70.8520	552	30 47 04	23.4947	74.2967
503	25 30 09	22.4277	70.9225	553	30 58 09	23.5160	74.3640
504	25 40 16	22.4499	70.9930	554	30 69 16	23.5372	74.4312
505	25 50 25	22.4722	71.0634	555	30 80 25	23.5584	74.4983
506	25 60 36	22.4944	71.1337	556	30 91 36	23.5797	74.5654
507	25 70 49	22.5167	71.2039	557	31 02 49	23.6008	74.6324
508	25 80 64	22.5389	71.2741	558	31 13 64	23.6220	74.6994
509	25 90 81	22.5610	71.3442	559	31 24 81	23.6432	74.7663
510	26 01 00	22.5832	71.4143	560	31 36 00	23.6643	74.8331
511	26 11 21	22.6053	71.4843	561	31 47 21	23.6854	74.8999
512	26 21 44	22.6274	71.5542	562	31 58 44	23.7065	74.9667
513	26 31 69	22.6495	71.6240	563	31 69 69	23.7276	75.0333
514	26 41 96	22.6716	71.6938	564	31 80 96	23.7487	75.0999
515	26 52 25	22.6936	71.7635	565	31 92 25	23.7697	75.1665
516	26 62 56	22.7156	71.8331	566	32 03 56	23.7908	75.2330
517	26 72 89	22.7376	71.9027	567	32 14 89	23.8118	75.2994
518	26 83 24	22.7596	71.9722	568	32 26 24	23.8328	75.3658
519	26 93 61	22.7816	72.0417	569	32 37 61	23.8537	75.4321
520	27 04 00	22.8035	72.1110	570	32 49 00	23.8747	75.4983
521	27 14 41	22.8254	72.1803	571	32 60 41	23.8956	75.5645
522	27 24 84	22.8473	72.2496	572	32 71 84	23.9165	75.6307
523	27 35 29	22.8692	72.3187	573	32 83 29	23.9374	75.6968
524	27 45 76	22.8910	72.3878	574	32 94 76	23.9583	75.7628
525	27 56 25	22.9129	72.4569	575	33 06 25	23.9792	75.8288
526	27 66 76	22.9347	72.5259	576	33 17 76	24.0000	75.8947
527	27 77 29	22.9565	72.5948	577	33 29 29	24.0208	75.9605
528	27 87 84	22.9783	72.6636	578	33 40 84	24.0416	76.0263
529	27 98 41	23.0000	72.7324	579	33 52 41	24.0624	76.0920
530	28 09 00	23.0217	72.8011	580	33 64 00	24.0832	76.1577
531	28 19 61	23.0434	72.8697	581	33 75 61	24.1039	76.2234
532	28 30 24	23.0651	72.9383	582	33 87 24	24.1247	76.2889
533	28 40 89	23.0868	73.0068	583	33 98 89	24.1454	76.3544
534	28 51 56	23.1084	73.0753	584	34 10 56	24.1661	76.4199
535	28 62 25	23.1301	73.1437	585	34 22 25	24.1868	76.4853
536	28 72 96	23.1517	73.2120	586	34 33 96	24.2074	76.5506
537	28 83 69	23.1733	73.2803	587	34 45 69	24.2281	76.6159
538	28 94 44	23.1948	73.3485	588	34 57 44	24.2487	76.6812
539	29 05 21	23.2164	73.4166	589	34 69 21	24.2693	76.7463
540	29 16 00	23.2379	73.4847	590	34 81 00	24.2899	76.8115
541	29 26 81	23.2594	73.5527	591	34 92 81	24.3105	76.8765
542	29 37 64	23.2809	73.6206	592	35 04 64	24.3311	76.9415
543	29 48 49	23.3024	73.6885	593	35 16 49	24.3516	77.0065
544	29 59 36	23.3238	73.7564	594	35 28 36	24.3721	77.0714
545	29 70 25	23.3452	73.8241	595	35 40 25	24.3926	77.1362
546	29 81 16	23.3666	73.8918	596	35 52 16	24.4131	77.2010
547	29 92 09	23.3880	73.9594	597	35 64 09	24.4336	77.2658
548	30 03 04	23.4094	74.0270	598	35 76 04	24.4540	77.3305
549	30 14 01	23.4307	74.0945	599	35 88 01	24.4745	77.3951
550	30 25 00	23.4521	74.1620	600	36 00 00	24.4949	77.4597

Table M (*Continued*)

N	N²	√N	√10N	N	N²	√N	√10N
601	36 12 01	24.5153	77.5242	651	42 38 01	25.5147	80.6846
602	36 24 04	24.5357	77.5887	652	42 51 04	25.5343	80.7465
603	36 36 09	24.5561	77.6531	653	42 64 09	25.5539	80.8084
604	36 48 16	24.5764	77.7174	654	42 77 16	25.5734	80.8703
605	36 60 25	24.5967	77.7817	655	42 90 25	25.5930	80.9321
606	36 72 36	24.6171	77.8460	656	43 03 36	25.6125	80.9938
607	36 84 49	24.6374	77.9102	657	43 16 49	25.6320	81.0555
608	36 96 64	24.6577	77.9744	658	43 29 64	25.6515	81.1172
609	37 08 81	24.6779	78.0385	659	43 42 81	25.6710	81.1788
610	37 21 00	24.6982	78.1025	660	43 56 00	25.6905	81.2404
611	37 33 21	24.7184	78.1665	661	43 69 21	25.7099	81.3019
612	37 45 44	24.7386	78.2304	662	43 82 44	25.7294	81.3634
613	37 57 69	24.7588	78.2943	663	43 95 69	25.7488	81.4248
614	37 69 96	24.7790	78.3582	664	44 08 96	25.7682	81.4862
615	37 82 25	24.7992	78.4219	665	44 22 25	25.7876	81.5475
616	37 94 56	24.8193	78.4857	666	44 35 56	25.8070	81.6088
617	38 06 89	24.8395	78.5493	667	44 48 89	25.8263	81.6701
618	38 19 24	24.8596	78.6130	668	44 62 24	25.8457	81.7313
619	38 31 61	24.8797	78.6766	669	44 75 61	25.8650	81.7924
620	38 44 00	24.8998	78.7401	670	44 89 00	25.8844	81.8535
621	38 56 41	24.9199	78.8036	671	45 02 41	25.9037	81.9146
622	38 68 84	24.9399	78.8670	672	45 15 84	25.9230	81.9756
623	38 81 29	24.9600	78.9303	673	45 29 29	25.9422	82.0366
624	38 93 76	24.9800	78.9937	674	45 42 76	25.9615	82.0975
625	39 06 25	25.0000	79.0569	675	45 56 25	25.9808	82.1584
626	39 18 76	25.0200	79.1202	676	45 69 76	26.0000	82.2192
627	39 31 29	25.0400	79.1833	677	45 83 29	26.0192	82.2800
628	39 43 84	25.0599	79.2465	678	45 96 84	26.0384	82.3408
629	39 56 41	25.0799	79.3095	679	46 10 41	26.0576	82.4015
630	39 69 00	25.0998	79.3725	680	46 24 00	26.0768	82.4621
631	39 81 61	25.1197	79.4355	681	46 37 61	26.0960	82.5227
632	39 94 24	25.1396	79.4984	682	46 51 24	26.1151	82.5833
633	40 06 89	25.1595	79.5613	683	46 64 89	26.1343	82.6438
634	40 19 56	25.1794	79.6241	684	46 78 56	26.1534	82.7043
635	40 32 25	25.1992	79.6869	685	46 92 25	26.1725	82.7647
636	40 44 96	25.2190	79.7496	686	47 05 96	26.1916	82.8251
637	40 57 69	25.2389	79.8123	687	47 19 69	26.2107	82.8855
638	40 70 44	25.2587	79.8749	688	47 33 44	26.2298	82.9458
639	40 83 21	25.2784	79.9375	689	47 47 21	26.2488	83.0060
640	40 96 00	25.2982	80.0000	690	47 61 00	26.2679	83.0662
641	41 08 81	25.3180	80.0625	691	47 74 81	26.2869	83.1264
642	41 21 64	25.3377	80.1249	692	47 88 64	26.3059	83.1865
643	41 34 49	25.3574	80.1873	693	48 02 49	26.3249	83.2466
644	41 47 36	25.3772	80.2496	694	48 16 36	26.3439	83.3067
645	41 60 25	25.3969	80.3119	695	48 30 25	26.3629	83.3667
646	41 73 16	25.4165	80.3714	696	48 44 16	26.3818	83.4266
647	41 86 09	25.4362	80.4363	697	48 58 09	26.4008	83.4865
648	41 99 04	25.4558	80.4984	698	48 72 04	26.4197	83.5464
649	42 12 01	25.4775	80.5605	699	48 86 01	26.4386	83.6062
650	42 25 00	25.4951	80.6226	700	49 00 00	26.4575	83.6660

Table M   (*Continued*)

N	N²	√N	√10N	N	N²	√N	√10N
701	49 14 01	26.4764	83.7257	751	56 40 01	27.4044	86.6603
702	49 28 04	26.4953	83.7854	752	56 55 04	27.4226	86.7179
703	49 42 09	26.5141	83.8451	753	56 70 09	27.4408	86.7756
704	49 56 16	26.5330	83.9047	754	56 85 16	27.4591	86.8332
705	49 70 25	26.5518	83.9643	755	57 00 25	27.4773	86.8907
706	49 84 36	26.5707	84.0238	756	57 15 36	27.4955	86.9483
707	49 98 49	26.5895	84.0833	757	57 30 49	27.5136	87.0057
708	50 12 64	26.6083	84.1427	758	57 45 64	27.5318	87.0632
709	50 26 81	26.6271	84.2021	759	57 60 81	27.5500	87.1206
710	50 41 00	26.6458	84.2615	760	57 76 00	27.5681	87.1780
711	50 55 21	26.6646	84.3208	761	57 91 21	27.5862	87.2353
712	50 69 44	26.6833	84.3801	762	58 06 44	27.6043	87.2926
713	50 83 69	26.7021	84.4393	763	58 21 69	27.6225	87.3499
714	50 97 96	26.7208	84.4985	764	58 36 96	27.6405	87.4071
715	51 12 25	26.7395	84.5577	765	58 52 25	27.6586	87.4643
716	51 26 56	26.7582	84.6168	766	58 67 56	27.6767	87.5214
717	51 40 89	26.7769	84.6759	767	58 82 89	27.6948	87.5785
718	51 55 24	26.7955	84.7349	768	58 98 24	27.7128	87.6356
719	51 69 61	26.8142	84.7939	769	59 13 61	27.7308	87.6926
720	51 84 00	26.8328	84.8528	770	59 29 00	27.7489	87.7496
721	51 98 41	26.8514	84.9117	771	59 44 41	27.7669	87.8066
722	52 12 84	26.8701	84.9706	772	59 59 84	27.7849	87.8635
723	52 27 29	26.8887	85.0294	773	59 75 29	27.8029	87.9204
724	52 41 76	26.9072	85.0882	774	59 90 76	27.8209	87.9773
725	52 56 25	26.9258	85.1469	775	60 06 25	27.8388	88.0341
726	52 70 76	26.9444	85.2056	776	60 21 76	27.8568	88.0909
727	52 85 29	26.9629	85.2643	777	60 37 29	27.8747	88.1476
728	52 99 84	26.9815	85.3229	778	60 52 84	27.8927	88.2043
729	53 14 41	27.0000	85.3815	779	60 68 41	27.9106	88.2610
730	53 29 00	27.0185	85.4400	780	60 84 00	27.9285	88.3176
731	53 43 61	27.0370	85.4985	781	60 99 61	27.9464	88.3742
732	53 58 24	27.0555	85.5570	782	61 15 24	27.9643	88.4308
733	53 72 89	27.0740	85.6154	783	61 30 89	27.9821	88.4873
734	53 87 56	27.0924	85.6738	784	61 46 56	28.0000	88.5438
735	54 02 25	27.1109	85.7321	785	61 62 25	28.0179	88.6002
736	54 16 96	27.1293	85.7904	786	61 77 96	28.0357	88.6566
737	54 31 69	27.1477	85.8487	787	61 93 69	28.0535	88.7130
738	54 46 44	27.1662	85.9069	788	62 09 44	28.0713	88.7694
739	54 61 21	27.1846	85.9651	789	62 25 21	28.0891	88.8257
740	54 76 00	27.2029	86.0233	790	62 41 00	28.1069	88.8819
741	54 90 81	27.2213	86.0814	791	62 56 81	28.1247	88.9382
742	55 05 64	27.2397	86.1394	792	62 72 64	28.1425	88.9944
743	55 20 49	27.2580	86.1974	793	62 88 49	28.1603	89.0505
744	55 35 36	27.2764	86.2554	794	63 04 36	28.1780	89.1067
745	55 50 25	27.2947	86.3134	795	63 20 25	28.1957	89.1628
746	55 65 16	27.3130	86.3713	796	63 36 16	28.2135	89.2188
747	55 80 09	27.3313	86.4292	797	63 52 09	28.2312	89.2749
748	55 95 04	27.3496	86.4870	798	63 68 04	28.2489	89.3308
749	56 10 01	27.3679	86.5448	799	63 84 01	28.2666	89.3868
750	56 25 00	27.3861	86.6025	800	64 00 00	28.2843	89.4427

Table M   (*Continued*)

N	N²	√N̄	√10N̄	N	N²	√N̄	√10N̄
801	64 16 01	28.3019	89.4986	851	72 42 01	29.1719	92.2497
802	64 32 04	28.3196	89.5545	852	72 59 04	29.1890	92.3038
803	64 48 09	28.3373	89.6103	853	72 76 09	29.2062	92.3580
804	64 64 16	28.3549	89.6660	854	72 93 16	29.2233	92.4121
805	64 80 25	28.3725	89.7218	855	73 10 25	29.2404	92.4662
806	64 96 36	28.3901	89.7775	856	73 27 36	29.2575	92.5203
807	65 12 49	28.4077	89.8332	857	73 44 49	29.2746	92.5743
808	65 28 64	28.4253	89.8888	858	73 61 64	29.2916	92.6283
809	65 44 81	28.4429	89.9444	859	73 78 81	29.3087	92.6823
810	65 61 00	28.4605	90.0000	860	73 96 00	29.3258	92.7362
811	65 77 21	28.4781	90.0555	861	74 13 21	29.3428	92.7901
812	65 93 44	28.4956	90.1110	862	74 30 44	29.3598	92.8440
813	66 09 69	28.5132	90.1665	863	74 47 69	29.3769	92.8978
814	66 25 96	28.5307	90.2219	864	74 64 96	29.3939	92.9516
815	66 42 25	28.5482	90.2774	865	74 82 25	29.4109	93.0054
816	66 58 56	28.5657	90.3327	866	74 99 56	29.4279	93.0591
817	66 74 89	28.5832	90.3881	867	75 16 89	29.4449	93.1128
818	66 91 24	28.6007	90.4434	868	75 34 24	29.4618	93.1665
819	67 07 61	28.6182	90.4986	869	75 51 61	29.4788	93.2202
820	67 24 00	28.6356	90.5539	870	75 69 00	29.4958	93.2738
821	67 40 41	28.6531	90.6091	871	75 86 41	29.5127	93.3274
822	67 56 84	28.6705	90.6642	872	76 03 84	29.5296	93.3809
823	67 73 29	28.6880	90.7193	873	76 21 29	29.5466	93.4345
824	67 89 76	28.7054	90.7744	874	76 38 76	29.5635	93.4880
825	68 06 25	28.7228	90.8295	875	76 56 25	29.5804	93.5414
826	68 22 76	28.7402	90.8845	876	76 73 76	29.5973	93.5949
827	68 39 29	28.7576	90.9395	877	76 91 29	29.6142	93.6483
828	68 55 84	28.7750	90.9945	878	77 08 84	29.6311	93.7017
829	68 72 41	28.7924	91.0494	879	77 26 41	29.6479	93.7550
830	68 89 00	28.8097	91.1043	880	77 44 00	29.6648	93.8083
831	69 05 61	28.8271	91.1592	881	77 61 61	29.6816	93.8616
832	69 22 24	28.8444	91.2140	882	77 79 24	29.6985	93.9149
833	69 38 89	28.8617	91.2688	883	77 96 89	29.7153	93.9681
834	69 55 56	28.8791	91.3236	884	78 14 56	29.7321	94.0213
835	69 72 25	28.8964	91.3783	885	78 32 25	29.7489	94.0744
836	69 88 96	28.9137	91.4330	886	78 49 96	29.7658	94.1276
837	70 05 69	28.9310	91.4877	887	78 67 69	29.7825	94.1807
838	70 22 44	28.9482	91.5423	888	78 85 44	29.7993	94.2338
839	70 39 21	28.9655	91.5969	889	79 03 21	29.8161	94.2868
840	70 56 00	28.9828	91.6515	890	79 21 00	29.8329	94.3398
841	70 72 81	29.0000	91.7061	891	79 38 81	29.8496	94.3928
842	70 89 64	29.0172	91.7606	892	79 56 64	29.8664	94.4458
843	71 06 49	29.0345	91.8150	893	79 74 49	29.8831	94.4987
844	71 23 36	29.0517	91.8695	894	79 92 36	29.8998	94.5516
845	71 40 25	29.0689	91.9239	895	80 10 25	29.9166	94.6044
846	71 57 16	29.0861	91.9783	896	80 28 16	29.9333	94.6573
847	71 74 09	29.1033	92.0326	897	80 46 09	29.9500	94.7101
848	71 91 04	29.1204	92.0869	898	80 64 04	29.9666	94.7629
849	72 08 01	29.1376	92.1412	899	80 82 01	29.9833	94.8156
850	72 25 00	29.1548	92.1954	900	81 00 00	30.0000	94.8683

Table M (*Continued*)

N	N²	$\sqrt{N}$	$\sqrt{10N}$	N	N²	$\sqrt{N}$	$\sqrt{10N}$
901	81 18 01	30.0167	94.9210	951	90 44 01	30.8383	97.5192
902	81 36 04	30.0333	94.9737	952	90 63 04	30.8545	97.5705
903	81 54 09	30.0500	95.0263	953	90 82 09	30.8707	97.6217
904	81 72 16	30.0666	95.0789	954	91 01 16	30.8869	97.6729
905	81 90 25	30.0832	95.1315	955	91 20 25	30.9031	97.7241
906	82 08 36	30.0998	95.1840	956	91 39 36	30.9192	97.7753
907	82 26 49	30.1164	95.2365	957	91 58 49	30.9354	97.8264
908	82 44 64	30.1330	95.2890	958	91 77 64	30.9516	97.8775
909	82 62 81	30.1496	95.3415	959	91 96 81	30.9677	97.9285
910	82 81 00	30.1662	95.3939	960	92 16 00	30.9839	97.9796
911	82 99 21	30.1828	95.4463	961	92 35 21	31.0000	98.0306
912	83 17 44	30.1993	95.4987	962	92 54 44	31.0161	98.0816
913	83 35 69	30.2159	95.5510	963	92 73 69	31.0322	98.1326
914	83 53 96	30.2324	95.6033	964	92 92 96	31.0483	98.1835
915	83 72 25	30.2490	95.6556	965	93 12 25	31.0644	98.2344
916	83 90 56	30.2655	95.7079	966	93 31 56	31.0805	98.2853
917	84 08 89	30.2820	95.7601	967	93 50 89	31.0966	98.3362
918	84 27 24	30.2985	95.8123	968	93 70 24	31.1127	98.3870
919	84 45 61	30.3150	95.8645	969	93 89 61	31.1288	98.4378
920	84 64 00	30.3315	95.9166	970	94 09 00	31.1448	98.4886
921	84 82 41	30.3480	95.9687	971	94 28 41	31.1609	98.5393
922	85 00 84	30.3645	96.0208	972	94 47 84	31.1769	98.5901
923	85 19 29	30.3809	96.0729	973	94 67 29	31.1929	98.6408
924	85 37 76	30.3974	96.1249	974	94 86 76	31.2090	98.6914
925	85 56 25	30.4138	96.1769	975	95 06 25	31.2250	98.7421
926	85 74 76	30.4302	96.2289	976	95 25 76	31.2410	98.7927
927	85 93 29	30.4467	96.2808	977	95 45 29	31.2570	98.8433
928	86 11 84	30.4631	96.3328	978	95 64 84	31.2730	98.8939
929	86 30 41	30.4795	96.3846	979	95 84 41	31.2890	98.9444
930	86 49 00	30.4959	96.4365	980	96 04 00	31.3050	98.9949
931	86 67 61	30.5123	96.4883	981	96 23 61	31.3209	99.0454
932	86 86 24	30.5287	96.5401	982	96 43 24	31.3369	99.0959
933	87 04 89	30.5450	96.5919	983	96 62 89	31.3528	99.1464
934	87 23 56	30.5614	96.6437	984	96 82 56	31.3688	99.1968
935	87 42 25	30.5778	96.6954	985	97 02 25	31.3847	99.2472
936	87 60 96	30.5941	96.7471	986	97 21 96	31.4006	99.2975
937	87 79 69	30.6105	96.7988	987	97 41 69	31.4166	99.3479
938	87 98 44	30.6268	96.8504	988	97 61 44	31.4325	99.3982
939	88 17 21	30.6431	96.9020	989	97 81 21	31.4484	99.4485
940	88 36 00	30.6594	96.9536	990	98 01 00	31.4643	99.4987
941	88 54 81	30.6757	97.0052	991	98 20 81	31.4802	99.5490
942	88 73 64	30.6920	97.0567	992	98 40 64	31.4960	99.5992
943	88 92 49	30.7083	97.1082	993	98 60 49	31.5119	99.6494
944	89 11 36	30.7246	97.1597	994	98 80 36	31.5278	99.6995
945	89 30 25	30.7409	97.2111	995	99 00 25	31.5436	99.7497
946	89 49 16	30.7571	97.2625	996	99 20 16	31.5595	99.7998
947	89 68 09	30.7734	97.3139	997	99 40 09	31.5753	99.8499
948	89 87 04	30.7896	97.3653	998	99 60 04	31.5911	99.8999
949	90 06 01	30.8058	97.4166	999	99 80 01	31.6070	99.9500
950	90 25 00	30.8221	97.4679	1,000	1 00 00 00	31.6228	100.0000

REFERENCES

Aitken, A. C., 1937: The Evaluation of a Certain Triple-product Matrix. *Proceedings of the Royal Society of Edinburgh,* 57:172–181.

Aspen, Alice A., 1949: Tables for Use in Comparisons Whose Accuracy Involves Two Variances, Separately Estimated. *Biometrika,* 36:290–291.

Bancroft, T. A., 1968: *Topics in Intermediate Statistical Methods.* Ames, Iowa: The Iowa State University Press.

Binder, A., 1955: The Choice of an Error Term in Analysis of Variance Designs. *Psychometrika,* 20:29–50.

Bradley, James V., 1968: *Distribution-free Statistical Tests.* Englewood Cliffs, N.J.: Prentice-Hall, Inc.

Carroll, John B., 1953: An Analytical Solution for Approximating Simple Structure in Factor Analysis. *Psychometrika,* 18:23–38.

Cochran, William G., and Gertrude M. Cox, 1960: *Experimental Designs.* New York: John Wiley & Sons, Inc.

Comrie, L. J. (ed.), 1965: *Barlow's Tables.* London: Science Paperbacks, E. and F. N. Spon.

Cornell, Francis G., 1956: *The Essentials of Educational Statistics.* New York: John Wiley & Sons, Inc.

Cox, D. R., 1961: *Planning of Experiments.* New York: John Wiley & Sons, Inc.

Cronbach, L. J., 1957: The Two Disciplines of Scientific Psychology. *The American Psychologist,* 12:671–684.

Dubois, Phillip H., 1965: *An Introduction to Psychological Statistics.* New York: Harper and Row, Publishers.

———, 1970: Varieties of Psychological Test Homogeneity. *The American Psychologist,* 25:532–536.

Duncan, D. B., 1955: Multiple Range and Multiple F-tests. *Biometrics,* 11:1–42.

———, 1957: Multiple Range Tests for Correlated and Heteroscedastic Means. *Biometrics,* 13:164–176.

Edwards, Allen L., 1967: *Statistical Methods,* 2d ed. New York: Holt, Rinehart and Winston, Inc.

———, 1968: *Experimental Design in Psychological Research,* 3d ed. New York: Holt, Rinehart and Winston, Inc.

Ferguson, George A., 1941: *The Reliability of Mental Tests.* London: University of London Press, Ltd.

———, 1951: A Note on the Kuder-Richardson Formula. *Educational and Psychological Measurement,* 11:612–615.

———, 1954: The Concept of Parsimony in Factor Analysis. *Psychometrika,* 19:281–290.

———, 1965: *Nonparametric Trend Analysis.* Montreal: McGill University Press.

Finney, D. J., 1944: The Application of Probit Analysis to the Results of Mental Tests. *Psychometrika*, 9:31–39.

———, 1947: *Probit Analysis.* New York: Cambridge University Press.

———, 1948: The Fisher-Yates Test of Significance in 2 × 2 Contingency Tables. *Biometrika*, 35:145–156.

———, 1960: *The Theory of Experimental Design.* Chicago: University of Chicago Press.

Fisher, R. A., 1948: *Statistical Methods for Research Workers*, 10th ed. Edinburgh: Oliver & Boyd, Ltd.

——— **and F. Yates,** 1963: *Statistical Tables for Biological, Agricultural, and Medical Research*, 4th ed. Edinburgh: Oliver & Boyd, Ltd.

Freund, John E., 1962: *Mathematical Statistics.* Englewood Cliffs, N.J.: Prentice-Hall, Inc.

———, 1967: *Modern Elementary Statistics*, 3d ed. Englewood Cliffs, N.J.: Prentice-Hall, Inc.

Friedman, M., 1937: The Use of Ranks to Avoid the Assumption of Normality Implict in the Analysis of Variance. *Journal of the American Statistical Association*, 32:675–701.

———, 1940: A Comparison of Alternative Tests of Significance for the Problem of *m* Rankings. *Annals of Mathematical Statistics*, 11:86–92.

Gourlay, Neil, 1955: F-test Bias for Experimental Designs in Educational Research. *Psychometrika*, 20:227–248.

Gronow, D. G. C., 1951: Test for the Significance of Differences between Means in Two Normal Populations Having Unequal Variances. *Biometrika*, 38:252–256.

Guilford, J. P., 1965: *Fundamental Statistics in Psychology and Educaion*, 4th ed. New York: McGraw-Hill Book Company.

Gulliksen, H., 1950: *Theory of Mental Tests.* New York: John Wiley & Sons, Inc.

Harman, Harry H., 1967: *Modern Factor Analysis.* Chicago: University of Chicago Press.

Hersberg, Paul A., 1969: *The Parameters of Cross-validation.* Psychometrika Monograph Supplement, no. 16.

Holzinger, Karl J., and Frances Swineford, 1937: The Bi-factor Method. *Psychometrika*, 2:41–54.

Jackson, R. W. B., and George A. Ferguson, 1941: *Studies on the Reliability of Tests.* Bulletin 12, University of Toronto, Department of Educational Research, Toronto.

——— **and** ———, 1942: *Manual of Educational Statistics.* University of Toronto, Department of Educational Research, Toronto.

Johnson, Palmer O., 1949: *Statistical Methods in Research.* Englewood Cliffs, N.J.: Prentice-Hall, Inc.

Kaiser, Henry F., 1958: The Varimax Criterion for Analytic Rotation in Factor Analysis. *Psychometrika*, 23:187–200.

————, 1960: Directional Statistical Decisions. *Psychological Review*, 67:160–167.

Keeping, E. S., 1962: *Introduction to Statistical Inference*. Princeton, N.J.:D. Van Nostrand Company, Inc.

Kendall, M. G., 1951: *The Advanced Theory of Statistics*. vol. 2, 3d ed. London: Charles Griffin and Company, Ltd.

————, 1952: *The Advanced Theory of Statistics*. vol. 1, 5th ed. London: Charles Griffin and Company, Ltd.

————, 1955: *Rank Correlation Methods*, 2d ed. London: Charles Griffin and Company, Ltd.

————, 1968: *The Advanced Theory of Statistics*, vol, 3, 2d ed. London: Charles Griffin and Company, Ltd.

Kenney, John F., and E. S. Keeping, 1951: *Mathematics of Statistics*, part 2, 2d ed. Princeton, N.J.: D. Van Nostrand Company, Inc.

———— **and** ————, 1954: *Mathematics of Statistics*, part 1, 3d ed. Princeton, N.J.: D. Van Nostrand Company, Inc.

Keuls, M., 1952: The Use of the Studentized Range in Connection with the Analysis of Variance. *Euphytica*, 1:112–122.

Kruskal, W. H., and W. A. Wallis, 1952: Use of Ranks in One-criterion Variance Analysis. *Journal of the American Statistical Association*, 47:583–621.

Kuder, G. F., and M. W. Richardson, 1937: The Theory and Estimation of Test Reliability. *Psychometrika* 2:151–160.

Lacey, John I., 1956: The Evaluation of Autonomic Responses: Towards a General Solution. *Annals of the New York Academy of Sciences*, 67:123–164.

Lawley, D. N., 1940: The Estimation of Factor Loadings by the Method of Maximum Likelihood. *Proceedings of the Royal Society of Edinburgh*, 60:44–82.

———— **and A. E. Maxwell,** 1963: *Factor Analysis as a Statistical Method*. London: Butterworth and Company, Ltd.

Lord, Frederic M., 1955a: Estimating Test Reliability. *Educational and Psychological Measurement*, 15:325–336.

————, 1955b: Sampling Fluctuations Resulting from the Sampling of Test Items. *Psychometrika*, 20:1–22.

————, 1957: Do Tests of the Same Length Have the Same Standard Errors of Measurement? *Educational and Psychological Measurement*, 17:510–521.

———— **and Melvin R. Novick,** 1968: *Statistical Theories of Mental Test Scores*. Reading, Mass.: Addison-Wesley Publishing Company.

Macmeeken, A. M., 1940: *The Intelligence of a Representative Group of Scottish Children*. London: University of London Press, Ltd.

Magnusson, David, 1967: *Test Theory*. Reading, Mass.: Addison-Wesley Publishing Company.

Mann, H. B., and D. R. Whitney, 1947: On a Test of Whether One of Two Random Variables is Stochastically Larger Than the Other. *Annals of Mathematical Statistics*, 18:50–60.

McNemar, Quinn, 1947: Note on the Sampling Error of the Differences betwe Correlated Proportions or Percentages. *Psychometrika*, 12:153–157.

————, 1969: *Psychological Statistics*, 4th ed. New York: John Wiley & Sons, In

Meredith, William M., 1967: *Basic Mathematical and Statistical Tables for Ps chology and Education*. New York: McGraw-Hill Book Company.

Nair, K. R., 1940: Tables of Confidence Intervals for the Median in Samples from Any Continuous Population. *Sankhya*, 4:551–558. (Not seen.)

Neuhaus, Jack O., and Charles Wrigley, 1954: The Quartimax Method: an Analytical Approach to Orthogonal Simple Structure. *British Journal of Statistical Psychology*, 7:81–91.

Newman, D., 1939: The Distribution of the Range in Samples from a Normal Population in Terms of an Independent Estimate of Standard deviation. *Biometrika*, 31:20–30.

Saunders, D. R., 1953: An Analytic Method for Rotation to Orthogonal Simple Structure. *Research Bulletin*, RB 53–10. Princeton, N.J.: Educational Testing Service.

Scheffé, H., 1953: A Method for Judging All Contrasts in the Analysis of Variance. *Biometrika*, 40:87–104.

————, 1959: *The Analysis of Variance*. New York: John Wiley & Sons, Inc.

Siegel, Sidney, 1956: *Nonparametric Statistics*. New York: McGraw-Hill Book Company.

Snedecor, George W., and William G. Cochran, 1967: *Statistical Methods*, 6th ed. Ames, Iowa: The Iowa State University Press.

Stevens, S. S., 1957: On the Psychophysical Law. *Psychological Review*, 64:153–181.

Thomson, Godfrey H., 1951: *The Factorial Analysis of Human Ability*, 5th ed. London: University of London Press, Ltd.

Thurstone, L. L., 1944: *A Factorial Study of Perception*. Chicago: University of Chicago Press.

Torgerson, Warren S., 1958: *Theory and Methods of Scaling*. New York: John Wiley & Sons, Inc.

Tsao, Fei, 1946: General Solution of the Analysis of Variance and Covariance in the Case of Unequal or Disproportionate Numbers of Observations in the Subclasses. *Psychometrika*, 11:107–128.

Tukey, John W., 1949: Comparing Individual Means in the Analysis of Variance. *Biometrics*, 5:99–114.

Walker, Helen M., and Joseph Lev, 1953: *Statistical Inference*. New York: Holt, Rinehart and Winston, Inc.

Welch, B. L., 1938: The Significance of the Differences between Two Means When the Population Variances are Unequal. *Biometrika*, 29:350–362.

————, 1947: The Generalization of Student's Problem When Several Different Population Variances Are Involved. *Biometrika*, 34:28–35.

Wilk, M. B., and O. Kempthorne, 1955: Fixed, Mixed, and Random Models. *Journal of the American Statistical Association,* 50:1144–1167.

Winer, B. J., 1962: *Statistical Principles in Experimental Design.* New York: McGraw-Hill Book Company.

Woo, T. L., 1928: Dextrality and Sinistrality of Hand and Eye, 2d memoir. *Biometrika,* 20A:79–148.

INDEX